T0336835

Random Graphs and Complex Networks

Volume 1

This rigorous introduction to network science presents random graphs as models for real-world networks. Such networks have distinctive empirical properties, and a wealth of new models have emerged to capture them. Classroom tested for over ten years, this text places recent advances in a unified framework to enable systematic study.

Designed for a master's-level course, where students may only have a basic background in probability, the text covers such important preliminaries as convergence of random variables, probabilistic bounds, coupling, martingales, and branching processes. Building on this base – and motivated by many examples of real-world networks, including the Internet, collaboration networks, and the World-Wide Web – it focuses on several important models for complex networks and investigates key properties, such as the connectivity of nodes. Numerous exercises allow students to develop intuition and experience in working with the models.

REMCO VAN DER HOFSTAD is full professor of probability at Eindhoven University of Technology and acting scientific director of the European Institute in Stochastics (Eurandom). He has authored over 100 research articles and has taught courses on random graphs at over 10 institutions. He received the Prix Henri Poincaré 2003 jointly with Gordon Slade, the Rollo Davidson Prize 2007, and is a laureate of the 'Innovative Research Vidi Scheme' 2003 and the 'Innovative Research Vici Scheme' 2008.

CAMBRIDGE SERIES IN STATISTICAL AND PROBABILISTIC MATHEMATICS

This series of high-quality upper-division textbooks and expository monographs covers all aspects of stochastic applicable mathematics. The topics range from pure and applied statistics to probability theory, operations research, optimization and mathematical programming. The books contain clear presentations of new developments in the field and also of the state of the art in classical methods. While emphasizing rigorous treatment of theoretical methods, the books also contain applications and discussions of new techniques made possible by advances in computational practice.

A complete list of books in the series can be found at www.cambridge.org/statistics.
Recent titles include the following:

17. *Elements of Distribution Theory*, by Thomas A. Severini
18. *Statistical Mechanics of Disordered Systems*, by Anton Bovier
19. *The Coordinate-Free Approach to Linear Models*, by Michael J. Wichura
20. *Random Graph Dynamics*, by Rick Durrett
21. *Networks*, by Peter Whittle
22. *Saddlepoint Approximations with Applications*, by Ronald W. Butler
23. *Applied Asymptotics*, by A. R. Brazzale, A. C. Davison and N. Reid
24. *Random Networks for Communication*, by Massimo Franceschetti and Ronald Meester
25. *Design of Comparative Experiments*, by R. A. Bailey
26. *Symmetry Studies*, by Marlos A. G. Viana
27. *Model Selection and Model Averaging*, by Gerda Claeskens and Nils Lid Hjort
28. *Bayesian Nonparametrics*, edited by Nils Lid Hjort *et al.*
29. *From Finite Sample to Asymptotic Methods in Statistics*, by Pranab K. Sen, Julio M. Singer and Antonio C. Pedrosa de Lima
30. *Brownian Motion*, by Peter Mörters and Yuval Peres
31. *Probability (Fourth Edition)*, by Rick Durrett
33. *Stochastic Processes*, by Richard F. Bass
34. *Regression for Categorical Data*, by Gerhard Tutz
35. *Exercises in Probability (Second Edition)*, by Loïc Chaumont and Marc Yor
36. *Statistical Principles for the Design of Experiments*, by R. Mead, S. G. Gilmour and A. Mead
37. *Quantum Stochastics*, by Mou-Hsiung Chang
38. *Nonparametric Estimation under Shape Constraints*, by Piet Groeneboom and Geurt Jongbloed
39. *Large Sample Covariance Matrices and High-Dimensional Data Analysis*, by Jianfeng Yao, Shurong Zheng and Zhidong Bai
40. *Mathematical Foundations of Infinite-Dimensional Statistical Models*, by Evarist Giné and Richard Nickl
41. *Confidence, Likelihood, Probability*, by Tore Schweder and Nils Lid Hjort
42. *Probability on Trees and Networks*, by Russell Lyons and Yuval Peres
43. *Random Graphs and Complex Networks*, by Remco van der Hofstad

Random Graphs and Complex Networks

Volume 1

Remco van der Hofstad
Technische Universiteit Eindhoven

Shaftesbury Road, Cambridge CB2 8EA, United Kingdom

One Liberty Plaza, 20th Floor, New York, NY 10006, USA

477 Williamstown Road, Port Melbourne, VIC 3207, Australia

314–321, 3rd Floor, Plot 3, Splendor Forum, Jasola District Centre, New Delhi – 110025, India

103 Penang Road, #05–06/07, Visioncrest Commercial, Singapore 238467

Cambridge University Press is part of Cambridge University Press & Assessment, a department of the University of Cambridge.

We share the University's mission to contribute to society through the pursuit of education, learning and research at the highest international levels of excellence.

www.cambridge.org
Information on this title: www.cambridge.org/9781107172876

First published 2017

A catalogue record for this publication is available from the British Library

ISBN 978-1-107-17287-6 Hardback

Aan Mad, Max en Lars
het licht in mijn leven

Ter nagedachtenis aan mijn ouders
die me altijd aangemoedigd hebben

Contents

Preface

In this book, we study *random graphs* as models for *real-world networks*. Since 1999, many real-world networks have been investigated. These networks turned out to have rather different properties than classical random graph models, for example in the number of connections that the elements in the network make. As a result, a wealth of new models were invented to capture these properties. This book summarizes the insights developed in this exciting period.

This book is intended to be used for a master-level course where students have a limited prior knowledge of special topics in probability. We have included many of the preliminaries, such as convergence of random variables, probabilistic bounds, coupling, martingales and branching processes. This book aims to be self-contained. When we do not give proofs of the preliminary results, we provide pointers to the literature.

The field of random graphs was initiated in 1959–1960 by Erdős and Rényi (1959; 1960; 1961a; 1961b). At first, the theory of random graphs was used to prove deterministic properties of graphs. For example, if we can show that a random graph with a positive probability has a certain property, then a graph must exist with this property. The method of proving deterministic statements using probabilistic arguments is called the *probabilistic method*, and goes back a long way. See among others the preface of a standard work in random graphs by Bollobás (2001), or the classic book devoted to *The Probabilistic Method* by Alon and Spencer (2000). Erdős was one of the pioneers of this method; see, e.g., Erdős (1947), where he proved that the Ramsey number $R(k)$ is at least $2^{k/2}$. The Ramsey number $R(k)$ is the value n for which any graph of size at least n or its complement contains a complete graph of size at least k. Erdős (1947) shows that for $n \leq 2^{k/2}$ the fraction of graphs for which the graph or its complement contains a complete graph of size k is bounded by $1/2$, so that there must be graphs of size $n \leq 2^{k/2}$ for which neither the graph nor its complement contains a complete graph of size k.

The initial work by Erdős and Rényi on random graphs has incited a great amount of work in the field, initially mainly in the combinatorics community. See the standard references on the subject by Bollobás (2001) and Janson, Łuczak and Ruciński (2000) for the state of the art. Erdős and Rényi (1960) give a rather complete picture of the various phase transitions that occur in the Erdős-Rényi random graph. An interesting quote (Erdős and Rényi, 1960, pages 2–3) is the following:

> It seems to us worthwhile to consider besides graphs also more complex structures from the same point of view, i.e. to investigate the laws governing their evolution in a similar spirit. This may be interesting not only from a purely mathematical point of view. In fact,

> the evolution of graphs can be seen as a rather simplified model of the evolution of certain communication nets...

This was an excellent prediction indeed! Later, interest in different random graphs arose due to the analysis of real-world networks. Many of these real-world networks turned out to share similar properties, such as the fact that they are *small worlds*, and are *scale free*, which means that they have degrees obeying *power laws*. The Erdős-Rényi random graph does not obey these properties, and therefore new random graph models needed to be invented. In fact, Erdős and Rényi (1960) already remark that

> Of course, if one aims at describing such a real situation, one should replace the hypothesis of equiprobability of all connections by some more realistic hypothesis.

See Newman (2003) and Albert and Barabási (2002) for two reviews of real-world networks and their properties to get an impression of what 'more realistic' could mean, and see the recent book by Newman (2010) for detailed accounts of properties of real-world networks and models for them. These other models are also partly covered in the classical works by Bollobás (2001) and Janson, Łuczak and Rucínsky (2000), but up until today there is no comprehensive text treating random graph models for complex networks. See Durrett (2007) for a recent book on random graphs, and, particularly, dynamical processes living on them. Durrett covers part of the material in the present book, and more, but the intended audience is different. Our goal is to provide a source for a 'Random Graphs' course at the master level.

We describe results for the Erdős-Rényi random graph as well as for random graph models for complex networks. Our aim is to give the simplest possible proofs for classical results, such as the phase transition for the largest connected component in the Erdős-Rényi random graph. Some proofs are more technical and difficult. The sections containing these proofs are indicated with a star * and can be omitted without losing the logic behind the results. We also give many exercises that help the reader to obtain a deeper understanding of the material by working on their solutions. These exercises appear in the last section of each of the chapters, and when applicable, we refer to them at the appropriate place in the text.

I have tried to give as many references to the literature as possible. However, the number of papers on random graphs is currently exploding. In MathSciNet (see `http://www.ams.org/mathscinet`), there were, on December 21, 2006, a total of 1,428 papers that contain the phrase 'random graphs' in the review text, on September 29, 2008, this number increased to 1,614, on April 9, 2013, to 2,346 and, on April 21, 2016, to 2,986. These are merely the papers on the topic in the math community. What is special about random graph theory is that it is extremely multidisciplinary, and many papers using random graphs are currently written in economics, biology, theoretical physics and computer science. For example, in Scopus (see `http://www.scopus.com/scopus/home.url`), again on December 21, 2006, there were 5,403 papers that contain the phrase 'random graph' in the title, abstract or keywords; on September 29, 2008, this increased to 7,928; on April 9, 2013, to 13,987 and, on April 21, 2016, to 19,841. It can be expected that these numbers will continue to increase, rendering it impossible to review all the literature.

In June 2014, I decided to split the preliminary version of this book up into two books. This has several reasons and advantages, particularly since the later part of the work is more tuned towards a research audience, while the first part is more tuned towards an audience of master students with varying backgrounds. For the latest version of Volume II, which focusses on connectivity properties of random graphs and their small-world behavior, we refer to

`http://www.win.tue.nl/~rhofstad/NotesRGCN.html`

For further results on random graphs, or for solutions to some of the exercises in this book, readers are encouraged to look there. Also, for a more playful approach to networks for a broad audience, including articles, videos and demos of many of the models treated in this book, we refer all readers to the Network Pages at `http://www.networkspages.nl`. The Network Pages are an interactive website developed by and for all those who are interested in networks. One can find demos for some of the models discussed here, as well as of network algorithms and processes on networks.

This book would not have been possible without the help and encouragement of many people. I thank Gerard Hooghiemstra for the encouragement to write it, and for using it at Delft University of Technology almost simultaneously while I used it at Eindhoven University of Technology in the Spring of 2006 and again in the Fall of 2008. I particularly thank Gerard for many useful comments, solutions to exercises and suggestions for improvements of the presentation throughout the book. Together with Piet Van Mieghem, we entered the world of random graphs in 2001, and I have tremendously enjoyed exploring this field together with you, as well as with Henri van den Esker, Dmitri Znamenski, Mia Deijfen and Shankar Bhamidi, Johan van Leeuwaarden, Júlia Komjáthy, Nelly Litvak and many others.

I thank Christian Borgs, Jennifer Chayes, Gordon Slade and Joel Spencer for joint work on random graphs that are like the Erdős-Rényi random graph, but do have geometry. This work has deepened my understanding of the basic properties of random graphs, and many of the proofs presented here have been inspired by our work in Borgs et al. (2005a, b, 2006). Special thanks go to Gordon Slade, who has introduced me to the world of percolation, which is closely linked to the world of random graphs (see also the classic on percolation by Grimmett (1999)). It is peculiar to see that two communities work on two so closely related topics with different methods and even different terminology, and that it has taken such a long time to build bridges between the subjects. I am very happy that these bridges are now rapidly appearing, and the level of communication between different communities has increased significantly. I hope that this book helps to further enhance this communication. Frank den Hollander deserves a special mention. Frank, you have been important as a driving force throughout my career, and I am very happy now to be working with you on fascinating random graph problems!

Further I thank Marie Albenque, Gianmarco Bet, Shankar Bhamidi, Finbar Bogerd, Marko Boon, Francesco Caravenna, Rui Castro, Kota Chisaki, Mia Deijfen, Michel Dekking, Henri van den Esker, Lucas Gerin, Jesse Goodman, Rajat Hazra, Markus Heydenreich, Frank den Hollander, Yusuke Ide, Lancelot James, Martin van Jole, Willemien Kets, Júlia Komjáthy, John Lapeyre, Nelly Litvak, Norio Konno, Abbas Mehrabian, Mislav Mišković, Mirko Moscatelli, Jan Nagel, Sidharthan Nair, Alex Olssen, Mariana Olvera-Cravioto, Helena Peña, Nathan Ross, Karoly Simon, Dominik Tomecki, Nicola Turchi, Thomas Vallier and Xiaotin Yu for remarks and ideas that have improved the content and presentation of these notes substantially. Wouter Kager has entirely read the February 2007 version of this book, giving many ideas for improvements of the arguments and of the methodology. Artëm Sapozhnikov, Maren Eckhoff and Gerard Hooghiemstra read and commented the October 2011 version. Sándor Kolumbán read large parts of the October 2015 version, and spotted many errors, typos and inconsistencies.

I especially thank Dennis Timmers, Eefje van den Dungen and Joop van de Pol, who, as, my student assistants, have been a great help in the development of this book, in making figures, providing solutions to the exercises, checking the proofs and keeping the references up to date. Maren Eckhoff and Gerard Hooghiemstra also provided many solutions to the exercises, for which I am grateful! Sándor Kolumbán and Robert Fitzner helped me to turn all pictures of real-world networks as well as simulations of network models into a unified style, a feat that is beyond my LaTeX skills. A big thanks for that! Also my thanks for suggestions and help with figures to Marko Boon, Alessandro Garavaglia, Dimitri Krioukov, Vincent Kusters, Clara Stegehuis, Piet Van Mieghem and Yana Volkovich.

This work would not have been possible without the generous support of the Netherlands Organisation for Scientific Research (NWO) through VIDI grant 639.032.304, VICI grant 639.033.806 and Gravitation Networks grant 024.002.003.

Course Outline

The relation between the chapters in Volumes I and II of this book is as follows:

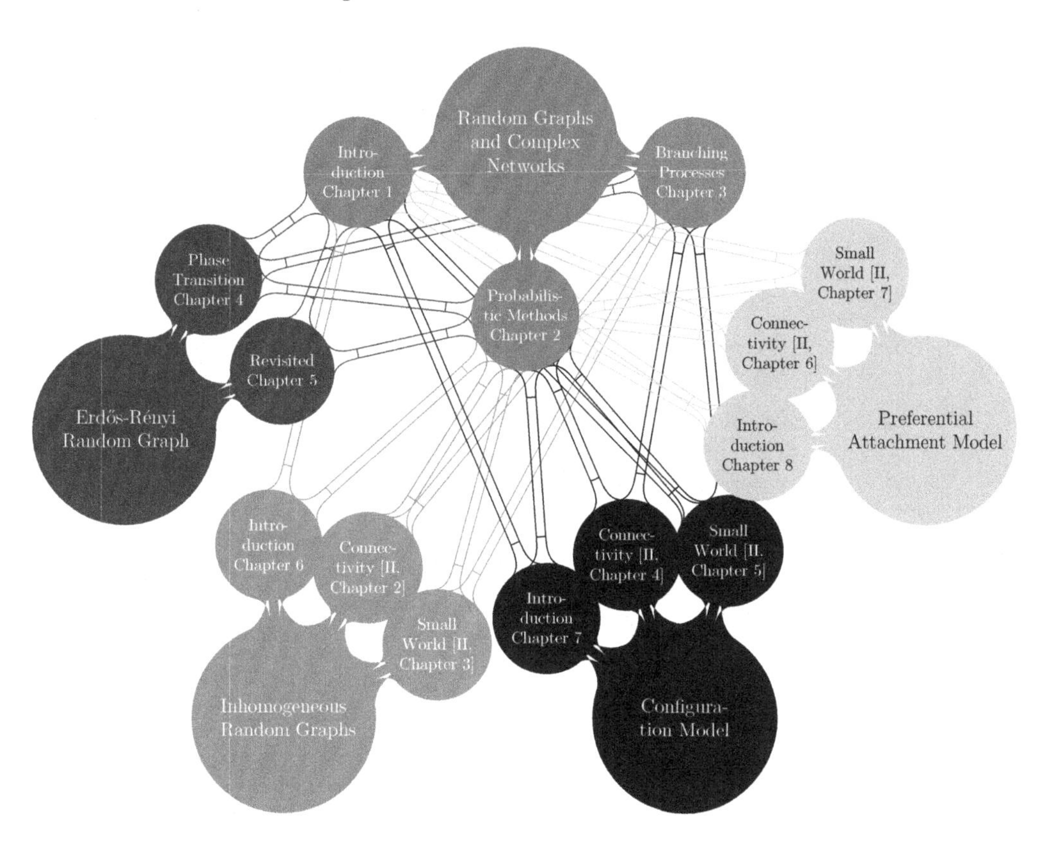

Here is some more explanation as well as a possible itinerary of a master course on random graphs. We include Volume II (see van der Hofstad (2015)) in the course outline:

Start with the introduction to real-world networks in Chapter 1, which forms the inspiration for what follows. Continue with Chapter 2, which gives the necessary probabilistic tools used in all later chapters, and pick those topics that your students are not familiar with and that are used in the later chapters that you wish to treat. Chapter 3 introduces branching processes, and is used in Chapters 4 and 5, as well as in most of Volume II.

After these preliminaries, you can start with the classical Erdős-Rényi random graph as covered in Chapters 4 and 5. Here you can choose the level of detail, and decide whether you wish to do the entire phase transition or would rather move on to the random graphs models for complex networks. It is possible to omit Chapter 5 before moving on.

After this, you can make your own choice of topics from the models for real-world networks. There are three classes of models for complex networks that are treated in this book. You can choose how much to treat in each of these models. You can either treat few models and discuss many aspects, or instead discuss many models at a less deep level. The introductory chapters about the three models, Chapter 6 for inhomogeneous random graphs, Chapter 7 for the configuration model, and Chapter 8 for preferential attachment models, provide a basic introduction to them, focussing on their degree structure. These introductory chapters need to be read in order to understand the later chapters about these models (particularly the ones in Volume II). The parts on the different models can be read independently.

1

Introduction

In this first chapter, we give an introduction to random graphs and complex networks. We discuss examples of real-world networks and their empirical properties, and give a brief introduction to the kinds of models that we investigate in the book. Further, we introduce the key elements of the notation used throughout the book.

1.1 Motivation

The advent of the computer age has incited an increasing interest in the fundamental properties of real-world networks. Due to the increased computational power, large data sets can now easily be stored and investigated, and this has had a profound impact on the empirical studies of large networks. As we explain in detail in this chapter, many real-world networks are small worlds and have large fluctuations in their degrees. These realizations have had fundamental implications for scientific research on networks. Network research is aimed to both to understand *why* many networks share these fascinating features, and also to investigate what the *properties* of these networks are in terms of the spread of diseases, routing information and ranking of the vertices present.

The study of complex networks plays an increasingly important role in science. Examples of such networks are electrical power grids and telecommunication networks, social relations, the World-Wide Web and Internet, collaboration and citation networks of scientists, etc. The structure of such networks affects their performance. For instance, the topology of social networks affects the spread of information or disease (see, e.g., Strogatz (2001)). The rapid evolution and success of the Internet have spurred fundamental research on the topology of networks. See Barabási (2002) and Watts (2003) for expository accounts of the discovery of network properties by Barabási, Watts and co-authors. In Newman et al. (2006), you can find some of the original papers detailing the empirical findings of real-world networks and the network models invented for them. The introductory book by Newman (2010) lists many of the empirical properties of, and scientific methods for, networks.

One common feature of complex networks is that they are *large*. As a result, a global description is utterly impossible, and researchers, both in the applications and in mathematics, have turned to their *local description*: how many vertices they have, by which *local rules* vertices connect to one another, etc. These local rules are *probabilistic*, reflecting the fact that there is a large amount of variability in how connections can be formed. Probability theory offers a highly effective way to deal with the *complexity* of networks, and leads us to consider *random graphs*. The simplest imaginable random graph is the Erdős-Rényi random graph, which arises by taking n vertices, and placing an edge between any pair of distinct vertices with a fixed probability p, independently for all pairs. We give an informal introduction to the classical Erdős-Rényi random graph in Section 1.8. We continue with a bit of graph theory.

1.2 Graphs and Their Degree and Connectivity Structure

This book describes random graphs, so we start by discussing graphs. A graph $G = (V, E)$ consists of a collection of vertices, called vertex set, V, and a collection of edges, called edge set, E. The vertices correspond to the objects that we model, the edges indicate some relation between pairs of these objects. In our settings, graphs are usually *undirected*. Thus, an edge is an unordered pair $\{u, v\} \in E$ indicating that u and v are directly connected. When G is undirected, if u is directly connected to v, then also v is directly connected to u. Thus, an edge can be seen as a pair of vertices. When dealing with social networks, the vertices represent the individuals in the population, while the edges represent the friendships among them.

Sometimes, we also deal with *directed* graphs, where edges are indicated by the *ordered* pair (u, v). In this case, when the edge (u, v) is present, the reverse edge (v, u) need not be present. One may argue about whether friendships in social networks are directed or not. In most applications, however, it is clear whether edges are directed or not. For example, in the World-Wide Web (WWW), where vertices represent web pages, an edge (u, v) indicates that the web page u has a hyperlink to the web page v, so the WWW is a directed network. In the Internet, instead, the vertices correspond to routers, and an edge $\{u, v\}$ is present when there is a physical cable linking u and v. This cable can be used in both directions, so that the Internet is undirected.

In this book, we only consider *finite* graphs. This means that V is a finite set of size, say, $n \in \mathbb{N}$. In this case, by numbering the vertices as $1, 2, \ldots, n$, we may as well assume that $V = [n] \equiv \{1, \ldots, n\}$, which we will do from now on. A special role is played by the *complete graph* denoted by K_n, for which the edge set is every possible pair of vertices, i.e., $E = \{\{i, j\}: 1 \leq i < j \leq n\}$. The complete graph K_n is the most highly connected graph on n vertices, and every other graph may be considered to be a subgraph of K_n obtained by keeping some edges and removing the rest. Of course, infinite graphs are also of interest, but since networks are finite, we stick to finite graphs.

The *degree* d_u of a vertex u is equal to the number of edges containing u, i.e.,

$$d_u = \#\{v \in V: \{u, v\} \in E\}. \tag{1.2.1}$$

Sometimes the degree is called the *valency*. In the social networks context, the degree of an individual is the number of her/his friends. We will often be interested in the structural properties of the degrees in a network, as indicated by the collection of degrees of all vertices or the degree sequence $\boldsymbol{d} = (d_v)_{v\in[n]}$. Such properties can be described nicely in terms of the *typical degree* denoted by $D_n = d_U$, where $U \in [n]$ is a vertex chosen uniformly at random from the collection of vertices. In turn, if we draw a histogram of the proportion of vertices having degree k for all k, then this histogram is precisely equal to the probability mass function $k \mapsto \mathbb{P}(D_n = k)$ of the random variable D_n, and it represents the *empirical distribution* of the degrees in the graph.

We continue by discussing certain related degrees. For example, when we draw an edge uniformly at random from E, and choose one of its two vertices uniformly at random, this corresponds to an individual engaged in a *random friendship*. Denote the degree of the corresponding random vertex by $D_n^\star$. Note that this vertex is *not* chosen uniformly at random from the collection of vertices! In the following theorem, we describe the law of $D_n^\star$ explicitly:

Theorem 1.1 (Friends in a random friendship) *Let $G = ([n], E)$ be a finite graph with degree sequence $\boldsymbol{d} = (d_v)_{v\in[n]}$. Let $D_n^\star$ be the degree of a random element in an edge that is drawn uniformly at random from E. Then*

$$\mathbb{P}(D_n^\star = k) = \frac{k}{\mathbb{E}[D_n]}\mathbb{P}(D_n = k). \tag{1.2.2}$$

In Theorem 1.1,

$$\mathbb{E}[D_n] = \frac{1}{n}\sum_{i\in[n]} d_i \tag{1.2.3}$$

is the average degree in the graph. The representation in (1.2.2) has a nice interpretation in terms of *size-biased random variables*. For a non-negative random variable X with $\mathbb{E}[X] > 0$, we define its size-biased version $X^\star$ by

$$\mathbb{P}(X^\star \leq x) = \frac{\mathbb{E}[X\mathbb{1}_{\{X\leq x\}}]}{\mathbb{E}[X]}. \tag{1.2.4}$$

Then, indeed, $D_n^\star$ is the size-biased version of D_n. In particular, note that, since the variance of a random variable $\mathrm{Var}(X) = \mathbb{E}[X^2] - \mathbb{E}[X]^2$ is non-negative,

$$\mathbb{E}[D_n^\star] = \frac{\mathbb{E}[D_n^2]}{\mathbb{E}[D_n]} = \mathbb{E}[D_n] + \frac{\mathrm{Var}(D_n)}{\mathbb{E}[D_n]} \geq \mathbb{E}[D_n], \tag{1.2.5}$$

the average number of friends of an individual in a random friendship is at least as large as that of a random individual. The inequality in (1.2.5) is strict whenever the degrees are not all equal, since then $\mathrm{Var}(D_n) > 0$. In Section 2.3, we further investigate the relation between D_n and $D_n^\star$. There, we show that, in some sense, $D_n^\star \geq D_n$ with probability 1. We will make the notion $D_n^\star \geq D_n$ perfectly clear in Section 2.3.

We next extend the above result to a random friend of a random individual. For this, we define the random vector (X_n, Y_n) by drawing an individual U uniformly at random from $[n]$, and then drawing a friend Z of U uniformly at random from the d_U friends of U and letting $X_n = d_U$ and $Y_n = d_Z$.

Theorem 1.2 (Your friends have more friends than you do) *Let $G = ([n], E)$ be a finite graph with degree sequence $\boldsymbol{d} = (d_v)_{v\in[n]}$. Assume that $d_v \geq 1$ for every $v \in [n]$. Let X_n be the degree of a vertex drawn uniformly at random from $[n]$, and Y_n be the degree of a uniformly drawn neighbor of this vertex. Then*

$$\mathbb{E}[Y_n] \geq \mathbb{E}[X_n], \tag{1.2.6}$$

the inequality being strict unless all degrees are equal.

When I view myself as the random individual, I see that Theorem 1.2 has the interpretation that, on average, a random friend of mine has more friends than I do! We now give a formal proof of this fact:

Proof We note that the joint law of (X_n, Y_n) is equal to

$$\mathbb{P}(X_n = k, Y_n = l) = \frac{1}{n} \sum_{(u,v)\in E'} \mathbb{1}_{\{d_u=k, d_v=l\}} \frac{1}{d_u}, \tag{1.2.7}$$

where the sum is over all *directed* edges E', i.e., we now consider (u, v) to be a different edge than (v, u) and we notice that, given that u is chosen as the uniform vertex, the probability that its neighbor v is chosen is $1/d_u$. Clearly, $|E'| = 2|E| = \sum_{i\in[n]} d_i$. Thus,

$$\mathbb{E}[X_n] = \frac{1}{n} \sum_{(u,v)\in E'} \sum_{k,l} k \mathbb{1}_{\{d_u=k, d_v=l\}} \frac{1}{d_u} = \frac{1}{n} \sum_{(u,v)\in E'} 1, \tag{1.2.8}$$

while

$$\mathbb{E}[Y_n] = \frac{1}{n} \sum_{(u,v)\in E'} \sum_{k,l} l \mathbb{1}_{\{d_u=k, d_v=l\}} \frac{1}{d_u} = \frac{1}{n} \sum_{(u,v)\in E'} \frac{d_v}{d_u}. \tag{1.2.9}$$

We will bound $\mathbb{E}[X_n]$ from above by $\mathbb{E}[Y_n]$. For this, we note that

$$1 \leq \frac{1}{2}(\frac{x}{y} + \frac{y}{x}) \tag{1.2.10}$$

for every $x, y > 0$, to obtain that

$$\mathbb{E}[X_n] \leq \frac{1}{n} \sum_{(u,v)\in E'} \frac{1}{2}(\frac{d_u}{d_v} + \frac{d_v}{d_u}) = \frac{1}{n} \sum_{(u,v)\in E'} \frac{d_u}{d_v} = \mathbb{E}[Y_n], \tag{1.2.11}$$

the penultimate equality following from the symmetry in (u, v). □

After the discussion of degrees in graphs, we continue with *graph distances*. For $u, v \in [n]$ and a graph $G = ([n], E)$, we let the graph distance $\mathrm{dist}_G(u, v)$ between u and v be equal to the minimal number of edges in a path linking u and v. When u and v are not in the same connected component, we set $\mathrm{dist}_G(u, v) = \infty$. We are interested in settings where G has a high amount of connectivity, so that many pairs of vertices are connected to one another by short paths. In order to describe how large distances between vertices typically are, we draw U_1 and U_2 uniformly at random from $[n]$ and we let

$$H_n = \mathrm{dist}_G(U_1, U_2). \tag{1.2.12}$$

Often, we will consider H_n conditionally on $H_n < \infty$. This means that we consider the typical number of edges between a uniformly chosen pair of *connected* vertices. As a result, H_n is sometimes referred to as the *typical distance*. Exercise 1.1 investigates the probability that $H_n < \infty$.

Just like the degree of a random vertex D_n, H_n is also a *random variable* even when the graph G is deterministic. The nice fact is that the distribution of H_n tells us something about *all* distances in the graph. An alternative and frequently used measure of distances in a graph is the *diameter* $\mathrm{diam}(G)$, defined as

$$\mathrm{diam}(G) = \max_{u,v\in[n]} \mathrm{dist}_G(u, v). \tag{1.2.13}$$

However, the diameter has several disadvantages. First, in many instances, the diameter is algorithmically more difficult to compute than the typical distances (since one has to measure the distances between all pairs of vertices and maximize over them). Second, it is a number instead of the distribution of a random variable, and therefore contains far less information than the distribution of H_n. Finally, the diameter is highly sensitive to small changes of the graph. For example, adding a small string of connected vertices to a graph may change the diameter dramatically, while it hardly influences the typical distances. As a result, in this book as well as Volume II, we put more emphasis on the typical distances. For many real-world networks, we will give plots of the distribution of H_n.

1.3 Complex Networks: the Infamous Internet Example

Complex networks have received a tremendous amount of attention in the past decades. In this section we use the Internet as an example of a real-world network to illustrate some of their properties. For an artistic impression of the Internet, see Figure 1.1.

Measurements have shown that many real-world networks share two fundamental properties: the *small-world phenomenon* roughly stating that distances in real-world networks are quite small and the *scale-free phenomenon* roughly stating that the degrees in real-world networks show an enormous amount of variability. We next discuss these two phenomena in detail.

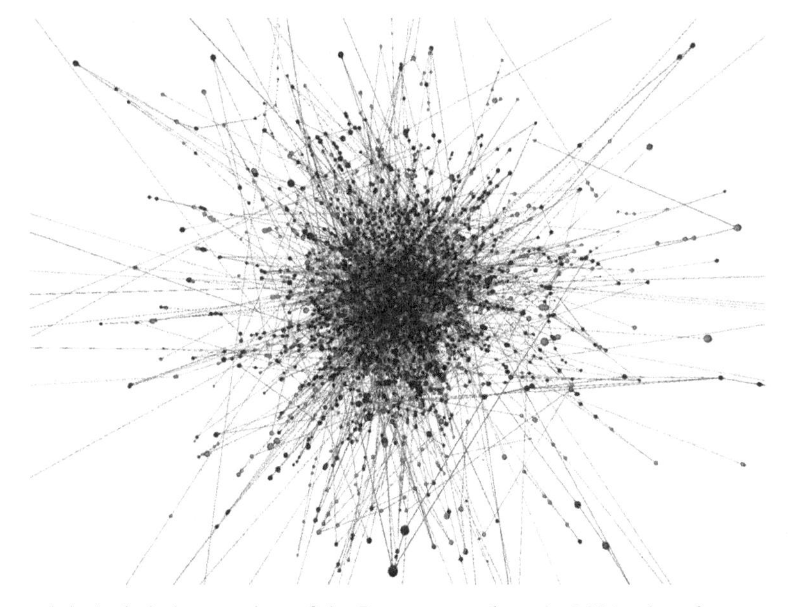

Figure 1.1 Artistic impression of the Internet topology in 2001 taken from `http://www.fractalus.com/steve/stuff/ipmap/`.

Small-World Phenomenon

The first fundamental network property is the fact that typical distances between vertices are small. This is called the 'small-world' phenomenon (see, e.g., the book by Watts (1999)). In particular, such networks are highly connected: their largest connected component contains a significant proportion of the vertices. Many networks, such as the Internet, even consist of *one* connected component, since otherwise e-mail messages could not be delivered. For example, in the Internet, IP-packets cannot use more than a threshold of physical links, and if distances in the Internet were larger than this threshold, then e-mail service would simply break down. Thus, the graph of the Internet has evolved in such a way that typical distances are relatively small, even though the Internet itself is rather large. For example, as seen in Figure 1.2(a), the number of Autonomous Systems (AS) traversed by an e-mail data set, sometimes referred to as the AS-count, is typically at most 7. In Figure 1.2(b), the proportion of routers traversed by an e-mail message between two uniformly chosen routers, referred to as the *hopcount*, is shown. It shows that the number of routers traversed is at most 27, while the distribution resembles a Poisson probability mass function.

Interestingly, various different data sets (focussing on different regional parts of the Internet) show roughly the same AS-counts. This shows that the AS-count is somewhat robust and it hints at the fact that the AS graph is relatively homogeneous. See Figure 1.3. For example, the AS-counts in North America and in Europe are quite close to the one in the entire AS graph. This implies that the dependence on geometry of the AS-count is rather weak, even though one would expect geometry to play a role. As a result, most of the models for the Internet, as well as for the AS graph, ignore geometry altogether.

Scale-Free Phenomenon

The second, maybe more surprising, fundamental property of many real-world networks is that the number of vertices with degree at least k decays slowly for large k. This implies that degrees are highly variable and that, even though the average degree is not so large, there exist vertices with extremely high degree. Often, the tail of the empirical degree distribution seems to fall off as an inverse power of k. This is called a 'power-law degree sequence', and resulting graphs often go under the name 'scale-free graphs'. It is visualized for the AS

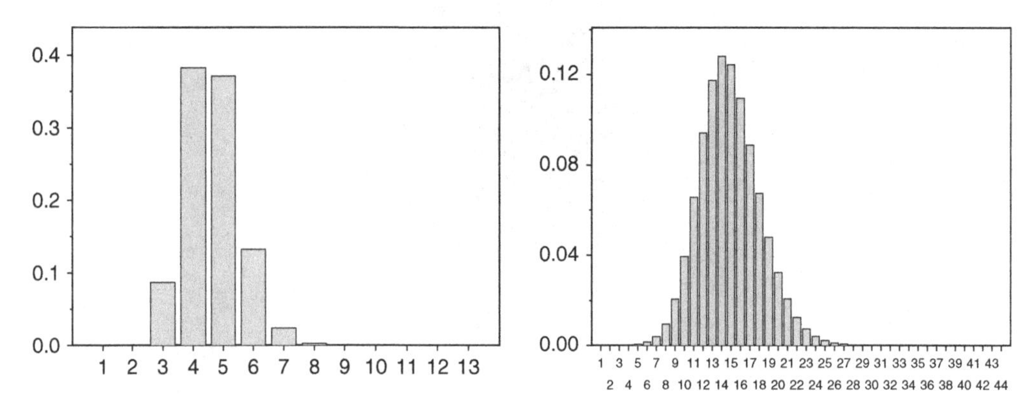

Figure 1.2 (a) Number of AS traversed in hopcount data. (b) Internet hopcount data. Courtesy of Hongsuda Tangmunarunkit.

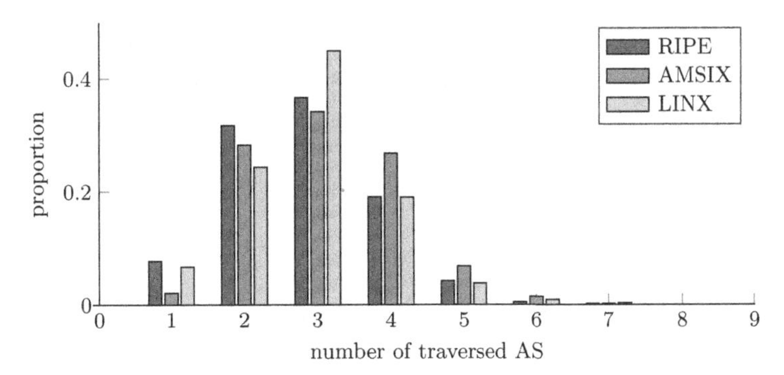

Figure 1.3 Number of AS traversed in various data sets. Data courtesy of Piet Van Mieghem.

graph in Figure 1.5, where the degree distribution of the AS graph is plotted on a log-log scale. Thus, we see a plot of $\log k \mapsto \log N_k$, where N_k is the number of vertices with degree k. When N_k is approximately proportional to an inverse power of k, i.e., when, for some normalizing constant c_n and some exponent τ,

$$N_k \approx c_n k^{-\tau}, \tag{1.3.1}$$

then

$$\log N_k \approx \log c_n - \tau \log k, \tag{1.3.2}$$

so that the plot of $\log k \mapsto \log N_k$ is close to a straight line. This is the reason why degree sequences in networks are often depicted in a log-log fashion, rather than in the more customary form of $k \mapsto N_k$. Here and in the remainder of this section, we write $\approx$ to denote an uncontrolled approximation. The power-law exponent τ can be estimated by the slope of the line in the log-log plot. Naturally, we must have that

$$\sum_k N_k = n < \infty, \tag{1.3.3}$$

so that it is reasonable to assume that $\tau > 1$.

For Internet, such log-log plots of the degrees first appeared in a paper by the Faloutsos brothers (1999) (see Figure 1.4 for the degree sequence in the Autonomous Systems graph). Here the power-law exponent is estimated as $\tau \approx 2.15 - 2.20$.

In recent years, many more Internet data sets have been collected. We particularly refer to the Center for Applied Internet Data Analysis (CAIDA) website for extensive measurements (see e.g., Krioukov et al. (2012) for a description of the data). See Figure 1.5, where the power-law exponent is now estimated as $\tau \approx 2.1$. See also Figure 1.7 for two examples of more recent measurements of the degrees of the Internet at the router or Internet Protocol (IP) level.

Measuring the Internet is quite challenging, particularly since the Internet is highly decentralized and distributed, so that a central authority is lacking. Huffaker et al. (2012) compare various data sets in terms of their coverage of the Internet and their accuracy.[1] The tool of the trade to obtain Internet data is called `traceroute`, an algorithm that allows you to send a

[1] See, in particular, `http://www.caida.org/research/topology/topo_comparison/`.

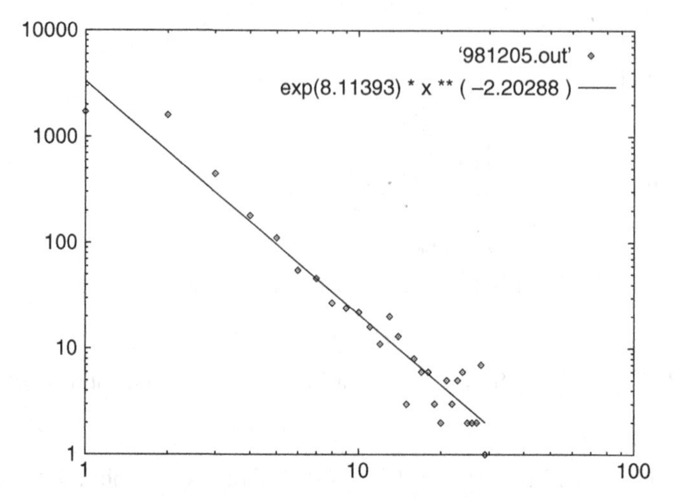

Figure 1.4 Degree sequence of Autonomous Systems (AS) on December 1998 on a log-log scale from Faloutsos, Faloutsos and Faloutsos (1999). These data suggest power-law degrees with exponent $\tau \approx 2.15 - 2.20$, the estimate on the basis of the data is 2.20288 with a multiplicative constant that is estimated as $e^{8.11393}$. This corresponds to c_n in (1.3.1).

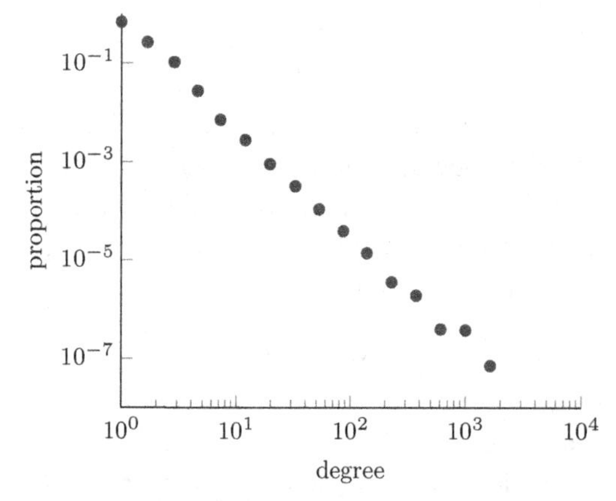

Figure 1.5 (a) Log-log plot of the probability mass function of the degree sequence of Autonomous Systems (AS) on April 2014 on a log-log scale from Krioukov et al. (2012) (data courtesy of Dmitri Krioukov). Due to the binning procedure that is applied, the figure looks smoother than many other log-log plots of degree sequences.

message between a source and a destination and to receive a list of the visited routers along the way. By piecing together many of such paths, one gets a picture of the Internet as a graph. This picture becomes more accurate when the number of sources and destinations increases, even though, as we describe in more detail below, it is not entirely understood how accurate these data sets are. In `traceroute`, also the direction of paths is obtained, and thus the graph reconstruction gives rise to a *directed* graph. The in- and out-degree sequences of this graph turn out to be quite different, as can be seen in Figure 1.6. It is highly interesting to explain such differences.

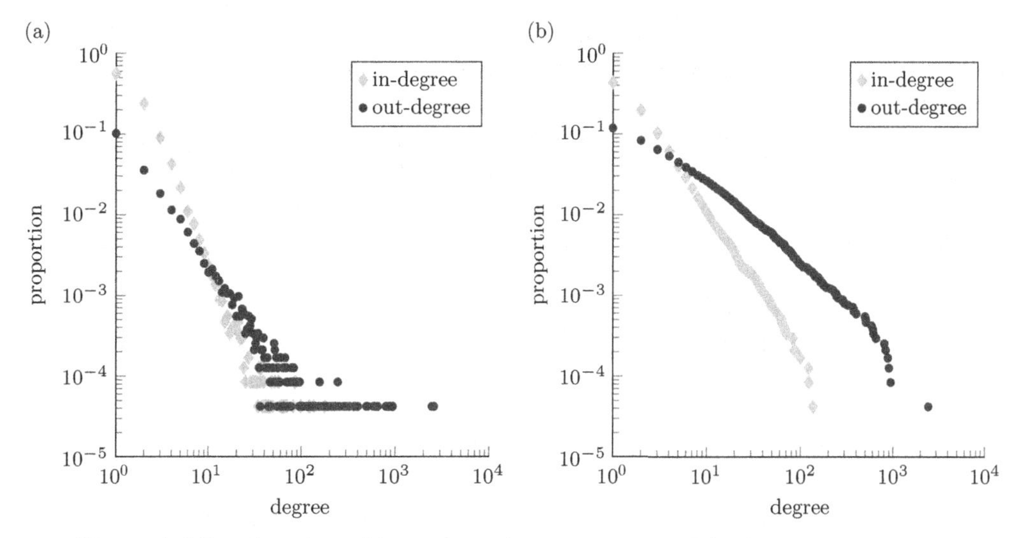

Figure 1.6 Log-log plots of in- and out-degree sequence of the Autonomous Systems graph. (a) Probability mass function. (b) Complementary cumulative distribution function.

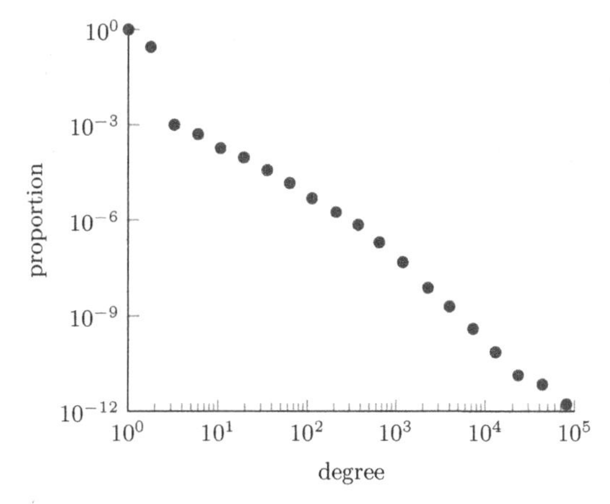

Figure 1.7 Log-log plot of the probability mass function of the degree distribution at router level from April 2014 (data courtesy of Dmitri Krioukov). The data set consists of 55,663,339 nodes and 59,685,901 edges. For a detailed explanation of how the data set was obtained, we refer to `http://www.caida.org/data/internet-topology-data-kit/`.

An interesting topic of research, receiving quite a bit of attention recently, is how the Internet behaves under malicious attacks or random breakdown (see, e.g., Albert et al. (2001) or Cohen et al. (2000, 2001)). The conclusion, based on various models for the Internet, is that the topology is critical for the vulnerability. When vertices with high degrees are taken out, the random graph models for the Internet cease to have the necessary connectivity properties. In particular, Albert et al. (2001) claim that when 2.5 percent of the Internet routers are *randomly* removed, the diameter of the Internet is unaffected, suggesting a remarkable tolerance to random attacks. Instead, when about 3 percent of the highest degree routers are *deterministically* removed, the Internet breaks down completely. Such results would have great implications for the resilience of networks, both to random and deliberate attacks.

A critical look at the proposed models for the Internet, and particularly, the claim of power-law degree sequences and the suggestion that attachment of edges in the Internet has a preference towards high-degree vertices, was given by Willinger, Govindan, Paxson and Shenker (2002). The authors conclude that the Barabási–Albert model (as described in more detail in Chapter 8) does not model the growth of the AS or IP graph appropriately, particularly since the degrees of the receiving vertices in the AS graph are even larger than for the Barabási–Albert model.

This criticism was most vehemently argued by Willinger, Anderson and Doyle (2009), with the suggestive title 'Mathematics and the internet: A source of enormous confusion and great potential'. In this view, the problem comes from the quality of the data. Indeed, the data set on which Faloutsos et al. (1999) base their analysis, and which is used again by Albert et al. (2001) to investigate the resilience properties of the Internet, was collected by Pansiot and Grad (1998) in order to study the efficiency of multicast versus unicast, which are different ways to send packages. The data was collected using `traceroute`, a tool that was not designed to be used to reconstruct the Internet as a graph. Pansiot and Grad realized that their way of reconstructing the Internet graph had some problems, and they write, 'We mention some problems we found in tracing routes, and we discuss the realism of the graph we obtained.' However, the Faloutsos brothers (1999) simply used the Pansiot–Grad data set, and took it at face value. This was then repeated by Albert, Jeong and Barabási (2001), which puts their results in a somewhat different light.

We now give some details of the problems with the data sets. Readers who are eager to continue can skip this part and move to the next section. Let us follow Willinger et al. (2009) to discuss the difficulties in using `traceroute` data to reconstruct a graph, which are threefold:

IP Alias Resolution Problem. A fundamental problem is that `traceroute` reports so-called *input interfaces*. Internet routers, the nodes of the Internet graph, may consist of several input interfaces, and it is a non-trivial problem to map these interfaces to routers. When errors are made in this procedure, the data does not truthfully represent the connectivity structure of the Internet routers.

Opaque Layer-2 Clouds. The Internet consists of different layers that facilitate the interoperability between heterogeneous network topologies. Since `traceroute` acts on layer-3, it is sometimes unable to trace through layer-2 clouds. This means that the internal connectivity structure of a larger unit of routers in layer-2 could be invisible for `traceroute`, so that `traceroute` shows connections between many, or even all, of these routers, even though most of these connections actually do not exist. This causes routers to be wrongfully assigned a very high degree.

Measuring Biases. Due to the way `traceroute` data is collected, an incomplete picture of the Internet is obtained, since only connections between routers that are actually being used by the data are reported. When this data set would be unbiased, a truthful picture of the Internet could still be obtained. Unfortunately, routers with a high degree are more likely to be used, which creates a so-called *sampling bias*. We return to these biases in Section 1.7.3.

Because of these problems with the data, Willinger, Anderson and Doyle (2009) doubt the conclusions by Faloutsos et al. (1999) and Albert et al. (2001) and all the subsequent work that is based on it. Given the fact that Albert et al. (2001) has close to 5,000 citations,[2] this clearly has significant impact. Further, Willinger, Anderson and Doyle also give engineering reasons, based on properties of the Internet, why it is unreasonable to expect the Internet to have a power-law degree sequence at IP level. This is a debate that will linger on for quite a while longer!

1.4 Scale-Free, Highly Connected and Small-World Graph Sequences

As mentioned in Section 1.3, many real-world networks are large. Mathematically, we can model this by taking the graph size to be equal to n, and study what happens when $n \to \infty$. Real-world networks share common features, e.g., that they have a relatively low average degree compared to the maximal possible degree $n - 1$ in a graph of size n, i.e., they are 'sparse'. Further, many real-world networks are 'small worlds' and 'scale free'. These notions are empirical, and inherently not mathematically precise. In this section, we explain what it means for a model of a real-world network to satisfy these properties. We start by discussing scale-free graph sequences.

1.4.1 Scale-Free Graph Sequences

Many real-world networks to be described later, such as the World-Wide Web and collaboration networks, grow in size with time. Therefore, it is reasonable to consider graphs of growing size, and to define the notions of scale-free, small-world and highly clustered random graphs as limiting statements when the size of the random graphs tends to infinity. This naturally leads us to study *graph sequences*. In this section, we look at graph sequences $(G_n)_{n\geq 1}$, where n denotes the size of the graph G_n, i.e., the number of vertices in G_n.

Denote the proportion of vertices with degree k in G_n by $P_k^{(n)}$, i.e.,

$$P_k^{(n)} = \frac{1}{n}\sum_{i=1}^{n} \mathbb{1}_{\{d_i^{(n)}=k\}}, \tag{1.4.1}$$

where $d_i^{(n)}$ denotes the degree of vertex $i \in [n] = \{1, \ldots, n\}$ in the graph G_n. We start by defining what it means for a random graph sequence $(G_n)_{n\geq 1}$ to be *sparse*:

Definition 1.3 (Sparse graph sequences) A graph sequence $(G_n)_{n\geq 1}$ is called *sparse* when

$$\lim_{n\to\infty} P_k^{(n)} = p_k, \qquad k \geq 0 \tag{1.4.2}$$

for some *deterministic* limiting probability distribution $(p_k)_{k\geq 0}$.

[2] From Google Scholar, on September 26, 2014.

We often apply Definition 1.3 to *random* graphs. Since the limit p_k in (1.4.2) is deterministic, the convergence in (1.4.2) must then be taken as convergence in probability or in distribution. Also, since $(p_k)_{k\geq 0}$ sums up to one, for large n most of the vertices have a bounded degree, which explains the phrase *sparse* graphs.

In some empirical work it is claimed that certain real-world networks show signs of *densification*, meaning that they become denser over time. Leskovec, Kleinberg and Faloutsos (2005; 2007) empirically find that $e(t) \propto n(t)^a$ as $t \to \infty$, where $e(t)$ and $n(t)$ denote the number of vertices and edges at time t, and $a \in (1, 2)$. For sparse graphs, $a = 1$. At the same time, they claim that the diameter shrinks over time.

We next define the notion of *scale-free graph sequences:*

Definition 1.4 (Scale-free graph sequences) We call a graph sequence $(G_n)_{n\geq 1}$ *scale-free with exponent* τ when it is sparse and

$$\lim_{k\to\infty} \frac{\log [1 - F(k)]}{\log (1/k)} = \tau - 1, \tag{1.4.3}$$

where $F(k) = \sum_{l\leq k} p_l$ denotes the cumulative distribution function corresponding to the probability mass function $(p_k)_{k\geq 0}$ defined in (1.4.2).

Thus, for a scale-free random graph process, its degree sequence converges to a limiting probability distribution as in (1.4.2), and the limiting distribution has asymptotic power-law tails described in (1.4.3). This gives a precise mathematical meaning to a graph sequence being scale free.

There are many other possible definitions. Indeed, in empirical work, often a log-log plot is given that plots $\log P_k^{(n)}$ versus $\log (1/k)$. Therefore, it may be reasonable to replace (1.4.3) by

$$\lim_{k\to\infty} \frac{\log p_k}{\log (1/k)} = \tau. \tag{1.4.4}$$

In some cases, the definition in (1.4.4) is a bit too restrictive, particularly when the probability mass function $k \mapsto p_k$ is not smooth. In specific models below, we use the version that is most appropriate in the setting under consideration. See Section 1.7 below for a more extensive discussion of power laws, as well as how to estimate them from real data.

The observation that many real-world networks have the above properties has incited a burst of activity in network modeling. Most of the models use random graphs as a way to model the *uncertainty* and the *lack of regularity* in real-world networks. In this book, we survey some of the proposed network models. These can be divided into two types: 'static' models, where we model a graph of a given size as a snapshot at a given time of a real-world network, and 'dynamic' models, where we model the growth of the network as time progresses. Static models aim to describe real-world networks and their topology at a given time instant. Dynamic models aim to *explain* how these networks came to be as they are. Such explanations often focus on the growth of the network as a way to explain the power-law degree sequences by means of simple and local growth rules.

When we try to model a power-law relationship between k and the number of vertices having degree k, the question is how to appropriately do so. In Chapters 6, 7 and 8, we discuss a number of models that have been proposed for graphs with a given degree sequence. For this, we let F_X be the distribution function of an integer-valued random variable X, and we denote its probability mass function by $(f_k)_{k\geq 0}$, so that

$$F_X(x) = \mathbb{P}(X \leq x) = \sum_{k\leq x} f_k. \tag{1.4.5}$$

We wish to obtain a random graph model in which N_k, the number of vertices of degree k, is roughly equal to nf_k, where we recall that n is the size of the network. For a power-law relationship as in (1.3.1), we should have that

$$N_k \approx nf_k. \tag{1.4.6}$$

Then, (1.4.4) turns into

$$f_k \propto k^{-\tau}, \tag{1.4.7}$$

where, to be able to make $f = (f_k)_{k\geq 0}$ a probability mass function, we need to require that $\tau > 1$, and $\propto$ in (1.4.7) denotes that the left-hand side is proportional to the right-hand side. Often (1.4.7) is too restrictive, and we wish to formulate a power-law relationship in a weaker sense. A different formulation could be to require that

$$1 - F_X(x) = \sum_{k>x} f_k \propto x^{1-\tau}. \tag{1.4.8}$$

Here $1 - F_X$ is often called the complementary cumulative distribution function, or simply the *tail*. Indeed, (1.4.8) is strictly weaker than (1.4.7), as investigated in Exercise 1.2.

An even weaker form of a power-law relation is to require that

$$1 - F_X(x) = L_X(x)x^{1-\tau}, \tag{1.4.9}$$

where the function $x \mapsto L_X(x)$ is *slowly varying*. Here, a function $x \mapsto \ell(x)$ is called *slowly varying* when, for all $c > 0$,

$$\lim_{x\to\infty} \frac{\ell(cx)}{\ell(x)} = 1. \tag{1.4.10}$$

In this case, $1 - F_X(x)$ is called *regularly varying with exponent* $1 - \tau$. This is formalized in the following definition:

Definition 1.5 (Regular and slow variation)

(a) A function $\ell\colon [0,\infty) \mapsto \mathbb{R}$ is called *slowly varying at infinity* when, for every $c > 0$,

$$\lim_{x\to\infty} \frac{\ell(cx)}{\ell(x)} = 1. \tag{1.4.11}$$

(b) A function $f\colon [0,\infty) \mapsto \mathbb{R}$ is called *regularly varying at infinity with exponent* α when

$$f(x) = x^{\alpha}\ell(x) \tag{1.4.12}$$

with ℓ slowly varying at infinity.

Exercise 1.3 investigates examples of slowly varying functions.

One should bear in mind that the notion of power-law degrees is an *asymptotic* one, and one can never be sure whether degrees follow a power law when one only observes a finite graph. One way to interpret power-law degree sequences is that they are a convenient *mathematical way* to model situations where a large amount of variability in the degrees in a real-world network is observed. We return to the question of power-law degree distributions and their tails in Section 1.7, where we speculate on why power laws might arise in real-world networks, as well as on the difficulties in estimating them.

In some networks, the degree distribution for large values is not quite a power law, while for small values it is. This is particularly appropriate when there are *costs* associated with having large degrees. For example, building and maintaining a large social network, or publishing papers with many people, requires time and energy, both of which are limited. Therefore, one cannot really expect to see the huge fluctuations that are predicted by power-law distributions. A related model is a power law with an *exponential cut off*, where the probability mass function is given by

$$p_k = ck^{-\tau}e^{-k/A}, \qquad k \geq 1, \tag{1.4.13}$$

for some large A. Thus, for k small compared to A the distribution looks like a power law, while for values that are large compared to A the exponential decay takes over.

We continue to discuss the notion of *highly connected* graph sequences.

1.4.2 Highly Connected Graph Sequences

The networks that we consider are often highly connected, in the sense that a large fraction of the vertices lies in a single connected component. For a graph $G = ([n], E)$ on n vertices and $v \in [n]$, let $\mathscr{C}(v)$ denote the *cluster* or *connected component* of $v \in [n]$, i.e., $\mathscr{C}(v) = \{u \in [n]\colon \mathrm{dist}_G(u, v) < \infty\}$, where we recall that $\mathrm{dist}_G(u, v)$ denotes the graph distance in G. Let $\mathscr{C}_{\max}$ denote the largest connected component, i.e., let $\mathscr{C}_{\max}$ satisfy $|\mathscr{C}_{\max}| = \max_{v\in[n]} |\mathscr{C}(v)|$, where $|\mathscr{C}(v)|$ denotes the number of vertices in $\mathscr{C}(v)$. Of course, the definition of $\mathscr{C}_{\max}$ might not be unambiguous, since there could be *two* or more maximal clusters, but this will often not occur in settings we are interested in. When there do exist two or more maximal clusters of the same size, then we let $\mathscr{C}_{\max}$ denote any of them with equal probability.

A graph sequence $(G_n)_{n\geq 1}$ is *highly connected* when $\mathscr{C}_{\max}$ is very large, and the maximal cluster is *unique* when all other clusters tend to be small. For the latter, we let $\mathscr{C}_{(2)}$ denote the second largest cluster. Then, the notion of highly connected graph sequences is formalized in the following definition:

Definition 1.6 (Highly connected graph sequences) A graph sequence $(G_n)_{n\geq 1}$ is called *highly connected* when

$$\liminf_{n\to\infty} |\mathscr{C}_{\max}|/n > 0, \tag{1.4.14}$$

in which case $\mathscr{C}_{\max}$ is called its *giant component*. Furthermore, for a highly connected graph sequence, its giant component is called *unique* when

$$\limsup_{n\to\infty} |\mathscr{C}_{(2)}|/n = 0. \tag{1.4.15}$$

The picture to have in mind is that highly connected graph sequences with a unique giant component consist of a 'huge blob' surrounded by 'dust'. In many random graph models, the second largest component is of size at most $\log n$.

Having defined highly connected graph sequences, we move on to define small-world graph sequences, for which we assume that the graphs in question are highly connected.

1.4.3 Small-World Graph Sequences

We continue to define what it means for a graph sequence $(G_n)_{n\geq 1}$ to be a small world. Intuitively, a small-world graph should have distances that are much smaller than those in a regular lattice or torus. When we consider the nearest-neighbor torus with width r in dimension d, and we draw two vertices uniformly at random, their distance is of order r. The size of the torus is $n = (2r+1)^d$, so that the typical distance between two uniform vertices is of order $n^{1/d}$, which grows as a positive power of n.

Recall that H_n denotes the distance between two uniformly chosen *connected* vertices, i.e., the graph distance between a pair of vertices drawn uniformly at random from all pairs of connected vertices in G_n. Here we recall that the graph distance $d_G(v_1, v_2)$ between the vertices v_1, v_2 denotes the minimal number of edges on paths connecting v_1 and v_2. Below, we deal with graph sequences $(G_n)_{n\geq 1}$ for which G_n is not necessarily connected, which explains why we condition on the two vertices being connected. We envision a situation where $(G_n)_{n\geq 1}$ is a highly connected graph sequence (recall Definition 1.6). We recall that H_n is the *typical distance of* G_n.

Definition 1.7 (Small-world graph sequences) We say that a graph sequence $(G_n)_{n\geq 1}$ is a *small world* when there exists a constant $K < \infty$ such that

$$\lim_{n\to\infty} \mathbb{P}(H_n \leq K \log n) = 1. \tag{1.4.16}$$

Further, we say that a graph sequence $(G_n)_{n\geq 1}$ is an *ultra-small world* when, for every $\varepsilon > 0$,

$$\lim_{n\to\infty} \mathbb{P}(H_n \leq \varepsilon \log n) = 1. \tag{1.4.17}$$

Note that, for a graph with a *bounded* degree $d_{\max}$, the typical distance is *at least* $(1-\varepsilon)\log n/\log d_{\max}$ with high probability (see Exercise 1.4). Thus, a random graph process with bounded degree is a small world *precisely* when the order of the typical distance is at most a constant times larger than the minimal value it can take on.

For a graph G, recall from (1.2.13) that $\mathrm{diam}(G)$ denotes the *diameter* of G, i.e., the maximal graph distance between any pair of connected vertices. Then, we could also have chosen to replace H_n in (1.4.16) by $\mathrm{diam}(G_n)$. However, as argued below (1.2.13), the diameter of a graph is a rather sensitive object which can easily be changed by making small changes to a graph in such a way that the scale-free nature and the typical distances H_n hardly change. This explains why we have a preference to work with the typical distance H_n rather than with the diameter $\mathrm{diam}(G_n)$.

We will see that in some models typical distances can be even much smaller than $\log n$, which explains the notion of an ultra-small world. Sometimes, there exists a constant K such that

$$\lim_{n\to\infty} \mathbb{P}(H_n \leq K \log\log n) = 1. \tag{1.4.18}$$

There are models for which (1.4.18) is satisfied, while $\mathrm{diam}(G_n)/\log n$ converges in probability to a positive limit due to the fact that there are long but thin strings of vertices that make the diameter large yet do not affect typical distances. This once more explains our preference for the typical graph distance H_n. We will also see cases for which H_n is of the order $\log n/\log\log n$.

In this section, we have given explicit definitions of sparse, scale-free, highly connected and small-world graph sequences. The remainder of this book, as well as Volume II, will discuss models for real-world networks, and will establish under which conditions the random graph sequences in these models satisfy these properties. In the next section, we discuss related empirical properties of real-world networks.

1.5 Further Network Statistics

In this section, we discuss a few other empirical properties of graph sequences: the clustering coefficient, degree-degree dependencies, centrality measures and community structure. These empirical and algorithmic properties of real-world networks will not be discussed in detail in this book. Needless to say, they are very interesting for the network models at hand.

Clustering in Networks

Clustering measures the extent to which neighbors of vertices are also neighbors of one another. For instance, in social networks, your friends tend to know each other, and therefore the amount of clustering is high. In general, clustering is high when the network has many triangles. The quantity that measures the amount of network clustering is the *clustering coefficient*. For a graph $G = ([n], E)$, we let

$$W_G = \sum_{1\leq i,j,k\leq n} \mathbb{1}_{\{ij,jk\in E\}} \tag{1.5.1}$$

denote two times the number of wedges in the graph G. The factor of two comes from the fact that the wedge ij, jk is the same as the wedge kj, ji, but it is counted twice in (1.5.1). We further let

$$\Delta_G = \sum_{1\leq i,j,k\leq n} \mathbb{1}_{\{ij,jk,ik\in E\}} \tag{1.5.2}$$

denote six times the number of triangles in G. Alternatively, we can write

$$W_G = 2\sum_{1\leq i,j,k\leq n:\, i<k} \mathbb{1}_{\{ij,jk\in E\}}, \qquad \Delta_G = 6\sum_{1\leq i<j<k\leq n} \mathbb{1}_{\{ij,jk,ik\in E\}}. \tag{1.5.3}$$

The clustering coefficient CC_G in G is defined as

$$\mathsf{CC}_G = \frac{\Delta_G}{W_G}. \tag{1.5.4}$$

Informally, the clustering coefficient measures the proportion of wedges for which the closing edge is also present. As such, it can be thought of as the probability that from a randomly drawn individual and two of its friends, the two friends are friends themselves.

Definition 1.8 (Highly clustered graph sequences) We say that a graph sequence $(G_n)_{n\geq 1}$ is *highly clustered* when

$$\liminf_{n\to\infty} \mathsf{CC}_{G_n} > 0. \tag{1.5.5}$$

As we will see in the remainder of this book, the models that we investigate tend *not* to be highly clustered, i.e., they tend to be *locally tree-like*. Throughout the book and Volume II, we do discuss extensions of network models that are highly clustered. These can, for example, be obtained by making local adaptations to the original models. Since the clustering coefficient only depends on the local neighborhoods of vertices, these random graph models are often closely related to the original models.

Sometimes the clustering coefficient is computed for fixed vertices. For this, the *clustering coefficient of vertex j* is defined to be

$$\mathsf{CC}_G(j) = \frac{1}{d_j(d_j-1)} \sum_{i,k} \mathbb{1}_{\{ij,jk,ik\in E\}}. \tag{1.5.6}$$

This allows one to define a different clustering coefficient, named after Watts and Strogatz (1998), as

$$\mathsf{CC}_{G,SW} = \frac{1}{n} \sum_{j\in[n]} \mathsf{CC}_G(j). \tag{1.5.7}$$

Degree-Degree Dependencies

Degree-degree dependencies measure whether vertices of high degree are more likely to connect to vertices of high degree, or rather to vertices of low degree. To define this, let E' denote the set of (directed) edges, so that $|E'| = 2|E|$ in the undirected setting. For the time being, assume that $G = ([n], E)$ is undirected. For each directed edge (u, v), we let the vector (d_u, d_v) denote the degrees at the ends of the edge. The vector $((d_u, d_v))_{(u,v)\in E'}$ is a collection of $|E'| = 2|E|$ two-dimensional integer variables. A network G is called *assortative* when a high value of d_u typically corresponds to a high value of d_v, and it is called *disassortative* otherwise. The consensus is that social networks tend to be assortative, while technological networks (like the Internet) tend to be disassortative.

Degree-degree dependencies deal with the dependency between the coordinates of a collection of two-dimensional variables. One way to measure this is to use the *correlation coefficient*. This results in the *assortativity coefficient* defined as

$$\rho_G = \frac{\sum_{i,j\in[n]} \left(\mathbb{1}_{\{ij\in E'\}} - d_i d_j/|E'|\right) d_i d_j}{\sum_{i,j\in[n]} \left(d_i \mathbb{1}_{\{i=j\}} - d_i d_j/|E'|\right) d_i d_j}. \tag{1.5.8}$$

Pick a (directed) edge uniformly at random from E', and let X and Y be the degrees of the vertices that the edge points to and from, respectively. Then, we can interpret ρ_G as the correlation coefficient of the random variables X and Y (see Exercise 1.5). Since ρ_G is a correlation coefficient, clearly $\rho_G \in [-1, 1]$. We call a network G assortative when $\rho_G > 0$ and disassortative when $\rho_G < 0$.

We can rewrite the assortativity coefficient ρ_G (see Exercise 1.6) as

$$\rho_G = \frac{\sum_{ij \in E'} d_i d_j - (\sum_{i \in [n]} d_i^2)^2/|E'|}{\sum_{i \in [n]} d_i^3 - (\sum_{i \in [n]} d_i^2)^2/|E'|}. \tag{1.5.9}$$

Equation (1.5.9) points us to a problem with the definition in (1.5.8). When dealing with an independent and identically distributed sample $((X_i, Y_i))_{i=1}^n$, the correlation coefficient is a valuable measure of dependence, but only when the *variances* of X_i and Y_i are bounded. In this case, the sample correlation coefficient ρ_n converges to the correlation coefficient ρ given by

$$\rho = \frac{\mathrm{Cov}(X, Y)}{\sqrt{\mathrm{Var}(X)\mathrm{Var}(Y)}}. \tag{1.5.10}$$

When the variances of X or Y do not exist, the correlation coefficient might not make much sense. In fact, Litvak and the author (see van der Hofstad and Litvak (2014), Litvak and van der Hofstad (2013)) proved that such convergence (even for an i.i.d. sample) can be to a proper random variable, that has support containing a subinterval of $[-1, 0]$ *and* a subinterval of $[0, 1]$, giving problems in the interpretation.

For networks, ρ_G in (1.5.9) is always well defined, and gives a value in $[-1, 1]$. However, also for networks there is a problem with this definition. Indeed, van der Hofstad and Litvak (2014) and Litvak and van der Hofstad (2013) prove that if a limiting value of ρ_G exists for a sequence of networks, then $\liminf_{n\to\infty} \rho_G \geq 0$, so no asymptotically disassortative graph sequences exist in this sense. Also, there is an example where the limit exists in distribution and the limit is a proper random variable. Naturally, other ways of classifying the degree-degree dependence can be proposed, such as the correlation of their *ranks*. Here, for a sequence of numbers $x_1, \ldots, x_n$ with ranks $r_1, \ldots, r_n$, x_i is the r_ith largest of $x_1, \ldots, x_n$. Ties tend to be broken by giving random ranks for equal values. For practical purposes, maybe a scatter plot of the values might be the most useful way to gain insight into degree-degree dependencies.

Closeness Centrality

In networks, one is often interested in how central or important vertices are. Of course, how important a vertex is depends on the notion of importance. See Boldi and Vigna (2014) for a survey of centrality measures. In the following, we discuss a few such notions, starting with *closeness centrality* as first introduced by Bavelas (1950). In this notion, vertices that are close to many other vertices are deemed to be important. This prompts the definition of closeness centrality of a vertex i as[3]

$$C_i = \frac{n}{\sum_{j \in [n]} \mathrm{dist}_G(i, j)}, \tag{1.5.11}$$

[3] Some authors remove $i = j$ from the sum and replace the factor n by $n - 1$.

where we recall that $\mathrm{dist}_G(i, j)$ denotes the graph distance between the vertices i and j. Thus, vertices with high closeness centrality are central in the network in terms of being close to most other vertices, while vertices with low closeness centrality are remote. One can expect that the closeness centrality of i plays an important role for the functionality of the network from the perspective of i, for example when a rumor is started at i or when i is the source of an infection. The above definition only makes sense for *connected* graphs, since $C_i = C_j = 0$ when i is not connected to j. This may be adapted by averaging only over connected components in (1.5.11), or by computing C_i only for the values of i in the largest connected component (which is often unique and contains a large proportion of the vertices; recall the notion of highly connected graph sequences in Definition 1.6).

Closeness centrality has the problem that pairs of disconnected vertices render its value hard to interpret. In a similar vein, pairs of vertices that are far apart also have a disproportionate effect on the closeness centrality of a graph. Boldi and Vigna (2014) propose to use *harmonic centrality* instead, which is defined as

$$\sum_{j \in [n]} \frac{1}{\mathrm{dist}_G(i, j)}. \tag{1.5.12}$$

In the harmonic centrality of i, the values $j \in [n]$ for which $\mathrm{dist}_G(i, j)$ is large contribute very little to the sum in (1.5.12). This prompts Boldi and Vigna to state that

> Harmonic centrality is strongly correlated to closeness centrality in simple networks, but naturally also accounts for nodes j that cannot reach i.

Betweenness Centrality

Another notion of centrality is *betweenness centrality* as independently invented by Anthonisse (1971) and Freeman (1977). While the closeness centrality of vertex i is supposed to measure how fast a message travels from vertex i, its betweenness centrality is supposed to measure how important vertex i is in sending the message quickly around. For vertices j, k, let n_{jk} denote the number of shortest paths between j and k. Further, let n^i_{jk} be the number of shortest paths between j and k that pass through i. Thus, $n^i_{jk} \geq 1$ when i lies on a shortest path from j to k, and 0 otherwise (which includes the case where j and k are not connected). The *betweenness centrality* of vertex i is defined as

$$B_i = \sum_{1 \leq j < k \leq n} n^i_{jk}/n_{jk}. \tag{1.5.13}$$

Note that this definition also makes sense for *directed* networks, where the sum should be over $j, k \in [n]$, $j \neq k$.

Vertices with high betweenness centrality serve as bottlenecks for the communication in the network. When under attack, they are the most likely to seriously diminish the network functionality for services that rely on the network's connectivity.

Community Structure of Networks

Often, networks consist of parts that are more highly connected than the entire network itself. These parts often go under the name of *communities*. In social networks, we can think of classes in schools, families, or groups of people interested in certain topics such as fans of soccer clubs or music bands. In scientific citation networks, where the vertices are papers and the (directed) edges are formed by references of one paper to another, we can think of the fields in science as giving rise to a community structure. However, many papers, particularly the important ones, are linking different subfields to one another by applying methodology that is common in one domain to a different domain. Exciting new developments in science often occur at the boundaries between fields.

Even though we have some intuitive idea about what makes a community, it is not clear exactly how it should be defined. For example, often people belong to several communities, and the boundaries of communities tend to be blurry. We thus have no precise *definition* of what makes a community, and many real-world networks do not have a clearly defined community structure. Of course, being able to detect communities is highly relevant, if only to be able to send the right advertisement to the people who are actually interested in it. As a result, *community detection* has become one of the important topics in network science of the past decades. Since network community structures are not clearly defined, such methodologies are often ill defined, but this only makes them more interesting and worthy to study. We refer to the extensive overview paper by Fortunato (2010) for a more detailed description of the problem, network models that have a community structure, as well as the available algorithms to detect them in real-world networks from a physics perspective. See Leskovec et al. (2010) for an empirical analysis of the performance of several community detection algorithms on real-world networks.

1.6 Other Real-World Network Examples

In this section, we describe a number of other examples of real-world networks that have been investigated in the literature, and where some of the empirical properties discussed in the previous sections, such as the small-world and scale-free phenomenon, are observed. Real-world networks are often divided into four main categories:

1. *Technological networks,* such as the Internet, transportation, telecommunication networks and power grids;
2. *Social networks,* such as friendship and virtual social networks, but also collaboration networks;
3. *Information networks,* such as the World-Wide Web and citation networks;
4. *Biological networks,* such as biochemical networks, protein interaction networks, metabolic and neural networks.

In this section, we discuss some of these examples. We discuss the 'Six Degrees of Separation' paradigm in social networks, the Facebook network, the Kevin Bacon Game and the movie actor network, Erdős numbers and collaboration networks and the World-Wide Web. We focus on some of the empirical findings in the above examples, and discuss some of the key publications on these empirical findings. Needless to say, each of these examples separately deserves its own book, but we do not dive too deep into the details. We try our best to give appropriate references to the literature; see also the notes and discussion in Section 1.10 for more details.

1.6.1 Six Degrees of Separation and Social Networks

In 1967, Stanley Milgram performed an interesting experiment.[4] In his experiment, Milgram sent 60 letters to various recruits in Wichita, Kansas, U.S.A., who were asked to deliver the letter to the wife of a divinity student living at a specified location in Cambridge, Massachusetts. The participants could only pass the letters (by hand) to personal acquaintances whom they thought might be able to reach the target, either directly, or via a 'friend of a friend'. While fifty people responded to the challenge, only three letters (or roughly 5 percent) eventually reached their destination. In later experiments, Milgram managed to increase the success rate to 35 percent and even 95 percent, by pretending that the value of the package was high, respectively, by adding more clues about the recipient, such as her/his occupation. See Milgram (1967); Travers and Milgram (1969) for more details.

The main conclusion from the work of Milgram was that most people in the world are connected by a chain of at most 6 'friends of friends', and this phrase was dubbed 'Six Degrees of Separation'. The idea itself was already proposed in 1929 by the Hungarian writer Frigyes Karinthy (1929) in a short story called 'Chains', see also Newman et al. (2006) where a translation of the story is reproduced.[5] Playwright John Guare popularized the phrase when he chose it as the title for his 1990 play. In it, Ousa, one of the main characters, says:

> Everybody on this planet is separated only by six other people. Six degrees of separation. Between us and everybody else on this planet. The president of the United states. A gondolier in Venice... It's not just the big names. It's anyone. A native in the rain forest. (...) An Eskimo. I am bound to everyone on this planet by a trail of six people. It is a profound thought.

The fact that any number of people can be reached by a chain of at most 6 intermediaries is indeed rather striking. It would imply that two people in as remote areas as Greenland and the Amazon could be linked by a sequence of at most 6 'friends of friends'. This makes the phrase 'It's a small world' very appropriate indeed! Another key reference in the small-world work in the social sciences is the paper by Pool and Kochen (1978), which was written in 1958, and was circulating around in the social sciences for twenty years before it was finally published in 1978.

The idea of Milgram was taken up afresh in 2001, with the added possibilities of the computer era. In 2001, Duncan Watts, a professor at Columbia University, together with colleagues Peter Dodds and Roby Muhamad redid Milgram's experiment using an e-mail message as the 'package' that needed to be delivered. Including initial senders and senders that were contacted along the chains of messages, data were recorded on 61,168 individuals from 166 countries, constituting 24,163 distinct message chains. Surprisingly, Watts found that the average number of intermediaries was again six. See Dodds et al. (2003) for details, including a much more extensive statistical analysis of the data compared to Milgram's analysis. Watts' research, and the advent of the computer age, has opened up new areas of

[4] See `http://www.stanleymilgram.com/milgram.php` for more background on the psychologist Milgram. Maybe Stanley Milgram is best known for his experiments on obedience, giving us a rather disturbing picture of how humans deal with authority. This experiment uses a highly controversial 'shock machine' to administer shocks to 'pupils' who do not obey.

[5] Possibly, Karinthy was in turn inspired by Guglielmo Marconi's 1909 Nobel Prize address.

inquiry related to six degrees of separation in diverse areas of network theory such as power grid analysis, disease transmission, graph theory, corporate communication, and computer circuitry. See Watts (1999) for a popular account of the small-world phenomenon.

To put the idea of a small world into network language, we define the vertices of the social graph to be the inhabitants of the world (so that $n \approx 6$ billion), and we draw an edge between two people when they know each other. Needless to say, we should make more precise what it means to 'know each other'. There are various possibilities. We could mean that the two people involved have shaken hands at some point, or that they know each other on a first name basis.

In a series of works, Jon Kleinberg (2000b; 2000a) investigates another aspect of Milgram's experiment, one that Milgram himself, and also Dodds et al. (2003), failed to recognize. Not only do social networks constitute a small world, but the people involved in the message passing also managed to *find* the short routes. Kleinberg called this the *navigability* of social networks. See also Easley and Kleinberg (2010) for a more algorithmic perspective to networks.

Our social connections play an important role in our lives, as made clear by the work of Mark Granovetter (1973; 1995), who argues that we find our jobs through our 'weak links' or 'weak ties'. Indeed, the people you know really well will be less likely to introduce you to a new employer, since they are few and they have already had plenty of occasions to do so. However, the many people whom you know less well, i.e., your weak ties, are an enormous source of information and contacts and thus may know of many more job opportunities. This is the reason why the word 'network' is often used as a verb rather than a noun (as we will mostly use it). Of course, in social networks one is not only interested in networks consisting of vertices and edges, but also in the properties of the links (for example, whether they are strong or weak ties). These properties are called *attributes*. The relevance of social networks and the value of their links is also made succinctly clear in the popular book on the 'tipping point' by Malcolm Gladwell (2006).

An interesting report about social networks that is worth mentioning is by Liljeros, Edling, Amaral and Stanley (2001), who investigated sexual networks in Sweden, where two people are connected when they have had a sexual relation in the previous year. They find that the degree distributions of males and females obey power laws, with estimated exponents of $\tau_{\text{fem}} \approx 2.5$ and $\tau_{\text{mal}} \approx 2.3$. When extending the data to the entire lifetime of the Swedish population, the estimated exponents decrease to $\tau_{\text{fem}} \approx 2.1$ and $\tau_{\text{mal}} \approx 1.6$. The latter only holds in the range between 20 and 400 contacts, after which it is exponentially truncated. Clearly, this has important implications for the transmittal of sexual diseases.

One of the main difficulties of social networks is that they are notoriously hard to measure. For instance, questionnaires are hard to interpret, since people have a different idea of what a certain social relation means. Also, questionnaires are physical and take time to collect. Finally, in a large population, one cannot ask all the people in the population, so that a list of the friends of a person surveyed does not bring out the social network as a whole, since her/his friends are unlikely to be surveyed as well. As a result, researchers are interested in examples of social networks that *can* be measured, for example due to the fact that they are *electronic*. Examples are e-mail networks or social networks such as Facebook.

In the next section, we study the prime example Facebook in some more detail.

1.6.2 Facebook

With 721 million active users, and 69 billion friendship links in 2011, Facebook was the largest virtual or online friendship network at the time. Since its creation in February 2004, Facebook has earned itself a prominent role in society, and has proven to be an enormous source of information about human (online) interaction. The Facebook graph has been investigated in detail by two teams of academic researchers around Lars Backstrom from Facebook (see Ugander et al. (Preprint (2011)); Backstrom et al. (2012)). Here, a user is defined to be active if he/she has logged in at least once in a 28 day time period in May 2011 (the period that was investigated in Ugander et al. (Preprint (2011)); Backstrom et al. (2012)) and has at least one Facebook friend. Studying a network this big gives rise to serious algorithmic challenges.

Ugander et al. (Preprint (2011)) find that 99.91 percent of the active Facebook users is in the giant component, so that Facebook is indeed very highly connected (recall Definition 1.6), the second largest connected component consisting of a meagre 2000 some users. The assortativity coefficient (recall (1.5.8)) is 0.226, quite large as one might expect from a social network. The complementary cumulative distribution function of the Facebook degrees is plotted in Figure 1.8. This distribution does not resemble a power law. Of course, the degree distribution is affected by the Facebook policy of limiting the number of friends by 5000 at the time of the measurements. While some social networks are reported to show densification, Facebook on the contrary is claimed to show sparsification in the sense that the proportion of existing edges to all possible edges decreases. On the other hand, the average degree does seem to increase, albeit slowly (see e.g., Table 1.1).

The typical distances in the Facebook graph, as well as their evolution, are studied in Backstrom et al. (2012); see Figure 1.9. We see that as time proceeds, distances shrink and stabilize at around 3–6. The average of the distance distribution, i.e., $\mathbb{E}[H_n]$ with H_n the hopcount, is 4.74, so that there are, on average, 3.74 degrees of separation. This was advertised broadly; for example, on November 22, 2011, the Telegraph posted a web article titled 'Facebook cuts six degrees of separation to four.'

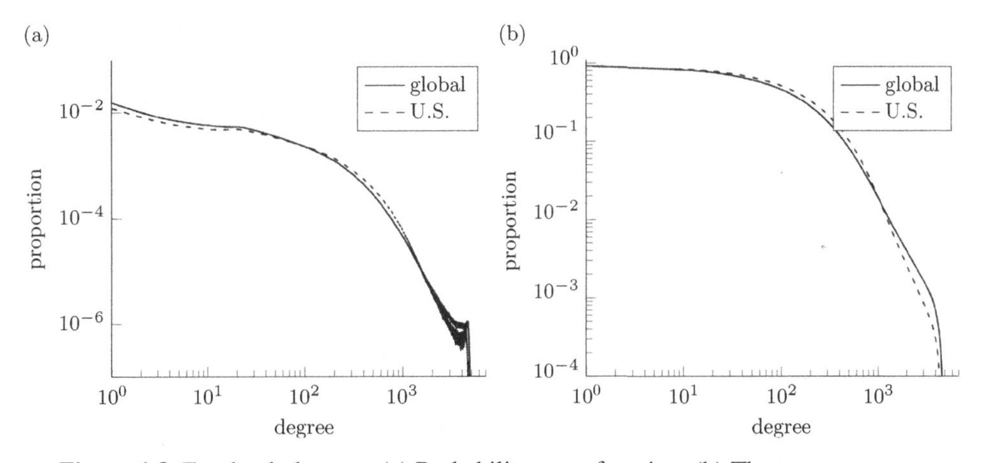

Figure 1.8 Facebook degrees. (a) Probability mass function. (b) The complementary cumulative distribution function (CCDF). The CCDF at degree k measures the fraction of users who have degree at least k.

Table 1.1 *Evolution of the Facebook degrees over time, with current denoting May 2011.*

Year	it	se	us	fb
2007	1.31	3.90	119.61	99.50
2008	5.88	46.09	106.05	76.15
2009	50.82	69.60	111.78	88.68
2010	122.92	100.85	128.95	113.00
2011	198.20	140.55	188.309	169.03
final	226.03	154.54	213.76	190.44

Note: `it` denotes the Italian subgraph of Facebook, `se` the Swedish, `us` the United States and `fb` the entire Facebook graph.

Table 1.2 *Evolution of the Facebook diameter.*

Year	it	se	us	fb
2007	41	17	13	14
2008	28	17	17	16
2009	21	16	16	15
2010	18	19	19	15
2011	17	20	18	35
final	19	19	20	58
	Exact diameter giant component			
final	25	23	30	41

Note: `it` denotes the Italian subgraph of Facebook, `se` the Swedish, `us` the United States and `fb` the entire Facebook graph.

To make the networks more amenable to analysis, also certain subgraphs have been investigated. In Table 1.1, `it` stands for Italy, `se` for Sweden, `us` for the United States and `fb` for the entire Facebook graph. While initially there are large differences in these subgraphs, after 2010 they appear to be highly similar. Only the distances in the entire Facebook graph appear to be somewhat larger than those in the respective subgraphs (see Figure 1.9(a) compared to 1.9(b)). Interestingly, it turns out that the variance of the hopcount is rather small, and in most cases (certainly after 2007, after which Facebook adoption in Sweden and Italy stopped, being much behind that in the U.S.), this variance is at most 0.5. Figure 1.10 shows how the average distance $\mathbb{E}[H_n]$ evolves over time.

Let us discuss one last feature of the Facebook data set. In Figure 1.11, we plot the average number of friends of friends of individuals with given degrees. We see that while the average number of friends of friends is larger than the average number of friends of the individual when the degree is at most 800, above this value it is lower. Theorem 1.2 of course states that, on *average*, your friends have more friends than you do. Figure 1.11 shows that this is only true up to a point. When you have a lot of friends yourself, you cannot expect your friends to have even more!

Facebook has sparked a tremendous research effort. Online social networks allow for a quantitative analysis of social networks that is often impossible for offline social networks.

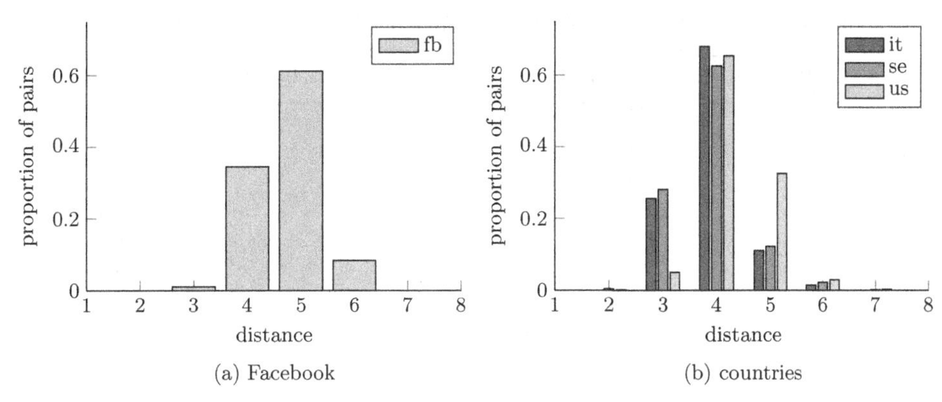

(a) Facebook (b) countries

Figure 1.9 (a) Distances in the Facebook graph (b) Distances in Facebook in different subgraphs.

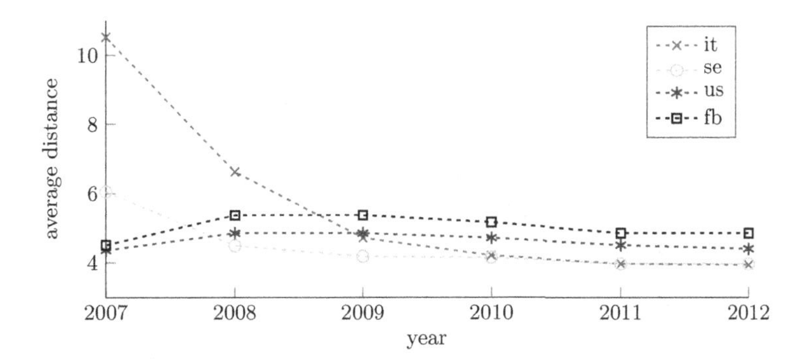

Figure 1.10 Evolution of Facebook average distance over time

However, this also raises the question of what an online friendship really means. In an extensive survey of the literature based on Facebook in various scientific domains, Wilson, Goslin and Graham (2012) summarize how Facebook data has been used in the literature to answer questions like 'Who is using Facebook and why?' and 'How are people presenting themselves on Facebook?' The answer to such questions may shed light on the relation between online and offline social networks, which are likely quite different.

1.6.3 Kevin Bacon Game and the Internet Movie Data Base

Another example of a large network is the *movie actor network*. In this example, the vertices are movie actors, and two actors share an edge when they have played in the same movie. Started as a student prank, this network has attracted some attention in connection to Kevin Bacon, who appears to be central in this network. The Computer Science Department at the University of Virginia has an interesting web site on this example.[6]

See Table 1.3(a) for the Kevin Bacon Numbers of all the actors in this network. There is one actor at distance 0 from Kevin Bacon (namely, Kevin Bacon himself), 2,769 actors have

[6] See The Oracle of Bacon at the University of Virginia website on `http://www.cs.virginia.edu/oracle/`.

Table 1.3 *(a) Kevin Bacon and (b) Sean Connery numbers (as of April 28, 2013).*

Kevin Bacon No.	# of actors
0	1
1	2,769
2	305,215
3	1,021,901
4	253,177
5	20,060
6	2,033
7	297
8	25
9	7

Sean Connery No.	# of actors
0	1
1	2,564
2	338,016
3	1,044,666
4	201,058
5	16,874
6	1,971
7	291
8	37
9	7

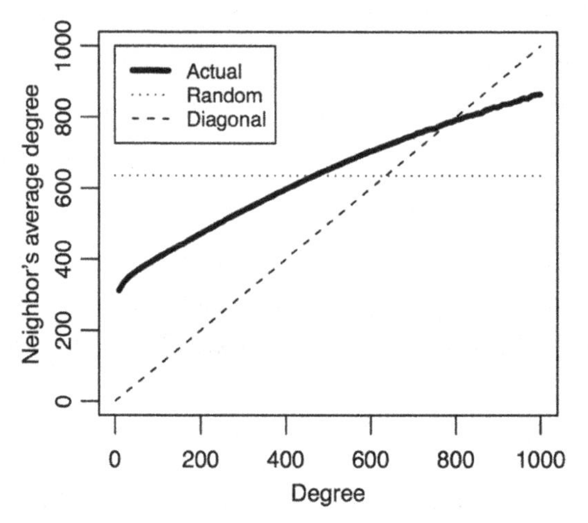

Figure 1.11 The average number of friends of friends of individuals of a fixed degree. 'Random' indicates the average degree, while 'Diagonal' indicates the line where the average degrees of neighbors of an individual equals that of the individual.

played in a movie starring Kevin Bacon, 338016 actors have played in a movie in which another movie star played who her/himself had played in a movie starring Kevin Bacon, etc. In total, the average Kevin Bacon number is 2.937, which is one over his closeness centrality. In search for 'Six Degrees of Separation', one could say that most pairs of actors are related by a chain of co-actors of length at most 6.

It turns out that Kevin Bacon is not the most central vertex in the graph. A more central actor is, for example, Sean Connery. See Table 1.3(b) for a table of the Sean Connery Numbers. By computing the average of these numbers we see that the average Connery Number is about 2.731, so that Connery is more central than Bacon. To quote the Oracle of Bacon: 'This is not to denigrate Mr. Bacon, and it should be noted that being the 370th best center out of 2.6 million people makes Bacon a better center than 99 percent of the people who have ever appeared in a feature film.' See Table 1.4 for the top 10 of most central actors, starring Harvey Keitel as the proud number 1! It is notable just how close the top 10 is,

Table 1.4 *Top 10 most central actors in the IMDb (as of April 28, 2013), and their average distance to all other actors (which is one over their closeness centrality). Remarkably, while some of these central actors are indeed well known, others seem less so.*

Number	Name	Average distance
1	Harvey Keitel	2.848635
2	Dennis Hopper	2.849329
3	Robert De Niro	2.855810
4	David Carradine	2.857729
5	Martin Sheen	2.858291
6	Udo Kier	2.859489
7	Michael Madsen	2.860010
8	Donald Sutherland	2.860447
9	Michael Caine	2.862189
10	Eric Roberts	2.867675

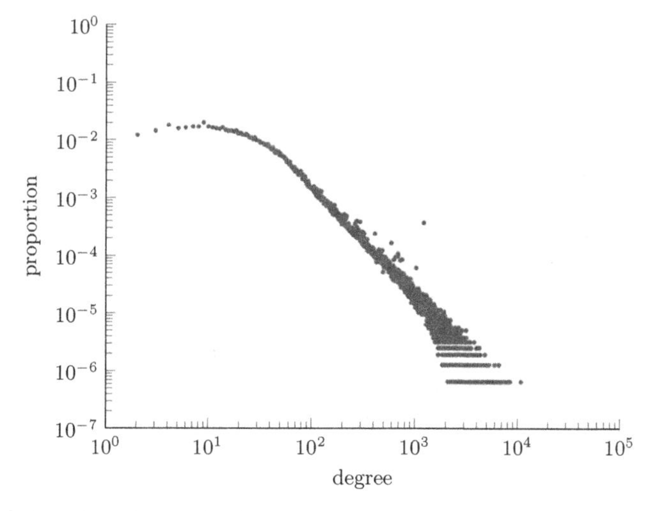

Figure 1.12 Log-log plot of the degree sequence in the Internet Movie Data base in 2007.

differing only in the third decimal. The worst possible closeness centrality is the actor who has inverse closeness centrality 10.105, which is more than three times as large as that of Kevin Bacon. On the web site `http://www.cs.virginia.edu/oracle/`, one can try out one's own favorite actors to see what Bacon number they have, or what the distance is between them.

In Figure 1.12, the degree sequence of the Internet Movie Data base is given in log-log format. We can see that a large part of the log-log plot is close to a straight line, but it appears as if the tails are slightly thinner. Indeed, while Barabási and Albert (1999) estimate the power-law exponent to be 2.3, Amaral et al. (2000) looked closer at the degree distribution to conclude that the power-law in fact has an exponential cut-off. One striking feature of the

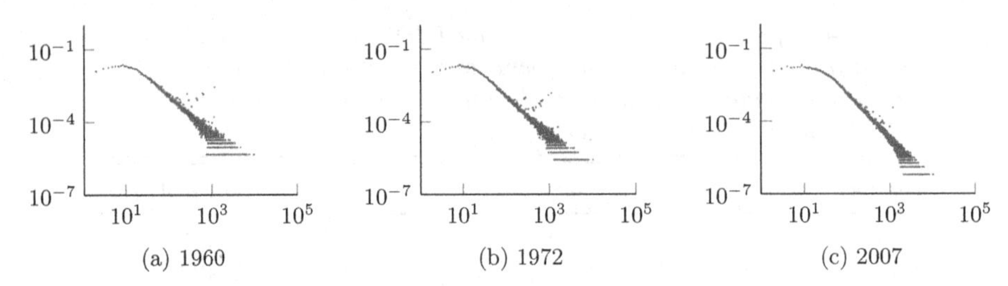

Figure 1.13 Log-log plot of the evolution of the degree sequence in the Internet Movie Data base in 1960, 1972 and 2007.

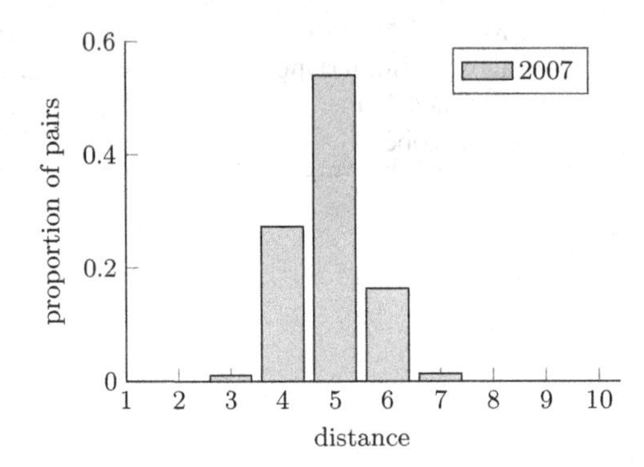

Figure 1.14 Typical distances in the Internet Movie Data base in 2007.

data is that it is quite *stable* in time. In Figure 1.13, the evolution of the degree sequence is shown. This stability is not due to the fact that the graph in, say, 1960 is contained in the one in 1972, since the number of movie actors grows too quickly. Further, the degrees of individual actors can only increase over time (and remain constant when actors retire) and new actors are added as time progresses. That the data set appears so stable must mean that there are organizing principles in how the data arises (or is acquired). How this occurs is unclear to us.

In Figure 1.14, the typical distances of actors in the Internet Movie Data base are given. We see that typical distances are relatively small, the vast majority of pairs of actors is within distance 7, so again 'Six Degrees of Separation'. Also the distances are stable, as can be seen by their evolution in Figure 1.15.

1.6.4 Erdős Numbers and Collaboration Networks

A further example of a complex network that has drawn attention is the collaboration graph in mathematics. This is popularized under the name 'Erdős number project'. In this network, the vertices are mathematicians, and there is an edge between two mathematicians when they have co-authored a paper. Thus, this network can be seen as the math equivalent of the more glamorous movie actor network.[7] The Erdős number of a mathematician is how many papers that mathematician is away from the legendary mathematician Paul Erdős,

[7] See `http://www.ams.org/msnmain/cgd/index.html` for more information.

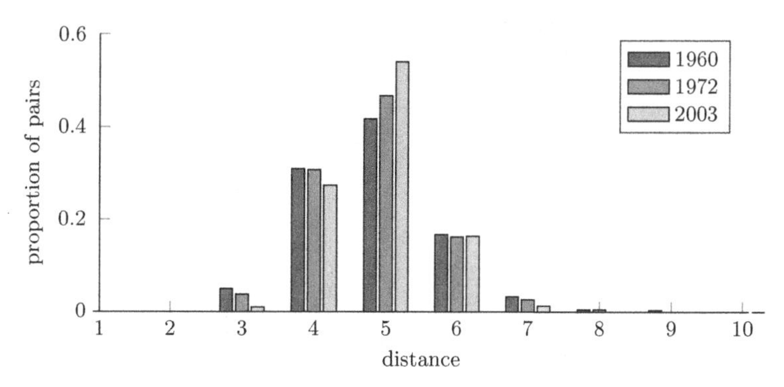

Figure 1.15 The evolution of the typical distances in the Internet Movie Data base in 1960, 1972 and 2003.

who was extremely prolific with around 1,500 papers and 504 collaborators. Thus, Paul Erdős has Erdős number 0, those who have published a paper with Paul Erdős have Erdős number 1, etc. Of those that are connected by a trail of collaborators to Erdős, the maximal Erdős number is claimed to be 15. On the above web site, one can see how far one's own professors are from Erdős. Also, it is possible to see the collaboration distance between any two mathematicians.

The Erdős numbers have also attracted attention in the literature. De Castro and Grossman (1999a,b) investigate the Erdős numbers of Nobel prize laureates, as well as Fields medal winners, to come to the conclusion that Nobel prize laureates have Erdős numbers of at most 8 with average 4–5, while Fields medal winners have Erdős numbers of at most 5 with average 3–4.[8] There the following summary of the collaboration graph is given. In July, 2004, the collaboration graph consisted of about 1.9 million authored papers in the Math Reviews database written by a total of about 401,000 different authors. Approximately 62.4 percent of these items are by a single author, 27.4 percent by two authors, 8.0 percent by three authors, 1.7 percent by four authors, 0.4 percent by five authors and 0.1 percent by six or more authors. The largest number of authors shown for a single item is in the 20s. Sometimes the author list includes 'et al.', so that the number of co-authors is not always known precisely.

The fraction of items authored by just one person has steadily decreased over time, starting out above 90 percent in the 1940s and currently standing at under 50 percent. The entire graph has about 676,000 edges, so that the average number of collaborators per person is 3.36. In the collaboration graph, there is one giant component consisting of about 268,000 vertices, so that the graph is highly connected (recall Definition 1.6). Of the remaining 133,000 authors, 84,000 of them have written no joint papers, and these authors correspond to isolated vertices. The average number of collaborators for people who have collaborated is 4.25. The average number of collaborators for people in the giant component is 4.73, while the average number of collaborators for people who have collaborated but are not in the giant component is 1.65. There are only 5 mathematicians with degree at least 200, the largest degree is for Erdős, who has 504 co-authors. The clustering coefficient of the collaboration graph is 1308045/9125801 = 0.14, so that it is highly clustered. The average path

[8] See also `http://www.oakland.edu/enp` for more information.

Table 1.5 *Erdős Numbers*

Erdős Number	# of Mathematicians
0	1
1	504
2	6,593
3	33,605
4	83,642
5	87,760
6	40,014
7	11,591
8	3,146
9	819
10	244
11	68
12	23
13	5

lengths are small, making this graph a *small-world* graph (recall Definition 1.7). Indeed, while the diameter of the largest connected component is 23, the average distance $\mathbb{E}[H_n]$ is estimated at 7.64, so there are seven degrees of separation (eight if you wish to include, say, 75 percent of the pairs).

For the Erdős numbers, we refer to Table 1.5. The median Erdős number is 5, the mean is 4.65 and the standard deviation is 1.21. We note that the Erdős number is finite if and only if the corresponding mathematician is in the largest connected component of the collaboration graph. Figure 1.16 shows the degree sequence of the collaboration graph amongst mathematicians, Figure 1.17 that of the induced subgraph of mathematicians with Erdős number equal to one. Figure 1.18 gives an artistic impression of the latter graph.[9]

Newman (2001) has studied several collaboration graphs. He finds that several database(s) are such that the degrees have power laws with exponential cut-offs. These data bases are various arXiv data bases in mathematics and theoretical physics, the MEDLINE data base in medicine and the data bases in high-energy physics and theoretical computer science. Also, the average distance between scientists is shown to be rather small, which is a sign of the small-world nature of these networks. Finally, the average distance is compared to $\log n / \log d$, where n is the size of the collaboration graph and d is the average degree (recall Definition 1.7). The fit shows that these are close. Newman (2000) gives further results.

Barabási et al. (2002) investigate the evolution of scientific collaboration graphs, taking the collaboration graph amongst mathematicians as one of their prime examples. The main conclusion is that scientists are more likely to write papers with other scientists who have written many papers, i.e., there is a tendency to write papers with others who have already written many. This preferential attachment is shown to be a possible explanation for the existence of power laws in collaboration networks (see Chapter 8). Given the fact that exponential cut-offs are also observed for collaboration networks, these claims should be taken with a grain of salt.

[9] This figure is taken from `http://www.orgnet.com/Erdos.html`.

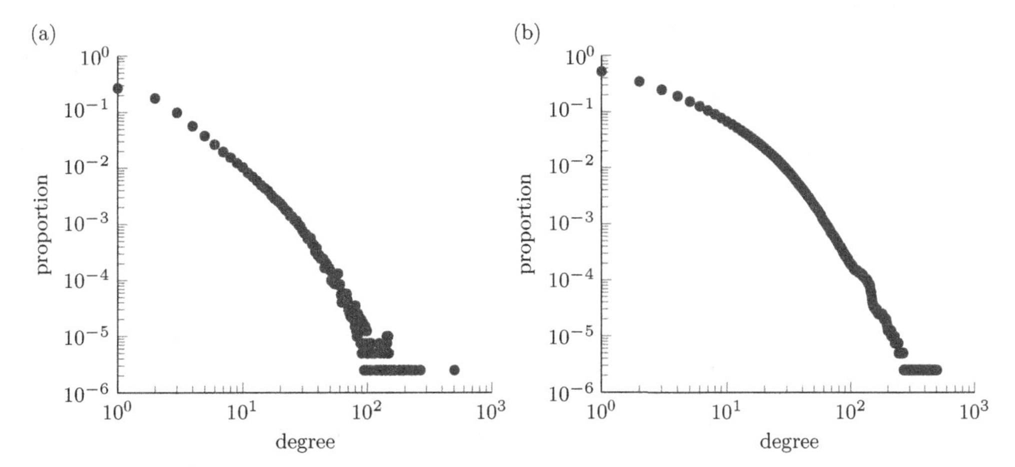

Figure 1.16 The degree sequence of the collaboration graph in mathematics from Grossman (2002). (a) Probability mass function. (b) The complementary cumulative distribution function.

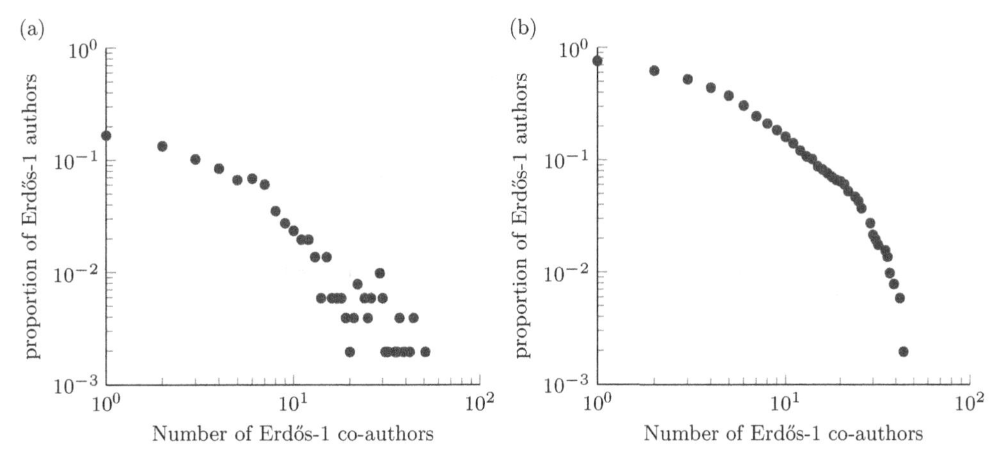

Figure 1.17 The degree sequence in the subgraph in the collaboration graph amongst mathematicians, of mathematicians with Erdős number one. (a) Probability mass function. (b) The cumulative complementary distribution function.

1.6.5 The World-Wide Web

A complex network that has attracted enormous attention is the World-Wide Web (WWW). Invented in the 1980s by Tim Berners-Lee and other scientists at the high-energy lab CERN in Geneva to exchange information between them, it has grown into *the* virtual source of information. Together with the Internet, the WWW has given rise to a network revolution since the 1980s.

There is some confusion about what the WWW really is, and what makes it different from the Internet. The elements of the WWW are web pages, and there is a (directed) connection between two web pages when the first links to the second. In the Internet, on the other hand, the edges correspond to physical cables between routers. Thus, while the WWW is virtual, the Internet is physical. With the world becoming ever more virtual, and the WWW growing

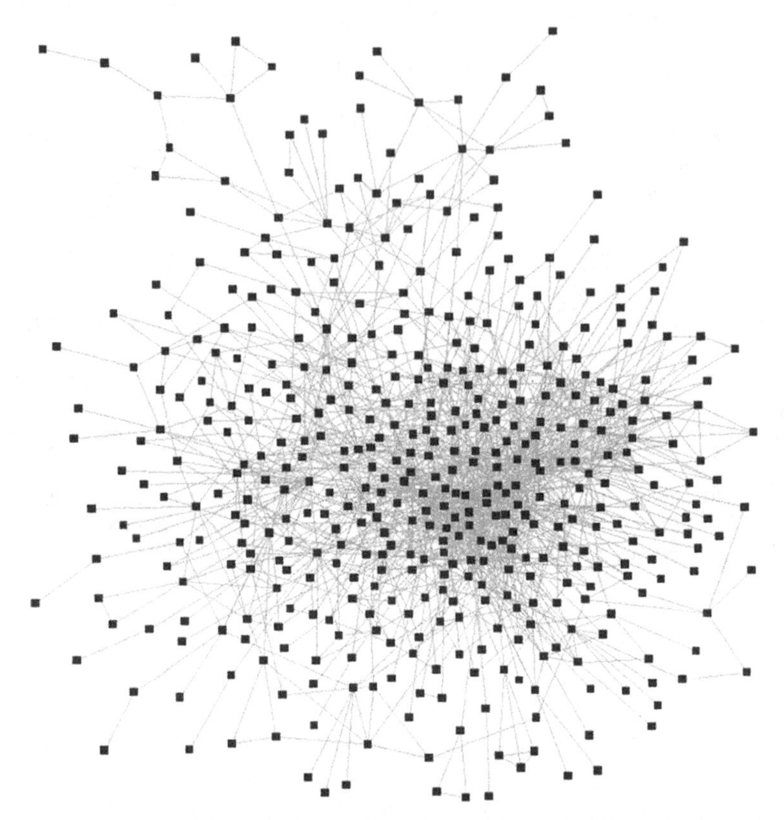

Figure 1.18 An artistic impression of the subgraph in the collaboration graph amongst mathematicians, of mathematicians with Erdős number one. This figure is taken from `http://www.orgnet.com/Erdos.html`.

at tremendous speed, the interest in properties of the WWW has grown as well. It is of great practical importance to know what the structure of the WWW is, for example, in order for search engines to be able to explore it efficiently.

It is hard to obtain data about the structure of the WWW. For example, it is already unclear how large the WWW really is. Hirate et al. (2008) estimated that the WWW consisted of 53.7 billion (=10^9) web pages in 2005, of which 34.7 billion web pages are indexed by Google. Early in 2005, Gulli and Signori estimated that it consists of 11.5 billion web pages. Thus, it is clear that the WWW is large, but it is hard to say precisely how large. This is partly due to the fact that large parts are unreachable, and other parts are not indexed by search engines. See Newman (2010, Section 4.1) for more background, the book by Bonato (2008) devoted to the WWW graph and Easley and Kleinberg (2010, Chapters 13–14) on the WWW and Web searching.

The most substantial early analysis of the WWW was performed by Broder et al. (2000), following up on earlier work by Kumar et al. (1999, 2000) in which the authors divide the WWW into several distinct parts; see also Figure 1.19. This division is into four main parts:

(a) The central core or Strongly Connected Component (SCC), consisting of those web pages that can reach each other along the directed links ($\approx$ 28 percent);

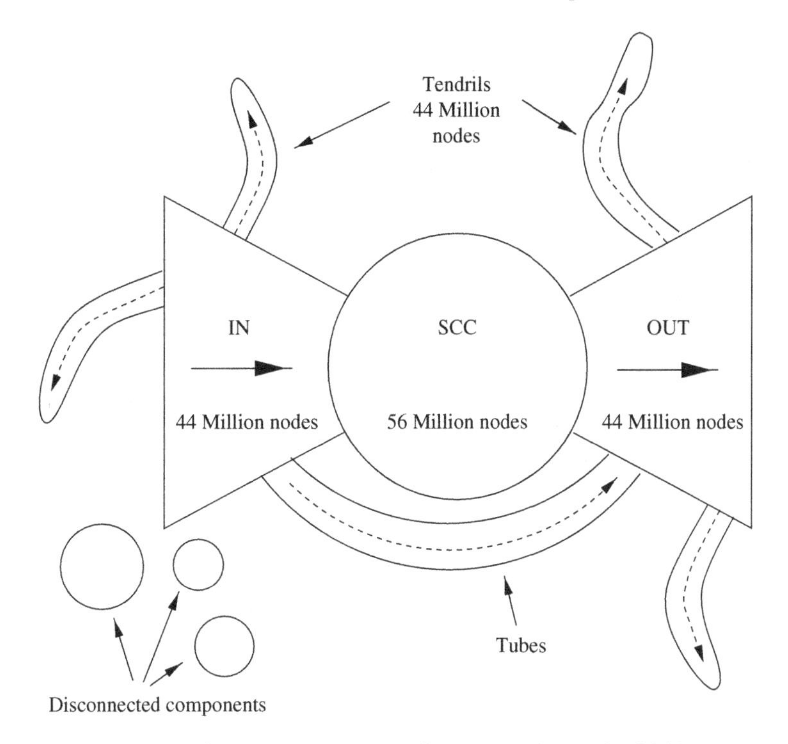

Figure 1.19 The WWW according to Broder et al. (2000).

(b) The IN part, consisting of pages that can reach the SCC, but cannot be reached from it (≈ 21 percent);
(c) The OUT part, consisting of pages that can be reached from the SCC, but do not link back into it (≈ 21 percent);
(d) The TENDRILS and other components, consisting of pages that can neither reach the SCC, nor be reached from it (≈ 30 percent).

Broder et al. (2000) also investigate the diameter of the WWW, finding that the SCC has diameter at least 28, while the WWW as a whole has diameter at least 500. This is partly due to the fact that the graph is *directed*. When the WWW is viewed as an *undirected* graph, the average distance between vertices decreases to around 7. Further, it was found that both the in- and out-degrees in the WWW follow a power-law, with power-law exponents estimated as $\tau_{\rm in} \approx 2.1$, $\tau_{\rm out} \approx 2.5$. This was first observed by Kumar et al. (1999). See also Figures 1.20–1.21, where the in- and out-degrees of two WWW data sets from Leskovec et al. (2009) are displayed.

Albert et al. (1999) also studied the degree distribution to find that the in-degrees obey a power law with exponent $\tau_{\rm in} \approx 2.1$ and the out-degrees obey a power law with exponent $\tau_{\rm out} \approx 2.45$, on the basis of several Web domains, such as `nd.edu`, `mit.edu` and `whitehouse.gov`, the Web domains of the home university of Barabási at Notre Dame, of MIT and of the White House, respectively. Further, they investigated the distances between vertices in these domains, to find that distances within domains grow linearly with the log of the size of the domains, with an estimated dependence of $h_n = 0.35 + 2.06 \log n$,

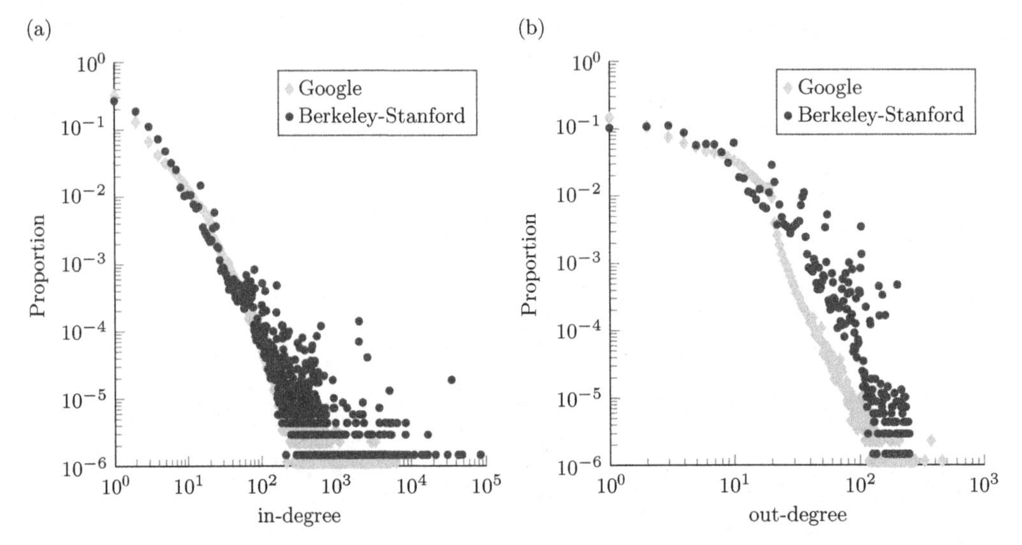

Figure 1.20 The probability mass function of the in- and out-degree sequences in the Berkeley-Stanford and Google competition graph data sets of the WWW in Leskovec et al. (2009). (a) in-degree; (b) out-degree.

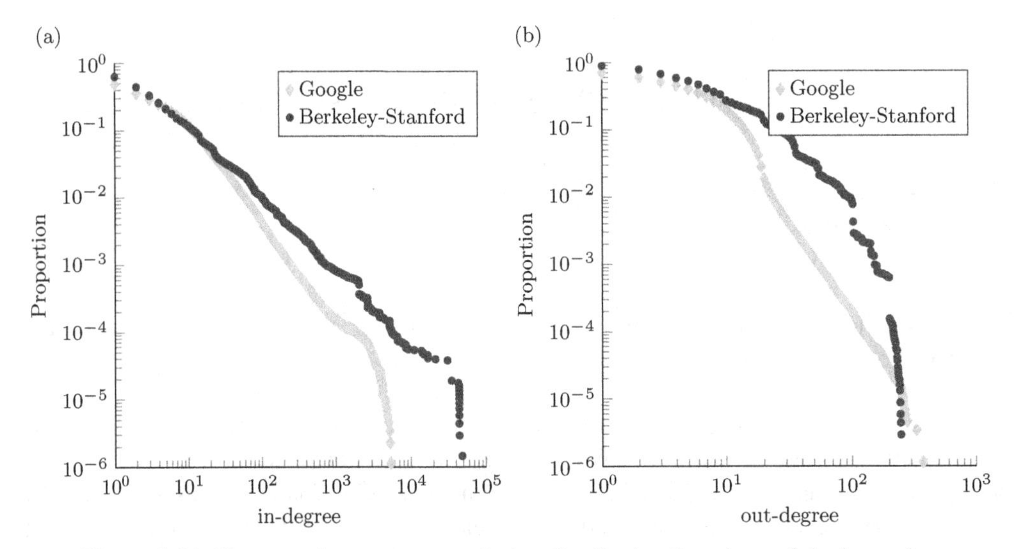

Figure 1.21 The complementary cumulative distribution functions of the in- and out-degree sequences in the Berkeley-Stanford and Google competition graph data sets of the WWW in Leskovec et al. (2009). (a) in-degree; (b) out-degree.

where h_n is the average distance between elements in the part of the WWW under consideration and n is the size of the subset of the WWW. Extrapolating this relation to the estimated size of the WWW at the time, $n = 8 \cdot 10^8$, Albert et al. (1999) concluded that the diameter of the WWW was 19 at the time, which prompted the authors to the following quote:

> Fortunately, the surprisingly small diameter of the web means that all information is just a few clicks away.

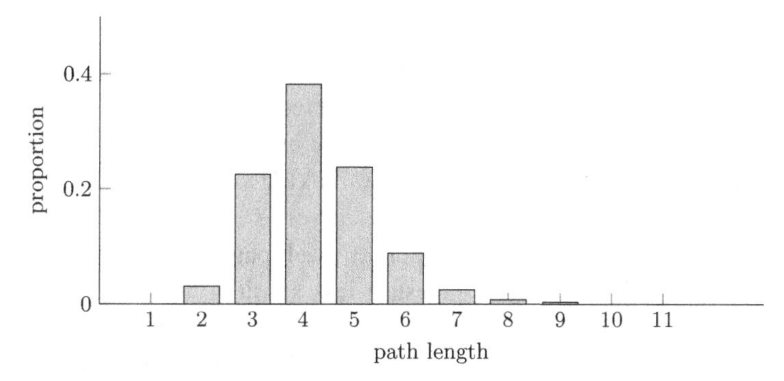

Figure 1.22 Average distances in the Strongly Connected Component of the WWW taken from Adamic (1999).

Adamic (1999) discusses distances in the WWW even further. When considering the WWW as an undirected graph, it is seen that the distances between most pairs of vertices within the SCC are small. See Figure 1.22 for a histogram of pairwise distances in the sample. Distances between pairs of vertices in the SCC tend to be at most 7: Six Degrees of Separation.

Barabási and Albert (1999) (see also Barabási et al. (2000)) argue that new web pages are more likely to attach to web pages that already have a high degree, giving a bias towards popular web pages. This is proposed as an explanation for the occurrences of power laws. We expand on this explanation in Section 1.8, and make the discussion rigorous in Chapter 8.

Kumar et al. (2000) propose models for the WWW where vertices are added that copy the links of older vertices in the graph; see also Krapivsky and Redner (2005). This is called an *evolving copying model*. In some cases, depending on the precise copying rules, the model is shown to lead to power-law degree sequences.

In Milgram's work discussed in Section 1.6.1, on the one hand, it is striking that short paths exist between most pairs of individuals; on the other hand, it may be even more striking that people actually succeed in *finding* them. Kleinberg (1999) addresses the problem of how 'authoritative sources' for the search on the WWW can be quantified. These authoritative sources can be found in an algorithmic way by relating them to the hubs in the network. This problem is intimately connected to the PageRank problem that we discuss now. Kleinberg et al. (1999) is a nice survey of measurements, models and methods for obtaining information on the WWW using Web crawling.

PageRank

A notoriously hard and fascinating problem for the Web is how to rank web pages on related topics such that the most important pages come first. Those of us who were around before Google revolutionized Web searching may remember the days when the list of output of search engines seemed not to be ordered at all. PageRank provides a simple algorithm to create such order, which is why its inventors, Sergey Brin and Larry Page, the founders of Google, gave their paper the catchy subtitle 'Bringing Order to the Web'. PageRank, it is claimed, is named after Larry Page.

PageRank is claimed as the main reason for the success of the search engine Google. Brin and Page put it to good use as founders of Google (see Brin and Page (1998) for the original reference). The problem with the Web is that, though by looking at a web page's HTML code we can easily tell which pages it *refers to*, we cannot easily tell which pages *refer to it*. Intuitively, one would think that a web page that is referenced by many other pages is more influential, and therefore is more likely to be the page one is looking for.

The basic idea behind search engines that use ranking schemes of the vertices, such as PageRank, is simple. Let R_i denote the rank of a vertex i, so that $(R_i)_{i\in[n]}$ denotes the vector of all ranks. Here the vertices correspond to web pages, and we normalize $(R_i)_{i\in[n]}$ such that $\sum_{i\in[n]} R_i = n$ to make the measure less sensitive to the size of the network. Vertices with a high value of R_i are the important ones, while vertices with a low value of R_i are deemed to be relatively unimportant. Here is a general description how ranking schemes can be used to 'bring order to the Web'. A web page has key words associated with it, indicating what the page is about. A ranking scheme allows us to order the web pages in order of their importance, as we explain in more detail below. When a person requests information about a certain query or topic, the search engine returns the ordered list of web pages that have the query or topic as a key word, that order being dictated by the ranks of the web pages involved. Therefore, to 'bring order to the Web', all we have to do is to provide an ordering of the importance of web pages regardless of their content and have a huge data base storing these ranks as well as the possible queries for web pages. PageRank provides precisely such an ordering.

The simplest possible ranking scheme for the importance of web pages is based on their *in-degrees*, the idea being that a popular web page has many other web pages linking to it. Unfortunately, this ranking scheme has serious flaws. For one, it is a measure that can easily be engineered in your favor if you create a 'link farm' with tons of hyperlinks to the web page whose rank you would like to boost. Also, it does not distinguish between links to a web page that come from an important page versus those that come from an irrelevant one. Of course, this seems to be a circular argument, since now we need to know the importance of another web page before we can decide whether the page to which it refers is important. PageRank resolves this circularity quite effectively.

The basic intuition behind PageRank is to define the PageRank of vertex j to satisfy the equation

$$R_j = \sum_{i\to j} \frac{R_i}{d_i^{(\mathrm{out})}}, \tag{1.6.1}$$

where the sum is over all $i \in [n]$ that have a directed edge to j, and $d_i^{(\mathrm{out})}$ is the out-degree of vertex i. The solution to (1.6.1) can even be found by repeated matrix iteration! Unfortunately, the definition in (1.6.1) is seriously flawed. Brin and Page themselves describe the problem as follows:

> Consider two web pages that point to each other but to no other page. And suppose there is some web page which points to one of them. Then, during iteration, this loop will accumulate rank but never distribute any rank (since there are no out-edges). The loop forms a sort of trap...

More generally, dangling ends, i.e., vertices for which $d_j^{(\mathrm{out})} = 0$, pose a problem. Bear in mind that these vertices *could* be quite important (examples are pdf files that are ubiquitous on the Web). Also when the graph can be partitioned into two connected pieces with no links between them, the solution to (1.6.1) is not unique, which again creates a serious problem, since an iteration scheme to approximate the solution to (1.6.1) will concentrate on only one of the connected components of the network, depending on the initial state of the iteration. PageRank is a relatively simple fix to these problems.

Let us first explain the solution in the absence of dangling ends. Let $G = ([n], E)$ denote the web graph, so that $[n]$ corresponds to the web pages and a directed edge $(i, j) \in E$ is present precisely when page i refers to j. Fix $\alpha \in (0, 1)$. Then, we let the vector of PageRanks $(R_i)_{i\in[n]}$ be the unique solution to the equation

$$R_i = \alpha \sum_{j \to i} \frac{R_j}{d_j^{(\mathrm{out})}} + 1 - \alpha, \tag{1.6.2}$$

which is a minor adaptation of (1.6.1). The parameter $\alpha \in (0, 1)$ is called the *damping factor*, and guarantees that (1.6.2) has a *unique* solution. The damping factor α is quite crucial. When $\alpha = 0$, all pages have PageRank 1. This is not very informative. On the other hand, PageRank concentrates on dangling ends when α is close to one, and this, too, is what we want. Experimentally, $\alpha = 0.85$ seems to work well and strikes a nice balance between these two extremes.

In the presence of dangling ends, we can just redistribute their mass equally over all vertices that are not dangling ends, so that (1.6.2) becomes

$$R_i = \alpha \sum_{j \to i} \frac{R_j}{d_j^{(\mathrm{out})}} + (1 - \alpha) \sum_{j \in \mathcal{D}} R_j + 1 - \alpha, \tag{1.6.3}$$

where $\mathcal{D} \subseteq [n]$ denotes the collection of dangling nodes (and, when $i \in \mathcal{D}$, the first term is zero by convention).

There is an interesting and useful representation of PageRank in terms of random walks on the graph. Let $\pi = (\pi_i)_{i\in[n]}$ be the stationary distribution of a random walk on G that, given that its location is i at time t, at time $t + 1$ jumps with probability $(1 - \alpha)/d_i^{(\mathrm{out})}$ to any neighbor j of i (if i is not a dangling end), and with probability α to a uniform vertex in $[n]$. Thus, letting $(X_t)_{t\geq 0}$ denote the random walk, its transition probabilities are given by

$$\mathbb{P}(X_{t+1} = j \mid X_t = i) = P_{ij} = \alpha \frac{\mathbb{1}_{\{(i,j)\in E\}}}{d_i^{(\mathrm{out})}} + \alpha \frac{\mathbb{1}_{\{i\in\mathcal{D}\}}}{n} + \frac{1-\alpha}{n}, \tag{1.6.4}$$

where again $\alpha \in (0, 1)$ is the damping factor from (1.6.3). Since $(X_t)_{t\geq 0}$ is an aperiodic and irreducible Markov chain, the stationary distribution π exists. When thinking about an edge $ij \in E$ as indicating that i finds j interesting, π_i measures how interesting i is, in the sense that π_i is the proportion of time the random walker spends in i. The random walker could be interpreted as a *bored surfer* who does not check the content of pages, but simply clicks a random hyperlink coming out of a web page, until he/she is bored and restarts uniformly. The latter also happens when the random walker happens to end up on a dangling end. The restarts are crucial, since otherwise the random walker would get stuck in the dangling ends of the Web (recall Figure 1.19).

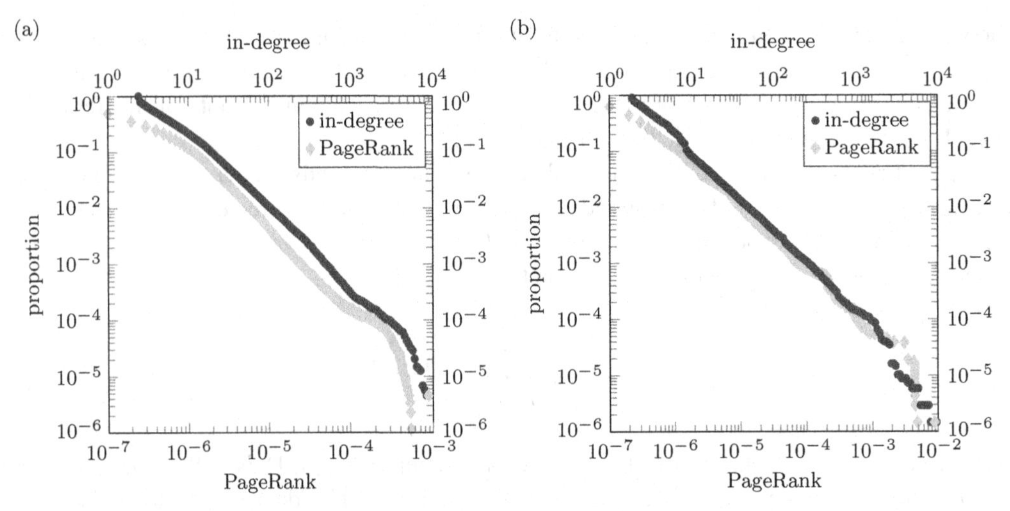

Figure 1.23 The in-degrees and PageRank in the WWW in data sets from Leskovec et al. (2009). (a) The Berkeley-Stanford data set. (b) The Google competition graph data set.

In terms of the above Markov chain, we define the PageRank R_i of a vertex i to be $R_i = n\pi_i$, so that the average of R_i is equal to

$$\frac{1}{n}\sum_{i\in[n]} R_i = \frac{1}{n}\sum_{i\in[n]} n\pi_i = \sum_{i\in[n]} \pi_i = 1, \tag{1.6.5}$$

since $(\pi_i)_{i\in[n]}$ is a probability distribution. Thus, the average PageRank of a page is 1 irrespective of the network size, the total PageRank of the network being n. Clearly, $(R_i)_{i\in[n]} = (n\pi_i)_{i\in[n]}$ solves (1.6.3).

Interestingly, empirical work on the Web graph has shown that the in-degrees in the WWW have a rather similar distribution to the PageRanks. Indeed, often not only the in-degree has an approximate power-law distribution, so do the PageRanks $(R_i)_{i\in[n]}$, and with the *same* power-law exponent. See Figure 1.23 where the similarity between in-degree and PageRank is shown for the two WWW data sets in Figures 1.20–1.21. This may explain why Google works so well. Indeed, since the ranks vary rather substantially, one would expect the most highly ranked web pages to be the most relevant ones, and the many web pages with small ranks to be, probably, not so relevant.

A matrix analytic viewpoint helps to compute the ranks $(R_i)_{i\in[n]}$. Indeed,

$$\sum_{j\in[n]} P_{ij}\pi_j = (1-\alpha)\sum_{j\in[n]} M_{ij}\pi_j + \frac{\alpha}{n}\sum_{j\in[n]} \pi_j = \alpha(M\pi)_i + \frac{1-\alpha}{n}, \tag{1.6.6}$$

where $M = (M_{ij})_{i,j\in[n]}$ is the stochastic matrix

$$M_{ij} = \frac{\mathbb{1}_{\{(i,j)\in E\}}}{d_i^{(\mathrm{out})}} + \frac{\mathbb{1}_{\{i\in\mathcal{D}\}}}{n}. \tag{1.6.7}$$

Since the second term in (1.6.6) is independent of $(\pi_i)_{i\in[n]}$, we can iterate and obtain that the vector π representing the stationary distribution equals

$$\pi = \frac{1-\alpha}{n}\sum_{k=0}^{\infty}\alpha^k M^k \vec{1}, \tag{1.6.8}$$

where $\vec{1}$ is the all-one vector. When we use the matrix-geometric series $\sum_{k=0}^{\infty} A^k = (I - A)^{-1}$, valid for any matrix with norm at most 1 and with I the identity matrix, we thus arrive at

$$\pi = \frac{\alpha}{n}[I - \alpha M]^{-1}\vec{1}, \tag{1.6.9}$$

so that the vector $R = (R_i)_{i\in[n]}$ equals

$$R = \alpha[I - \alpha M]^{-1}\vec{1}. \tag{1.6.10}$$

One could say that Google's birth was due to giant matrix inversions!

1.6.6 The Brain

The brain is arguably the largest complex network. Consisting of 10^{11} neurons connected to one another by axons interacting in an intricate way, it is a marvel how well it functions. See Figure 1.24 for an example of neurons and their axons. To a large extent it is unknown precisely what makes the brain work so well. When considering the brain as a network, different perspectives are relevant. First of all, the brain at the level of neurons constitutes an enormous network. However, compared to many other networks, there is a high amount of hierarchy or community structure present in the brain, where functional units consisting of thousands or even millions of neurons are responsible for specific tasks. This structure is sometimes called the *functional brain network*. For many brain regions it is known which functionality they have; for other regions, this is less well understood. Mapping the functional network of the brain is a highly active research area going under the name *connectomics*. For example, the *Human Connectome Project* is sponsored by sixteen components of the National Institutes of Health in the U.S., and aims to map the anatomical and functional connectivity within the healthy human brain. In Europe, the *Human Brain Project* combines research in over 80 European and international research institutions (see `https://www.humanbrainproject.eu`).

When viewing the brain as a network of interconnecting neurons, the interaction between neurons is highly intricate. The interaction between excitatory neurons is such that one neuron firing enhances the likelihood of the other connected neuron firing. Inhibitory neurons that form some 20–25 percent of all neurons in the cortex, instead, decrease the likelihood of firing of the neurons connected to them. See e.g., Lengler et al. (2013) who investigate the importance of heterogeneity in simple simulation settings of neurons formed by a random graph, with realistic dynamics between them. Further, the network of brain neurons changes over time, as the paradigm 'use it or lose it' describes. Indeed, connections that are heavily used are strengthened, while those that are infrequently used are being pruned. Thus, the topology of the brain network depends on the activity on it. Bressloff (2014)

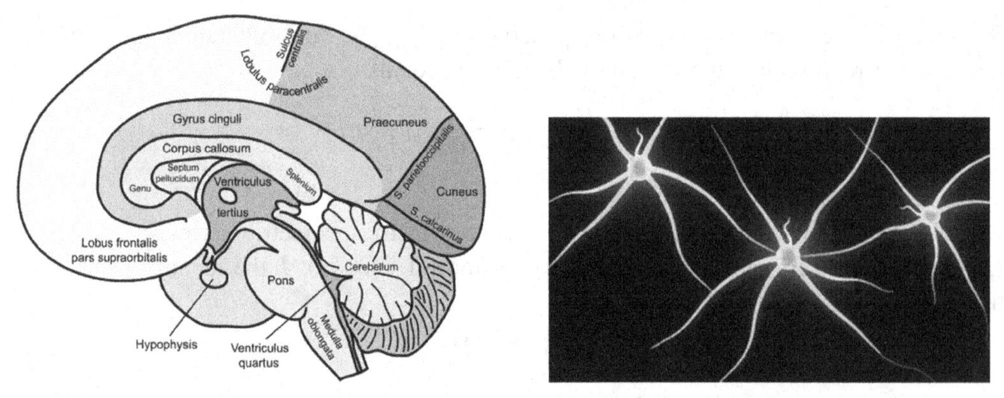

Figure 1.24 Functional brain network (image courtesy of commons.wikimedia.org) and neurons (image courtesy of dream designs at FreeDigitalPhotos.net).

gives a highly readable account of mathematical neuroscience, where the famous model by Hodgins–Huxley describing the firing patterns of a single neuron is explained in detail, and also some network aspects are being treated.

Another type of network that is commonly considered in neuroscience relates to the *functional network*. Data on it is often obtained from Electroencephalography (EEG) or fMRI data. In an EEG, electrical activity along the scalp is recorded in several spots. The activity at these spots can be summarized in a weighted network, where the vertices describe the spots on the scalp, whereas the weights along an edge describe the correlation between the data signals observed at the pair of spots in the edge. Sometimes these weighted networks are turned into unweighted networks by truncating the weights, i.e., edges with a high amount of correlation are kept and those with low correlation are ignored. It is for such networks that network science has been used the most extensively so far. The small-world, scale-free and highly clustered features of such networks have received substantial attention. Let us mention the fact that the hubs in the network are more likely to be connected to one another, an empirical fact that is dubbed the 'rich-club' phenomenon (see van den Heuvel and Sporns (2011)). We refer the reader to the book by Olaf Sporns (2011), called *Networks of the Brain*, as well as the references therein, for an accessible introduction to brain networks and their functionality.

1.7 Tales of Tails

In this section, we take a closer look at the occurrence of power laws. In Section 1.7.1, we discuss the literature on this topic, which dates back to the 1920s. In Section 1.7.2, we describe the new viewpoints on power laws in real-world networks. In Section 1.7.3, we consider the statistical issues of power laws, focussing on the estimation of the power-law exponent. Finally, in Section 1.7.4, we discuss where network data can be found on the web.

1.7.1 Old Tales of Tails

Mathematicians are drawn to simple relations, believing that they explain the rules that gave rise to them. Thus, finding such relations uncovers the hidden structure behind the often chaotic appearance of data. A power-law relationship is such a simple relation. We say that there is a power-law relationship between two variables when one is proportional to a power of the other. Or, in more mathematical language, the variable k and the characteristic $f(k)$ are in a power-law relation when $f(k)$ is proportional to a power of k, i.e., for some number $\tau > 0$ and constant $C > 0$,

$$f(k) = Ck^{-\tau}. \tag{1.7.1}$$

Power laws are intimately connected to so-called '80/20 rules'. For example, when studying the wealth in populations, Pareto (1896) observed a huge variability. Most individuals do not earn so much, but there are those rare individuals that earn a substantial part of the total income. Pareto's principle became best known under the name '80/20 rule', indicating, for example, that 20 percent of the people earn 80 percent of the total income. This law appears to be true much more generally. For example, 20 percent of the people are claimed to own 80 percent of the land, 20 percent of the employees are claimed to earn 80 percent of the profit of large companies and maybe it is even true that 20 percent of the scientists write 80 percent of the papers. In each of these cases, no typical size exists due to the high amount of variability present, which explains why these properties are called 'scale free'.

Intuitively, when an 80/20 rule holds, it seems that a power law must be hidden in the background! Power laws play a crucial role in mathematics, as well as in many applications and have a long history. Zipf (1929) was one of the first to find a power law in the frequencies of occurrence of words in large pieces of text. He found that the relative frequency of a word is roughly inversely proportional to its rank in the frequency table of all words. Thus, the most frequent word is about twice as frequent as the second most frequent word, and about three times as frequent as the third most frequent word, etc. In short, with k the rank of the word and $f(k)$ the relative frequency of the kth most frequent word, $f(k) \propto k^{-\tau}$ with τ close to 1. This is called *Zipf's law*.

Already in the 1920s, several other examples of power laws were found. Lotka (1926) investigated papers that were referred to in the Chemical Abstracts in the period from 1901–1916, and found that the number of scientists appearing with two entries is close to $1/2^2 = 1/4$ of the number of scientists with just one entry. The number of scientists appearing with three entries is close to $1/3^2 = 1/9$ times the number of scientists appearing with one entry, etc. Again, with $f(k)$ denoting the number of scientists appearing in k entries, we have that $f(k) \propto k^{-\tau}$, where τ now is close to 2. This is dubbed *Lotka's Law*. Recently, effort has been put into explaining power laws using the notion of 'self-organization'. Per Bak (1996), one of the central figures in this area, called his book on the topic 'How Nature Works'.

Power-law relations are one-step extensions of linear relations. Conveniently, even when one does not understand the mathematical definition of a power law too well, one can still *observe* one in a simple way: in a log-log plot, power laws are turned into straight lines: we take the log of the power-law relationship (1.7.1) to obtain

$$\log f(k) = \log C - \tau \log k, \tag{1.7.2}$$

so that $\log f(k)$ is in a linear relationship with $\log k$, with coefficient equal to $-\tau$. Not only does this allow us to visually inspect whether $f(k)$ is in a power-law relationship to k, it also allows us to estimate the exponent τ. This is precisely what has been done in order to obtain the power-law exponents in the examples in Section 1.3. An account of the history of power-laws can be found in the survey of Mitzenmacher (2004), where also possible explanations why power laws arise so frequently are discussed.

1.7.2 New Tales of Tails

In this section, we discuss the occurrence of power-law degree sequences in real-world networks. We start by giving a heuristic explanation for the occurrence of power-law degree sequences, in the setting of exponentially growing graphs. This heuristic is based on some assumptions that we formulate now.

We assume that

(1) The number of vertices $V(t)$ grows *exponentially* in t at some rate $\rho > 0$, i.e., $V(t) \approx e^{\rho t}$.

(2) The number $N(t)$ of links *into* a vertex at a time t after its creation is $N(t) \approx e^{\beta t}$. (Note that we must have that $\beta \leq \rho$, since the number of links into a vertex is bounded from above by the number of vertices.) Thus, also the number of links into a vertex grows exponentially with time.

We note that assumption (1) is equivalent to the assumption that

(1') The lifetime T of a random vertex, i.e., the time since its birth, has law

$$\mathbb{P}(T > x) = e^{-\rho x}, \tag{1.7.3}$$

so the density of the lifetime of a random vertex is equal to

$$f_T(x) = \rho e^{-\rho x}. \tag{1.7.4}$$

Using the above assumptions, we see that the number of links *into* a random vertex X equals

$$\mathbb{P}(X > i) = i^{-\rho/\beta}, \tag{1.7.5}$$

since

$$\begin{aligned}
\mathbb{P}(X > i) &= \int_0^\infty f_T(x)\mathbb{P}(X > i \mid T = x)dx \\
&= \int_0^\infty \rho e^{-\rho x}\mathbb{P}(X > i \mid T = x)dx \\
&= \int_{(\log i)/\beta}^\infty \rho e^{-\rho x}dx = e^{-(\log i)\rho/\beta} = i^{-\rho/\beta}.
\end{aligned}$$

where we use (2) to conclude that $\mathbb{P}(X > i \mid T = x) = 1$ precisely when $x > (\log i)/\beta$ since the number of links of the vertex grows exponentially with time in the last line. Stretching the above heuristic a bit further, we get

$$\mathbb{P}(X = i) = \mathbb{P}(X > i - 1) - \mathbb{P}(X > i) \sim i^{-(\rho/\beta+1)}. \tag{1.7.6}$$

This heuristic suggests a power law for the in-degrees of the graph, with power-law exponent $\tau = \rho/\beta + 1 \geq 2$. This heuristic explains the occurrence of power laws with exponents that are at least 2.

An alternative reason why power laws occur so generally will be given in Chapter 8. Interestingly, so far, also in this explanation, only power laws that are at least 2 are obtained.

1.7.3 Power Laws, Their Estimation and Criticism

While power-law degree sequences are claimed to occur quite generally in real-world networks, there are also some critical observations, particularly about the measurements that produce power laws in the Internet, as already discussed in some detail in Section 1.3. Here we return to so-called *sampling biases*. It is argued that `traceroute`-measurements, by which the Internet-topology is uncovered, could be partially responsible for the fact that power-law degree sequences are observed in the Internet. Indeed, Lakhina et al. (2003) show that when we perform `traceroute`-measurements to certain subgraphs of the Erdős-Rényi random graph, the data seems to exhibit power-law degree sequences. As we argue in more detail in the next section, the Erdős-Rényi random graph does *not* have power-law degree sequences, so these power-law observations must be an *artifact* of the way the `traceroute` measurements are performed. We will informally introduce the Erdős-Rényi random graph in the next section.

In Internet measurements, subgraphs are typically obtained by exploring paths between sets of pairs of vertices. Indeed, fix a subset S of the vertex set V in a graph $G = (V, E)$. We obtain a subgraph of G by computing the shortest paths between all pairs of vertices in S, and piecing these paths together. The subgraph that we obtain is the part of the network that appears along a shortest path between the starting points and destinations in S, and this is the way `traceroute` is used in the Internet. For shortest-path routing, vertices with a high degree turn out to be more likely to appear than vertices with low degrees. Therefore, such data sets tend to *overestimate* the tail distribution of the degrees in the complete network. This bias in `traceroute` data was further studied by Clauset, Moore and collaborators (see Achlioptas et al. (2005); Clauset and Moore (2005)), in which both for Erdős-Rényi random graphs and for random regular graphs it was shown that subgraphs appear to obey a power law.

While the above criticism may be serious for the Internet, and possibly for the World-Wide Web, where degree distributions are investigated using web-crawling potentially leading to sampling biases, there are many networks whose edge and vertex sets are completely available that also are reported to show power-law degree sequences. In this case, the observed power laws cannot be so easily dismissed. However, one needs to be careful in using and analyzing data confirming power-law degree sequences. Particularly, we cannot rule out that some estimates of the power-law degree exponent τ are *biased*, and that the true values of τ are substantially larger. Possibly, this criticism may explain why so often power laws are observed with exponents in the interval $(2, 3)$.

When all degrees in a network are observed, one can visually check whether a power law is present for the degree sequence by making the log-log plot of the frequency of occurrence of vertices with degree k versus k, and verifying whether this is close to a straight line. Of course, this is not very precise, and one would clearly prefer an estimation procedure of the power-law exponent of the degree sequence. This brings us into the statistical arena. Suppose that we have a data set of size m given by $X_1, \ldots, X_m$, and we order this data as

$X_{(1)} \leq X_{(2)} \leq \cdots \leq X_{(m)}$. One way to estimate τ is to use the *Hill-estimator* $H_{k,m}$ for $1/(\tau-1)$ given by

$$H_{k,m} = \frac{1}{k}\sum_{i=1}^{k} \log(X_{(i)}/X_{(k+1)}), \tag{1.7.7}$$

where $k \geq 1$ is a parameter that needs to be chosen carefully. Statistical theory, as for example presented in the book by Resnick (2007), states that $H_{k,m}$ converges in probability to $1/(\tau-1)$ when $m \to \infty$ and $k/m \to 0$ when the data $(X_i)_{i=1}^m$ are i.i.d. random variables. This is closely related to the fact that the maximum of m i.i.d. random variables having a power-law distribution with exponent τ is of the order $n^{1/(\tau-1)}$. See Section 2.6 for more details about extremes of random variables.[10]

The difficulty is how to choose k. It turns out that, in many data sets, the Hill estimator varies substantially with k, so that choosing the 'right' value of k is non-trivial. A Hill plot can help. In a Hill plot, one plots $(k, 1/H_{k,m})$, and takes the maximal value k^* for which the curve is close to a straight line below k^*. Again, this is not very precise! The procedure often works nicely though when dealing with a pure power law, i.e. when $1 - F(x) = \mathbb{P}(X_1 > x) = (c/x)^{\tau-1}$ for $x \geq c$ and 0 otherwise. However, when slowly varying functions as in (1.4.9) are included, the accuracy drops substantially. The above discussion applies to the 'nice' case where the data comes from an i.i.d. sample of random variables. In real-world networks, the degrees are not even i.i.d., and thus one can only expect worse convergence properties. We conclude that it is hard to estimate the power-law exponent τ, and thus that the reported values appearing in various more applied papers need to be regarded with caution. Of course, this does not mean that we should not try to estimate τ. For example, Newman and collaborators (see e.g., Newman (2005); Clauset et al. (2009)) give sensible suggestions on how to do this in practice.

1.7.4 Network Data

There are many network data sets available on the web, for you to investigate or just to use as illustrating examples. The data for the Internet Movie Data base (IMDb) can be downloaded from `http://www.imdb.com/interfaces`, where the conditions of use can also be found. For the Internet, the CAIDA website[11] gives many data sets, both at IP as well as AS level, of the Internet topology.

There are also a few websites collecting various network data sets. The website of the Stanford Network Analysis Project (SNAP) [12] by Jure Leskovec has many network data sets from various origins that can be downloaded.[13] Also on the Pajek wiki by Vladimir Batagelj and Andrej Mrvar,[14] which focusses on the Pajek software to analyze data, one can find example data sets.[15] Finally, some of the pioneers in network theory have data sets

[10] In more detail, $H_{k,m}$ has the alternative form $H_{m,k} = \frac{1}{k}\sum_{i=1}^{k} i \log(X_{(i)}/X_{(i+1)})$. The random variables $i \log(X_{(i)}/X_{(i+1)})$ converge to independent exponential random variables with mean $1/(\tau-1)$. Exchanging the limits gives that we can estimate $1/(\tau-1)$ by $H_{m,k}$.

[11] `http://www.caida.org/data/`

[12] `http://snap.stanford.edu/`

[13] `http://snap.stanford.edu/data/index.html`

[14] `http://pajek.imfm.si/doku.php`

[15] `http://pajek.imfm.si/doku.php?id=data:index`

available on their web pages. We specifically refer to the web pages of Mark Newman[16] and of Albert-Laszló Barabási.[17]

1.8 Random Graph Models for Complex Networks

In the previous sections, we have described properties of real-world networks. These networks are quite large, and in most cases it is virtually impossible to describe them in detail or to give an accurate model for how they came to be. To circumvent this problem, random graphs have been considered as network models. These random graphs describe by which *local* and *probabilistic* rules vertices are connected to one another. The use of probabilistic rules is to be able to describe the *complexity* of the networks. In deterministic models, often too much structure is present, making the resulting networks unsuitable to describe real-world networks. In general, probability theory can be thought as a powerful way to model complexity. This approach introduces *randomness* in network theory, and leads us to consider random graphs as network models. However, it does not tell us a priori *what* these random graph models should look like.

The field of random graphs was established in the late 1950s and early 1960s. While there were a few papers appearing around (and even before) that time, the paper by Erdős and Rényi (1960) is generally considered to have founded the field. The authors Erdős and Rényi studied the simplest imaginable random graph. Their graph has n vertices, and each pair of vertices is independently connected with a fixed probability p. When we think of this graph as describing a social network, the elements denote the individuals, while two individuals are connected when they know one another. The probability for elements to be connected is sometimes called the *edge probability*. Let $\mathrm{ER}_n(p)$ denote the resulting random graph. Note that the precise model above is introduced by Gilbert (1959), and Erdős and Rényi (1960) instead formulated a model with a fixed number of edges (rather than a binomial number). It is not hard to see that the two models are intimately related (see e.g., Section 4.6, where the history is explained in a bit more detail). The Erdős-Rényi random graph was named after Erdős and Rényi due to the deep and striking results proved in Erdős and Rényi (1960), which opened up an entirely new field. Earlier papers investigating random graphs are by Erdős (1947), who uses the probabilistic method to prove graph properties, and by Solomonoff and Rapoport (1951), where the Erdős-Rényi random graph is introduced as a model for neurons and investigated non-rigorously.

Despite the fact that $\mathrm{ER}_n(p)$ is the simplest imaginable model of a random network, it has a fascinating *phase transition* when p varies. Phase transitions are important in physics. The paradigm example is the solid-fluid transition of water, which occurs when we move the temperature from below 0° Celsius to above 0° Celsius. Similar phase transitions occur in various real-world phenomena, such as magnetism or conductance of porous materials. Many models have been invented that describe and explain such phase transitions, and we will see some examples in this book. The Erdős-Rényi random graph exhibits a phase transition in the size of the maximal connected component, as well as in its connectivity.

The degree of a vertex, say vertex 1, in $\mathrm{ER}_n(\lambda/n)$ has a binomial distribution with success probability $p = \lambda/n$ and $n-1$ trials. In Chapter 5, we will see that the proportion of vertices

[16] `http://www-personal.umich.edu/~mejn/netdata/`

[17] `http://www3.nd.edu/~networks/resources.html`

with degree k is also close to $\mathbb{P}(\mathsf{Bin}(n-1, \lambda/n) = k)$, where we write $\mathsf{Bin}(m, q)$ to denote a binomial random variable with m trials and success probability q. It is well known that, for large n, the binomial distribution with parameters n and $p = \lambda/n$ is close to the Poisson distribution with parameter λ. More precisely,

$$\lim_{n\to\infty} \mathbb{P}\big(\mathsf{Bin}(n, \lambda/n) = k\big) = \mathrm{e}^{-\lambda}\frac{\lambda^k}{k!}, \qquad k = 0, 1, \ldots. \tag{1.8.1}$$

The probability mass function $p_k = \mathrm{e}^{-\lambda}\frac{\lambda^k}{k!}$ is the probability mass function of the Poisson distribution with parameter λ. This result can be strengthened to a law of large numbers stating that the proportion of vertices with degree k converges in probability to p_k. This in particular implies that $\mathrm{ER}(n, \lambda/n)$ is a *sparse random graph sequence.* For k large, the Poisson probability mass function p_k is much smaller than $k^{-\tau}$ for any τ, so the Erdős-Rényi random graph is *not* scale free in contrast to many real-world networks, as explained in more detail in Section 1.3. In Chapters 6, 7 and 8, we describe three scale-free random graph models. In Chapter 6, we describe the *generalized random graph.* The philosophy of this model is simple: we adapt the Erdős-Rényi random graph in such a way that it becomes scale free. In Chapter 7, we describe the *configuration model*, a random graph model where the degrees of vertices are fixed. In Chapter 8, we discuss the *preferential attachment model*, a dynamical random graph model.

As mentioned above, the degrees of the Erdős-Rényi random graph with edge probability $p = \lambda/n$ are close to a Poisson random variable with mean λ and these are *not* scale free. However, we can make these degrees scale free by letting the parameter λ vary over the vertices, in such a way that the distribution of weights over the vertices approaches a power law. Thus, to each vertex i, we associate a weight w_i such that the edges emanating from i are occupied with a probability depending on w_i. There are many ways in which this can be done. For example, in the generalized random graph introduced by Britton, Deijfen and Martin-Löf (2006), the probability that an edge exists between vertices i and j is equal to

$$p_{ij} = \frac{w_i w_j}{w_i w_j + \ell_n}, \tag{1.8.2}$$

where $\ell_n = \sum_{i\in[n]} w_i$ is the total weight of the graph and different edges are independent. In Chapter 6, we will prove that when the weights $(w_i)_{i\in[n]}$ satisfy an approximate power law, this inhomogeneous version of the Erdős-Rényi random graph does lead to scale-free random graphs. One way to quantify the fact that the weights $(w_i)_{i\in[n]}$ satisfy an approximate power law is to assume that there exists a random variable W such that

$$\lim_{n\to\infty} \frac{1}{n}\sum_{i\in[n]} \mathbb{1}_{\{w_i > x\}} = \mathbb{P}(W > x), \tag{1.8.3}$$

where $\mathbb{P}(W > x)$ is regularly varying at $x = \infty$ with exponent $\tau - 1$. Here, we write $\mathbb{1}_{\mathcal{E}}$ to denote the indicator of the event $\mathcal{E}$. Equation (1.8.3) states that the empirical weight distribution converges to a limiting distribution function. This in turn implies that the degree sequence also has an asymptotic power-law distribution, which can be understood by noting that the degree of vertex i is close to a Poisson distribution with parameter w_i, which concentrates around w_i when w_i is large. This heuristic will be made precise in Chapter 6.

There are various other possibilities to create an inhomogeneous version of the Erdős-Rényi random graph, some of which will be discussed also. See the papers by Chung and Lu (2002b) or by Norros and Reittu (2006) for two specific examples, and the impressive paper by Bollobás, Janson and Riordan (2007) for the most general set-up of generalized random graphs.

In the second scale-free random graph model, which we discuss in detail in Chapter 7, we take the degrees as a start for the model. Thus, to each vertex i, we associate a degree d_i, and in some way connect up the different edges. Clearly, we need that the sum of the degrees $\sum_{i\in[n]} d_i$ be *even*, and we assume this from now on. We can connect up the vertices by imagining there to be d_i *half-edges* incident to vertex i, and pairing up the half-edges in some random way. One way to do this is to pair all the half-edges *uniformly*, and this leads to the *configuration model*. Naturally, it is possible that the above procedure does not lead to a simple graph, since self-loops and multiple edges can occur. As it turns out, when the degrees are not too large, or more precisely, when they asymptotically have *finite variance* in the sense that $\frac{1}{n}\sum_{i\in[n]} d_i^2$ remains bounded in n, the graph is with positive probability simple, meaning that there are no self-loops nor multiple edges. By conditioning on the graph being simple, we end up with a uniform graph with the specified degrees. Sometimes this is also referred to as the *repeated configuration model*, since we can think of conditioning as repeatedly forming the graph until it is simple, which happens with strictly positive probability. A second approach to dealing with self-loops and multiple edges is to simply remove them, leading to the so-called *erased configuration model*. In Chapter 7, we investigate these two models, and show that the two degree distributions are close to each other when the graph size n tends to infinity. Thus, even the erasing does not alter the degrees too much. Further, the configuration model leads to a scale-free random graph when the degrees obey an asymptotic power law, i.e., when there exists a limiting power-law random variable D such that the empirical degree distribution converges to the distribution of D, i.e., for every $x \in \mathbb{R}$,

$$\lim_{n\to\infty} \frac{1}{n} \sum_{i\in[n]} \mathbb{1}_{\{d_i > x\}} = \mathbb{P}(D > x). \tag{1.8.4}$$

Scale-free behavior of the configuration model then corresponds to $\mathbb{P}(D > x)$ being regularly varying at $x = \infty$ with exponent $\tau - 1$.

The generalized random graph and configuration models *describe* networks and, in some sense, do so quite satisfactorily. Indeed, they give rise to models with degrees that can be matched to degree distributions found in real-world networks. However, they do not *explain* how the networks were formed so as to have power-law degrees. A possible explanation for the occurrence of scale-free behavior was given by Barabási and Albert (1999), by a feature called *preferential attachment*. Most real-world networks grow. For example, the WWW has increased from a few web pages in 1990 to an estimated size of a few billion in 2014. *Growth* is an aspect that is not taken into account in Erdős-Rényi random graphs, although it would not be hard to reformulate them as a growth process where elements are successively added and connections are randomly added or removed. Thus, growth by itself is not enough to explain the occurrence of power laws. However, viewing real-world networks as evolving in time does give us the possibility to investigate *how* they grow.

So, how do real networks grow? Think of a social network describing a certain population in which a newcomer arrives, increasing the network by one vertex. She/he will start to socialize with people in the population, and this process is responsible for the newly created connections to the newly arrived person. In an Erdős-Rényi random graph, the connections of the newcomer will be spread uniformly over the population. Is this realistic? Is the newcomer not more likely to get to know people who are socially active and, therefore, already have a larger degree? Probably so! We do not live in a perfectly egalitarian world. Rather, we live in a self-reinforcing world, where people who are known are more likely to become even more well known! Therefore, rather than equal probabilities for our newcomer to get acquainted with other individuals in the population, there is a bias towards individuals who already know many people. Alternatively, when we think of the degrees of elements as describing the *wealth* of the individuals in the population, we live in a world where the rich get richer! This is the idea behind preferential attachment models as introduced in Chapter 8.

Phrased in a more mathematical way, preferential attachment models are such that new elements are more likely to attach to elements with high degrees compared to elements with small degree. For example, suppose that new elements are born with a fixed number of edges to the older elements. Each edge is connected to a specific older element with a probability that is proportional to the degree of that older element at the time the connection is created. This phenomenon is now mostly called *preferential attachment*, and was first described informally by Barabási and Albert (1999). See also the book by Barabási (2002) for a highly accessible and enthusiastic personal account by Barabási. Albert and Barabási have been two of the major players in the investigation of the similarities between real-world networks. The notion of preferential attachment in networks has led the theoretical physics and the mathematics communities to study the structure of preferential attachment models in numerous papers. For some of the references, see Chapter 8. In particular, we will see that, as in (1.8.4), the empirical degree distribution of the preferential attachment model does converge and that the limiting degree distribution satisfies an asymptotic power law. Note that now this power law is *not* simply imposed in the model, but is the result of a simple and attractive local growth rule.

While the above explanation works for social networks, in other examples some form of preferential attachment is also likely to be present. For example, in the WWW, when a new web page is created it is more likely to link to an already popular site, such as Google, than, for instance, to my personal web page. For the Internet, it may be profitable for new routers to be connected to highly connected routers, since these give rise to short distances. Even in biological networks, a subtle form of preferential attachment is claimed to exist. For example, in preferential attachment models the *oldest* vertices tend to have the highest degrees. In protein interaction networks, there is some evidence that the proteins that interact with the most other proteins, and thus have a large degree in the network, are also the ones that have been active in cells of organisms the longest.

In Chapter 8, we introduce and study preferential attachment models, and show that preferential attachment leads to scale-free random graphs. The power-law exponent of the degrees depends sensitively on the precise parameters of the model, such as the number of added edges and how dominant the preferential attachment effect is, in a similar way as the suggested power-law exponent in the heuristic derivation in (1.7.6) depends on the parameters of that model.

In Chapters 6, 7 and 8, we investigate the *degrees* of the proposed random graph models and explain the scale-free nature of these models. In Volume II van der Hofstad (2015), we investigate further properties of these models, focussing on the *connected components* and the *distances* within these graphs. We frequently refer to Volume II, and write, e.g., [II, Chapter 2] to refer to Chapter 2 in Volume II. As observed in Section 1.3, most real-world networks are *small worlds*. As a result, one would hope that random graph models for real-world networks are such that distances between their elements are small. In [II, Chapters 2, 3, 4, 5 and 6], we quantify this and relate graph distances to properties of degrees. A further property that we will investigate is the *phase transition* of the largest connected component, as described in detail for the Erdős-Rényi random graph in Chapter 4. We also discuss *universality*, a notion that is strongly rooted in statistical physics and that describes why and when models with distinct *local* rules have the same *global* properties. In terms of network modeling this is crucial, since the various models are all invented with the goal to describe the same real-world networks, even though the local rules that define the graphs can be quite different. As such, it would be undesirable if they would have entirely distinct properties.

The aim of this book is to study random graph models for real-world complex networks in a *rigorous mathematical* way and to put the study of random graphs on a firm theoretical basis. The level is aimed at master-level students, even though there are topics of interest to specialists as well. The text contains many exercises presented in the last section of each chapter, and we encourage the reader to work them through so as to get optimally acquainted with the notions used in this book.

Of course, the literature on random graphs is extensive, and we are not able to discuss all of it. Particularly, we do not discuss the *functionality* of networks. Functionality is about processes that live on the network. For example, the purpose of a road network is to facilitate traffic traveling on the roads in the network. Internet facilitates e-mail, the WWW search and social networks diffusion of ideas and infectious diseases. The behavior of such processes has attracted tremendous attention in the literature, but this is beyond the scope of this book. We close this chapter with a summary of some standard notation used in this book, as well as some notes and discussion.

1.9 Notation

In this book, we denote events by calligraphic letters, such as $\mathcal{A}, \mathcal{B}, \mathcal{C}$ and $\mathcal{E}$. We use $\mathbb{1}_{\mathcal{E}}$ to denote the indicator function of the event $\mathcal{E}$. We use capital letters, such as X, Y, Z, U, V, W, to denote random variables. There are some exceptions, for example, F_X and M_X denote the distribution function and moment generating function of a random variable X, respectively, and we emphasize this by writing the subscript X explicitly. We say that a sequence of events $(\mathcal{E}_n)_{n\geq 1}$ occurs *with high probability* when $\lim_{n\to\infty} \mathbb{P}(\mathcal{E}_n) = 1$. We often abbreviate this as 'whp'. We call a sequence of random variables $(X_i)_{i\geq 1}$ *independent and identically distributed* (i.i.d.) when they are independent and X_i has the same distribution as X_1 for every $i \geq 1$.

We use special notation for certain random variables. We write $X \sim \mathsf{Be}(p)$ when X has a Bernoulli distribution with success probability p, i.e., $\mathbb{P}(X = 1) = 1 - \mathbb{P}(X = 0) = p$. We write $X \sim \mathsf{Bin}(n, p)$ when the random variable X has a binomial distribution with parameters n and p, and we write $X \sim \mathsf{Poi}(\lambda)$ when X has a Poisson distribution with parameter λ. We sometimes abuse notation, and write $\mathbb{P}(\mathsf{Bin}(n, p) = k)$ to denote $\mathbb{P}(X = k)$ when $X \sim \mathsf{Bin}(n, p)$.

We also rely on Landau notation to denote asymptotics. We write $f(n) = o(g(n))$ as $n \to \infty$ when $g(n) > 0$ and $\lim_{n\to\infty} |f(n)|/g(n) = 0$. We write $f(n) = O(g(n))$ as $n \to \infty$ when $g(n) > 0$ and $\limsup_{n\to\infty} |f(n)|/g(n) < \infty$. Finally, we write $f(n) = \Theta(g(n))$ as $n \to \infty$ when $f(n) = O(g(n))$ and $g(n) = O(f(n))$.

We extend Landau notation to sequences of random variables. For a sequence of random variables $(X_n)_{n\geq 1}$ and a limiting variable X, we write that $X_n \xrightarrow{d} X$ when X_n converges in distribution to X, $X_n \xrightarrow{\mathbb{P}} X$ when X_n converges in probability to X and $X_n \xrightarrow{a.s.} X$ when X_n converges almost surely to X. We say that a sequence of random variables $(X_n)_{n\geq 1}$ is *tight* when $\limsup_{K\to\infty} \limsup_{n\to\infty} \mathbb{P}(X_n \leq K) = 0$. For two sequences of random variables $(X_n)_{n\geq 1}$ and $(Y_n)_{n\geq 1}$, where Y_n is non-negative, we write $X_n = O_{\mathbb{P}}(Y_n)$ when $(X_n/Y_n)_{n\geq 1}$ is a tight sequence of random variables, and $X_n = o_{\mathbb{P}}(Y_n)$ when $X_n/Y_n \xrightarrow{\mathbb{P}} 0$. We give precise definitions for these notions in Definition 2.1 below. We refer the reader to Janson (2011) for an excellent discussion of notation for asymptotics in probability.

We let $\mathbb{N} = \{1, 2, 3, \ldots\}$ denote the positive integers, and $\mathbb{N}_0 = \mathbb{N} \cup \{0\}$ the non-negative integers. We say that a random variable X is *integer valued* when X takes values in the integers. We will often be concerned with non-negative integer-valued random variables, such as Bernoulli, binomial and Poisson random variables.

1.10 Notes and Discussion

Notes on Section 1.1

There are many nice accounts of networks and their role in science. The book Newman (2010) is a great resource for definitions, empirical findings in network data and discussions of the various network models and processes on them, from a physics perspective. The present book is far less ambitious, but instead intends to be fully rigorous. The Lectures on Complex Networks by Dorogovtsev (2010) form an introduction to the rich work of theoretical physics on the subject, aimed at researchers rather than students as Newman (2010) does. There are also many popular accounts, such as the books by Watts (1999, 2003) focussing on the small-world properties of real-world networks and the book by Barabási (2002) about the 'new science of networks'. These sources are a great inspiration for why we actually do study networks!

Notes on Section 1.2

Graph theory is an old topic, first invented by Euler to solve the infamous Königsberg bridge problem (see e.g., (Harary, 1969, Section 1.1)). We recommend the books by Harary (1969) and Bollobás (1998). Theorem 1.2 is inspired by the paper of Feld (1991) called 'Why your friends have more friends than you do', even though our proof is, as far as we know, original.

Notes on Section 1.3

The physics perspective to the Internet is summarized in the book by Pastor-Satorras and Vespignani (2007), where the criticism on the scale-free properties of the Internet is also addressed. We refer to (Pastor-Satorras and Vespignani, 2007,

Section 4.4) for these arguments. We do realize that this debate has not been concluded.

Siganos, Faloutsos, Faloutsos and Faloutsos (2003) take up where Faloutsos et al. (1999) has left, and further study power laws arising in the Internet. Jin and Bestavros (2006) summarize various Internet measurements and study how the small-world properties of the AS graph can be obtained from the degree properties and a suitable way of connecting vertices.

Yook, Jeong and Barabási (2002) find that the Internet topology depends on geometry, and find that the fractal dimension is equal to $D_f = 1.5$. They continue to propose a model for the Internet growth that predicts this behavior combining *preferential attachment* with geometry. We discuss preferential attachment models in more detail in Chapter 8. Bollobás and Riordan (2003b) analyze how preferential attachment models behave under malicious attacks or random breakdown.

Willinger, Anderson and Doyle claim that measurement biases only form a minor problem in Internet data. However, because these problems may be analyzed in a mathematical way, they have attracted some attention in the applied mathematics and computer science domains. See, in particular, Achlioptas et al. (2005), Clauset and Moore (2003) and Lakhina et al. (2003). Willinger, Anderson and Doyle (2009) also present another model for the Internet, based on so-called *highly optimized trade-off* (HOT), a feature they claim makes more sense from an Internet engineering perspective.

Notes on Section 1.4

We have given precise mathematical definitions for the notions of graph sequences being highly connected, small worlds and scale free, extending earlier definitions in van der Hofstad (2010). Our definitions are based upon a summary of the relevant results proven for random graph models.

We restrict ourselves to *sparse* random graphs, i.e., random graphs where the average degree remains bounded as the network size grows (recall Definition 1.3). In recent years, there has been an intensive and highly successful effort to describe asymptotic properties of graphs in the *dense* setting, where the average degree grows proportionally to the network size. This theory is described in terms of graph limits or *graphons*, which can be thought of as describing the limit of the rescaled adjacency matrix of the graph. The key ingredient in this theory is Szemerédi's regularity lemma (see Szemerédi (1978)), which states roughly that one can partition a graph into several more or less homogeneous parts with homogeneous edge probabilities for the edges in between. See the book by Lovász (2012) for more details. We refrain from discussing the dense setting in more detail.

Notes on Section 1.5

For a detailed account on various network statistics and metrics, we refer the reader to (Newman, 2010, Chapters 7–8). Newman (2010, Chapter 10) also discusses several algorithms for graph partitioning and community detection.

Notes on Section 1.6

There are many references to the social science literature on social networks in the book by Watts (2003), who later obtained a position in the social sciences. Here we provide some pointers to the literature where more details can be found. Newman, Watts and Strogatz

(2002) survey various models for social networks that have appeared in their papers. Many of the original references can also be found in the collection Newman et al. (2006), along with an introduction explaining their relevance. We refer to Newman (2010, Section 3.1) for many examples of social networks that have been investigated empirically. For an overview of the relevance of social networks in economy, we refer to Vega-Redondo (2007).

Amaral, Scala, Bartélémy and Stanley (2000) calculated degree distributions of several networks, among others a friendship network of 417 junior high school students and a social network of friendships between Mormons in Provo, Utah. For these examples, the degree distributions surprisingly turn out to be closer to a normal distribution than to a power law.

Ebel, Mielsch and Bornholdt (2002) investigate the topology of an e-mail network at the University of Kiel over a period of 112 days. They conclude that the degree sequence obeys a power law, with an exponential cut-off for degrees larger than 200. The estimated degree exponent is 1.81. The authors note that since this data set is gathered at a server, the observed degree of the *external* vertices underestimates their true degree. When only the *internal* vertices are taken into account, the estimate for the power-law exponent decreases to 1.32. When taking into account that the network is in fact *directed*, the power-law exponent of the in-degree is estimated at 1.49, while the out-degrees have an exponent estimated at 2.03. Watts and Strogatz (1998) investigate the small-world nature of the movie-actor network, finding that it has more clustering and shorter distances than a random graph with equal edge density.

Notes on Section 1.7

I learned of the heuristic explanation for power laws in Section 1.7.2 from a talk by Aldous (n.d.).

1.11 Exercises for Chapter 1

Exercise 1.1 (Connectivity in G) Let G have connected components $C_1, C_2, \ldots, C_m$. Thus, $(C_i)_{i=1}^m$ is a partition of $[n]$, every pair of vertices in C_i is connected and no vertices in C_i are connected to vertices in C_j for $i \neq j$. Show that

$$\mathbb{P}(H_n < \infty) = \frac{1}{n^2} \sum_{i=1}^{m} |C_i|^2, \tag{1.11.1}$$

where $|C_i|$ denotes the number of vertices in C_i.

Exercise 1.2 (Relations of notions of power-law distributions) Show that when (1.4.7) holds, then (1.4.8) holds. Find an example where (1.4.8) holds in the form that there exists a constant C such that, as $x \to \infty$,

$$1 - F_X(x) = Cx^{1-\tau}(1 + o(1)), \tag{1.11.2}$$

but (1.4.7) fails.

Exercise 1.3 (Examples of slowly varying functions) Show that $x \mapsto \log x$ and $x \mapsto \mathrm{e}^{(\log x)^\gamma}$, $\gamma \in (0, 1)$, are slowly varying, but that $x \mapsto \mathrm{e}^{(\log x)^\gamma}$ is *not* slowly varying when $\gamma = 1$.

Exercise 1.4 (Lower bound typical distance) Let the connected graph sequence $(G_n)_{n\geq 1}$ have a bounded degree, i.e., $\max_{v\in[n]} d_v = d_{\max} \leq K$ and G_n is connected for every $n \geq 1$. Show that, for every $\varepsilon > 0$,

$$\lim_{n\to\infty} \mathbb{P}(H_n \leq (1-\varepsilon)\log n/\log d_{\max}) = 0. \tag{1.11.3}$$

Exercise 1.5 (Assortativity coefficient as a correlation coefficient) Pick a (directed) edge uniformly at random from E', and let X and Y be the degrees of the vertices that the edge points to and from, respectively. Prove that

$$\rho_G = \frac{\mathrm{Cov}(X, Y)}{\sqrt{\mathrm{Var}(X)}\sqrt{\mathrm{Var}(Y)}} \tag{1.11.4}$$

can be interpreted as the correlation coefficient of the random variables X and Y.

Exercise 1.6 (Rewrite assortativity coefficient) Prove (1.5.9).

Part I

Preliminaries

2

Probabilistic Methods

In this chapter, we describe basic results in probability theory that we rely on in this book. We discuss convergence of random variables, the first and second moment methods, large deviations, coupling, martingales and extreme value theory. We give some proofs, but not all, and focus on the ideas behind the results.

Organization of this Chapter

We describe convergence of random variables in Section 2.1, coupling in Section 2.2 and stochastic domination in Section 2.3. In Section 2.4, we describe bounds on random variables, namely the Markov inequality, the Chebychev inequality and the Chernoff bound. Particular attention will be given to binomial random variables, as they play a crucial role throughout this book. In Section 2.5, we describe a few results on martingales. Finally, in Section 2.6, we describe extreme value theory of random variables. Not all proofs are given in this chapter. We do provide references to the literature, as well as intuition where possible.

2.1 Convergence of Random Variables

In the Erdős-Rényi random graph with $p = \lambda/n$ for some $\lambda > 0$, the degree of a vertex is distributed as a $\mathsf{Bin}(n-1, p)$ random variable. When n is large and $np = \lambda$ is fixed, a $\mathsf{Bin}(n-1, p)$ random variable is close to a Poisson random variable with mean λ. In Chapter 4, we make heavy use of this convergence result, and a quantitative version of it is stated in Theorem 2.10 below.

In order to formalize that

$$\mathsf{Bin}(n, p) \approx \mathsf{Poi}(np), \tag{2.1.1}$$

we need to introduce the notion of convergence of random variables. Random variables are defined to be functions on a sample space. There are several possible notions for convergence of functions on continuous spaces. In a similar fashion, there are several notions of convergence of random variables, three of which we state in the following definition. For more background on the convergence of random variables, we refer the reader to the classic book Billingsley (1968).

Definition 2.1 (Convergence of random variables)

(a) A sequence X_n of random variables *converges in distribution* to a limiting random variable X when

$$\lim_{n\to\infty} \mathbb{P}(X_n \leq x) = \mathbb{P}(X \leq x), \tag{2.1.2}$$

for every x for which $F(x) = \mathbb{P}(X \leq x)$ is continuous. We write this as $X_n \xrightarrow{d} X$.

(b) A sequence X_n of random variables *converges in probability* to a limiting random variable X when, for every $\varepsilon > 0$,

$$\lim_{n\to\infty} \mathbb{P}(|X_n - X| > \varepsilon) = 0. \tag{2.1.3}$$

We write this as $X_n \xrightarrow{\mathbb{P}} X$.

(c) A sequence X_n of random variables *converges almost surely* to a limiting random variable X when

$$\mathbb{P}(\lim_{n\to\infty} X_n = X) = 1. \tag{2.1.4}$$

We write this as $X_n \xrightarrow{a.s.} X$.

It is not hard to see that convergence in probability implies convergence in distribution. The notion of convergence almost surely may be the most difficult to grasp. It turns out that convergence almost surely implies convergence in probability, making it the strongest version of convergence to be discussed here. In this book, we mainly work with convergence in distribution and in probability. There are examples where convergence in distribution holds, but convergence in probability fails; see Exercise 2.1. There are other notions of convergence that we do not discuss, such as convergence in L^1 or L^2. We again refer to Billingsley (1968), or to introductory books in probability, such as the ones by Billingsley (1995), Feller (1968, 1971) and Grimmett and Stirzaker (2001).

We next state some theorems that give convenient criteria by which we can conclude that random variables converge in distribution. In their statement, we make use of a number of functions of random variables that we introduce now:

Definition 2.2 (Generating functions of random variables) Let X be a random variable. Then

(a) The *characteristic function* of X is the function

$$\phi_X(t) = \mathbb{E}[\mathrm{e}^{\mathrm{i}tX}], \qquad t \in \mathbb{R}, \tag{2.1.5}$$

where $\mathrm{i}^2 = -1$ is the imaginary unit.

(b) The *probability generating function* of an integer-valued random variable X is the function

$$G_X(t) = \mathbb{E}[t^X], \qquad t \in \mathbb{R}, \tag{2.1.6}$$

so that $G_X(0) = \mathbb{P}(X = 0)$.

(c) The *moment generating function* of X is the function

$$M_X(t) = \mathbb{E}\big[e^{tX}\big], \qquad t \in \mathbb{R}. \tag{2.1.7}$$

We note that the characteristic function exists for every random variable X, since $|e^{itX}| = 1$ for every t. The moment-generating function, however, does not always exist, as you are asked to show in Exercise 2.3.

Theorem 2.3 (Criteria for convergence in distribution) *The sequence of random variables $(X_n)_{n\geq 1}$ converges in distribution to a random variable X*

(a) if and only if the characteristic functions $\phi_n(t)$ of X_n converge to the characteristic function $\phi_X(t)$ of X for all $t \in \mathbb{R}$. This is guaranteed when $\phi_n(t)$ converges to $\phi(t)$ and $\phi(t)$ is continuous at $t = 0$.
(b) when, for some $\varepsilon > 0$, the moment generating functions $M_n(t)$ of X_n converge to the moment generating function $M_X(t)$ of X for all $|t| < \varepsilon$.
(c) when, for some $\varepsilon > 0$, the probability generating functions $G_n(t)$ of X_n converge to the probability generating function $G_X(t)$ of X for all $|t| < 1 + \varepsilon$.
(d) when the moments $\mathbb{E}[X_n^r]$ converge to the moments $\mathbb{E}[X^r]$ of X for each $r = 1, 2, \ldots$, provided the moments of X satisfy

$$\limsup_{r\to\infty} \frac{|\mathbb{E}[X^r]|^{1/r}}{r} < \infty. \tag{2.1.8}$$

(e) when the moments $\mathbb{E}[X_n^r]$ converge to the moments $\mathbb{E}[X^r]$ of X for each $r = 1, 2, \ldots$, and $M_X(t)$, the moment generating function of X, is finite for t in some neighborhood of the origin.

We finally discuss a special case of convergence in distribution, namely, when we deal with a sum of indicators, and the limit is a Poisson random variable. This is a particularly important example in this book. See Exercises 2.4, 2.5 and 2.6 for properties of the Poisson distribution. Below, we write $(X)_r = X(X-1)\cdots(X-r+1)$, so that $\mathbb{E}[(X)_r]$ is the *rth factorial moment of X*.

For a random variable X taking values in $\{0, 1, \ldots, n\}$, the factorial moments of X uniquely determine the probability mass function, since

$$\mathbb{P}(X = k) = \sum_{r=k}^{n} (-1)^{k+r} \frac{\mathbb{E}[(X)_r]}{(r-k)!k!}, \tag{2.1.9}$$

see e.g. (Bollobás, 2001, Corollary 1.11). To see (2.1.9), we write

$$\mathbb{1}_{\{X=k\}} = \binom{X}{k}\Big(1-1\Big)^{X-k}, \tag{2.1.10}$$

using the convention that $0^0 = 1$. Then, by Newton's binomial,

$$\mathbb{1}_{\{X=k\}} = \binom{X}{k} \sum_{i=0}^{X-k} (-1)^i \binom{X-k}{i} = \sum_{i=0}^{\infty} (-1)^i \binom{X}{k}\binom{X-k}{i}, \tag{2.1.11}$$

where, by convention, $\binom{n}{k} = 0$ when $k < 0$ or $k > n$. Rearranging the binomials gives

$$\mathbb{1}_{\{X=k\}} = \sum_{r=k}^{\infty} (-1)^{k+r} \frac{(X)_r}{(r-k)!k!}, \tag{2.1.12}$$

where $r = k + i$. Taking expectations yields

$$\mathbb{P}(X = k) = \sum_{r=k}^{\infty} (-1)^{k+r} \frac{\mathbb{E}[(X)_r]}{(r-k)!k!}, \tag{2.1.13}$$

which is (2.1.9). Similar results also hold for unbounded random variables, since the sum

$$\sum_{r=k}^{n} (-1)^{k+r} \frac{\mathbb{E}[(X)_r]}{(r-k)!k!} \tag{2.1.14}$$

is alternatingly larger than $\mathbb{P}(X = k)$ (for $n + k$ even) and smaller than $\mathbb{P}(X = k)$ (for $n + k$ odd), see, e.g., Bollobás (2001, Corollary 1.13). This implies the following result:

Theorem 2.4 (Convergence to a Poisson random variable) *A sequence of integer-valued random variables $(X_n)_{n\geq 1}$ converges in distribution to a Poisson random variable with parameter λ when, for all $r = 1, 2, \ldots,$*

$$\lim_{n\to\infty} \mathbb{E}[(X_n)_r] = \lambda^r. \tag{2.1.15}$$

Exercises 2.7 and 2.8 investigate properties of factorial moments, while Exercise 2.9 asks you to prove Theorem 2.4. Theorem 2.4 is particularly convenient when dealing with sums of indicators, i.e., when

$$X_n = \sum_{i\in\mathcal{I}_n} I_{i,n}, \tag{2.1.16}$$

where $I_{i,n}$ takes the values 0 and 1 only, since the following result gives an explicit description of the factorial moments of such random variables:

Theorem 2.5 (Factorial moments of sums of indicators) *When $X = \sum_{i\in\mathcal{I}} I_i$ is a sum of indicators, then*

$$\mathbb{E}[(X)_r] = \sum^{*}_{i_1,\ldots,i_r\in\mathcal{I}} \mathbb{E}\Big[\prod_{l=1}^{r} I_{i_l}\Big] = \sum^{*}_{i_1,\ldots,i_r\in\mathcal{I}} \mathbb{P}\big(I_{i_1} = \cdots = I_{i_r} = 1\big), \tag{2.1.17}$$

where $\sum^{*}_{i_1,\ldots,i_r\in\mathcal{I}}$ denotes a sum over distinct indices.

Exercise 2.10 investigates (2.1.17) for $r = 2$, while in Exercise 2.11, you are asked to use Theorem 2.5 to show that binomial random variables with parameters n and $p = \lambda/n$ converge in distribution to a Poisson random variable with parameter λ.

Proof We prove (2.1.17) by induction on $r \geq 1$ and for all probability measures $\mathbb{P}$ and corresponding expectations $\mathbb{E}$. For $r = 1$, $(X)_1 = X$ and (2.1.17) follows from the fact that the expectation of a sum of random variables is the sum of expectations of the indicators, which, in turn, is equal to $\sum_{i\in\mathcal{I}} \mathbb{P}(I_i = 1)$. This initializes the induction hypothesis.

In order to advance the induction hypothesis, we first note that it suffices to prove the statement for indicators I_i for which $\mathbb{P}(I_i = 1) > 0$. Then, for $r \geq 2$, we write out

$$(X)_r = \sum_{i_1\in\mathcal{I}} I_{i_1}(X-1)\cdots(X-r+1). \tag{2.1.18}$$

Denote by $\mathbb{P}_{i_1}$ the conditional distribution given that $I_{i_1} = 1$, i.e., for any event $\mathcal{E}$,

$$\mathbb{P}_{i_1}(\mathcal{E}) = \frac{\mathbb{P}(\mathcal{E} \cap \{I_{i_1} = 1\})}{\mathbb{P}(I_{i_1} = 1)}. \tag{2.1.19}$$

Then we can rewrite

$$\mathbb{E}\big[I_{i_1}(X-1)\cdots(X-r+1)\big] = \mathbb{P}(I_{i_1} = 1)\mathbb{E}_{i_1}\big[(X-1)\cdots(X-r+1)\big]. \tag{2.1.20}$$

We define

$$Y = X - I_{i_1} = \sum_{j\in\mathcal{I}\setminus\{i_1\}} I_j, \tag{2.1.21}$$

and note that $X = Y + 1$ conditionally on $I_{i_1} = 1$. As a result,

$$\mathbb{E}_{i_1}\big[(X-1)\cdots(X-r+1)\big] = \mathbb{E}_{i_1}\big[Y\cdots(Y-r+2)\big] = \mathbb{E}_{i_1}\big[(Y)_{r-1}\big]. \tag{2.1.22}$$

We now apply the induction hypothesis to $\mathbb{E}_{i_1}\big[(Y)_{r-1}\big]$, to obtain

$$\mathbb{E}_{i_1}\big[(Y)_{r-1}\big] = \sum_{i_2,\ldots,i_r\in\mathcal{I}\setminus\{i_1\}}^{*} \mathbb{P}_{i_1}\big(I_{i_2} = \cdots = I_{i_r} = 1\big). \tag{2.1.23}$$

As a result, we arrive at

$$\mathbb{E}[(X)_r] = \sum_{i_1\in\mathcal{I}} \mathbb{P}(I_{i_1} = 1) \sum_{i_2,\ldots,i_r\in\mathcal{I}\setminus\{i_1\}}^{*} \mathbb{P}_{i_1}\big(I_{i_2} = \cdots = I_{i_r} = 1\big). \tag{2.1.24}$$

We complete the proof by noting that

$$\mathbb{P}(I_{i_1} = 1)\mathbb{P}_{i_1}\big(I_{i_2} = \cdots = I_{i_r} = 1\big) = \mathbb{P}\big(I_{i_1} = I_{i_2} = \cdots = I_{i_r} = 1\big), \tag{2.1.25}$$

and

$$\sum_{i_1\in\mathcal{I}} \sum_{i_2,\ldots,i_r\in\mathcal{I}\setminus\{i_1\}}^{*} = \sum_{i_1,\ldots,i_r\in\mathcal{I}}^{*}. \tag{2.1.26}$$

□

For later use, we define multidimensional versions of Theorems 2.4 and 2.5:

Theorem 2.6 (Convergence to independent Poisson random variables) *A vector of integer-valued random variables $\big((X_{1,n},\ldots,X_{d,n})\big)_{n\geq 1}$ converges in distribution to a vector of independent Poisson random variables with parameters $\lambda_1,\ldots,\lambda_d$ when, for all $r_1,\ldots,r_d \in \mathbb{N}$,*

$$\lim_{n\to\infty} \mathbb{E}[(X_{1,n})_{r_1}\cdots(X_{d,n})_{r_d}] = \lambda_1^{r_1}\cdots\lambda_d^{r_d}. \tag{2.1.27}$$

Theorem 2.7 (Factorial moments of sums of indicators) *When $X_{\ell,n} = \sum_{i\in\mathcal{I}_\ell} I_{i,\ell}$ for all $\ell = 1,\ldots,d$ are sums of indicators, then*

$$\mathbb{E}[(X_{1,n})_{r_1}\cdots(X_{d,n})_{r_d}] \tag{2.1.28}$$
$$= \sum^*_{i_1^{(1)},\ldots,i_{r_1}^{(1)}\in\mathcal{I}_1} \cdots \sum^*_{i_1^{(d)},\ldots,i_{r_d}^{(d)}\in\mathcal{I}_d} \mathbb{P}\big(I^{(\ell)}_{i_s} = 1 \text{ for all } \ell = 1,\ldots,d,\ s = 1,\ldots,r_\ell\big).$$

Exercise 2.12 asks you to prove Theorem 2.7 using Theorem 2.5. The fact that the convergence of moments as in Theorems 2.3, 2.4 and 2.6 yields convergence in distribution is sometimes called the *method of moments*, and is often a convenient way of proving convergence in distribution.

2.2 Coupling

For any λ fixed, it is well known (see, e.g., Exercise 2.11) that, when $n \to \infty$,

$$\mathsf{Bin}(n,\lambda/n) \xrightarrow{d} \mathsf{Poi}(\lambda). \tag{2.2.1}$$

To prove this convergence, as well as to quantify the difference between $\mathsf{Bin}(n,\lambda/n)$ and $\mathsf{Poi}(\lambda)$, we will use a *coupling* proof. Couplings will be quite useful in what follows, so we discuss them, as well as the related topic of *stochastic orderings*, in detail. An excellent treatment of coupling theory is given in the book by Thorisson (2000), to which we refer for more details.

Two random variables X and Y are *coupled* when they are defined on the same probability space in such a way that they have the correct marginal distributions. X and Y are defined on the same probability space when there is one probability law $\mathbb{P}$ such that $\mathbb{P}(X \in \mathcal{E}, Y \in \mathcal{F})$ is defined for all events $\mathcal{E}$ and $\mathcal{F}$. Coupling is formalized in the following definition, where it is also generalized to more than one random variable:

Definition 2.8 (Coupling of random variables) The random variables $(\hat{X}_1,\ldots,\hat{X}_n)$ are a *coupling* of the random variables $X_1,\ldots,X_n$ when $(\hat{X}_1,\ldots,\hat{X}_n)$ are defined on the same probability space, and are such that the marginal distribution of $\hat{X}_i$ is the same as that of X_i for all $i = 1,\ldots,n$, i.e., for all measurable subsets $\mathcal{E}$ of $\mathbb{R}$,

$$\mathbb{P}(\hat{X}_i \in \mathcal{E}) = \mathbb{P}(X_i \in \mathcal{E}). \tag{2.2.2}$$

The key point of Definition 2.8 is that while the random variables $X_1,\ldots,X_n$ may not be defined on *one* probability space, the coupled random variables $(\hat{X}_1,\ldots,\hat{X}_n)$ are defined on the *same* probability space. The coupled random variables $(\hat{X}_1,\ldots,\hat{X}_n)$ are related to the original random variables $X_1,\ldots,X_n$ by the fact that the marginal distributions of $(\hat{X}_1,\ldots,\hat{X}_n)$ are equal to those of the random variables $X_1,\ldots,X_n$. Note that one coupling arises by taking $(\hat{X}_1,\ldots,\hat{X}_n)$ to be independent, with $\hat{X}_i$ having the same distribution as X_i. However, in our proofs, we often make use of more elaborate couplings that give rise to stronger results.

Couplings are useful to prove that random variables are related. We now describe a general coupling between two random variables that makes them *equal* with high probability. We let X and Y be two discrete random variables with

$$\mathbb{P}(X = x) = p_x, \qquad \mathbb{P}(Y = y) = q_y, \qquad x \in \mathcal{X}, y \in \mathcal{Y}, \tag{2.2.3}$$

where $(p_x)_{x\in\mathcal{X}}$ and $(q_y)_{y\in\mathcal{Y}}$ are two probability mass functions on two subsets $\mathcal{X}$ and $\mathcal{Y}$ of the same space. A convenient distance between discrete probability distributions $(p_x)_{x\in\mathcal{X}}$ and $(q_y)_{y\in\mathcal{Y}}$ in (2.2.3) is the *total variation distance* between the discrete probability mass functions $(p_x)_{x\in\mathcal{X}}$ and $(q_y)_{y\in\mathcal{Y}}$. In general, for two probability measures μ and ν, the total variation distance is given by

$$d_{\mathrm{TV}}(\mu, \nu) = \sup_A |\mu(A) - \nu(A)|, \tag{2.2.4}$$

where $\mu(A)$ and $\nu(A)$ are the probabilities of the event A under the measures μ and ν and the supremum is over all (Borel) subsets A of $\mathbb{R}$. For the discrete probability distributions $(p_x)_{x\in\mathcal{X}}$ and $(q_y)_{y\in\mathcal{Y}}$, these measures are given by

$$\mu(A) = \sum_{a\in A} p_a, \qquad \nu(A) = \sum_{a\in A} q_a, \tag{2.2.5}$$

where we abuse notation and take sums over $\mathcal{X} \cup \mathcal{Y}$. We let $d_{\mathrm{TV}}(p, q) = d_{\mathrm{TV}}(\mu, \nu)$ where μ and ν are defined in (2.2.5).

For discrete probability mass functions, it is not hard to see that

$$d_{\mathrm{TV}}(p, q) = \frac{1}{2}\sum_x |p_x - q_x|. \tag{2.2.6}$$

When F and G are the distribution functions corresponding to two continuous densities $f = (f(x))_{x\in\mathbb{R}}$ and $g = (g(x))_{x\in\mathbb{R}}$, so that for every measurable $A \subseteq \mathbb{R}$,

$$\mu(A) = \int_A f(x)dx, \qquad \nu(A) = \int_A g(x)dx, \tag{2.2.7}$$

then we obtain

$$d_{\mathrm{TV}}(f, g) = \frac{1}{2}\int_{-\infty}^{\infty} |f(x) - g(x)|dx. \tag{2.2.8}$$

You are asked to prove the identities (2.2.6) and (2.2.8) in Exercise 2.13. The main result linking the total variation distance between two discrete random variables and a coupling of them is the following theorem:

Theorem 2.9 (Maximal coupling) *For any two discrete random variables X and Y, there exists a coupling $(\hat{X}, \hat{Y})$ of X and Y such that*

$$\mathbb{P}(\hat{X} \neq \hat{Y}) = d_{\mathrm{TV}}(p, q), \tag{2.2.9}$$

while, for any coupling $(\hat{X}, \hat{Y})$ of X and Y,

$$\mathbb{P}(\hat{X} \neq \hat{Y}) \geq d_{\mathrm{TV}}(p, q). \tag{2.2.10}$$

Proof We start by defining the coupling that achieves (2.2.9). For this, we define the random vector $(\hat{X}, \hat{Y})$ by

$$\mathbb{P}(\hat{X} = \hat{Y} = x) = (p_x \wedge q_x), \tag{2.2.11}$$

$$\mathbb{P}(\hat{X} = x, \hat{Y} = y) = \frac{(p_x - (p_x \wedge q_x))(q_y - (p_y \wedge q_y))}{\frac{1}{2}\sum_z |p_z - q_z|}, \qquad x \neq y, \tag{2.2.12}$$

where $(a \wedge b) = \min\{a, b\}$. First of all, this is a probability distribution, since

$$\sum_x (p_x - (p_x \wedge q_x)) = \sum_x (q_x - (p_x \wedge q_x)) = \frac{1}{2}\sum_x |p_x - q_x|. \tag{2.2.13}$$

By (2.2.11)–(2.2.13),

$$\mathbb{P}(\hat{X} = x) = p_x, \qquad \mathbb{P}(\hat{Y} = y) = q_y, \tag{2.2.14}$$

so that $\hat{X}$ and $\hat{Y}$ have the right marginal distributions as required in Definition 2.8. Moreover, by (2.2.13),

$$\begin{aligned}\mathbb{P}(\hat{X} \neq \hat{Y}) &= \sum_{x,y} \frac{(p_x - (p_x \wedge q_x))(q_y - (p_y \wedge q_y))}{\frac{1}{2}\sum_z |p_z - q_z|} \\ &= \frac{1}{2}\sum_x |p_x - q_x| = d_{\mathrm{TV}}(p, q).\end{aligned} \tag{2.2.15}$$

This proves (2.2.9).

The bound in (2.2.10) shows that this is an optimal or maximal coupling. Indeed, for all x, and any coupling $(\hat{X}, \hat{Y})$ of X, Y,

$$\mathbb{P}(\hat{X} = \hat{Y} = x) \leq \mathbb{P}(\hat{X} = x) = \mathbb{P}(X = x) = p_x, \tag{2.2.16}$$

and also

$$\mathbb{P}(\hat{X} = \hat{Y} = x) \leq \mathbb{P}(\hat{Y} = x) = \mathbb{P}(Y = x) = q_x. \tag{2.2.17}$$

Therefore, any coupling satisfies

$$\mathbb{P}(\hat{X} = \hat{Y} = x) \leq (p_x \wedge q_x), \tag{2.2.18}$$

and thus

$$\mathbb{P}(\hat{X} = \hat{Y}) = \sum_x \mathbb{P}(\hat{X} = \hat{Y} = x) \leq \sum_x (p_x \wedge q_x). \tag{2.2.19}$$

As a result, for *any* coupling,

$$\mathbb{P}(\hat{X} \neq \hat{Y}) \geq 1 - \sum_x (p_x \wedge q_x) = \frac{1}{2}\sum_x |p_x - q_x|. \tag{2.2.20}$$

The coupling in (2.2.11)–(2.2.12) attains this equality, which makes it the best coupling possible, in the sense that it maximizes $\mathbb{P}(\hat{X} = \hat{Y})$. This proves (2.2.10). □

In this book, we often work with binomial random variables that we wish to compare to Poisson random variables. We will make use of the following theorem, which will be proved using a coupling argument:

Theorem 2.10 (Poisson limit for binomial random variables) *Let $(I_i)_{i=1}^n$ be independent with $I_i \sim \mathsf{Be}(p_i)$, and let $\lambda = \sum_{i=1}^n p_i$. Let $X = \sum_{i=1}^n I_i$ and let Y be a Poisson random variable with parameter λ. Then, there exists a coupling $(\hat{X}, \hat{Y})$ of X and Y such that*

$$\mathbb{P}(\hat{X} \neq \hat{Y}) \leq \sum_{i=1}^n p_i^2. \tag{2.2.21}$$

Consequently, for every $\lambda \geq 0$ and $n \in \mathbb{N}$, there exists a coupling $(\hat{X}, \hat{Y})$ of X and Y, where $X \sim \mathsf{Bin}(n, \lambda/n)$ and $Y \sim \mathsf{Poi}(\lambda)$, such that

$$\mathbb{P}(\hat{X} \neq \hat{Y}) \leq \frac{\lambda^2}{n}. \tag{2.2.22}$$

Exercise 2.15 investigates consequences of Theorem 2.10.

Proof Throughout the proof, we let $I_i \sim \mathsf{Be}(p_i)$ and we assume that $(I_i)_{i=1}^n$ are independent. Further, we let $J_i \sim \mathsf{Poi}(p_i)$ and assume that $(J_i)_{i=1}^n$ are independent. We write

$$p_{i,x} = \mathbb{P}(I_i = x) = p_i \mathbb{1}_{\{x=1\}} + (1 - p_i)\mathbb{1}_{\{x=0\}}, \tag{2.2.23}$$

$$q_{i,x} = \mathbb{P}(J_i = x) = \mathrm{e}^{-p_i} \frac{p_i^x}{x!} \tag{2.2.24}$$

for the Bernoulli and Poisson probability mass functions.

We next let $(\hat{I}_i, \hat{J}_i)$ be a coupling of I_i, J_i, where $(\hat{I}_i, \hat{J}_i)$ are independent for different i. For each pair I_i, J_i, the maximal coupling $(\hat{I}_i, \hat{J}_i)$ described above satisfies

$$\mathbb{P}(\hat{I}_i = \hat{J}_i = x) = \min\{p_{i,x}, q_{i,x}\} = \begin{cases} 1 - p_i & \text{for } x = 0, \\ p_i \mathrm{e}^{-p_i} & \text{for } x = 1, \\ 0 & \text{for } x \geq 2, \end{cases} \tag{2.2.25}$$

where we have used the inequality $1 - p_i \leq \mathrm{e}^{-p_i}$ for $x = 0$. Thus, since $1 - \mathrm{e}^{-p_i} \leq p_i$,

$$\mathbb{P}(\hat{I}_i \neq \hat{J}_i) = 1 - \mathbb{P}(\hat{I}_i = \hat{J}_i) = 1 - (1 - p_i) - p_i \mathrm{e}^{-p_i} = p_i(1 - \mathrm{e}^{-p_i}) \leq p_i^2. \tag{2.2.26}$$

Next, let $\hat{X} = \sum_{i=1}^n \hat{I}_i$ and $\hat{Y} = \sum_{i=1}^n \hat{J}_i$. Then, $\hat{X}$ has the same distribution as $X = \sum_{i=1}^n I_i$, and $\hat{Y}$ has the same distribution as $Y = \sum_{i=1}^n J_i \sim \mathsf{Poi}(p_1 + \cdots + p_n)$. Finally, by Boole's inequality and (2.2.26),

$$\mathbb{P}(\hat{X} \neq \hat{Y}) \leq \mathbb{P}\Big(\bigcup_{i=1}^n \{\hat{I}_i \neq \hat{J}_i\}\Big) \leq \sum_{i=1}^n \mathbb{P}(\hat{I}_i \neq \hat{J}_i) \leq \sum_{i=1}^n p_i^2. \tag{2.2.27}$$

This completes the proof of Theorem 2.10. □

For $p = (p_x)_{x \in \mathcal{X}}$ and $q = (q_x)_{x \in \mathcal{X}}$ on the same (discrete) space, the total variation distance $d_{\mathrm{TV}}(p, q)$ is obviously larger than any term $\frac{1}{2}|p_x - q_x|$ for any $x \in \mathcal{X}$, so that convergence in total variation distance of $p^{(n)} = (p_x^{(n)})_{x \in \mathcal{X}}$ to a probability mass function $p = (p_x)_{x \in \mathcal{X}}$ implies pointwise convergence of the probability mass functions $\lim_{n\to\infty} p_x^{(n)} = p_x$ for every $x \in \mathcal{X}$. Interestingly, for discrete random variables which all take values in the same discrete space, these notions turn out to be equivalent; see Exercise 2.16.

2.3 Stochastic Ordering

To compare random variables, we will further rely on the notion of *stochastic ordering*, which is defined as follows:

Definition 2.11 (Stochastic domination) Let X and Y be two random variables, not necessarily living on the same probability space. The random variable X is *stochastically smaller* than the random variable Y when, for every $x \in \mathbb{R}$, the inequality

$$\mathbb{P}(X \leq x) \geq \mathbb{P}(Y \leq x) \tag{2.3.1}$$

holds. We denote this by $X \preceq Y$.

A nice coupling reformulation of stochastic ordering is presented in the following lemma, which sometimes goes under the name of *Strassen's Theorem* named after Strassen (1965), who extended it to partially ordered sets:

Lemma 2.12 (Coupling definition of stochastic domination) *The random variable X is stochastically smaller than the random variable Y if and only if there exists a coupling $(\hat{X}, \hat{Y})$ of X, Y such that*

$$\mathbb{P}(\hat{X} \leq \hat{Y}) = 1. \tag{2.3.2}$$

Proof When $\mathbb{P}(\hat{X} \leq \hat{Y}) = 1$,

$$\begin{aligned} \mathbb{P}(Y \leq x) &= \mathbb{P}(\hat{Y} \leq x) = \mathbb{P}(\hat{X} \leq \hat{Y} \leq x) \\ &\leq \mathbb{P}(\hat{X} \leq x) = \mathbb{P}(X \leq x), \end{aligned} \tag{2.3.3}$$

so that X is stochastically smaller than Y.

For the other direction, suppose that X is stochastically smaller than Y. We define the generalized inverse of a distribution function F by

$$F^{-1}(u) = \inf\{x \in \mathbb{R} \colon F(x) \geq u\}, \tag{2.3.4}$$

where $u \in [0, 1]$. If U is a uniform random variable on $[0, 1]$, then it is well known that the random variable $F^{-1}(U)$ has distribution function F. This follows since the function F^{-1} is such that

$$F^{-1}(u) > x \qquad \text{precisely when} \qquad u > F(x). \tag{2.3.5}$$

Therefore, we obtain that

$$\mathbb{P}(F^{-1}(U) \leq x) = \mathbb{P}(U \leq F(x)) = F(x), \tag{2.3.6}$$

since the distribution function F_U of U equals $F_U(u) = u$ for $u \in [0, 1]$ and in the first equality we have used (2.3.5). We conclude that indeed $F^{-1}(U)$ has distribution function F.

Denote the marginal distribution functions of X and Y by F_X and F_Y. Then (2.3.1) is equivalent to

$$F_X(x) \geq F_Y(x) \tag{2.3.7}$$

for all x. It follows that, for all $u \in [0, 1]$,

$$F_X^{-1}(u) \leq F_Y^{-1}(u). \tag{2.3.8}$$

Therefore, since $\hat{X} = F_X^{-1}(U)$ and $\hat{Y} = F_Y^{-1}(U)$ have the same marginal distributions as X and Y, respectively, it follows that

$$\mathbb{P}(\hat{X} \leq \hat{Y}) = \mathbb{P}(F_X^{-1}(U) \leq F_Y^{-1}(U)) = 1. \tag{2.3.9}$$

□

2.3.1 Examples of Stochastically Ordered Random Variables

There are many examples of pairs of random variables that are stochastically ordered, and we now describe a few of them:

Binomial Random Variables

A simple example of random variables that are stochastically ordered is as follows. Let $m, n \in \mathbb{N}$ be integers such that $m \leq n$. Let $X \sim \mathsf{Bin}(m, p)$ and $Y \sim \mathsf{Bin}(n, p)$. Then, we claim that $X \preceq Y$. To see this, let $\hat{X} = \sum_{i=1}^{m} I_i$ and $\hat{Y} = \sum_{i=1}^{n} I_i$, where $(I_i)_{i \geq 1}$ is an i.i.d. sequence of Bernoulli random variables, i.e.,

$$\mathbb{P}(I_i = 1) = 1 - \mathbb{P}(I_i = 0) = p, \qquad i = 1, \ldots, n, \tag{2.3.10}$$

and $I_1, I_2, \ldots, I_n$ are mutually independent. Then, since $I_i \geq 0$ for each i,

$$\mathbb{P}(\hat{X} \leq \hat{Y}) = 1. \tag{2.3.11}$$

Therefore, $X \preceq Y$.

The stochastic domination above also holds when, conditionally on the integer-valued random variable Z, $X \sim \mathsf{Bin}(n - Z, p)$ and $Y \sim \mathsf{Bin}(n, p)$, as investigated in Exercise 2.17. This domination result will prove to be useful in the investigation of the Erdős-Rényi random graph.

Poisson Random Variables

Another example of random variables that are stochastically ordered is as follows. Let $\lambda, \mu \geq 0$ be such that $\lambda \leq \mu$. Let $X \sim \mathsf{Poi}(\lambda)$ and $Y \sim \mathsf{Poi}(\mu)$. Then, $X \preceq Y$. To see this, let $\hat{X} \sim \mathsf{Poi}(\lambda)$, $\hat{Z} \sim \mathsf{Poi}(\mu - \lambda)$, where $\hat{X}$ and $\hat{Z}$ are independent, and let $\hat{Y} = \hat{X} + \hat{Z}$. Then, $\hat{Y} \sim \mathsf{Poi}(\mu)$. Moreover, since $\hat{Z} \geq 0$ for each i,

$$\mathbb{P}(\hat{X} \leq \hat{Y}) = 1. \tag{2.3.12}$$

Therefore, $X \preceq Y$.

2.3.2 Stochastic Ordering and Size-Biased Random Variables

Recall from (1.2.4) that, for a non-negative random variable X with $\mathbb{E}[X] > 0$, we define its size-biased version $X^\star$ by

$$\mathbb{P}(X^\star \leq x) = \frac{\mathbb{E}[X \mathbb{1}_{\{X \leq x\}}]}{\mathbb{E}[X]}. \tag{2.3.13}$$

Theorem 1.1 states that the vertex incident to a uniformly chosen edge has the size-biased degree distribution, which, by (1.2.5), has a larger mean than that of the degree distribution itself. The following result shows that in fact $X^\star \succeq X$:

Proposition 2.13 (Stochastic ordering and size-biasing) *Let X be a non-negative random variable with $\mathbb{E}[X] > 0$, and let $X^\star$ be its size-biased version. Then, $X^\star \succeq X$.*

In the proof of Proposition 2.13, we make use of the following correlation inequality that is of independent interest:

Lemma 2.14 (Correlation inequality for monotone functions) *Let X be a random variable and $f, g\colon \mathbb{R} \mapsto \mathbb{R}$ two non-decreasing functions. Then*

$$\mathbb{E}[f(X)g(X)] \geq \mathbb{E}[f(X)]\mathbb{E}[g(X)]. \tag{2.3.14}$$

Lemma 2.14 states that $f(X)$ and $g(X)$ are non-negatively correlated.

Proof Let X_1, X_2 be two independent copies of X. We use that

$$\mathbb{E}[f(X)g(X)] = \mathbb{E}[f(X_1)g(X_1)] = \mathbb{E}[f(X_2)g(X_2)], \tag{2.3.15}$$

and, by independence,

$$\mathbb{E}[f(X)]\mathbb{E}[g(X)] = \mathbb{E}[f(X_1)g(X_2)] = \mathbb{E}[f(X_2)g(X_1)]. \tag{2.3.16}$$

Therefore,

$$\begin{aligned} &\mathbb{E}[f(X)g(X)] - \mathbb{E}[f(X)]\mathbb{E}[g(X)] \\ &= \tfrac{1}{2}\mathbb{E}[f(X_1)g(X_1)] + \tfrac{1}{2}\mathbb{E}[f(X_2)g(X_2)] - \tfrac{1}{2}\mathbb{E}[f(X_1)g(X_2)] - \tfrac{1}{2}\mathbb{E}[f(X_2)g(X_1)] \\ &= \mathbb{E}[(f(X_1) - f(X_2))(g(X_1) - g(X_2))] \geq 0, \end{aligned} \tag{2.3.17}$$

where we use that $(f(X_1) - f(X_2))(g(X_1) - g(X_2)) \geq 0$ a.s., since f and g are both non-decreasing. □

Proof of Proposition 2.13 By Lemma 2.14 applied to $f(y) = y$ and $g(y) = g_x(y) = \mathbb{1}_{\{y \geq x\}}$, which are both non-decreasing in y,

$$\mathbb{E}[X\mathbb{1}_{\{X>x\}}] \geq \mathbb{E}[X]\mathbb{E}[\mathbb{1}_{\{X>x\}}] = \mathbb{E}[X]\mathbb{P}(X > x), \tag{2.3.18}$$

so that

$$\mathbb{P}(X^\star > x) \geq \mathbb{P}(X > x). \tag{2.3.19}$$

This is equivalent to the statement that $\mathbb{P}(X^\star \leq x) \leq \mathbb{P}(X \leq x)$, so that $X^\star \succeq X$. □

2.3.3 Consequences of Stochastic Domination

In this section, we discuss a number of consequences of stochastic domination, such as the fact that the means of a stochastically ordered pair of random variables is ordered as well:

Theorem 2.15 (Ordering of means for stochastically ordered random variables) *Suppose $X \preceq Y$. Then*

$$\mathbb{E}[X] \leq \mathbb{E}[Y]. \tag{2.3.20}$$

Proof We apply Lemma 2.12. Let $(\hat{X}, \hat{Y})$ be a coupling of X and Y such that $\hat{X} \leq \hat{Y}$ with probability 1. Then

$$\mathbb{E}[X] = \mathbb{E}[\hat{X}] \leq \mathbb{E}[\hat{Y}] = \mathbb{E}[Y]. \tag{2.3.21}$$

□

Theorem 2.16 (Preservation of ordering under monotone functions) *Suppose $X \preceq Y$, and $g\colon \mathbb{R} \to \mathbb{R}$ is non-decreasing. Then $g(X) \preceq g(Y)$.*

Proof Let $(\hat{X}, \hat{Y})$ be a coupling of X and Y such that $\hat{X} \leq \hat{Y}$ with probability 1 (see Lemma 2.12). Then, $g(\hat{X})$ and $g(\hat{Y})$ have the same distributions as $g(X)$ and $g(Y)$, and $g(\hat{X}) \leq g(\hat{Y})$ with probability one, by the fact that g is non-decreasing. Therefore, again by Lemma 2.12 the claim follows. □

2.4 Probabilistic Bounds

We often make use of a number of probabilistic bounds, which we summarize and prove in this section.

2.4.1 First and Second Moment Methods

We start with the *Markov inequality*, sometimes also called the *first moment method*:

Theorem 2.17 (Markov inequality) *Let X be a non-negative random variable with $\mathbb{E}[X] < \infty$. Then,*

$$\mathbb{P}(X \geq a) \leq \frac{\mathbb{E}[X]}{a}. \tag{2.4.1}$$

In particular, when X is non-negative and integer valued with $\mathbb{E}[X] \leq m$, then

$$\mathbb{P}(X = 0) \geq 1 - m. \tag{2.4.2}$$

By (2.4.2), when the integer-valued random variable has a small mean it must be equal to 0 with high probability. This is called the *first moment method* and is a powerful tool.

Proof Equation (2.4.1) follows by

$$a\mathbb{P}(X \geq a) \leq \mathbb{E}[X \mathbb{1}_{\{X \geq a\}}] \leq \mathbb{E}[X]. \tag{2.4.3}$$

□

We continue with the *Chebychev inequality*, which can often be used to show that $X > 0$ with high probability, for an integer-valued random variable X:

Theorem 2.18 (Chebychev inequality) *Assume that the random variable X has variance $\mathrm{Var}(X) = \sigma^2$. Then,*

$$\mathbb{P}\Big(\big|X - \mathbb{E}[X]\big| \geq a\Big) \leq \frac{\sigma^2}{a^2}. \tag{2.4.4}$$

In particular, when X is non-negative and integer valued with $\mathbb{E}[X] \geq m$ and $\mathrm{Var}(X) = \sigma^2$,

$$\mathbb{P}(X = 0) \leq \frac{\sigma^2}{m^2}. \tag{2.4.5}$$

By (2.4.5), if the integer random variable has a large mean and a variance that is small compared to the square of the mean, then it must be positive with high probability. This is called the *second moment method*.

Proof For (2.4.4), we note that

$$\mathbb{P}\Big(|X - \mathbb{E}[X]| \geq a\Big) = \mathbb{P}\Big((X - \mathbb{E}[X])^2 \geq a^2\Big), \tag{2.4.6}$$

and apply the Markov inequality. For (2.4.5), we note that

$$\mathbb{P}(X = 0) \leq \mathbb{P}\big(|X - \mathbb{E}[X]| \geq \mathbb{E}[X]\big) \leq \frac{\mathrm{Var}(X)}{\mathbb{E}[X]^2} \leq \frac{\sigma^2}{m^2}. \tag{2.4.7}$$

□

2.4.2 Large Deviation Bounds

We sometimes rely on bounds on the probability that a sum of independent random variables is larger than its expectation. For such probabilities, *large deviation theory* gives good bounds. We describe these bounds here. For more background on large deviations, we refer the reader to the books on the subject by Dembo and Zeitouni (1998), den Hollander (2000) and Olivieri and Vares (2005).

Theorem 2.19 (Cramér's upper bound, Chernoff bound) *Let $(X_i)_{i\geq 1}$ be a sequence of i.i.d. random variables. Then, for all $a \geq \mathbb{E}[X_1]$, there exists a rate function $a \mapsto I(a)$ such that*

$$\mathbb{P}\Big(\sum_{i=1}^{n} X_i \geq na\Big) \leq \mathrm{e}^{-nI(a)}, \tag{2.4.8}$$

while, for all $a \leq \mathbb{E}[X_1]$,

$$\mathbb{P}\Big(\sum_{i=1}^{n} X_i \leq na\Big) \leq \mathrm{e}^{-nI(a)}. \tag{2.4.9}$$

The rate function $a \mapsto I(a)$ can be computed, for $a \geq \mathbb{E}[X_1]$, as

$$I(a) = \sup_{t\geq 0}\Big(ta - \log \mathbb{E}[\mathrm{e}^{tX_1}]\Big), \tag{2.4.10}$$

while, for $a \leq \mathbb{E}[X_1]$,

$$I(a) = \sup_{t\leq 0}\Big(ta - \log \mathbb{E}[\mathrm{e}^{tX_1}]\Big). \tag{2.4.11}$$

Note that the function $t \mapsto ta - \log \mathbb{E}\big[\mathrm{e}^{tX_1}\big]$ is concave, and the derivative in 0 is $a - \mathbb{E}[X_1] \geq 0$ for $a \geq \mathbb{E}[X_1]$. Therefore, for $a \geq \mathbb{E}[X_1]$, the supremum of $t \mapsto (ta - \log \mathbb{E}[\mathrm{e}^{tX_1}])$ is attained for a $t \geq 0$ when $\mathbb{E}[\mathrm{e}^{tX_1}]$ exists in a neighborhood of $t = 0$. As a result, (2.4.10)–(2.4.11) can be combined as

$$I(a) = \sup_{t}\Big(ta - \log \mathbb{E}[\mathrm{e}^{tX_1}]\Big). \tag{2.4.12}$$

Further, by the same concavity and when $\mathbb{E}\big[\mathrm{e}^{tX_1}\big] < \infty$ for all $|t| < \varepsilon$ for some $\varepsilon > 0$, $I(a) = 0$ only when $a = \mathbb{E}[X_1]$. Thus, the probabilities in (2.4.8)–(2.4.9) decay exponentially unless $a = \mathbb{E}[X_1]$. We now prove Theorem 2.19:

Proof We only prove (2.4.8); the proof of (2.4.9) is identical when we replace X_i by $-X_i$. We rewrite, for every $t \geq 0$,

$$\mathbb{P}\Big(\sum_{i=1}^{n} X_i \geq na\Big) = \mathbb{P}\Big(\mathrm{e}^{t\sum_{i=1}^{n} X_i} \geq \mathrm{e}^{tna}\Big) \leq \mathrm{e}^{-nta}\mathbb{E}\Big[\mathrm{e}^{t\sum_{i=1}^{n} X_i}\Big], \tag{2.4.13}$$

where we have used Markov's inequality (Theorem 2.17). Since $(X_i)_{i\geq 1}$ is a sequence of i.i.d. random variables,

$$\mathbb{E}\Big[\mathrm{e}^{t\sum_{i=1}^{n} X_i}\Big] = \mathbb{E}[\mathrm{e}^{tX_1}]^n, \tag{2.4.14}$$

so that, for every $t \geq 0$,

$$\mathbb{P}\Big(\sum_{i=1}^{n} X_i \geq na\Big) \leq \Big(\mathrm{e}^{-ta}\mathbb{E}[\mathrm{e}^{tX_1}]\Big)^n. \tag{2.4.15}$$

Minimizing the right-hand side over $t \geq 0$ gives that

$$\mathbb{P}\Big(\sum_{i=1}^{n} X_i \geq na\Big) \leq \mathrm{e}^{-n\sup_{t\geq 0}\big(ta-\log\mathbb{E}[\mathrm{e}^{tX_1}]\big)}. \tag{2.4.16}$$

This proves (2.4.8). □

In Exercise 2.20, the rate function of Poisson random variables is investigated.

2.4.3 Bounds on Binomial Random Variables

In this section, we investigate the tails of the binomial distribution. We start by a corollary of Theorem 2.19:

Corollary 2.20 (Large deviations for binomial distributions) *Let X_n be a binomial random variable with parameters p and n. Then, for all $a \in (p, 1]$,*

$$\mathbb{P}\big(X_n \geq na\big) \leq \mathrm{e}^{-nI(a)}, \tag{2.4.17}$$

where

$$I(a) = a\log\Big(\frac{a}{p}\Big) + (1-a)\log\Big(\frac{1-a}{1-p}\Big). \tag{2.4.18}$$

Moreover,

$$I(a) \geq I_p(a) \tag{2.4.19}$$

where

$$I_p(a) = p - a - a\log(p/a). \tag{2.4.20}$$

We can recognize (2.4.20) as the rate function of a Poisson random variable with mean p (see Exercise 2.20). Thus, Corollary 2.20 suggests that the upper tail of a binomial random variable is thinner than the one of a Poisson random variable. In (2.4.18) and from here onward, we define $x\log x = 0$ for $x = 0$. This arises when $a = 1$ in (2.4.18).

Proof We start by proving (2.4.17) using (2.4.8). We note that, by (2.4.10), we obtain a bound with $I(a)$ instead of $I_p(a)$, where, with $X_1 \sim \mathsf{Be}(p)$,

$$I(a) = \sup_{t \geq 0} \left(ta - \log \mathbb{E}[e^{tX_1}]\right) = \sup_t \Big(ta - \log\big(pe^t + (1-p)\big)\Big) \tag{2.4.21}$$
$$= a \log\Big(\frac{a}{p}\Big) + (1-a)\log\Big(\frac{1-a}{1-p}\Big).$$

We note that, since $1 + x \leq e^x$ for every x,

$$pe^t + (1-p) = 1 + p(e^t - 1) \leq e^{p(e^t-1)}, \tag{2.4.22}$$

so that

$$I(a) \geq \sup_t \big(ta - p(e^t - 1)\big) = p - a - a\log\big(p/a\big) = I_p(a). \tag{2.4.23}$$

□

We continue to study tails of the binomial distribution, following the classic book on random graphs by Janson, Łuczak and Rucinski (2000). The main bounds are the following:

Theorem 2.21 *Let $X_i \sim \mathsf{Be}(p_i)$, $i = 1, 2, \ldots, n$, be independent Bernoulli distributed random variables, and write $X = \sum_{i=1}^n X_i$ and $\lambda = \mathbb{E}[X] = \sum_{i=1}^n p_i$. Then*

$$\mathbb{P}(X \geq \mathbb{E}[X] + t) \leq \exp\left(-\frac{t^2}{2(\lambda + t/3)}\right), \qquad t \geq 0; \tag{2.4.24}$$

$$\mathbb{P}(X \leq \mathbb{E}[X] - t) \leq \exp\left(-\frac{t^2}{2\lambda}\right), \qquad t \geq 0. \tag{2.4.25}$$

Exercise 2.21 extends Theorem 2.21 to Poisson random variables.

Proof Let $Y \sim \mathsf{Bin}(n, \lambda/n)$ where we recall that $\lambda = \sum_{i=1}^n p_i$. Since $x \mapsto \log x$ is concave, for every $x_1, \ldots, x_n > 0$,

$$\sum_{i=1}^n \frac{1}{n}\log(x_i) \leq \log\Big(\frac{1}{n}\sum_{i=1}^n x_i\Big). \tag{2.4.26}$$

As a result, for every real u, upon taking the logarithm,

$$\mathbb{E}\big[e^{uX}\big] = \prod_{i=1}^n (1 + (e^u - 1)p_i) = e^{n\sum_{i=1}^n \frac{1}{n}\log(1+(e^u-1)p_i)} \tag{2.4.27}$$
$$\leq e^{n\log(1+(e^u-1)\lambda/n)} = \Big(1 + (e^u - 1)\lambda/n\Big)^n = \mathbb{E}\big[e^{uY}\big].$$

Then we compute that, for all $u \geq 0$, by the Markov inequality (Theorem 2.17),

$$\mathbb{P}(X \geq \mathbb{E}[X] + t) \leq e^{-u(\mathbb{E}[X]+t)}\mathbb{E}\big[e^{uX}\big] \leq e^{-u(\mathbb{E}[X]+t)}\mathbb{E}\big[e^{uY}\big] \tag{2.4.28}$$
$$= e^{-u(\lambda+t)}(1 - p + pe^u)^n,$$

where $p = \lambda/n$ and using that $\mathbb{E}[X] = \lambda$.

When $t > n - \lambda$, the left-hand side of (2.4.28) equals 0 and there is nothing to prove. For $\lambda + t < n$, the right-hand side of (2.4.28) attains its minimum for u satisfying $e^u = (\lambda + t)(1 - p)/(n - \lambda - t)p$. This yields, for $0 \leq t \leq n - \lambda$,

$$\mathbb{P}(X \geq \lambda + t) \leq \left(\frac{\lambda}{\lambda + t}\right)^{\lambda + t} \left(\frac{n - \lambda}{n - \lambda - t}\right)^{n - \lambda - t}. \tag{2.4.29}$$

For $0 \leq t \leq n - \lambda$, we can rewrite (2.4.29) as

$$\mathbb{P}(X \geq \lambda + t) \leq \exp\left(-\lambda\varphi\left(\frac{t}{\lambda}\right) - (n - \lambda)\varphi\left(-\frac{t}{n - \lambda}\right)\right), \tag{2.4.30}$$

where $\varphi(x) = (1 + x)\log(1 + x) - x$ for $x \geq -1$ (and $\varphi(x) = \infty$ for $x < -1$).

Replacing X by $n - X$, we also obtain, for $0 \leq t \leq \lambda$,

$$\begin{aligned}\mathbb{P}(X \leq \lambda - t) &= \mathbb{P}(n - X \geq n - \lambda + t) \\ &\leq \exp\left(-\lambda\varphi\left(-\frac{t}{\lambda}\right) - (n - \lambda)\varphi\left(\frac{t}{n - \lambda}\right)\right).\end{aligned} \tag{2.4.31}$$

Since $\varphi(x) \geq 0$ for every x we can ignore the second term in the exponent. Furthermore, $\varphi(0) = 0$ and $\varphi'(x) = \log(1 + x) \leq x$, so that $\varphi(x) \geq x^2/2$ for $x \in [-1, 0]$, which proves (2.4.25). Similarly, $\varphi(0) = \varphi'(0) = 0$ and, for $x \in [0, 1]$,

$$\varphi''(x) = \frac{1}{1 + x} \geq \frac{1}{(1 + x/3)^3} = \left(\frac{x^2}{2(1 + x/3)}\right)'', \tag{2.4.32}$$

so that $\varphi(x) \geq x^2/(2(1 + x/3))$, which proves (2.4.24). □

2.5 Martingales

In this section, we state and prove some useful results concerning martingales. We assume some familiarity with conditional expectations. For readers who are unfamiliar with filtrations and conditional expectations given a σ-algebra, we start by giving the simplest case of a martingale:

Definition 2.22 (Martingale) A stochastic process $(M_n)_{n \geq 0}$ is a *martingale* when

$$\mathbb{E}[|M_n|] < \infty \qquad \text{for all } n \geq 0, \tag{2.5.1}$$

and, a.s.,

$$\mathbb{E}[M_{n+1} \mid M_0, M_1, \ldots, M_n] = M_n \qquad \text{for all } n \geq 0. \tag{2.5.2}$$

As can be seen from (2.5.2), a martingale can be interpreted as a fair game. Indeed, when M_n denotes the profit after the nth game has been played, (2.5.2) tells us that the expected profit at time $n + 1$ given the profit up to time n is equal to the profit at time n. Exercise 2.22 states that the mean of a martingale is constant. We now give a second definition, which we will need in Chapter 8 where a martingale is defined with respect to a more general filtration:

Definition 2.23 (Martingale definition, general) A stochastic process $(M_n)_{n\geq 0}$ is a *martingale* with respect to $(X_n)_{n\geq 1}$ if

$$\mathbb{E}[|M_n|] < \infty \qquad \text{for all } n \geq 0, \tag{2.5.3}$$

M_n is measurable with respect to the σ-algebra generated by $(X_0, \dots, X_n)$, and, a.s.,

$$\mathbb{E}[M_{n+1} \mid X_0, \dots, X_n] = M_n \qquad \text{for all } n \geq 0. \tag{2.5.4}$$

Similarly, $(M_n)_{n\geq 0}$ is a *submartingale*, when (2.5.3) holds and M_n is measurable with respect to the σ-algebra generated by $(X_0, \dots, X_n)$, but (2.5.4) is replaced with

$$\mathbb{E}[M_{n+1} \mid X_0, \dots, X_n] \geq M_n \qquad \text{for all } n \geq 0. \tag{2.5.5}$$

For $X_n = M_n$, the definitions in (2.5.2) and (2.5.4) coincide. Exercises 2.23, 2.24 and 2.25 give examples of martingales.

In the following two sections, we state and prove two key results for martingales, the *martingale convergence theorem* and the *Azuma-Hoeffding inequality*. These results are an indication of the power of martingales. Martingale techniques play a central role in modern probability theory due to these results.

2.5.1 Martingale Convergence Theorem

We start with the martingale convergence theorem:

Theorem 2.24 (Martingale convergence theorem) *Let $(M_n)_{n\geq 0}$ be a submartingale with respect to $(X_n)_{n\geq 0}$ satisfying*

$$\mathbb{E}[|M_n|] \leq B \qquad \textit{for all } n \geq 0. \tag{2.5.6}$$

Then, $M_n \xrightarrow{a.s.} M_\infty$ for some limiting random variable M_∞ satisfying $\mathbb{E}[|M_\infty|] < \infty$.

Martingale convergence theorems come in various forms. There is also an L^2-version, for which it is assumed that $\mathbb{E}[M_n^2] \leq B$ uniformly for all $n \geq 1$. In this case, one also obtains the convergence $\lim_{n\to\infty} \mathbb{E}[M_n^2] = \mathbb{E}[M_\infty^2]$:

Theorem 2.25 (L^2-Martingale convergence theorem) *Let $(M_n)_{n\geq 0}$ be a martingale process with respect to $(X_n)_{n\geq 0}$ satisfying*

$$\mathbb{E}[M_n^2] \leq B \qquad \textit{for all } n \geq 0. \tag{2.5.7}$$

Then, $M_n \xrightarrow{a.s.} M_\infty$, for some limiting random variable M_∞ which is finite with probability 1, and also $\mathbb{E}[(M_n - M_\infty)^2] \to 0$.

Exercises 2.26 and 2.27 investigate examples of convergent martingales. Exercise 2.28 investigates how to construct submartingales by taking maxima of martingales.

The key step in the classical probabilistic proof of Theorem 2.24 is 'Snell's up-crossings inequality'. Suppose that $\{m_n : n \geq 0\}$ is a real sequence and $[a, b]$ is a real interval. An *up-crossing* of $[a, b]$ is defined to be a crossing by m of $[a, b]$ in the upwards direction.

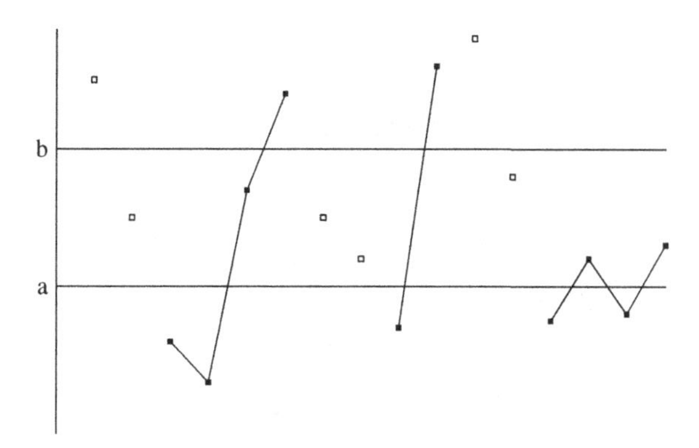

Figure 2.1 Up-crossings

More precisely, let $T_1 = \min\{n\colon m_n \le a\}$, the first time m hits the interval $(-\infty, a]$ and $T_2 = \min\{n > T_1\colon m_n \ge b\}$, the first subsequent time when m hits $[b, \infty)$; we call the interval $[T_1, T_2]$ an up-crossing of $[a, b]$. In addition, for $k > 1$, define the stopping times T_n by

$$T_{2k-1} = \min\{n > T_{2k-2}\colon m_n \le a\}, \qquad T_{2k} = \min\{n > T_{2k-1}\colon m_n \ge b\}, \tag{2.5.8}$$

so that the number of up-crossings of $[a, b]$ is equal to the number of intervals $[T_{2k-1}, T_{2k}]$ for $k \ge 1$. Let $U_n(a, b; m)$ be the number of up-crossings of $[a, b]$ by the sequence m up to time n, and let $U(a, b; m) = \lim_{n\to\infty} U_n(a, b; m)$ be the total number of up-crossings of $[a, b]$ by m.

Let $(M_n)_{n\ge 0}$ be a submartingale and let $U_n(a, b; M)$ be the number of up-crossings of $[a, b]$ by M up to time n. Then Snell's up-crossing inequality gives a bound on the expected number of up-crossings of an interval $[a, b]$:

Proposition 2.26 (Up-crossing inequality) *If $a < b$ then*

$$\mathbb{E}[U_n(a, b; M)] \le \frac{\mathbb{E}[(M_n - a)_+]}{b - a},$$

where $(M_n - a)_+ = \max\{0, M_n - a\}$.

Proof Setting $Z_n = (M_n - a)_+$, we have that Z_n is a non-negative submartingale. Indeed, $\mathbb{E}[|Z_n|] \le \mathbb{E}[|M_n|] + |a| < \infty$. Further, since for every random variable X and $a \in \mathbb{R}$,

$$\mathbb{E}[(X - a)_+] \ge \mathbb{E}[X - a]_+, \tag{2.5.9}$$

it holds that

$$\begin{aligned}\mathbb{E}[Z_{n+1} \mid X_0, \dots, X_n] &= \mathbb{E}[(M_{n+1} - a)_+ \mid X_0, \dots, X_n] \\ &\ge \big(\mathbb{E}[M_{n+1} \mid X_0, \dots, X_n] - a\big)_+ \ge Z_n,\end{aligned} \tag{2.5.10}$$

where the first inequality is (2.5.9), and the last inequality follows from the submartingale property $\mathbb{E}[M_{n+1} \mid X_0, \dots, X_n] \ge M_n$. Up-crossings of $[a, b]$ by M correspond to up-crossings of $[0, b - a]$ by Z, so that $U_n(a, b; M) = U_n(0, b - a; Z)$.

Let $[T_{2k-1}, T_{2k}]$, for $k \geq 1$, be the up-crossings of Z of $[0, b-a]$, and define the indicator functions

$$I_i = \begin{cases} 1 & \text{if } i \in (T_{2k-1}, T_{2k}] \text{ for some } k, \\ 0 & \text{otherwise} \end{cases} \tag{2.5.11}$$

Note that the event $\{I_i = 1\}$ depends on $M_0, M_1, \ldots, M_{i-1}$ only. Since $M_0, M_1, \ldots, M_{i-1}$ are measurable with respect to the σ-algebra generated by $(X_0, \ldots, X_{i-1})$, also I_i is measurable with respect to the σ-algebra generated by $(X_0, \ldots, X_{i-1})$. Now

$$(b-a)U_n(0, b-a; Z) \leq \sum_{i=1}^{n} (Z_i - Z_{i-1}) I_i \tag{2.5.12}$$

since each up-crossing of $[0, b-a]$ by Z contributes an amount of at least $b-a$ to the summation. The expectation of the summands on the right-hand side of (2.5.12) is equal to

$$\begin{aligned} \mathbb{E}[(Z_i - Z_{i-1})I_i] &= \mathbb{E}\Big[\mathbb{E}\big[(Z_i - Z_{i-1})I_i \mid X_0, \ldots, X_{i-1}\big]\Big] \\ &= \mathbb{E}[I_i(\mathbb{E}[Z_i \mid X_0, \ldots, X_{i-1}] - Z_{i-1})] \\ &\leq \mathbb{E}[\mathbb{E}[Z_i \mid X_0, \ldots, X_{i-1}] - Z_{i-1}] = \mathbb{E}[Z_i] - \mathbb{E}[Z_{i-1}], \end{aligned} \tag{2.5.13}$$

where we use that I_i is measurable with respect to the σ-algebra generated by $(X_0, \ldots, X_{i-1})$ for the second equality and we use that Z is a submartingale and $0 \leq I_i \leq 1$ to obtain the inequality. Summing over i and taking expectations on both sides of (2.5.12), we obtain

$$(b-a)\mathbb{E}[U_n(0, b-a; Z)] \leq \mathbb{E}[Z_n] - \mathbb{E}[Z_0] \leq \mathbb{E}[Z_n], \tag{2.5.14}$$

which completes the proof of Proposition 2.26. □

Now we have the tools to complete the proof of Theorem 2.24:

Proof of Theorem 2.24 Suppose $(M_n)_{n \geq 0}$ is a submartingale and $\mathbb{E}[|M_n|] \leq B$ for all n. Let Λ be defined as

$$\Lambda = \{\omega \colon M_n(\omega) \text{ does not converge to a limit in } [-\infty, \infty]\}.$$

The claim of almost sure convergence of M_n follows when we show that $\mathbb{P}(\Lambda) = 0$. The set Λ has an equivalent definition

$$\begin{aligned} \Lambda &= \{\omega \colon \liminf M_n(\omega) < \limsup M_n(\omega)\} \\ &= \bigcup_{a,b \in \mathbb{Q} \colon a<b} \{\omega \colon \liminf M_n(\omega) < a < b < \limsup M_n(\omega)\} \\ &= \bigcup_{a,b \in \mathbb{Q} \colon a<b} \Lambda_{a,b}. \end{aligned}$$

However,

$$\Lambda_{a,b} \subseteq \{\omega \colon U(a, b; M) = \infty\},$$

so that, by Proposition 2.26, $\mathbb{P}(\Lambda_{a,b}) = 0$ for every $a < b$. Since Λ is a countable union of sets $\Lambda_{a,b}$, it follows that $\mathbb{P}(\Lambda) = 0$. This concludes the first part of the proof that M_n converges almost surely to a limit M_∞.

To show that the limit is bounded, we use Fatou's lemma (see Theorem A.7 in the appendix) to conclude

$$\mathbb{E}[|M_\infty|] = \mathbb{E}[\liminf_{n\to\infty} |M_n|] \le \liminf_{n\to\infty} \mathbb{E}[|M_n|] \le \sup_{n\ge 0} \mathbb{E}[|M_n|] < \infty.$$

In particular, by Markov's inequality (recall Theorem 2.17),

$$\mathbb{P}(|M_\infty| < \infty) = 1.$$

This completes the proof of Theorem 2.24. □

2.5.2 Azuma–Hoeffding Inequality

We continue with the Azuma–Hoeffding inequality, which provides exponential bounds for the tails of a special class of martingales:

Theorem 2.27 (Azuma-Hoeffding inequality) *Let $(M_n)_{n\ge 0}$ be a martingale process with the property that there exist constants $K_n \ge 0$ such that almost surely*

$$|M_n - M_{n-1}| \le K_n \qquad \text{for all } n \ge 0, \tag{2.5.15}$$

where, by convention, we define $M_{-1} = \mu = \mathbb{E}[M_n]$ (recall also Exercise 2.22). Then, for every $a \ge 0$ and $n \ge 1$,

$$\mathbb{P}(|M_n - \mu| \ge a) \le 2\exp\Big\{-\frac{a^2}{2\sum_{i=0}^n K_i^2}\Big\}. \tag{2.5.16}$$

Theorem 2.27 is very powerful, as it provides bounds on the tails on the distribution of M_n. In many cases, the bounds are close to optimal. The particular strength of Theorem 2.27 is that the bound is valid *for all $n \ge 1$*.

Proof For $\psi > 0$, the function $g(y) = \mathrm{e}^{\psi y}$ is convex, so that, for all y with $|y| \le 1$,

$$\mathrm{e}^{\psi y} \le \frac{1}{2}(1-y)\mathrm{e}^{-\psi} + \frac{1}{2}(1+y)\mathrm{e}^{\psi}. \tag{2.5.17}$$

Applying this with $y = Y$ to a random variable Y having mean 0 and satisfying $\mathbb{P}(|Y| \le 1) = 1$, we obtain

$$\mathbb{E}[\mathrm{e}^{\psi Y}] \le \mathbb{E}[\frac{1}{2}(1-Y)\mathrm{e}^{-\psi} + \frac{1}{2}(1+Y)\mathrm{e}^{\psi}] = \frac{1}{2}(\mathrm{e}^{-\psi} + \mathrm{e}^{\psi}). \tag{2.5.18}$$

We can use that $(2n)! \ge 2^n n!$ for all $n \ge 0$ to obtain

$$\frac{1}{2}(\mathrm{e}^{-\psi} + \mathrm{e}^{\psi}) = \sum_{n\ge 0} \frac{\psi^{2n}}{(2n)!} \le \sum_{n\ge 0} \frac{\psi^{2n}}{2^n n!} = \mathrm{e}^{\psi^2/2}. \tag{2.5.19}$$

By Markov's inequality (Theorem 2.17), for any $\psi > 0$,

$$\mathbb{P}(M_n - \mu \ge x) = \mathbb{P}(\mathrm{e}^{\psi(M_n-\mu)} \ge \mathrm{e}^{\psi x}) \le \mathrm{e}^{-\psi x}\mathbb{E}[\mathrm{e}^{\psi(M_n-\mu)}]. \tag{2.5.20}$$

Writing $Y_n = M_n - M_{n-1}$, we obtain

$$\mathbb{E}[\mathrm{e}^{\psi(M_n-\mu)}] = \mathbb{E}[\mathrm{e}^{\psi(M_{n-1}-\mu)}\mathrm{e}^{\psi Y_n}].$$

Conditioning on $X_0, \ldots, X_{n-1}$ yields

$$\mathbb{E}[\mathrm{e}^{\psi(M_n-\mu)} \mid X_0, \ldots, X_{n-1}] = \mathrm{e}^{\psi(M_{n-1}-\mu)}\mathbb{E}[\mathrm{e}^{\psi Y_n} \mid X_0, \ldots, X_{n-1}] \le \mathrm{e}^{\psi(M_{n-1}-\mu)} \exp\big(\frac{1}{2}\,\psi^2 K_n^2\big), \tag{2.5.21}$$

where (2.5.18) and (2.5.19) are applied to the random variable Y_n/K_n which satisfies

$$\mathbb{E}[Y_n \mid X_0, \ldots, X_{n-1}] = \mathbb{E}[M_n \mid X_0, \ldots, X_{n-1}] - \mathbb{E}[M_{n-1} \mid X_0, \ldots, X_{n-1}] = M_{n-1} - M_{n-1} = 0. \tag{2.5.22}$$

Take expectations on both sides of (2.5.21) and iterate to find

$$\mathbb{E}\big[\mathrm{e}^{\psi(M_n-\mu)}\big] \le \mathbb{E}\big[\mathrm{e}^{\psi(M_{n-1}-\mu)}\big]\exp\big(\frac{1}{2}\,\psi^2 K_n^2\big) \le \exp\Big(\frac{1}{2}\,\psi^2 \sum_{i=0}^{n} K_i^2\Big). \tag{2.5.23}$$

Therefore, by (2.5.20), for all $\psi > 0$,

$$\mathbb{P}(M_n - \mu \ge x) \le \exp\Big(- \psi x + \frac{1}{2}\,\psi^2 \sum_{i=0}^{n} K_i^2\Big). \tag{2.5.24}$$

The exponential is minimized with respect to ψ by setting $\psi = x/\sum_{i=0}^{n} K_i^2$. Hence,

$$\mathbb{P}(M_n - \mu \ge x) \le \exp\left(-\frac{x^2}{2\sum_{i=0}^{n} K_i^2}\right). \tag{2.5.25}$$

Using that $-M_n$ is also a martingale and by symmetry, we obtain that

$$\mathbb{P}(M_n - \mu \le -x) \le \exp\left(-\frac{x^2}{2\sum_{i=0}^{n} K_i^2}\right). \tag{2.5.26}$$

Adding the two bounds completes the proof. □

Exercises 2.29 and 2.30 investigate consequences of the Azuma–Hoeffding inequality. We close this section with a version of the *optional stopping theorem,* which is a key result in probability theory:

Theorem 2.28 (Optional stopping theorem) *Let $(M_n)_{n\ge 0}$ be a martingale for the increasing σ-fields $(\mathcal{F}_n)_{n\ge 0}$ and suppose that τ_1, τ_2 are stopping times with $0 \le \tau_1 \le \tau_2$. If the process $(M_{n\wedge\tau_2})_{n\ge 0}$ is bounded, then $\mathbb{E}[M_{\tau_1}] = \mathbb{E}[M_{\tau_2}]$. If, instead, $(M_n)_{n\ge 0}$ is a submartingale, then under the same boundedness condition, the bound $\mathbb{E}[M_{\tau_1}] \le \mathbb{E}[M_{\tau_2}]$ holds.*

2.6 Order Statistics and Extreme Value Theory

In this section, we study the largest values of a sequence of i.i.d. random variables. For more background on order statistics, we refer the reader to the standard reference on the subject by Embrechts, Klüppelberg and Mikosch (1997). We are particularly interested in the case where the random variables in question have heavy tails. We let $(X_i)_{i=1}^n$ be an i.i.d. sequence, and introduce the *order statistics* of $(X_i)_{i=1}^n$ by

$$X_{(1)} \leq X_{(2)} \leq \cdots \leq X_{(n)}, \tag{2.6.1}$$

so that $X_{(1)} = \min\{X_1, \ldots, X_n\}$, $X_{(2)}$ is the second smallest of $(X_i)_{i=1}^n$, etc. In the notation in (2.6.1), we ignore the fact that the distribution of $X_{(i)}$ depends on n. Sometimes the notation $X_{(1:n)} \leq X_{(2:n)} \leq \cdots \leq X_{(n:n)}$ is used instead to make the dependence on n explicit. In this section, we mainly investigate $X_{(n)}$, i.e., the maximum of $X_1, \ldots, X_n$. We note that the results immediately translate to $X_{(1)}$, by changing X_i to $-X_i$.

We denote the distribution function of the random variables $(X_i)_{i=1}^n$ by

$$F_X(x) = \mathbb{P}(X_1 \leq x). \tag{2.6.2}$$

Before stating the results, we introduce a number of special distributions. We say that the random variable Y has a *Fréchet distribution* with parameter $\alpha > 0$ if

$$\mathbb{P}(Y \leq y) = \begin{cases} 0, & y \leq 0, \\ \mathrm{e}^{-y^{-\alpha}} & y > 0. \end{cases} \tag{2.6.3}$$

We say that the random variable Y has a *Weibull distribution* with parameter $\alpha > 0$ if[1]

$$\mathbb{P}(Y \leq y) = \begin{cases} \mathrm{e}^{-(-y)^{\alpha}}, & y \leq 0, \\ 1 & y > 0. \end{cases} \tag{2.6.4}$$

We say that the random variable Y has a *Gumbel distribution* if

$$\mathbb{P}(Y \leq y) = \mathrm{e}^{-\mathrm{e}^{-y}}, \qquad y \in \mathbb{R}. \tag{2.6.5}$$

One of the fundamental results in extreme value theory is the following characterization of possible limit distributions of $X_{(n)}$:

Theorem 2.29 (Fisher–Tippett Three Types Theorem for maxima) *Let $(X_n)_{n\geq 1}$ be a sequence of i.i.d. random variables. If there exist norming constants $c_n > 0$ and $d_n \in \mathbb{R}$ and some non-degenerate random variable Y such that*

$$\frac{X_{(n)} - c_n}{d_n} \xrightarrow{d} Y, \tag{2.6.6}$$

then Y has a Fréchet, Weibull or Gumbel distribution.

A fundamental role in extreme value statistics is played by approximate solutions u_n of $[1 - F_X(u)] = 1/n$. More precisely, we define u_n by

$$u_n = \sup\{u \colon 1 - F_X(u) \geq 1/n\}. \tag{2.6.7}$$

We often deal with random variables that have a power-law distribution. For such random variables, the following theorem identifies the Fréchet distribution as the only possible

[1] Sometimes, the Weibull distribution is also defined as a non-negative random variable. We follow Embrechts et al. (1997) in our definition.

extreme value limit. In its statement, we recall the definition of slowly and regularly varying functions in Definition 1.5 in Section 1.3:

Theorem 2.30 (Maxima of heavy-tailed random variables) *Let $(X_n)_{n\geq 1}$ be a sequence of i.i.d. unbounded random variables satisfying*

$$1 - F_X(x) = x^{1-\tau} L_X(x), \tag{2.6.8}$$

where $x \mapsto L_X(x)$ is a slowly varying function *and where $\tau > 1$. Then*

$$\frac{X_{(n)}}{u_n} \xrightarrow{d} Y, \tag{2.6.9}$$

where Y has a Fréchet distribution with parameter $\alpha = \tau - 1$ and u_n is defined in (2.6.7).

Exercise 2.31 shows that u_n is regularly varying for power-law distributions.

For completeness, we also state two theorems identifying when the Weibull distribution or Gumbel distribution occur as the limiting distribution in extreme value theory. We start with the case of random variables that are bounded from above, for which the Weibull distribution arises as the limit distribution for the maximum:

Theorem 2.31 (Maxima of bounded random variables) *Let $(X_n)_{n\geq 1}$ be a sequence of i.i.d. random variables satisfying that $F_X(x_X) = 1$ for some $x_X \in \mathbb{R}$ and*

$$1 - F_X(x_X - x^{-1}) = x^{-\alpha} L_X(x), \tag{2.6.10}$$

where $x \mapsto L_X(x)$ is a slowly varying function and where $\alpha > 1$. Then

$$\frac{X_{(n)} - x_X}{d_n} \xrightarrow{d} Y, \tag{2.6.11}$$

where Y has a Weibull distribution with parameter α, and $d_n = x_X - u_n$ where u_n is defined in (2.6.7).

Theorem 2.31 is the reason why the Weibull distribution is chosen to have support in $(-\infty, 0)$ in (2.6.4). Alternatively, when we would have considered minima of random variables that are bounded from below, it would have been more natural to define the Weibull distribution on the interval $(0, \infty)$ instead.

We continue with the case of random variables that possibly have an unbounded support, but have a thin upper tail, for which the Gumbel distribution arises as the limit distribution for the maximum:

Theorem 2.32 (Maxima of random variables with thin tails) *Let $(X_n)_{n\geq 1}$ be a sequence of i.i.d. bounded random variables satisfying that $F_\infty(x_F) = 1$ for some $x_F \in [0, \infty]$, and*

$$\lim_{x \uparrow x_F} \frac{1 - F_\infty(x + ta(x))}{1 - F_\infty(x)} = \mathrm{e}^{-t}, \qquad t \in \mathbb{R}, \tag{2.6.12}$$

where $x \mapsto a(x)$ is given by

$$a(x) = \int_x^{x_F} \frac{1 - F_\infty(t)}{1 - F_\infty(x)} dt. \tag{2.6.13}$$

Then

$$\frac{X_{(n)} - u_n}{d_n} \xrightarrow{d} Y, \tag{2.6.14}$$

where Y has a Gumbel distribution, and $d_n = a(u_n)$ where u_n is defined in (2.6.7).

In our analysis, the joint convergence of maximum and sum appears in the case where the random variables $(X_i)_{i=1}^n$ have *infinite mean*. It is well known that the order statistics of the random variables, as well as their sum, are governed by u_n in the case $\tau \in (1, 2)$. The following theorem shows this in detail. In the theorem below, $E_1, E_2, \ldots$ is an i.i.d. sequence of exponential random variables with unit mean and $\Gamma_j = E_1 + E_2 + \cdots + E_j$, so that Γ_j has a Gamma distribution with parameters j and 1.

It is well known that when the distribution function F_∞ of $(X_i)_{i=1}^n$ satisfies (2.6.8), then $\sum_{i=1}^n X_i$ has size approximately $n^{1/(\tau-1)}$, just as holds for the maximum, and the rescaled sum $n^{-1/(\tau-1)} \sum_{i=1}^n X_i$ converges to a stable distribution. (See, e.g., Durrett, 2010, Section 3.7.) The next result generalizes this statement to convergence of the sum together with the first order statistics:

Theorem 2.33 (Convergence in distribution of order statistics and sum) *$(X_n)_{n\geq 0}$ be a sequence of i.i.d. random variables satisfying* (2.6.8) *for some $\tau \in (1, 2)$. Then, for any $k \in \mathbb{N}$, and with $L_n = X_1 + \cdots + X_n$,*

$$\frac{1}{u_n}\left(L_n, \left(X_{(n+1-i)}\right)_{i=1}^n\right) \xrightarrow{d} \left(\eta, (\xi_i)_{i\geq 1}\right), \quad as\ n \to \infty, \tag{2.6.15}$$

where $(\eta, (\xi_i)_{i\geq 1})$ is a random vector which can be represented by

$$\eta = \sum_{j=1}^{\infty} \Gamma_j^{-1/(\tau-1)}, \qquad \xi_i = \Gamma_i^{-1/(\tau-1)}, \tag{2.6.16}$$

and where u_n is slowly varying with exponent $1/(\tau-1)$ (recall Exercise 2.31). Moreover,

$$\xi_k k^{1/(\tau-1)} \xrightarrow{\mathbb{P}} 1 \text{ as } k \to \infty. \tag{2.6.17}$$

Proof We only sketch the proof for the pure power-law case. Then, $1 - F_X(x) = x^{-(\tau-1)}$ for $x \geq 1$ and $1 - F_X(x) = 1$ for $x < 1$. Then $X_i \sim F_X^{-1}(U) = F_X^{-1}(\mathrm{e}^{-E})$, where U has a uniform distribution on $(0, 1)$ and $E \sim \mathsf{Exp}(1)$ has a standard exponential distribution. We compute that

$$F_X^{-1}(u) = (1-u)^{-1/(\tau-1)}, \tag{2.6.18}$$

so that

$$X_i \sim (1 - \mathrm{e}^{-E_i})^{-1/(\tau-1)}. \tag{2.6.19}$$

Therefore, $(X_{(i)})_{i=1}^n$ has the same distribution as $\left((1-\mathrm{e}^{-E_i})^{-1/(\tau-1)}\right)_{i=1}^n$. It is well known that

$$n\left(E_{(n+1-i)}\right)_{i=1}^n \xrightarrow{d} (\Gamma_i)_{i\geq 1}, \tag{2.6.20}$$

for example since the joint law of $\left(E_{(n+1-i)}\right)_{i=1}^{n}$ is the same as that of $(\Gamma_{i,n})_{i=1}^{n}$, where

$$\Gamma_{i,n} = \sum_{j=1}^{i} \frac{E_j}{n+1-j}. \tag{2.6.21}$$

In Exercise 2.32, you are asked to prove (2.6.21). By (2.6.21), we have the joint distributional equality

$$\frac{1}{u_n}\left(L_n, \left(X_{(n+1-i)}\right)_{i=1}^{n}\right) \tag{2.6.22}$$
$$\stackrel{d}{=} n^{1/(\tau-1)}\Big(\sum_{i=1}^{n}(1-\mathrm{e}^{-\Gamma_{i,n}})^{-1/(\tau-1)}, \left(\Gamma_{i,n}^{-1/(\tau-1)}\right)_{i=1}^{n}\Big).$$

It is not hard to see that this implies the claim. □

Interestingly, much can be said about the random probability distribution $P_i = \xi_i/\eta$, which is called the *Poisson–Dirichlet distribution* (see, e.g., Pitman and Yor (1997)). For example, Pitman and Yor, (1997, Eqn. (10)) proves that for any $f\colon [0,1] \to \mathbb{R}$, and with $\alpha = \tau - 1 \in (0,1)$,

$$\mathbb{E}\big[\sum_{i=1}^{\infty} f(P_i)\big] = \frac{1}{\Gamma(\alpha)\Gamma(1-\alpha)} \int_0^1 f(u)u^{-\alpha-1}(1-u)^{\alpha-1}du. \tag{2.6.23}$$

We do not delve into this subject further.

2.7 Notes and Discussion

Notes on Section 2.1

Theorem 2.3(a) can be found in the book by Breiman (see Breiman, 1992, Theorem 8.28). Theorem 2.3(c) is Breiman, 1992, Problem 25 page 183. Theorem 2.3(d) is Breiman, 1992, Theorem 8.48 and Proposition 8.49. For a thorough discussion on convergence issues of integer-valued random variables including Theorems 2.4–2.6 and much more, see the classic book on random graphs by Bollobás (see Bollobás, 2001, Section 1.4).

Notes on Sections 2.2 and 2.3

The classic texts on coupling and stochastic domination are by Thorisson (2000) and Lindvall (2002).

Notes on Section 2.4

Theorem 2.19 has a long history. See, e.g., Dembo and Zeitouni (1998, Theorem 2.2.3) for a more precise version of Cramér's Theorem, which states that (2.4.8)–(2.4.9) are sharp, in the sense that $-\frac{1}{n}\log \mathbb{P}(\frac{1}{n}\sum_{i=1}^{n} X_i \leq a)$ converges to $I(a)$. See Olivieri and Vares (2005, Theorem 1.1) for a version of Cramér's Theorem that also includes the Chernoff bound. Similar bounds as in Theorem 2.21 under the same conditions and, even more generally, for independent random variables X_i such that $0 \leq X_i \leq 1$, are given, for example, in Bennet (1962); Hoeffding (1963) and Alon and Spencer (2000, Appendix A). The bound (2.4.29) is implicit in Chernoff (1952) and is often called the *Chernoff bound*, appearing for the first time explicitly in Okamoto (1958).

Notes on Section 2.5

For more details on martingales, we refer the reader to the books by Grimmett and Stirzaker (2001) or Williams (1991). Our proof of the martingale convergence theorem (Theorem 2.24) follows (Grimmett and Stirzaker, 2001, Section 12.3). For interesting examples of martingale arguments, as well as adaptations of the Azuma-Hoeffding inequality in Theorem 2.27, see Chung and Lu (2006b). The Azuma–Hoeffding inequality was first proved by Azuma (1967) and Hoeffding (1963). The Optional Stopping Time Theorem (Theorem 2.28) can be found in Durrett (2010, Theorem 5.7.4) or Williams (1991).

Notes on Section 2.6

For a thorough discussion of extreme value results, as well as many examples, we refer to the standard work on the topic by Embrechts, Klüppelberg and Mikosch (1997), where also the main results of this section can be found. In more detail, Theorem 2.29 is Embrechts et al. (1997, Theorem 3.2.3), Theorem 2.30 is Embrechts et al. (1997, Theorem 3.3.7), Theorem 2.31 is Embrechts et al. (1997, Theorem 3.3.12) and Theorem 2.32 is Embrechts et al. (1997, Theorem 3.3.27).

2.8 Exercises for Chapter 2

Exercise 2.1 (Convergence in distribution, but not in probability) Find an example of a sequence of random variables where convergence in distribution occurs, but convergence in probability does not.

Exercise 2.2 (Example of a converging sequence of random variables) Show that the sequence of independent random variables $(X_n)_{n\geq 1}$, where X_n takes the value n with probability $1/n$ and 0 with probability $1 - 1/n$ converges both in distribution and in probability to 0. Does X_n converge to zero almost surely?

Exercise 2.3 (Random variable without moment generating function) Find a random variable for which the moment generating function is equal to $+\infty$ for every $t \neq 0$.

Exercise 2.4 (Moment bounds for Poisson distribution) Show that a Poisson random variable satisfies the moment condition in (2.1.8).

Exercise 2.5 (Factorial moments of Poisson distribution) Prove that when X is a Poisson random variable with mean λ,

$$\mathbb{E}[(X)_r] = \lambda^r. \tag{2.8.1}$$

Exercise 2.6 (Recursion formula for Poisson moments) Show that the moments of a Poisson random variable X with mean λ satisfy the recursion

$$\mathbb{E}[X^m] = \lambda\mathbb{E}[(X + 1)^{m-1}]. \tag{2.8.2}$$

Exercise 2.7 (Probability mass function in terms of factorial moments) Show that if

$$\sum_{r\geq 0} \frac{\mathbb{E}[(X)_r]}{(r-k)!} < \infty, \tag{2.8.3}$$

then also

$$\mathbb{P}(X=k) = \sum_{r=k}^{\infty} (-1)^{k+r} \frac{\mathbb{E}[(X)_r]}{(r-k)!k!}. \tag{2.8.4}$$

Exercise 2.8 (Convergence using factorial moments) Use Exercise 2.7 to conclude that when $\lim_{n\to\infty} \mathbb{E}[(X_n)_r] = \mathbb{E}[(X)_r]$ for all $r \geq 1$, where X_n and X are all integer-valued non-negative random variables, then also $X_n \xrightarrow{d} X$.

Exercise 2.9 (Proof of Theorem 2.4) Use Exercise 2.8 to prove Theorem 2.4.

Exercise 2.10 (Proof of Theorem 2.5 for $r = 2$) Prove (2.1.17) for $r = 2$.

Exercise 2.11 (Convergence of $\mathsf{Bin}(n, \lambda/n)$ to Poisson) Compute the factorial moments of a binomial random variable with parameters n and $p = \lambda/n$. Use this together with Exercise 2.5 to conclude that a binomial random variable with parameters n and $p = \lambda/n$ converges in distribution to a Poisson random variable with mean λ.

Exercise 2.12 (Proof of Theorem 2.7) Prove Theorem 2.7 using Theorem 2.5.

Exercise 2.13 (Total variation and L^1-distances) Prove (2.2.6) and (2.2.8).

Exercise 2.14 (Coupling and total variation distance) Prove (2.2.13).

Exercise 2.15 (Consequences of maximal coupling) Let $X \sim \mathsf{Bin}(n, \lambda/n)$ and $Y \sim \mathsf{Poi}(\lambda)$. Write $f_i = \mathbb{P}(X = i)$ and $g_i = \mathbb{P}(Y = i)$. Prove that Theorem 2.10 implies that $d_{\mathrm{TV}}(f, g) \leq \lambda^2/n$. Conclude also that, for every $i \in \mathbb{N}$,

$$\big|\mathbb{P}(X=i) - \mathbb{P}(Y=i)\big| \leq \lambda^2/n. \tag{2.8.5}$$

Exercise 2.16 (Equivalence of pointwise and total variation convergence) Show that if the probability mass function $p^{(n)} = (p_x^{(n)})_{x\in\mathcal{X}}$ satisfies that $\lim_{n\to\infty} p_x^{(n)} = p_x$ for every $x \in \mathcal{X}$, and $p = (p_x)_{x\in\mathcal{X}}$ is a probability mass function, then also $\lim_{n\to\infty} d_{\mathrm{TV}}(p^{(n)}, p) = 0$.

Exercise 2.17 (Stochastic domination of binomials) Let $Z \geq 0$ be an integer-valued random variable such that $\mathbb{P}(Z \in \{0, \dots, n\}) = 1$. Let $X \sim \mathsf{Bin}(n - Z, p)$ and $Y \sim \mathsf{Bin}(n, p)$, where $X \sim \mathsf{Bin}(n - Z, p)$ means that $X \sim \mathsf{Bin}(n - z, p)$ conditionally on $Z = z$. Prove that $X \preceq Y$.

Exercise 2.18 (Stochastic domination of normals with ordered means) Let X and Y be normal distributions with equal variances σ^2 and means $\mu_X \leq \mu_Y$. Is $X \preceq Y$?

Exercise 2.19 (Stochastic domination of normals with ordered variances) Let X and Y be normal distributions with variances $\sigma_X^2 < \sigma_Y^2$ and equal means μ. Is $X \preceq Y$?

Exercise 2.20 (Large deviation for Poisson variables) Compute $I(a)$ for $(X_i)_{i\geq 1}$ being independent Poisson random variables with mean λ. Show that, for $a > \lambda$,

$$\mathbb{P}\Big(\sum_{i=1}^{n} X_i \geq na\Big) \leq \mathrm{e}^{-nI_\lambda(a)}, \tag{2.8.6}$$

where $I_\lambda(a) = a(\log(a/\lambda) - 1) + \lambda$. Show also that, for $a < \lambda$,

$$\mathbb{P}\Big(\sum_{i=1}^{n} X_i \leq na\Big) \leq \mathrm{e}^{-nI_\lambda(a)}. \tag{2.8.7}$$

Prove that $I_\lambda(a) > 0$ for all $a \neq \lambda$.

Exercise 2.21 Prove that Theorem 2.21 also holds for the Poisson distribution by a suitable limiting argument.

Exercise 2.22 (The mean of a martigale is constant) Show that when $(M_n)_{n\geq 0}$ is a martingale, then $\mu = \mathbb{E}[M_n]$ is independent of n.

Exercise 2.23 (Product martingale) Let $(X_i)_{i\geq 1}$ be an independent sequence of random variables with $\mathbb{E}[|X_i|] < \infty$ and $\mathbb{E}[X_i] = 1$. Let $M_0 = 1$. Show that, for $n \geq 1$,

$$M_n = \prod_{i=1}^{n} X_i \tag{2.8.8}$$

is a martingale.

Exercise 2.24 (Sum martingale) Let $(X_i)_{i\geq 1}$ be an independent sequence of random variables with $\mathbb{E}[|X_i|] < \infty$ and $\mathbb{E}[X_i] = 0$. Let $M_0 = 0$. Show that, for $n \geq 1$,

$$M_n = \sum_{i=1}^{n} X_i \tag{2.8.9}$$

is a martingale.

Exercise 2.25 (Doob martingale) Let $M_n = \mathbb{E}[Y \mid X_0, \ldots, X_n]$ for some random variables $(X_n)_{n\geq 0}$ and Y with $\mathbb{E}[|Y|] < \infty$. Show that $(M_n)_{n\geq 0}$ is a martingale process with respect to $(X_n)_{n\geq 0}$. $(M_n)_{n\geq 0}$ is called a *Doob* martingale.

Exercise 2.26 (Convergence of non-negative martingales) Use Theorem 2.24 to prove that when the martingale $(M_n)_{n\geq 0}$ is *non-negative*, i.e., when $M_n \geq 0$ with probability 1 for all $n \geq 1$, then $M_n \xrightarrow{a.s.} M_\infty$ to some limiting random variable M_∞ with $\mathbb{E}[M_\infty] < \infty$.

Exercise 2.27 (Convergence of product martingale) Let $(X_i)_{i\geq 1}$ be an independent sequence of random variables with $\mathbb{E}[X_i] = 1$ and for which $X_i \geq 0$ with probability 1. Show that the martingale $M_0 = 1$ and, for $n \geq 1$,

$$M_n = \prod_{i=1}^{n} X_i \tag{2.8.10}$$

converges in probability to a random variable which is finite with probability 1.
Hint: Prove that $\mathbb{E}[|M_n|] = \mathbb{E}[M_n] = 1$, and apply Exercise 2.26 or Theorem 2.24.

Exercise 2.28 (Maximum of martingales is submartingale) For $i = 1, \ldots, m$, let $(M_n^{(i)})_{n\geq 0}$ be a sequence of martingales with respect to $(X_n)_{n\geq 1}$. Show that

$$M_n = \max_{i=1}^{m} M_n^{(i)} \tag{2.8.11}$$

is a submartingale with respect to $(X_n)_{n\geq 1}$.

Exercise 2.29 (Azuma–Hoeffding for binomials) Show that Theorem 2.27 implies that for $X \sim \mathsf{Bin}(n, p)$ with $p \leq 1/2$

$$\mathbb{P}(|X - np| \geq a) \leq 2\exp\Big\{-\frac{a^2}{2n(1-p)^2}\Big\}. \tag{2.8.12}$$

Exercise 2.30 (Azuma–Hoeffding and the CLT) Let $(X_i)_{i\geq 0}$ be an independent identically distributed sequence of random variables with $\mathbb{E}[X_i] = 0$ and $|X_i| \leq 1$, and define the martingale $(M_n)_{n\geq 0}$ by

$$M_n = \sum_{i=1}^{n} X_i, \qquad (M_0 = 0). \tag{2.8.13}$$

Use the Azuma–Hoeffding Inequality (Theorem 2.27) to show that

$$\mathbb{P}(|M_n| \geq a) \leq 2\exp\Big(-\frac{a^2}{2n}\Big). \tag{2.8.14}$$

Take $a = x\sqrt{n}$, and use the central limit theorem to prove that $\mathbb{P}(|M_n| \geq a)$ converges. Compare the limit to the bound in (2.8.14).

Exercise 2.31 (Regular variation of u_n) Show that when (2.6.8) holds, then u_n is regularly varying at $n = \infty$ with exponent $1/(\tau - 1)$.

Exercise 2.32 (Order statistics of exponential sample) Prove (2.6.21) using the memoryless property of the exponential distribution.

3

Branching Processes

Branching processes appear throughout this book to describe the connected components of various random graphs. To prepare for this, in this chapter we describe branching processes in quite some detail. Special attention will be given to branching processes with a Poisson offspring distribution, as well as to their relation to branching processes with a binomial offspring distribution.

Organization of This Chapter

We start by describing the survival versus extinction transition in Section 3.1, and compute moments of the total family size in Section 3.2. We provide a useful random-walk perspective on branching processes in Section 3.3. In Section 3.4, we restrict to supercritical branching processes, and in Section 3.5, we use the random-walk perspective to derive the probability distribution of the branching process total progeny. In Section 3.6, we discuss Poisson branching processes, and in Section 3.7, we compare binomial branching processes to Poisson branching processes.

3.1 Survival versus Extinction

A branching process is the simplest possible model for a population evolving in time. Suppose that each individual independently gives birth to a number of children with the same distribution, independently across the different individuals. We denote the offspring distribution by $(p_i)_{i\geq 0}$, where

$$p_i = \mathbb{P}\,(\text{individual has } i \text{ children}). \tag{3.1.1}$$

We denote by Z_n the number of individuals in the nth generation, where, by convention, we let $Z_0 = 1$. Then Z_n satisfies the recursion relation

$$Z_n = \sum_{i=1}^{Z_{n-1}} X_{n,i}, \tag{3.1.2}$$

where $(X_{n,i})_{n,i\geq 1}$ is a doubly infinite array of i.i.d. random variables. We will often write X for the offspring distribution, so that $(X_{n,i})_{n,i\geq 1}$ is a doubly infinite array of i.i.d. random variables with $X_{n,i} \sim X$ for all $n, i \geq 0$. In this case, the law $(p_i)_{i\geq 0}$ of X is called the *offspring distribution* of the branching process.

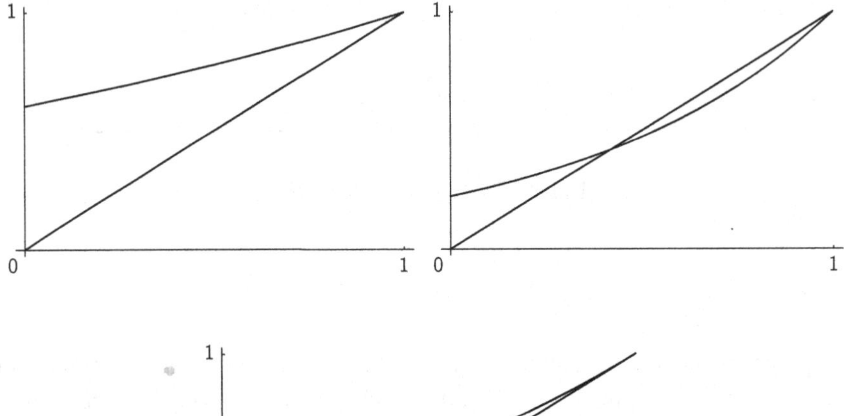

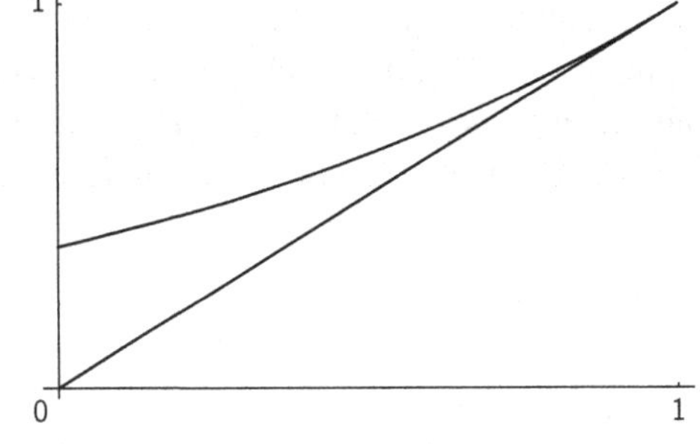

Figure 3.1 The solution of $s = G_X(s)$ when $\mathbb{E}[X] < 1$, $\mathbb{E}[X] > 1$ and $\mathbb{E}[X] = 1$, respectively. Note that $\mathbb{E}[X] = G'_X(1)$, and the solution η to $\eta = G_X(\eta)$ satisfies $\eta < 1$ precisely when $G'_X(1) > 1$.

One of the major results of branching processes is that when $\mathbb{E}[X] \leq 1$, the population dies out with probability one (unless $X_{1,1} = 1$ with probability one), while when $\mathbb{E}[X] > 1$, there is a non-zero probability that the population will survive forever. In order to state the result, we denote the *extinction probability* by

$$\eta = \mathbb{P}(\exists n\colon Z_n = 0). \tag{3.1.3}$$

Then the main result on the phase transition for branching processes is the following theorem:

Theorem 3.1 (Survival vs. extinction for branching processes) *For a branching process with i.i.d. offspring X, $\eta = 1$ when $\mathbb{E}[X] < 1$, while $\eta < 1$ when $\mathbb{E}[X] > 1$. Further, $\eta = 1$ when $\mathbb{E}[X] = 1$ and $\mathbb{P}(X = 1) < 1$. The extinction probability η is the smallest solution in $[0, 1]$ of*

$$\eta = G_X(\eta), \tag{3.1.4}$$

with $s \mapsto G_X(s)$ the probability generating function of the offspring distribution X, i.e.,

$$G_X(s) = \mathbb{E}[s^X]. \tag{3.1.5}$$

Proof We write

$$\eta_n = \mathbb{P}(Z_n = 0). \tag{3.1.6}$$

Because $\{Z_n = 0\} \subseteq \{Z_{n+1} = 0\}$, we have that $\eta_n \uparrow \eta$. Let

$$G_n(s) = \mathbb{E}[s^{Z_n}] \tag{3.1.7}$$

denote the probability generating function of the number of individuals in the nth generation. Then, since $\mathbb{P}(X = 0) = G_X(0)$ for an integer-valued random variable X,

$$\eta_n = G_n(0). \tag{3.1.8}$$

By conditioning on the first generation and using that $\mathbb{P}(Z_1 = i) = \mathbb{P}(X = i) = p_i$, we obtain that, for $n \geq 2$,

$$G_n(s) = \mathbb{E}[s^{Z_n}] = \sum_{i=0}^{\infty} p_i \mathbb{E}[s^{Z_n} \mid Z_1 = i] = \sum_{i=0}^{\infty} p_i G_{n-1}(s)^i. \tag{3.1.9}$$

In the last step, we have used the fact that each of the i individuals produces offspring in the nth generation independently and in an identical way, with probability generating function equal to $s \mapsto G_{n-1}(s)$, where Z_n is the sum of the contributions of the i individuals in the first generation. Therefore, writing $G_X(s) = G_1(s)$ for the generating function of $X_{1,1}$,

$$G_n(s) = G_X(G_{n-1}(s)). \tag{3.1.10}$$

When we substitute $s = 0$, we obtain that η_n satisfies the recurrence relation

$$\eta_n = G_X(\eta_{n-1}). \tag{3.1.11}$$

See Figure 3.2 for the evolution of $n \mapsto \eta_n$.

When $n \to \infty$, we have that $\eta_n \uparrow \eta$, so that by continuity of $s \mapsto G_X(s)$,

$$\eta = G_X(\eta). \tag{3.1.12}$$

When $\mathbb{P}(X = 1) = 1$, then $Z_n = 1$ a.s., and there is nothing to prove. When, further, $\mathbb{P}(X \leq 1) = 1$, but $p = \mathbb{P}(X = 0) > 0$, then $\mathbb{P}(Z_n = 0) = 1 - (1 - p)^n \to 1$, so again there is nothing to prove. Therefore, for the remainder of this proof, we assume that $\mathbb{P}(X \leq 1) < 1$.

Suppose that $\psi \in [0, 1]$ satisfies $\psi = G_X(\psi)$. We claim that $\eta \leq \psi$. We use induction to prove that $\eta_n \leq \psi$ for all n. Indeed, $\eta_0 = 0 \leq \psi$, which initializes the induction hypothesis.

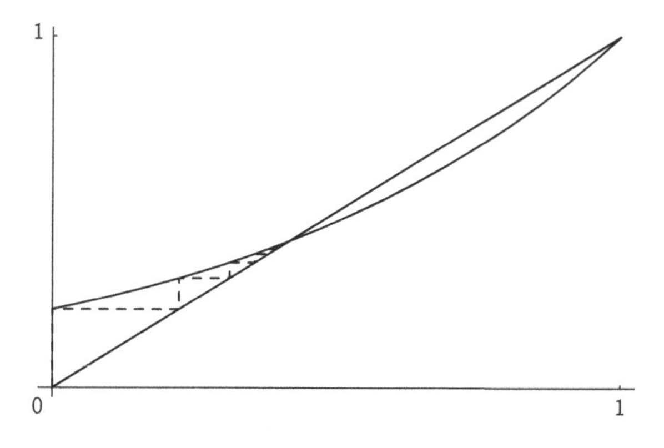

Figure 3.2 The iteration for $n \mapsto \eta_n$ in (3.1.11).

To advance the induction, we use (3.1.11), the induction hypothesis, as well as the fact that $s \mapsto G_X(s)$ is increasing on [0, 1], to see that

$$\eta_n = G_X(\eta_{n-1}) \leq G_X(\psi) = \psi, \tag{3.1.13}$$

where the final conclusion comes from the fact that ψ is a solution of $\psi = G_X(\psi)$. Therefore, $\eta_n \leq \psi$, which advances the induction. Since $\eta_n \uparrow \eta$, we conclude that $\eta \leq \psi$ for all solutions ψ of $\psi = G_X(\psi)$. Therefore, η is the smallest such solution.

We note that $s \mapsto G_X(s)$ is increasing and convex for $s \geq 0$, since

$$G_X''(s) = \mathbb{E}[X(X-1)s^{X-2}] \geq 0. \tag{3.1.14}$$

When $\mathbb{P}(X \leq 1) < 1$, we have that $\mathbb{E}[X(X-1)s^{X-2}] > 0$, so that $s \mapsto G_X(s)$ is strictly increasing and strictly convex for $s > 0$. Therefore, there can be at most two solutions of $s = G_X(s)$ in [0, 1]. Note that $s = 1$ is always a solution of $s = G_X(s)$, since G_X is a probability generating function. Since $G_X(0) > 0$, there is precisely one solution when $G_X'(1) < 1$, while there are two solutions when $G_X'(1) > 1$. The former implies that $\eta = 1$ when $G_X'(1) < 1$, while the latter implies that $\eta < 1$ when $G_X'(1) > 1$. When $G_X'(1) = 1$, again there is precisely one solution, except when $G_X(s) = s$, which is equivalent to $\mathbb{P}(X = 1) = 1$. Since $G_X'(1) = \mathbb{E}[X]$, this proves the claim. □

We call a branching process *subcritical* when $\mathbb{E}[X] < 1$, *critical* when $\mathbb{E}[X] = 1$ and $\mathbb{P}(X = 1) < 1$ and *supercritical* when $\mathbb{E}[X] > 1$. The case where $\mathbb{P}(X = 1) = 1$ is uninteresting, and is omitted in this definition.

In many cases, we are interested in the *survival probability*, denoted by $\zeta = 1 - \eta$, which is the probability that the branching process survives forever, i.e.,

$$\zeta = \mathbb{P}(Z_n > 0 \; \forall n \geq 0). \tag{3.1.15}$$

See Figure 3.3 for the survival probability of a Poisson branching process with parameter λ as a function of λ. Exercises 3.1–3.5 investigate aspects of branching processes and discuss an example.

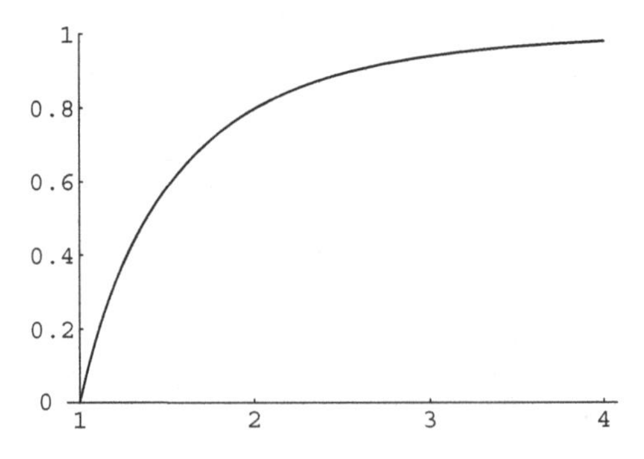

Figure 3.3 The survival probability $\zeta = \zeta_\lambda$ for a Poisson branching process with mean offspring equal to λ for $\lambda \geq 1$. The survival probability equals $\zeta = 1 - \eta$, where η is the extinction probability. Note that $\zeta_\lambda \equiv 0$ for $\lambda \in [0, 1]$.

We continue by studying the total progeny T of the branching process, which is defined as

$$T = \sum_{n=0}^{\infty} Z_n. \tag{3.1.16}$$

We denote by $G_T(s)$ the probability-generating function of T, i.e.,

$$G_T(s) = \mathbb{E}[s^T]. \tag{3.1.17}$$

The main result is the following:

Theorem 3.2 (Total progeny probability generating function) *For a branching process with i.i.d. offspring X having probability generating function $G_X(s) = \mathbb{E}[s^X]$, the probability generating function of the total progeny T satisfies the relation that, for all $s \in [0, 1)$,*

$$G_T(s) = sG_X(G_T(s)). \tag{3.1.18}$$

Theorem 3.2 requires some care when the branching process has a positive survival probability. We note that when the branching process survives with positive probability, i.e., $\eta < 1$ with η the smallest solution to $\eta = G_X(\eta)$, then $G_T(s) = \mathbb{E}[s^T]$ only receives a contribution from $T < \infty$ when $|s| < 1$. Further, $G_T(1) \equiv \lim_{s\nearrow 1} G_T(s)$ satisfies $G_T(1) = G_X(G_T(1))$, so that either $G_T(1) = 1$ or $G_T(1) = \eta$. Since, for each $s \in [0, 1)$,

$$G_T(s) = \mathbb{E}[s^T \mathbb{1}_{\{T<\infty\}}] \leq \mathbb{P}(T < \infty), \tag{3.1.19}$$

we thus obtain that $\lim_{s\nearrow 1} G_T(s) = \eta$.

Proof We again condition on the size of the first generation, and use that when $Z_1 = i$, for $j = 1, \ldots, i$, the total progeny T_j of the jth child of the initial individual satisfies that $(T_j)_{j=1}^{i}$ is an i.i.d. sequence of random variables with law equal to that of T. Therefore, using also that

$$T = 1 + \sum_{j=1}^{i} T_j, \tag{3.1.20}$$

where, by convention, the empty sum, arising when $i = 0$, is equal to zero, we obtain

$$\begin{aligned} G_T(s) &= \sum_{i=0}^{\infty} p_i \mathbb{E}[s^T \mid Z_1 = i] = s \sum_{i=0}^{\infty} p_i \mathbb{E}[s^{T_1+\cdots+T_i}] \\ &= s \sum_{i=0}^{\infty} p_i G_T(s)^i = sG_X(G_T(s)). \end{aligned} \tag{3.1.21}$$

This completes the proof. □

3.2 Family Moments

In this section, we compute the mean generation size of a branching process and use this to compute the mean family size or the mean total progeny. The main result is the following theorem:

Theorem 3.3 (Moments of generation sizes) *For all $n \geq 0$, and with $\mu = \mathbb{E}[Z_1] = \mathbb{E}[X]$ the expected offspring of a given individual,*

$$\mathbb{E}[Z_n] = \mu^n. \tag{3.2.1}$$

Proof Recall that

$$Z_n = \sum_{i=1}^{Z_{n-1}} X_{n,i}, \tag{3.2.2}$$

where $(X_{n,i})_{n,i\geq 1}$ is a doubly infinite array of i.i.d. random variables. In particular, $(X_{n,i})_{i\geq 1}$ is independent of Z_{n-1}. The completion of the proof of Theorem 3.3 is left to the reader as Exercise 3.12. □

Theorem 3.4 (Markov inequality for subcritical branching processes) *Fix $n \geq 0$. Let $\mu = \mathbb{E}[Z_1] = \mathbb{E}[X]$ be the expected offspring of a given individual, and assume that $\mu < 1$. Then,*

$$\mathbb{P}(Z_n > 0) \leq \mu^n. \tag{3.2.3}$$

Theorem 3.4 implies that in the subcritical regime, i.e., when the expected offspring μ satisfies $\mu < 1$, the probability that the population survives up to time n is exponentially small in n. In particular, the expected total progeny is finite, as the next theorem shows:

Theorem 3.5 (Expected total progeny) *For a branching process with i.i.d. offspring X having mean offspring $\mu < 1$,*

$$\mathbb{E}[T] = \frac{1}{1-\mu}. \tag{3.2.4}$$

3.3 Random-Walk Perspective to Branching Processes

In branching processes, it is common to study the number of descendants of each individual in a given generation. For random graph purposes, it is often convenient to use a different construction of a branching process by sequentially investigating the number of children of each member of the population. This picture leads to a *random-walk formulation of branching processes*. For more background on random walks, we refer the reader to the classic book by Spitzer (1976) or the book by Grimmett and Stirzaker (2001, Section 5.3).

We now give the random-walk representation of a branching process. Let $X_1', X_2', \ldots$ be independent and identically distributed random variables with the same distribution as $X_{1,1}$ in (3.1.2). Define $S_0', S_1', \ldots$ by the recursion

$$\begin{aligned} S_0' &= 1, \\ S_i' &= S_{i-1}' + X_i' - 1 = X_1' + \cdots + X_i' - (i-1). \end{aligned} \tag{3.3.1}$$

Let T' be the smallest t for which $S_t' = 0$, i.e.,

$$T' = \inf\{t \colon S_t' = 0\} = \inf\{t \colon X_1' + \cdots + X_t' = t - 1\}. \tag{3.3.2}$$

In particular, if such a t does not exist, then we define $T' = +\infty$.

The above description turns out to be *equivalent* to the normal definition of a branching process, but records the branching process tree in a different manner. For example, in the random-walk picture, it is slightly more difficult to extract the distribution of the generation sizes, and the answer depends on how we explore the branching process tree. Usually, this exploration is performed in either a breadth-first or a depth-first order. For the distribution of $(S'_t)_{t\geq 0}$ this makes no distinction. To see that the two pictures agree, we will show that the distribution of the random variable T' is equal in distribution to the *total progeny* of the branching process as defined in (3.1.16), and it is equal to the total number of individuals in the family tree of the initial individual. We will also see that $(X'_k)_{k\geq 1}$ in (3.3.1) and $(X_{i,n})_{i,n\geq 1}$ in (3.1.2) are related through $X'_k = X_{i_k,n_k}$ for a unique i_k, n_k that depends on the order in which the tree is explored.

The branching process belonging to the recursion in (3.3.1) is the following. The population starts with one active individual. At time i, we select one of the active individuals in the population, and give it X'_i children. The children (if any) are set to active, and the individual itself becomes inactive. See Figure 3.4 for a graphical representation of this exploration in a breadth-first order.

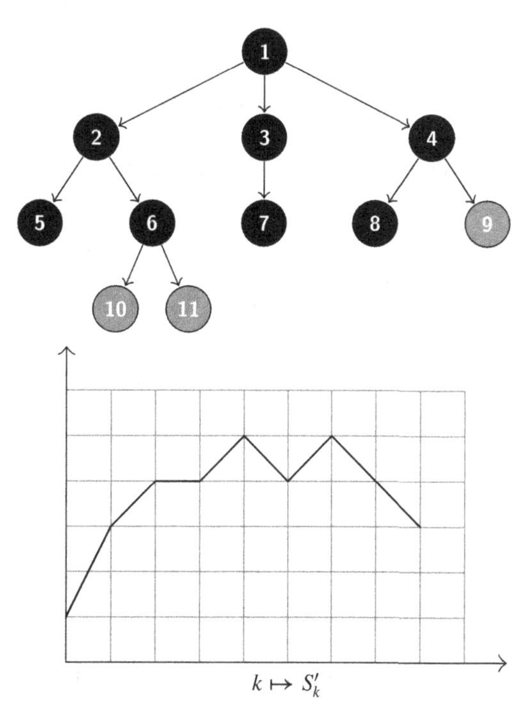

Figure 3.4 The exploration of a Galton–Watson or branching process tree as formulated in (3.3.1). We explore the tree in the breadth-first order. The black vertices have been explored, the grey vertices have not yet been explored. We have explored 8 vertices, and $X'_1 = 3$, $X'_2 = 2$, $X'_3 = 1$, $X'_4 = 2$, $X'_5 = 0$, etc. The next vertex to be explored is vertex 9, and there are 2 more vertices to be explored (i.e., $S'_8 = 3$). Below, we plot the random-walk path arising from the exploration of the branching process tree.

This exploration process is continued as long as there are active individuals in the population. Then, the process S'_i describes the number of active individuals after the first i individuals have been explored. The process stops when $S'_t = 0$, but the recursion can be defined for all t since this leaves the value of T' unaffected. Note that, for a branching process, (3.3.1) only makes sense as long as $i \leq T'$, since only then $S'_i \geq 0$ for all $i \leq T$. However, (3.3.1) itself can be defined for *all* $i \geq 0$, also when $S'_i < 0$. This fact will be useful in the sequel. The equivalence is proved formally in the following lemma:

Lemma 3.6 (Interpretation of $(S'_i)_{i\in[T']}$) *The random process* $(S'_i)_{i\in[T']}$ *in (3.3.1) has the same distribution as the random process* $(S_i)_{i\in[T]}$*, where* S_i *denotes the number of unexplored individuals in the exploration of a branching process population after exploring* i *individuals successively, and* T *denotes its total progeny.*

Proof We prove Lemma 3.6 by induction on i. The induction hypothesis is that $(S'_0, \ldots, S'_i)$ has the same law as $(S_0, \ldots, S_i)$, where $(S'_i)_{i\in[T']}$ and $(S_i)_{i\in[T]}$ have been defined in Lemma 3.6.

Clearly, the statement is correct when $i = 0$ where $S_0 = S'_0 = 0$, which initiates the induction hypothesis. We next advance the induction hypothesis. For this, suppose that the statement is true for $i-1$. We are done when $S_{i-1} = 0$, since then all individuals have been explored and the total number of explored individuals is equal to the size of the family tree, which is T by definition. Thus, assume that $S_{i-1} > 0$. Then we pick an arbitrary unexplored individual and denote the number of its children by X'_i. By the independence property of the offspring of different individuals in a branching process, conditionally on $(S_0, \ldots, S_{i-1})$, the distribution of X'_i is equal to the distribution of $X_{1,1}$, say, independently of $(S_0, \ldots, S_{i-1})$. Also, after exploring the children of the ith individual, we have added X_i individuals that still need to be explored, and have explored a single individual, so that now the total number of unexplored individuals is equal to $S_{i-1} + X'_i - 1 = S'_{i-1} + X'_i - 1$. By (3.3.1), S'_i satisfies the same recursion. We conclude that also $(S'_0, \ldots, S'_i)$ has the same law as $(S_0, \ldots, S_i)$. This completes the proof using induction. □

Lemma 3.6 gives a nice interpretation of the random process $(S'_i)_{i\geq 0}$ in (3.3.1). Since the branching process total progeny is explored *precisely* at the moment that all of its individuals have been explored, it follows that T in (3.3.2) has the same distribution as the total progeny of the branching process. From now on and with Lemma 3.6 in mind, we will slightly abuse notation and write $(S_i)_{i\geq 0}$ and $(X_i)_{i\geq 1}$ instead of $(S'_i)_{i\geq 0}$ and $(X'_i)_{i\geq 1}$. In Exercises 3.13 and 3.14, you are asked to investigate the random-walk description for branching processes in more detail.

We next investigate the possible random-walk trajectories compatible with a total progeny $T = t$. Denote by $H = (X_1, \ldots, X_T)$ the history of the process up to time T. We include the case where $T = \infty$, in which case the vector H has infinite length. A sequence $(x_1, \ldots, x_t)$ with $x_i \geq 0$ for every $i = 1, \ldots, t$ is a possible history if and only if the sequence x_i satisfies (3.3.1), i.e., when $s_i > 0$ for all $i < t$, while $s_t = 0$, where $s_i = x_1 + \cdots + x_i - (i-1)$.

Then, for any $t < \infty$,

$$\mathbb{P}(H = (x_1, \ldots, x_t)) = \prod_{i=1}^{t} p_{x_i}. \tag{3.3.3}$$

We next use the random-walk perspective in order to describe the distribution of a branching process *conditioned on extinction*. Call the distributions $(p_k)_{k\geq 0}$ and $(p'_k)_{k\geq 0}$ a *conjugate pair* if

$$p'_k = \eta^{k-1} p_k, \tag{3.3.4}$$

where η is the extinction probability belonging to the offspring distribution $(p_k)_{k\geq 0}$, so that $\eta = G_X(\eta)$. In Exercise 3.15, you are asked to prove that the conjugate distribution $(p'_k)_{k\geq 0}$ defined in (3.3.4) is a probability distribution.

The relation between a supercritical branching process conditioned on extinction and its conjugate branching process is as follows:

Theorem 3.7 (Duality principle for branching processes) *Let $(p_k)_{k\geq 0}$ and $(p'_k)_{k\geq 0}$ be a conjugate pair of offspring distributions. The branching process with distribution $(p_k)_{k\geq 0}$ conditioned on extinction, has the same distribution as the branching process with offspring distribution $(p'_k)_{k\geq 0}$.*

The duality principle takes a particularly appealing form for Poisson branching processes (see Theorem 3.15 below).

Proof It suffices to show that for every finite history $H = (x_1, \ldots, x_t)$, the probability (3.3.3) is the same for the branching process with offspring distribution $(p_k)_{k\geq 0}$, when conditioned on extinction, and the branching process with offspring distribution $(p'_k)_{k\geq 0}$. Fix a $t < \infty$. First observe that

$$\begin{aligned}\mathbb{P}(H = (x_1, \ldots, x_t) \mid \text{extinction}) &= \frac{\mathbb{P}(\{H = (x_1, \ldots, x_t)\} \cap \text{extinction})}{\mathbb{P}(\text{extinction})} \\ &= \eta^{-1}\mathbb{P}(H = (x_1, \ldots, x_t)),\end{aligned} \tag{3.3.5}$$

since a finite history implies that the population becomes extinct. Then, we use (3.3.3), together with the fact that

$$\prod_{i=1}^{t} p_{x_i} = \prod_{i=1}^{t} p'_{x_i}\eta^{-(x_i-1)} = \eta^{t-\sum_{i=1}^{t} x_i} \prod_{i=1}^{t} p'_{x_i} = \eta \prod_{i=1}^{t} p'_{x_i}, \tag{3.3.6}$$

since $x_1 + \cdots + x_t = t - 1$. Substitution into (3.3.5) yields that

$$\mathbb{P}(H = (x_1, \ldots, x_t) \mid \text{extinction}) = \mathbb{P}'(H = (x_1, \ldots, x_t)), \tag{3.3.7}$$

where $\mathbb{P}'$ is the distribution of the branching process with offspring distribution $(p'_k)_{k\geq 0}$. □

Exercises 3.16 and 3.17 investigate properties of the dual branching process, such as the generating function and mean of the offspring distribution.

Another convenient feature of the random-walk perspective for branching processes is that it allows one to study what the probability of extinction is when the family tree has at least a given size. The main result in this respect is given below:

Theorem 3.8 (Extinction probability with large total progeny) *For a branching process with i.i.d. offspring X having mean $\mu = \mathbb{E}[X] > 1$,*

$$\mathbb{P}(k \leq T < \infty) \leq \frac{e^{-Ik}}{1 - e^{-I}}, \tag{3.3.8}$$

where the exponential rate I is given by

$$I = \sup_{t \leq 0} \big(t - \log \mathbb{E}[e^{tX}]\big) > 0. \tag{3.3.9}$$

Theorem 3.8 can be reformulated by saying that when the total progeny is large, then the branching process will survive with high probability. Note that when $\mu = \mathbb{E}[X] > 1$, then we can also write

$$I = \sup_{t} \big(t - \log \mathbb{E}[e^{tX}]\big) \tag{3.3.10}$$

(see also (2.4.12)). In Theorem 3.8, it is *not* assumed that $\mathbb{E}[e^{tX}] < \infty$ for all $t \in \mathbb{R}$! Since $X \geq 0$, we clearly *do* have that $\mathbb{E}[e^{tX}] < \infty$ for all $t \leq 0$. Therefore, since also the derivative of $t \mapsto t - \log \mathbb{E}[e^{tX}]$ in $t = 0$ is equal to $1 - \mathbb{E}[X] < 0$, the supremum is attained at a negative t, and, therefore, we obtain that $I > 0$ under no assumptions on the existence of the moment generating function of the offspring distribution. We now give the full proof:

Proof We use the fact that $T = s$ implies that $S_s = 0$, which in turn implies that $X_1 + \cdots + X_s = s - 1 \leq s$. Therefore,

$$\mathbb{P}(k \leq T < \infty) \leq \sum_{s=k}^{\infty} \mathbb{P}(S_s = 0) \leq \sum_{s=k}^{\infty} \mathbb{P}(X_1 + \cdots + X_s \leq s). \tag{3.3.11}$$

For the latter probability, we use (2.4.9) and (2.4.11) in Theorem 2.19 with $a = 1 < \mathbb{E}[X]$. Then, we arrive at

$$\mathbb{P}(k \leq T < \infty) \leq \sum_{s=k}^{\infty} e^{-sI} = \frac{e^{-Ik}}{1 - e^{-I}}, \tag{3.3.12}$$

as required. □

3.4 Supercritical Branching Processes

In this section, we prove a convergence result for the size Z_n of the nth generation in a supercritical branching process. Clearly, in the (sub)critical case, $Z_n \xrightarrow{\mathbb{P}} 0$, and there is nothing to prove. In the supercritical case, when the expected offspring is equal to $\mu > 1$, it is also known that (see, e.g., Athreya and Ney (1972, Theorem 2, p. 8)) $\lim_{n\to\infty} \mathbb{P}(Z_n = k) = 0$ unless $k = 0$, and $\mathbb{P}(\lim_{n\to\infty} Z_n = 0) = 1 - \mathbb{P}(\lim_{n\to\infty} Z_n = \infty) = \eta$, where η is the extinction probability of the branching process. In particular, the branching process population cannot stabilize. It remains to investigate what happens when $\eta < 1$, in which case $\lim_{n\to\infty} Z_n = \infty$ with probability $1 - \eta > 0$. We prove the following convergence result:

Theorem 3.9 (Convergence for supercritical branching processes) *For a branching process with i.i.d. offspring X having mean $\mu = \mathbb{E}[X] > 1$, $\mu^{-n} Z_n \xrightarrow{a.s.} W_\infty$ for some random variable W_∞ which is finite with probability 1.*

Proof We use the martingale convergence theorem (Theorem 2.24), and, in particular, its consequence formulated in Exercise 2.26. Denote $M_n = \mu^{-n} Z_n$, and recall that $(M_n)_{n \geq 1}$ is a martingale by Exercise 3.9. Theorem 3.3 implies that $\mathbb{E}[|M_n|] = \mathbb{E}[M_n] = 1$, so that Theorem 2.24 gives the result. □

Unfortunately, not much is known about the limiting random variable W_∞. Exercise 3.18 shows that its probability generating function $G_W(s) = \mathbb{E}[s^{W_\infty}]$ satisfies the implicit relation, for $s \in [0, 1]$,

$$G_W(s) = G_X\big(G_W(s^{1/\mu})\big). \tag{3.4.1}$$

In particular, it could be the case that $W_\infty = 0$ a.s., in which case Theorem 3.9 is rather uninformative. We next investigate when $\mathbb{P}(W_\infty > 0) = 1 - \eta = \zeta$:

Theorem 3.10 (Kesten–Stigum Theorem) *For a branching process with i.i.d. offspring X having mean $\mathbb{E}[X] > 1$ and extinction probability η, $\mathbb{P}(W_\infty = 0) = \eta$ precisely when $\mathbb{E}[X \log X] < \infty$. When $\mathbb{E}[X \log X] < \infty$, also $\mathbb{E}[W_\infty] = 1$, while, when $\mathbb{E}[X \log X] = \infty$, $\mathbb{P}(W_\infty = 0) = 1$.*

Theorem 3.10 implies that $\mathbb{P}(W_\infty > 0) = 1 - \eta$ precisely when $\mathbb{E}[X \log X] < \infty$. Here η is the extinction probability of the branching process, so that conditionally on survival, $W_\infty > 0$ a.s. It is remarkable that the precise condition when $W_\infty = 0$ a.s. can be so easily expressed in terms of a moment condition on the offspring distribution. Exercises 3.19 and 3.20 investigate further properties of the limiting variable W_∞.

We omit the proof of Theorem 3.10.

Theorem 3.10 leaves us with the question what happens when $\mathbb{E}[X \log X] = \infty$. In this case, Seneta (1969) has shown that there always exists a proper renormalization, i.e., there exists a sequence $(c_n)_{n \geq 1}$ with $\lim_{n \to \infty} c_n^{1/n} = \mu$ such that Z_n / c_n converges to a non-degenerate limit. However, $c_n = o(\mu^n)$, so that $\mathbb{P}(W_\infty = 0) = 1$.

We continue to study the number of individuals with an *infinite line of descent*, i.e., the individuals whose family tree survives forever. Interestingly, conditionally on survival, these individuals form a branching process again, as we describe now. In order to state the result, we let $Z_n^{(\infty)}$ denote those individuals from the nth generation of $(Z_k)_{k \geq 0}$ whose descendants survive forever. Then, the main result concerning $(Z_n^{(\infty)})_{n \geq 0}$ is as follows:

Theorem 3.11 (Individuals with an infinite line of descent) *Conditionally on survival, the process $(Z_n^{(\infty)})_{n \geq 0}$ is again a branching process with offspring distribution $(p_k^{(\infty)})_{k \geq 0}$ given by $p_0^{(\infty)} = 0$ and, for $k \geq 1$,*

$$p_k^{(\infty)} = \frac{1}{\zeta} \sum_{j=k}^{\infty} \binom{j}{k} \eta^{j-k} (1 - \eta)^k p_j. \tag{3.4.2}$$

Moreover,

$$\mu^{(\infty)} = \mathbb{E}[Z_1^{(\infty)}] = \mu = \mathbb{E}[Z_1]. \tag{3.4.3}$$

In particular, the branching process $(Z_n^{(\infty)})_{n\geq 0}$ *is supercritical with the same expected offspring as* $(Z_n)_{n\geq 0}$ *itself.*

Comparing Theorem 3.11 to Theorem 3.7, we see that in the supercritical regime, the branching process conditioned on extinction is a branching process with the conjugate or dual (subcritical) offspring distribution, while, conditionally on survival, the individuals with an infinite line of descent form a (supercritical) branching process. Exercise 3.21 shows that $(p_k^{(\infty)})_{k\geq 0}$ is a probability distribution.

Proof We let $\mathcal{A}_\infty = \{Z_n > 0 \ \forall n \geq 0\}$ be the event that $(Z_n)_{n\geq 0}$ survives forever. We will prove by induction on $n \geq 0$ that the distribution of $(Z_k^{(\infty)})_{k=0}^n$ conditionally on $\mathcal{A}_\infty$ is equal to that of $(\hat{Z}_k)_{k=0}^n$, where $(\hat{Z}_k)_{k\geq 0}$ is a branching process with offspring distribution $(p_k^{(\infty)})_{k\geq 0}$ given in (3.4.2). We start by initializing the induction hypothesis. For this, we note that $Z_0^{(\infty)} = 1$ on $\mathcal{A}_\infty$, whereas, by convention, $\hat{Z}_0 = 1$. This initializes the induction hypothesis.

To advance the induction hypothesis, we argue as follows. Suppose that, conditionally on $\mathcal{A}_\infty$, the distribution of $(Z_k^{(\infty)})_{k=0}^n$ is equal to that of $(\hat{Z}_k)_{k=0}^n$. Then, we show that, conditionally on $\mathcal{A}_\infty$, the distribution of $(Z_k^{(\infty)})_{k=0}^{n+1}$ is also equal to that of $(\hat{Z}_k)_{k=0}^{n+1}$. By the induction hypothesis, this immediately follows if the *conditional* distribution of $Z_{n+1}^{(\infty)}$ given $(Z_k^{(\infty)})_{k=0}^n$ is equal to the conditional distribution of $\hat{Z}_{n+1}$ given $(\hat{Z}_k)_{k=0}^n$.

The law of $\hat{Z}_{n+1}$ given $(\hat{Z}_k)_{k=0}^n$ is that of a sum of $\hat{Z}_n$ i.i.d. random variables with law $(p_k^{(\infty)})_{k\geq 0}$. The law of $Z_{n+1}^{(\infty)}$ given $(Z_k^{(\infty)})_{k=0}^n$ is equal to the law of $Z_{n+1}^{(\infty)}$ given $Z_n^{(\infty)}$, by the Markov property of the branching process. Further, each individual with infinite line of descent in the nth generation gives rise to a random and i.i.d. number of individuals with infinite line of descent in the $(n+1)$st generation with the same law as $Z_1^{(\infty)}$ conditionally on $\mathcal{A}_\infty$. Indeed, both are equal in distribution to the number of children with infinite line of descent of an individual whose offspring survives. As a result, to complete the proof of (3.4.2), we must show that

$$\mathbb{P}\big(Z_1^{(\infty)} = k \mid \mathcal{A}_\infty\big) = p_k^{(\infty)}. \tag{3.4.4}$$

For $k = 0$, this is trivial, since $Z_1^{(\infty)} \geq 1$ conditionally on $\mathcal{A}_\infty$, so that both sides are equal to 0. For $k \geq 1$, on the other hand, the proof follows by conditioning on Z_1. Indeed, $Z_1^{(\infty)} = k$ implies that $Z_1 \geq k$ and that $\mathcal{A}_\infty$ occurs, so that

$$\begin{aligned}\mathbb{P}\big(Z_1^{(\infty)} = k \mid \mathcal{A}_\infty\big) &= \zeta^{-1}\mathbb{P}\big(Z_1^{(\infty)} = k\big) = \zeta^{-1}\sum_{j\geq k}\mathbb{P}\big(Z_1^{(\infty)} = k \mid Z_1 = j\big)\mathbb{P}(Z_1 = j)\\ &= \zeta^{-1}\sum_{j\geq k}\binom{j}{k}\eta^{j-k}(1-\eta)^k p_j = p_k^{(\infty)},\end{aligned} \tag{3.4.5}$$

since each of the j individuals has infinite line of descent with probability $\zeta = 1-\eta$, so that $\mathbb{P}\big(Z_1^{(\infty)} = k \mid Z_1 = j\big) = \mathbb{P}(\mathsf{Bin}(j, 1-\eta) = k)$. Equation (3.4.5) proves (3.4.4).

We complete the proof of Theorem 3.11 by proving (3.4.3). We start by proving (3.4.3) when $\mu = \mathbb{E}[X] < \infty$. Note that (3.4.2) also holds for $k = 0$. Then, we write

$$\mu^{(\infty)} = \sum_{k=0}^{\infty} k p_k^{(\infty)} = \sum_{k=0}^{\infty} k \frac{1}{\zeta} \sum_{j=k}^{\infty} \binom{j}{k} \eta^{j-k} (1-\eta)^k p_j$$
$$= \frac{1}{\zeta} \sum_{j=0}^{\infty} p_j \sum_{k=0}^{j} k \binom{j}{k} \eta^{j-k} (1-\eta)^k = \frac{1}{\zeta} \sum_{j=0}^{\infty} p_j (\zeta j) = \sum_{j=0}^{\infty} j p_j = \mu. \tag{3.4.6}$$

This proves (3.4.3) when $\mu < \infty$. When $\mu = \infty$, on the other hand, we only need to show that $\mu^{(\infty)} = \infty$ as well. This can easily be seen by an appropriate truncation argument, and is left to the reader in Exercise 3.22. □

With Theorems 3.11 and 3.9 in hand an interesting picture of the structure of supercritical branching processes emerges. Indeed, by Theorem 3.9, $Z_n \mu^{-n} \xrightarrow{a.s.} W_\infty$, where, if the $X \log X$-condition in Theorem 3.10 is satisfied, $\mathbb{P}(W_\infty > 0) = \zeta$, the branching process survival probability. On the other hand, by Theorem 3.11 and conditionally on $\mathcal{A}_\infty$, $(Z_n^{(\infty)})_{n \geq 0}$ is also a branching process with expected offspring $\mu = \mathbb{E}[X]$, which survives with probability 1. Furthermore, the $X \log X$-condition for $(Z_n^{(\infty)})_{n \geq 0}$ as formulated in the Kesten–Stigum Theorem (Theorem 3.10) clearly follows from that of $(Z_n)_{n \geq 0}$. As a result, $Z_n^{(\infty)} \mu^{-n} \xrightarrow{a.s.} W_\infty^{(\infty)}$, where $\mathbb{P}(W_\infty^{(\infty)} > 0 \mid \mathcal{A}_\infty) = 1$. However, $Z_n^{(\infty)} \leq Z_n$ for all $n \geq 0$, by definition. This raises the question, what is the relative size of $Z_n^{(\infty)}$ and Z_n, conditionally on $\mathcal{A}_\infty$. This question is answered in the following theorem:

Theorem 3.12 (Proportion of particles with infinite line of descent) *Conditionally on survival,*

$$\frac{Z_n^{(\infty)}}{Z_n} \xrightarrow{a.s.} \zeta. \tag{3.4.7}$$

Theorem 3.12 is useful to transfer results on branching processes that survive with probability 1, such as $(Z_n^{(\infty)})_{n \geq 0}$ conditionally on survival, to branching processes which have a non-zero extinction probability, such as $(Z_n)_{n \geq 0}$.

Proof of Theorem 3.12 We only give the proof in the case where the mean offspring $\mu = \mathbb{E}[X]$ is finite and the $X \log X$-condition holds, so that $\mathbb{E}[X \log X] < \infty$. Applying Theorem 3.11 together with Theorem 3.9 and the fact that, conditionally on survival, $\mathbb{E}[Z_1^{(\infty)}] = \mu$ (see (3.4.3)), we obtain that there exists $W^{(\infty)}$ such that $Z_n^{(\infty)} \mu^{-n} \to W^{(\infty)}$. Moreover, by Theorem 3.10 and the fact that the survival probability of the branching process $(Z_n^{(\infty)})_{n \geq 0}$ equals 1 (recall Exercise 3.1), we have that $\mathbb{P}(W^{(\infty)} > 0) = 1$. Further, again by Theorem 3.9, now applied to $(Z_n)_{n \geq 0}$, conditionally on survival, Z_n / μ^n converges in distribution to the conditional distribution of W_∞ conditionally on $W_\infty > 0$. Thus, we obtain that $Z_n^{(\infty)} / Z_n$ converges a.s. to a finite and positive limiting random variable that we denote by R.

In order to see that this limit in fact equals ζ, we use that the distribution of $Z_n^{(\infty)}$ given that $Z_n = k$ is binomial with parameter k and probability of success ζ, conditionally on being at least 1. (Indeed, $Z_n^{(\infty)} \geq 1$ since we condition on the survival event.) Since $Z_n \to \infty$ as

$n \to \infty$ conditionally on survival, $Z_n^{(\infty)}/Z_n$ converges in probability to ζ. This implies that $R = \zeta$ a.s., as required. □

3.5 Hitting-Time Theorem and the Total Progeny

In this section, we derive a general result for the law of the total progeny for branching processes, by making use of the hitting-time theorem for random walks. The main result is the following:

Theorem 3.13 (Law of total progeny) *For a branching process with i.i.d. offspring distribution $Z_1 = X$,*

$$\mathbb{P}(T = n) = \frac{1}{n}\mathbb{P}(X_1 + \cdots + X_n = n - 1), \tag{3.5.1}$$

where $(X_i)_{i=1}^n$ are i.i.d. copies of X.

Exercises 3.23 and 3.24 investigate the special cases of binomial and geometric offspring distributions.

We prove Theorem 3.13 below. In fact, we prove a more general version of Theorem 3.13, which states that

$$\mathbb{P}(T_1 + \cdots + T_k = n) = \frac{k}{n}\mathbb{P}(X_1 + \cdots + X_n = n - k), \tag{3.5.2}$$

where $T_1, \ldots, T_k$ are k independent random variables with the same distribution as T. Alternatively, we can think of $T_1 + \cdots + T_k$ as being the total progeny of a branching process starting with k individuals, i.e., a branching process for which $Z_0 = k$.

The proof is based on the random-walk representation of a branching process, together with the random-walk hitting-time theorem. In its statement, we write $\mathbb{P}_k$ for the law of a random walk starting in k, we let $(Y_i)_{i \geq 1}$ be the i.i.d. steps of the random walk and we let $S_n = k + Y_1 + \cdots + Y_n$ be the position of the walk, starting in k, after n steps. We finally let

$$H_0 = \inf\{n \colon S_n = 0\} \tag{3.5.3}$$

denote the first hitting time of the origin of the walk. Then, the hitting-time theorem is the following result:

Theorem 3.14 (Hitting-time theorem) *For a random walk with i.i.d. steps $(Y_i)_{i \geq 1}$ satisfying that Y_i is integer valued and*

$$\mathbb{P}(Y_i \geq -1) = 1, \tag{3.5.4}$$

the distribution of H_0 is given by

$$\mathbb{P}_k(H_0 = n) = \frac{k}{n}\mathbb{P}_k(S_n = 0). \tag{3.5.5}$$

Theorem 3.14 is a remarkable result, since it states that, conditionally on the event $\{S_n = 0\}$, and regardless of the precise distribution of the steps of the walk $(Y_i)_{i \geq 1}$ satisfying (3.5.4), the probability of the walk to be at 0 *for the first time* at time n is equal to k/n.

Equation (3.5.2) follows from Theorem 3.14 since the law of $T_1 + \cdots + T_k$ is that of the hitting time of zero of a random walk starting in k with step distribution $Y_i = X_i - 1$, where $(X_i)_{i \geq 1}$ are the offspring of the vertices. Indeed, by the random-walk Markov property, the law of H_0 when the random walk starts from k is the same as $H_{0,1} + \cdots + H_{0,k}$, where $(H_{0,i})_{i=1}^k$ are i.i.d. random variables having the same law as H_0 with $k = 1$. Thus, since $(H_{0,i})_{i=1}^k$ have the same law as $(T_i)_{i=1}^k$, $H_{0,1} + \cdots + H_{0,k}$ also has the same law as $T_1 + \cdots + T_k$. Since $X_i \geq 0$, we have that $Y_i = X_i - 1 \geq -1$, which completes the proof of (3.5.2) and hence of Theorem 3.13. The details are left as Exercise 3.25. Exercise 3.26 investigates whether the condition $\mathbb{P}(Y_i \geq -1) = 1$ is necessary.

Proof of Theorem 3.14 We prove (3.5.5) for all $k \geq 0$ by induction on $n \geq 1$. When $n = 1$, both sides are equal to 0 when $k > 1$ and $k = 0$, and are equal to $\mathbb{P}(Y_1 = -1)$ when $k = 1$. This initializes the induction.

To advance the induction, we take $n \geq 2$ and note that both sides are equal to 0 when $k = 0$. Thus, we may assume that $k \geq 1$. We condition on the first step to obtain

$$\mathbb{P}_k(H_0 = n) = \sum_{s=-1}^{\infty} \mathbb{P}_k(H_0 = n \mid Y_1 = s)\mathbb{P}(Y_1 = s). \tag{3.5.6}$$

By the random-walk Markov property,

$$\mathbb{P}_k(H_0 = n \mid Y_1 = s) = \mathbb{P}_{k+s}(H_0 = n - 1) = \frac{k+s}{n-1}\mathbb{P}_{k+s}(S_{n-1} = 0), \tag{3.5.7}$$

where in the last equality we used the induction hypothesis, which is allowed since $k \geq 1$ and $s \geq -1$, so that $k + s \geq 0$. It is here that we use the main assumption that $Y \geq -1$ a.s. This leads to

$$\mathbb{P}_k(H_0 = n) = \sum_{s=-1}^{\infty} \frac{k+s}{n-1}\mathbb{P}_{k+s}(S_{n-1} = 0)\mathbb{P}(Y_1 = s). \tag{3.5.8}$$

We undo the law of total probability, using that $\mathbb{P}_{k+s}(S_{n-1} = 0) = \mathbb{P}_k(S_n = 0 \mid Y_1 = s)$, to arrive at

$$\begin{aligned} \mathbb{P}_k(H_0 = n) &= \sum_{s=-1}^{\infty} \frac{k+s}{n-1}\mathbb{P}_k(S_n = 0 \mid Y_1 = s)\mathbb{P}(Y_1 = s) \\ &= \frac{1}{n-1}\mathbb{P}_k(S_n = 0)\Big(k + \mathbb{E}_k[Y_1 \mid S_n = 0]\Big), \end{aligned} \tag{3.5.9}$$

where $\mathbb{E}_k[Y_1 \mid S_n = 0]$ is the conditional expectation of Y_1 given that $S_n = 0$ occurs. We next note that $\mathbb{E}_k[Y_i \mid S_n = 0]$ is independent of i, so that

$$\mathbb{E}_k[Y_1 \mid S_n = 0] = \frac{1}{n}\sum_{i=1}^{n} \mathbb{E}_k[Y_i \mid S_n = 0] = \frac{1}{n}\mathbb{E}_k\Big[\sum_{i=1}^{n} Y_i \mid S_n = 0\Big] = -\frac{k}{n}, \tag{3.5.10}$$

since $\sum_{i=1}^n Y_i = S_n - k = -k$ when $S_n = 0$. Therefore, we arrive at

$$\mathbb{P}_k(H_0 = n) = \frac{1}{n-1}\Big[k - \frac{k}{n}\Big]\mathbb{P}_k(S_n = 0) = \frac{k}{n}\mathbb{P}_k(S_n = 0). \tag{3.5.11}$$

This advances the induction, and completes the proof of Theorem 3.14. □

3.6 Properties of Poisson Branching Processes

In this section, we specialize to branching processes with Poisson offspring distributions. We denote the distribution of a Poisson branching process by $\mathbb{P}^*_\lambda$. We also write T^* for the total progeny of the Poisson branching process, and X^* for a Poisson random variable.

For a Poisson random variable X^* with mean λ, the probability generating function of the offspring distribution is equal to

$$G^*_\lambda(s) = \mathbb{E}^*_\lambda[s^{X^*}] = \sum_{i=0}^{\infty} s^i \mathrm{e}^{-\lambda} \frac{\lambda^i}{i!} = \mathrm{e}^{\lambda(s-1)}. \tag{3.6.1}$$

Therefore, the relation for the extinction probability η in (3.1.4) becomes

$$\eta_\lambda = \mathrm{e}^{\lambda(\eta_\lambda - 1)}, \tag{3.6.2}$$

where we add the subscript λ and write $\eta = \eta_\lambda$ to make the dependence on λ explicit.

For $\lambda \leq 1$, the equation (3.6.2) has the unique solution $\eta_\lambda = 1$, which corresponds to almost-sure extinction. For $\lambda > 1$ there are two solutions, of which the smallest satisfies $\eta_\lambda \in (0, 1)$.

Conditionally on extinction, a Poisson branching process has law $(p'_k)_{k\geq 0}$ given by

$$p'_k = \eta_\lambda^{k-1} p_k = \mathrm{e}^{-\lambda\eta_\lambda} \frac{(\lambda\eta_\lambda)^k}{k!}, \tag{3.6.3}$$

where we have used (3.6.2). Note that this offspring distribution is again Poisson, but now with mean

$$\mu_\lambda = \lambda\eta_\lambda. \tag{3.6.4}$$

Again by (3.6.2),

$$\mu_\lambda \mathrm{e}^{-\mu_\lambda} = \lambda\eta_\lambda \mathrm{e}^{-\lambda\eta_\lambda} = \lambda \mathrm{e}^{-\lambda}. \tag{3.6.5}$$

This motivates the following definition of conjugate pairs. We call $\mu < 1 < \lambda$ a *conjugate pair* when

$$\mu \mathrm{e}^{-\mu} = \lambda \mathrm{e}^{-\lambda}. \tag{3.6.6}$$

Then, λ and $\mu_\lambda = \lambda\eta_\lambda$ are conjugate pairs. Since $x \mapsto x\mathrm{e}^{-x}$ is first increasing and then decreasing, with a maximum of e^{-1} at $x = 1$, the equation $x\mathrm{e}^{-x} = \lambda\mathrm{e}^{-\lambda}$ for $\lambda > 1$ has precisely two solutions: the interesting solution $x = \mu_\lambda < 1$ and the trivial solution $x = \lambda > 1$. Therefore, for Poisson offspring distributions, the duality principle in Theorem 3.7 can be reformulated as follows:

Theorem 3.15 (Poisson duality principle) *Let $\mu < 1 < \lambda$ be conjugate pairs. The Poisson branching process with mean λ, conditioned on extinction, has the same distribution as a Poisson branching process with mean μ.*

We further describe the law of the total progeny of a Poisson branching process:

Theorem 3.16 (Total progeny for Poisson branching processes) *For a branching process with i.i.d. offspring X^*, where X^* has a Poisson distribution with mean λ,*

$$\mathbb{P}^*_\lambda(T^* = n) = \frac{(\lambda n)^{n-1}}{n!} e^{-\lambda n}, \qquad (n \geq 1). \tag{3.6.7}$$

Exercises 3.28–3.30 investigate properties of the total progeny distribution in Theorem 3.16.

We use Theorem 3.16 to prove Cayley's Theorem on the number of labeled trees (Cayley, 1889). In its statement, we define a *labeled tree* on $[n]$ to be a tree of size n where all vertices have a label in $[n]$ and each label occurs precisely once. We now make this definition precise. An *edge* of a labeled tree is a pair $\{v_1, v_2\}$, where v_1 and v_2 are the labels of two connected vertices in the tree. The *edge set* of a tree of size n is the collection of its $n-1$ edges. Two labeled trees are equal if and only if they consist of the same edge sets. There is a bijection between labeled trees of n vertices and spanning trees of the complete graph K_n on the vertices $[n]$. Cayley's Theorem reads as follows:

Theorem 3.17 (Cayley's Theorem) *The number of labeled trees of size n is equal to n^{n-2}. Equivalently, the number of spanning trees of the complete graph of size n equals n^{n-2}.*

Proof Let T_n denote the number of labeled trees of size n. Our goal is to prove that $T_n = n^{n-2}$. We give two proofs, one based on double counting and one on a relation to Poisson branching processes.

Double Counting Proof

This beautiful proof is due to Jim Pitman, and gives Cayley's formula and a generalization by counting in two different ways. We follow Aigner and Ziegler (2014). We start with some notation.

A *forest* is a collection of disjoint trees. A *rooted forest* on $[n]$ is a forest together with a choice of a root in each component tree that covers $[n]$, i.e., each vertex is in one of the trees of the forest. Let $\mathcal{F}_{n,k}$ be the set of all rooted forests that consist of k rooted trees. Thus $\mathcal{F}_{n,1}$ is the set of all rooted trees. Note that $|\mathcal{F}_{n,1}| = nT_n$, since there are n choices for the root of a tree of size n. We now regard $F_{n,k} \in \mathcal{F}_{n,k}$ as a directed graph with all edges directed away from the roots. Say that a forest F contains another forest F' if F contains F' as a directed graph. Clearly, if F properly contains F' , then F has fewer components than F'.

Here is the crucial idea. Call a sequence $F_1, \ldots, F_k$ of forests a fragmentation sequence if $F_i \in \mathcal{F}_{n,i}$ and F_i contains F_{i+1} for all i. This means that F_{i+1} can be obtained from F_i by removing an edge and letting the new root of the newly created tree having no root be the vertex in the removed edge. Now let F_k be a fixed forest in $\mathcal{F}_{n,k}$ and let $N(F_k)$ be the number of rooted trees containing F_k, and $N^\star(F_k)$ the number of fragmentation sequences ending in F_k.

We count $N^\star(F_k)$ in two ways, first by starting at a tree and secondly by starting at F_k. Suppose $F_1 \in \mathcal{F}_{n,1}$ contains F_k. Since we may delete the $k-1$ edges of $F_1 \setminus F_k$ in any possible order to get a refining sequence from F_1 to F_k, we find

$$N^\star(F_k) = N(F_k)(k-1)!. \tag{3.6.8}$$

Let us now start at the other end. To produce from F_k an F_{k-1}, we have to add a directed edge, from any vertex a, to any of the $k-1$ roots of the trees that do not contain a. Thus we have $n(k-1)$ choices. Similarly, for F_{k-1}, we may produce a directed edge from any vertex b to any of the $k-2$ roots of the trees not containing b. For this we have $n(k-2)$ choices. Continuing this way, we arrive at

$$N^{\star}(F_k) = n^{k-1}(k-1)!, \tag{3.6.9}$$

and out comes, with (3.6.8), the unexpectedly simple relation

$$N(F_k) = n^{k-1} \qquad \text{for any} \qquad F_k \in \mathcal{F}_{n,k}. \tag{3.6.10}$$

For $k = n$, F_n consists just of n isolated vertices. Hence $N(F_n)$ counts the number of all rooted trees, and we obtain $|F_{n,1}| = n^{n-1}$, which is Cayley's formula. □

Poisson Branching Processes Proof

This proof makes use of *family trees*. It is convenient to represent the vertices of a family tree in terms of *words*. These words arise inductively as follows. The root is the word $\varnothing$. The children of the root are the words $1, 2, \ldots, d_{\varnothing}$, where, for a word w, we let d_w denote the number of children of w. The children of 1 are $11, 12, \ldots, 1d_1$, etc. A family tree is then uniquely represented by its set of words. For example, the word 1123 represents the third child of the second child of the first child of the first child of the root. This representation is sometimes called the *Ulam–Harris representation of trees.*

Two family trees are the same if and only if they are represented by the same set of words. We obtain a branching process when the variables $(d_w)_w$ are equal to a collection of i.i.d. random variables. For a word w, we let $|w|$ be its length, where $|\varnothing| = 0$. The length of a word w is the number of steps the word is away from the root, and equals its generation.

Let $\mathcal{T}$ denote the family tree of a branching process with Poisson offspring distribution with parameter 1. We compute the probability of obtaining a given tree t as

$$\mathbb{P}(\mathcal{T} = t) = \prod_{w \in t} \mathbb{P}(\xi = d_w), \tag{3.6.11}$$

where ξ is a Poisson random variable with parameter 1 and d_w is the number of children of the word w in the tree t. For a Poisson branching process, $\mathbb{P}(\xi = d_w) = \mathrm{e}^{-1}/d_w!$, so that

$$\mathbb{P}(\mathcal{T} = t) = \prod_{w \in t} \frac{\mathrm{e}^{-1}}{d_w!} = \frac{\mathrm{e}^{-n}}{\prod_{w \in t} d_w!}, \tag{3.6.12}$$

where n denotes the number of vertices in t. We note that the above probability is the same for each tree with the same number of vertices of degree k for each k.

Conditionally on having total progeny $T^* = n$ and a family tree $\mathcal{T}$, we introduce a labeling on $\mathcal{T}$ as follows. We give the root label 1, and give all other vertices a label from the set $\{2, \ldots, n\}$ uniformly at random without replacement, giving a labeled tree $\mathcal{L}$ on n vertices. We will prove that $\mathcal{L}$ is a *uniform* labeled tree on n vertices.

We start by introducing some notation. For a family tree t and a labeled tree ℓ, we write $t \sim \ell$ when ℓ can be obtained from t by adding labels. Given a labeled tree ℓ and any family tree t such that $t \sim \ell$, let $L(\ell)$ be the number of ways to label t in a different way (such that the root is labeled 1) such that the labeled tree obtained is isomorphic to ℓ as a rooted unordered labeled tree. It is not hard to see that $L(\ell)$ does not depend on the choice of t as long as $t \sim \ell$. Here $L(\ell)$ reflects the amount of symmetry present in any family tree t such that $t \sim \ell$.

Given a labeled tree ℓ, there are precisely

$$\#\{t : t \sim \ell\} = \frac{1}{L(\ell)} \prod_{w \in \ell} d_w! \tag{3.6.13}$$

family trees compatible with ℓ, since permuting the children of any vertex does not change the labeled tree. Here, we let $d_\varnothing$ be the degree of the vertex labelled 1 in ℓ, while d_w is the degree minus one for $w \in \ell$ such that $w \neq \varnothing$.

Further, the probability that, for all $w \in \mathcal{T}$, w receives label i_w with $i_\varnothing = 1$ is precisely equal to $1/(n-1)!$, where $n = |\mathcal{T}|$. Noting that there are $L(\ell)$ different assignments of labels in t that give rise to the labeled tree ℓ, we obtain, for any $t \sim \ell$,

$$\mathbb{P}(t \text{ receives labels } \ell) = \frac{L(\ell)}{(|\ell|-1)!}. \tag{3.6.14}$$

As a result, when labeling a family tree of a Poisson branching process with mean 1 offspring, the probability of obtaining a given labeled tree ℓ of arbitrary size equals

$$\begin{aligned} \mathbb{P}(\mathcal{L} = \ell) &= \sum_{t_\ell \sim \ell} \mathbb{P}(\mathcal{T} = t_\ell)\mathbb{P}(t_\ell \text{ receives labels } \ell) \\ &= \sum_{t_\ell \sim \ell} \frac{\mathrm{e}^{-|\ell|}}{\prod_{w \in t_\ell} d_w!} \frac{L(\ell)}{(|\ell|-1)!} \\ &= \#\{t : t \sim \ell\} \frac{\mathrm{e}^{-|\ell|}}{\prod_{w \in \ell} d_w!} \frac{L(\ell)}{(|\ell|-1)!} \\ &= \frac{1}{L(\ell)} \prod_{w \in \ell} d_w! \frac{\mathrm{e}^{-|\ell|}}{\prod_{w \in t_\ell} d_w!} \frac{L(\ell)}{(|\ell|-1)!} = \frac{\mathrm{e}^{-|\ell|}}{(|\ell|-1)!}. \end{aligned} \tag{3.6.15}$$

Therefore, conditionally on $T^* = n$, the probability of a given labeled tree $\mathcal{L}$ of size n equals

$$\mathbb{P}(\mathcal{L} = \ell \mid |\mathcal{L}| = n) = \frac{\mathbb{P}(\mathcal{L} = \ell)}{\mathbb{P}(|\mathcal{L}| = n)}. \tag{3.6.16}$$

By Theorem 3.16,

$$\mathbb{P}(|\mathcal{L}| = n) = \mathbb{P}(T^* = n) = \frac{\mathrm{e}^{-n} n^{n-2}}{(n-1)!}. \tag{3.6.17}$$

As a result, for each labeled tree $\mathcal{L}$ of size $|\mathcal{L}| = n$,

$$\mathbb{P}(\mathcal{L} = \ell \mid |\mathcal{L}| = n) = \frac{\mathbb{P}(\mathcal{L} = \ell)}{\mathbb{P}(|\mathcal{L}| = n)} = \frac{\mathrm{e}^{-n}}{(n-1)!} \frac{(n-1)!}{\mathrm{e}^{-n} n^{n-2}} = \frac{1}{n^{n-2}}. \tag{3.6.18}$$

The obtained probability is *uniform* over all labeled trees. Therefore, the number of labeled trees equals

$$\mathbb{P}(\mathcal{L} = \ell \mid |\mathcal{L}| = n)^{-1} = n^{n-2}. \tag{3.6.19}$$

The above not only proves Cayley's Theorem, but also gives an explicit construction of a uniform labeled tree from a Poisson branching process. □

We next investigate the asymptotics of the probability mass function of the total progeny of Poisson branching processes:

Theorem 3.18 (Asymptotics for total progeny for Poisson branching processes) *For a branching process with i.i.d. offspring X^*, where X^* has a Poisson distribution with mean λ, as $n \to \infty$,*

$$\mathbb{P}^*_\lambda(T^* = n) = \frac{1}{\lambda\sqrt{2\pi n^3}} \mathrm{e}^{-I_\lambda n}(1 + O(1/n)), \tag{3.6.20}$$

where

$$I_\lambda = \lambda - 1 - \log\lambda. \tag{3.6.21}$$

In particular, when $\lambda = 1$,

$$\mathbb{P}^*_1(T^* = n) = (2\pi)^{-1/2} n^{-3/2}[1 + O(n^{-1})]. \tag{3.6.22}$$

Proof By Theorem 3.16,

$$\mathbb{P}^*_\lambda(T^* = n) = \frac{1}{\lambda}\lambda^n \mathrm{e}^{-(\lambda-1)n}\mathbb{P}^*_1(T^* = n) = \frac{1}{\lambda}\mathrm{e}^{-I_\lambda n}\mathbb{P}^*_1(T^* = n), \tag{3.6.23}$$

so that (3.6.20) follows from (3.6.22). Again using Theorem 3.16,

$$\mathbb{P}^*_1(T^* = n) = \frac{n^{n-1}}{n!}\mathrm{e}^{-n}. \tag{3.6.24}$$

By Stirling's formula,

$$n! = \sqrt{2\pi n}\mathrm{e}^{-n} n^n(1 + O(1/n)). \tag{3.6.25}$$

Substitution of (3.6.25) into (3.6.24) yields (3.6.22). □

Equation (3.6.22) is an example of a *power-law* relationship that often holds at criticality. The above $n^{-3/2}$ behavior is associated more generally with the distribution of the total progeny whose offspring distribution has finite variance (see, e.g., Aldous (1993, Proposition 24)).

In Chapter 4, we will investigate the behavior of the Erdős–Rényi random graph by making use of couplings to branching processes. There, we also need the fact that, for $\lambda > 1$, the survival probability is sufficiently smooth (see, e.g., Section 4.4):

Corollary 3.19 (Differentiability of the survival probability) *Let η_λ denote the extinction probability of a branching process with a mean λ Poisson offspring distribution, and $\zeta_\lambda = 1 - \eta_\lambda$ its survival probability. Then, for all $\lambda > 1$,*

$$\frac{d}{d\lambda}\zeta_\lambda = \frac{\zeta_\lambda(1 - \zeta_\lambda)}{1 - \mu_\lambda} \in (0, \infty), \tag{3.6.26}$$

where μ_λ is the conjugate of λ defined in (3.6.6). When $\lambda \searrow 1$,

$$\zeta_\lambda = 2(\lambda - 1)(1 + o(1)). \tag{3.6.27}$$

Proof We prove the equivalent statement $\frac{d}{d\lambda}\eta_\lambda = -\frac{\eta_\lambda(1-\eta_\lambda)}{1-\mu_\lambda}$. By (3.6.2), η_λ satisfies $\eta_\lambda = \mathrm{e}^{\lambda(1-\eta_\lambda)}$ with $\eta_\lambda < 1$ for $\lambda > 1$.

Consider the function $f\colon (0,1) \to (1,\infty)$ defined by

$$f(x) = \frac{\log x}{x-1}. \tag{3.6.28}$$

It is straightforward to verify that $f\colon (0,1) \to (1,\infty)$ is strictly decreasing on $(0,1)$ with $\lim_{x\searrow 0} f(x) = \infty$ and $\lim_{x\nearrow 1} f(x) = 1$. As a result, f is a bijective map from $(0,1)$ to $(1,\infty)$ and is continuously differentiable in its domain. Therefore, its inverse is also continuously differentiable, and $f^{-1}(\lambda) = \eta_\lambda$ for $\lambda > 1$. Thus also, $\lambda \mapsto \eta_\lambda$ is continuously differentiable with respect to λ on $(1,\infty)$.

Further, taking the logarithm on both sides of (3.6.2) yields $\log \eta_\lambda = \lambda(\eta_\lambda - 1)$. Differentiating with respect to λ leads to

$$\frac{1}{\eta_\lambda}\frac{d}{d\lambda}\eta_\lambda = (\eta_\lambda - 1) + \lambda\frac{d}{d\lambda}\eta_\lambda. \tag{3.6.29}$$

Solving for $\frac{d}{d\lambda}\eta_\lambda$ leads to

$$\frac{d}{d\lambda}\eta_\lambda = -\frac{\eta_\lambda(1-\eta_\lambda)}{1-\lambda\eta_\lambda}, \tag{3.6.30}$$

which, using that $\lambda\eta_\lambda = \mu_\lambda$, completes the proof of (3.6.26).

To prove (3.6.27), we note that by (3.6.2),

$$\zeta_\lambda = 1 - \mathrm{e}^{-\lambda\zeta_\lambda}. \tag{3.6.31}$$

We know that $\zeta_\lambda \searrow 0$ as $\lambda \searrow 1$. Therefore, we Taylor expand $x \mapsto \mathrm{e}^{-\lambda x}$ around $x = 0$ to obtain

$$\zeta_\lambda = \lambda\zeta_\lambda - \tfrac{1}{2}(\lambda\zeta_\lambda)^2 + o(\zeta_\lambda^2). \tag{3.6.32}$$

Dividing through by $\zeta_\lambda > 0$ and rearranging terms yields

$$\tfrac{1}{2}\zeta_\lambda = \lambda - 1 + o(\lambda - 1). \tag{3.6.33}$$

This completes the proof of (3.6.27). □

3.7 Binomial and Poisson Branching Processes

In the following theorem, we relate the total progeny of a Poisson branching process to that of a binomial branching process with parameters n and success probability λ/n. In its statement, we write $\mathbb{P}_{n,p}$ for the law of a binomial branching process with parameters n and success probability p.

Theorem 3.20 (Poisson and binomial branching processes) *For a branching process with binomial offspring distribution with parameters n and p, and the branching process with Poisson offspring distribution with parameter $\lambda = np$, for each $k \geq 1$,*

$$\mathbb{P}_{n,p}(T \geq k) = \mathbb{P}^*_\lambda(T^* \geq k) + e_n(k), \tag{3.7.1}$$

where T and T^* are the total progenies of the binomial and Poisson branching processes, respectively, and where

$$|e_n(k)| \leq \frac{\lambda^2}{n} \sum_{s=1}^{k-1} \mathbb{P}_\lambda^*(T^* \geq s). \tag{3.7.2}$$

In particular, $|e_n(k)| \leq k\lambda^2/n$.

Proof We use a coupling proof. The branching processes are described by their offspring distributions, which are binomial and Poisson random variables respectively. We use the coupling in Theorem 2.10 for each of the random variables X_i and X_i^* determining the branching processes as in (3.3.1), where $X_i \sim \mathsf{Bin}(n, \lambda/n)$, $X_i^* \sim \mathsf{Poi}(\lambda)$, and where

$$\mathbb{P}(X_i \neq X_i^*) \leq \frac{\lambda^2}{n}. \tag{3.7.3}$$

Further, the vectors $(X_i, X_i^*)_{i \geq 1}$ are i.i.d. We use $\mathbb{P}$ to denote the joint probability distributions of the binomial and Poisson branching processes, where all the offsprings are coupled in the above way.

We start by noting that

$$\mathbb{P}_{n,p}(T \geq k) = \mathbb{P}(T \geq k, T^* \geq k) + \mathbb{P}(T \geq k, T^* < k), \tag{3.7.4}$$

and

$$\mathbb{P}_\lambda^*(T^* \geq k) = \mathbb{P}(T \geq k, T^* \geq k) + \mathbb{P}(T^* \geq k, T < k). \tag{3.7.5}$$

Subtracting the two probabilities yields

$$|\mathbb{P}_{n,p}(T \geq k) - \mathbb{P}_\lambda^*(T^* \geq k)| \leq \max\big\{\mathbb{P}(T \geq k, T^* < k), \mathbb{P}(T^* \geq k, T < k)\big\}. \tag{3.7.6}$$

We then use Theorem 2.10, as well as the fact that the event $\{T \geq k\}$ is determined by the values of $X_1, \ldots, X_{k-1}$ only. Thus, T is a stopping time with respect to the filtration $\sigma(X_1, \ldots, X_k)$. Indeed, by (3.3.1), when we know $X_1, \ldots, X_{k-1}$, then we can also verify whether there exists a $t < k$ such that $X_1 + \cdots + X_t = t - 1$, implying that $T < k$. When there is no such t, then $T \geq k$. Similarly, by investigating $X_1^*, \ldots, X_{k-1}^*$, we can verify whether there exists a $t < k$ such that $X_1^* + \cdots + X_t^* = t - 1$, implying that $T^* < k$.

When $T \geq k$ and $T^* < k$, or when $T^* \geq k$ and $T < k$, there must be a value of $s < k$ for which $X_s \neq X_s^*$. Therefore, we can bound, by splitting depending on the first value $s < k$ where $X_s \neq X_s^*$,

$$\mathbb{P}(T \geq k, T^* < k) \leq \sum_{s=1}^{k-1} \mathbb{P}(X_i = X_i^* \; \forall i \leq s-1, X_s \neq X_s^*, T \geq k), \tag{3.7.7}$$

where the condition $X_i = X_i^* \; \forall i \leq s - 1$ is absent for $s = 1$.

Now we note that when $X_i = X_i^*$ for all $i \leq s - 1$ and $T \geq k$, this implies in particular that $X_1^* + \cdots + X_i^* \geq i$ for all $i \leq s - 1$, which in turn implies that $T^* \geq s$. Moreover, the event $\{T^* \geq s\}$ depends only on $X_1^*, \ldots, X_{s-1}^*$, and, therefore, is independent of the event that $X_s \neq X_s^*$. Thus, we arrive at the fact that

$$\mathbb{P}(T \geq k, T^* < k) \leq \sum_{s=1}^{k-1} \mathbb{P}(T^* \geq s, X_s \neq X_s^*)$$
$$= \sum_{s=1}^{k-1} \mathbb{P}(T^* \geq s)\mathbb{P}(X_s \neq X_s^*). \tag{3.7.8}$$

By Theorem 2.10,

$$\mathbb{P}(X_s \neq X_s^*) \leq \frac{\lambda^2}{n}, \tag{3.7.9}$$

so that

$$\mathbb{P}(T \geq k, T^* < k) \leq \frac{\lambda^2}{n} \sum_{s=1}^{k-1} \mathbb{P}(T^* \geq s). \tag{3.7.10}$$

A similar argument proves that also

$$\mathbb{P}(T^* \geq k, T < k) \leq \frac{\lambda^2}{n} \sum_{s=1}^{k-1} \mathbb{P}(T^* \geq s). \tag{3.7.11}$$

We conclude from (3.7.6) that

$$|\mathbb{P}_{n,p}(T \geq k) - \mathbb{P}_\lambda^*(T^* \geq k)| \leq \frac{\lambda^2}{n} \sum_{s=1}^{k-1} \mathbb{P}_\lambda^*(T^* \geq s). \tag{3.7.12}$$

This completes the proof of Theorem 3.20. □

3.8 Notes and Discussion

Notes on Sections 3.1–3.2

The results on the phase transition for branching processes are classical. For more information about branching processes, we refer to the classical books on the subject, by Athreya and Ney (1972), by Harris (1963) and by Jagers (1975).

Notes on Section 3.3

We learned of the random-walk representation of branching processes from the book of Alon and Spencer (see Alon and Spencer, 2000, Section 10.4).

Notes on Section 3.4

Theorem 3.10 was first proved by Kesten and Stigum (1966a,b, 1967). A proof of Theorem 3.10 is given in Athreya and Ney (1972, Pages 24–26), while in Lyons et al. (1995) a conceptual proof is given. See Durrett (2007, Proof of Theorem 2.16) for a simple proof of the statement under the stronger condition that $\mathbb{E}[X^2] < \infty$, using the L^2-martingale convergence theorem (see also below Theorem 2.24).

Notes on Section 3.5

The current proof of Theorem 3.14 is taken from van der Hofstad and Keane (2008), where an extension is also proved by conditioning on the numbers of steps of various sizes. The first proof of the special case of Theorem 3.14 for $k = 1$ can be found in Otter (1949). The extension to $k \geq 2$ is by Kemperman (1961), or by Dwass (1969) using a result in Dwass (1968).

Most of these proofs make unnecessary use of generating functions, in particular, the Lagrange inversion formula, which the simple proof given here does not employ. See also Grimmett and Stirzaker (2001, Page 165–167) for a more recent version of the generating function proof. Wendel (1975) gives various proofs of the hitting-time theorem, including a combinatorial proof making use of a relation by Dwass (1962). A proof for random walks making only steps of size ± 1 using the reflection principle can for example be found in Grimmett and Stirzaker (2001, Page 79).

A very nice proof of Theorem 3.14 can be found in the classic book by Feller, see Feller (1971, Lemma 1, page 412), which has come to be known as *Spitzer's combinatorial* lemma (Spitzer, 1956). This lemma states that in a sequence $(y_1, \ldots, y_n)$ with $y_1 + \cdots + y_n = -k$ and $y_i \geq -1$ for every $i \in [n]$, out of all n possible cyclical reorderings, precisely k are such that all partial sums $s_i = y_1 + \cdots + y_i$ satisfy $s_i > -k$ for every $i \leq n-1$. Extensions of this result to Lévy processes can be found in Alili et al. (2005) and to two-dimensional walks in Járai and Kesten (2004).

The hitting-time theorem is closely related to the ballot theorem, which has a long history dating back to Bertrand in 1887 (see Konstantopoulos (1995) for an excellent overview of the history and literature). The version of the ballot theorem by Konstantopoulos (1995) states that, for a random walk $(S_n)_{n\geq 0}$ starting at 0, with exchangeable, non-negative steps, the probability that $S_m < m$ for all $m = 1, \ldots, n$, conditionally on $S_n = k$, equals k/n. This proof borrows upon queueing theory methodology, and is related to, yet slightly different from, our proof.

The ballot theorem for random walks with independent steps is the following result:

Theorem 3.21 (Ballot theorem) *Consider a random walk with i.i.d. steps $(X_i)_{i\geq 1}$ taking non-negative integer values. Then, with $S_m = X_1 + \cdots + X_m$ the position of the walk after m steps,*

$$\mathbb{P}_0(S_m < m \text{ for all } 1 \leq m \leq n \mid S_n = n-k) = \frac{k}{n}. \tag{3.8.1}$$

You are asked to prove the ballot theorem in Theorem 3.21 in Exercise 3.31.

Notes on Section 3.6

An alternative proof of Theorem 3.17 can be found in van Lint and Wilson (2001). Theorem 3.16, together with (3.6.2), can also be proved by making use of Lambert's W function. Indeed, we use that the probability generating function of the total progeny in (3.1.18), for Poisson branching process, reduces to

$$G_T(s) = s\mathrm{e}^{\lambda(G_T(s)-1)}. \tag{3.8.2}$$

Equation (3.8.2) actually defines a function analytic in $\mathbb{C}\backslash[1,\infty)$, and we are taking the principal branch. Equation (3.8.2) can be written in terms of the Lambert W function, which is defined by $W(x)\mathrm{e}^{W(x)} = x$, as $G_T(s) = -W(-s\lambda\mathrm{e}^{-\lambda})/\lambda$. The branches of W are described by Corless et al. (1996), where the fact that

$$W(x) = -\sum_{n=1}^{\infty} \frac{n^{n-1}}{n!}(-x)^n. \tag{3.8.3}$$

is also derived. Theorem 3.17 follows immediately from this equation upon substituting $x = \lambda\mathrm{e}^{-\lambda}$ and using that the coefficient of s^n in $G_T(s)$ equals $\mathbb{P}(T = n)$. Also, $\eta_\lambda = \lim_{s\uparrow 1} G_T(s) = -W(-\lambda\mathrm{e}^{-\lambda})/\lambda$. This also allows for a more direct proof of Corollary 3.19, since

$$\frac{d}{d\lambda}\eta_\lambda = -\frac{d}{d\lambda}\Big[\frac{W(-\lambda\mathrm{e}^{-\lambda})}{\lambda}\Big], \tag{3.8.4}$$

and where, since $W(x)\mathrm{e}^{W(x)} = x$,

$$W'(x) = \frac{1}{x}\frac{W(x)}{1+W(x)}. \tag{3.8.5}$$

We omit the details of this proof, taking a more combinatorial approach instead.

The proof of (3.6.26) in Corollary 3.19 is due to Souvik Ray and replaces a slightly more involved previous proof.

3.9 Exercises for Chapter 3

Exercise 3.1 (Almost sure survival) Show that the branching process extinction probability η satisfies $\eta = 0$ precisely when $p_0 = 0$.

Exercise 3.2 (Binary branching) When the offspring distribution is given by

$$p_x = (1-p)\mathbb{1}_{\{x=0\}} + p\mathbb{1}_{\{x=2\}}, \tag{3.9.1}$$

we speak of binary branching. Prove that $\eta = 1$ when $p \leq 1/2$ and, for $p > 1/2$,

$$\eta = \frac{1-p}{p}. \tag{3.9.2}$$

Exercise 3.3 (Geometric branching (see Athreya and Ney, 1972, Pages 6–7)) Let the probability distribution $(p_k)_{k\geq 0}$ be given by

$$\begin{cases} p_k = b(1-p)^{k-1} & \text{for } k = 1, 2, \ldots; \\ p_0 = 1 - b/p & \text{for } k = 0, \end{cases} \tag{3.9.3}$$

so that, for $b = p$, the offspring distribution has a geometric distribution with success probability p. Show that the extinction probability η is given by $\eta = 1$ if $\mu = \mathbb{E}[X] = b/p^2 \leq 1$, while, with the abbreviation $q = 1 - p$ and for $b/p^2 > 1$,

$$\eta = \frac{1-\mu p}{q}. \tag{3.9.4}$$

More generally, let 1 and s_0 be the two solutions to $G_X(s) = s$. Show that $s_0 = \frac{1-\mu p}{q}$.

Exercise 3.4 (Exercise 3.3 cont.) Let the probability distribution $(p_k)_{k\geq 0}$ be given by (3.9.3). Show that $G_n(s)$, the generating function of Z_n, is given by

$$G_n(s) = \begin{cases} 1-\mu^n \dfrac{1-s_0}{\mu^n - s_0} + \dfrac{\mu^n \left(\frac{1-s_0}{\mu^n - s_0}\right)^2 s}{1-\left(\frac{\mu^n-1}{\mu^n-s_0}\right)} & \text{when } b \neq p^2; \\ \dfrac{nq-(nq-p)s}{p+nq-nps} & \text{when } b = p^2. \end{cases} \tag{3.9.5}$$

Exercise 3.5 (Exercise 3.4 cont.) Conclude from Exercise 3.4 that, for $(p_k)_{k\geq 0}$ in (3.9.3),

$$\mathbb{P}(Z_n > 0,\ \exists m > n \text{ such that } Z_m = 0) = \begin{cases} \mu^n \dfrac{1-s_0}{\mu^n - s_0} & \text{when } b < p^2; \\ \dfrac{p}{p+nq} & \text{when } b = p^2; \\ \dfrac{(1-\eta)\eta}{\mu^n - \eta} & \text{when } b > p^2. \end{cases} \tag{3.9.6}$$

Exercise 3.6 (Exercise 3.2 cont.) In the case of binary branching, i.e., when $(p_k)_{k\geq 0}$ is given by (3.9.1), show that the probability generating function of the total progeny satisfies

$$G_T(s) = \frac{1-\sqrt{1-4s^2pq}}{2sp}. \tag{3.9.7}$$

Exercise 3.7 (Exercise 3.5 cont.) Show, using Theorem 3.2, that, for $(p_k)_{k\geq 0}$ in (3.9.3), the probability generating function of the total progeny satisfies

$$G_T(s) = \frac{\sqrt{(p+s(b-pq))^2 - 4pqs(p-b)} - (p+sbq)}{2pq}. \tag{3.9.8}$$

Exercise 3.8 (Mean generation size completed) Complete the proof of Theorem 3.3 by conditioning on Z_{n-1} and showing that

$$\mathbb{E}\Big[\sum_{i=1}^{Z_{n-1}} X_{n,i} \,\Big|\, Z_{n-1} = m\Big] = m\mu, \tag{3.9.9}$$

so that

$$\mathbb{E}[Z_n] = \mu \mathbb{E}[Z_{n-1}]. \tag{3.9.10}$$

Exercise 3.9 (Generation-size martingale) Prove that the stochastic process $(\mu^{-n} Z_n)_{n\geq 1}$ is a martingale.

Exercise 3.10 (Subcritical branching processes) Prove Theorem 3.4 by using Theorem 3.3, together with the Markov inequality in Theorem 2.17.

Exercise 3.11 (Critical branching processes) Show that $Z_n \xrightarrow{\mathbb{P}} 0$ when the branching process is critical and $\mathbb{P}(X = 1) < 1$. [Recall page 90 for the definition of a critical branching process.] On the other hand, conclude that $\mathbb{E}[Z_n] = 1$ for all $n \geq 1$ for critical branching processes.

Exercise 3.12 (Proof expected total progeny) Prove the formula for the expected total progeny size in subcritical branching processes in Theorem 3.5.

Exercise 3.13 (Verification of law of total progeny as random-walk hitting time) Compute $\mathbb{P}(T = k)$ for T in (3.3.2) and $\mathbb{P}(T = k)$ for T in (3.1.16) explicitly, for $k = 1, 2$ and 3.

Exercise 3.14 (Exercise 3.2 cont.) In the case of binary branching, i.e., when the offspring distribution is given by (3.9.1), show that

$$\mathbb{P}(T = k) = \frac{1}{p}\mathbb{P}\big(S_0 = S_{k+1} = 0, S_i > 0 \;\forall 1 \leq i \leq k\big), \tag{3.9.11}$$

where $(S_i)_{i\geq 1}$ is a simple random walk, i.e.,

$$S_i = Y_1 + \cdots + Y_i, \tag{3.9.12}$$

where $(Y_i)_{i\geq 1}$ are i.i.d. random variables with distribution

$$\mathbb{P}(Y_1 = 1) = 1 - \mathbb{P}(Y_1 = -1) = p. \tag{3.9.13}$$

This gives a one-to-one relation between random walk excursions and the total progeny of a binary branching process.

Exercise 3.15 (Conjugate pair) Prove that $(p'_x)_{x\geq 0}$ defined in (3.3.4) is a probability distribution.

Exercise 3.16 (Generating function of dual offspring distribution) Let $G_d(s) = \mathbb{E}'[s^{X_1}]$ be the probability generating function of the offspring of the conjugate branching process defined in (3.3.4). Show that

$$G_d(s) = \frac{1}{\eta}G_X(\eta s). \tag{3.9.14}$$

Exercise 3.17 (Dual of supercritical branching process is subcritical) Let X' have probability mass function $(p'_k)_{k\geq 0}$ defined in (3.3.4). Show that $\eta < 1$ implies that

$$\mathbb{E}[X'] < 1. \tag{3.9.15}$$

Thus, the branching process with offspring distribution $(p'_k)_{k\geq 0}$ is *subcritical* when $(p_k)_{k\geq 0}$ is *supercritical*.

Exercise 3.18 (Proof of implicit relation for martingale limit) Prove (3.4.1).

Exercise 3.19 (Positivity of martingale limit on survival event) Prove that $\mathbb{P}(W_\infty > 0) = 1 - \eta$ implies that $\mathbb{P}(W_\infty > 0 \mid \text{survival}) = 1$.

Exercise 3.20 (Mean of martingale limit) Prove, using Fatou's lemma (Theorem A.7), that $\mathbb{E}[W_\infty] \leq 1$ always holds.

Exercise 3.21 Prove that $(p_k^{(\infty)})_{k\geq 0}$ defined in (3.4.2) is a probability distribution.

Exercise 3.22 (Proof of Theorem 3.11 in the infinite-mean case) Prove (3.4.3) when $\mu = \infty$.

Exercise 3.23 (Total progeny for binomial branching processes) Compute the probability mass function of the total progeny of a branching process with a binomial offspring distribution using Theorem 3.13.

Exercise 3.24 (Total progeny for geometric branching processes) Compute the probability mass function of the total progeny of a branching process with a geometric offspring distribution using Theorem 3.13.
Hint: note that when $(X_i)_{i=1}^n$ are i.i.d. geometric, then $X_1+\cdots+X_n$ has a negative binomial distribution.

Exercise 3.25 (The total progeny from the hitting-time theorem) Prove that Theorem 3.14 implies (3.5.2).

Exercise 3.26 (The condition in the hitting-time theorem) Is Theorem 3.14 still true when the restriction that $\mathbb{P}(Y_i \geq -1) = 1$ is dropped?

Exercise 3.27 (Extension of Theorem 3.14) Extend the hitting-time theorem Theorem 3.14 to the case where $(Y_i)_{i=1}^n$ is an exchangeable sequence rather than an i.i.d. sequence. Here a sequence $(Y_i)_{i=1}^n$ is called exchangeable when its distribution is the same as the distribution of any permutation of the sequence.
Hint: if $(Y_i)_{i=1}^n$ is exchangeable, then so is $(Y_i)_{i=1}^n$ conditioned on $\sum_{i=1}^n Y_i = -k$.

Exercise 3.28 (Verification of (3.6.7)) Verify (3.6.7) for $n = 1, 2$ and $n = 3$.

Exercise 3.29 (The total progeny of a Poisson branching process) Prove Theorem 3.16 using Theorem 3.13.

Exercise 3.30 (Large, but finite, Poisson total progeny) Use Theorem 3.16 to show that, for any λ, and for k sufficiently large,

$$\mathbb{P}_\lambda^*(k \leq T^* < \infty) \leq \mathrm{e}^{-I_\lambda k}, \tag{3.9.16}$$

where $I_\lambda = \lambda - 1 - \log \lambda$.

Exercise 3.31 (Proof of the ballot theorem) Prove the ballot theorem (Theorem 3.21) using the random-walk hitting-time theorem (Theorem 3.14).
Hint: Let $S'_m = k + (S_n - n) - (S_{n-m} - n + m)$, and note that $S_m < m$ for all $1 \leq m \leq n$ precisely when $S'_m > 0$ for all $0 \leq m < n$, and $(S'_m)_{m\geq 0}$ is a random walk taking steps $Y_m = S'_m - S'_{m-1} = X_{n-m} - 1$.

Part II

Basic Models

4

Phase Transition for the Erdős–Rényi Random Graph

In this chapter, we study the connected components of the Erdős–Rényi random graph. Such connected components can be described in terms of branching processes. As we have seen in Chapter 3, branching processes have a phase transition: when the expected offspring is below 1, the branching process dies out almost surely, while when the expected offspring exceeds 1, then it survives with positive probability. The Erdős–Rényi random graph has a related phase transition. Indeed, when the expected degree is smaller than 1, the components are small, the largest one being of order $\log n$. On the other hand, when the expected degree exceeds 1, there is a giant connected component that contains a positive proportion of all vertices. The aim of this chapter is to quantify and prove these facts.

Organization of this Chapter

The chapter is organized as follows. In the introduction in Section 4.1 we argue that connected components in the Erdős–Rényi random graph can be described by branching processes. The link between the Erdős–Rényi random graph and branching processes is described in more detail in Section 4.2, where we prove upper and lower bounds for the tails of the cluster size (or connected component size) distribution. The *connected component containing* v is denoted by $\mathscr{C}(v)$, and consists of all vertices that can be reached from v using occupied edges. We sometimes also call $\mathscr{C}(v)$ the *cluster of* v. The connection between branching processes and clusters is used extensively in Sections 4.3–4.5. In Section 4.3, we study the subcritical regime of the Erdős–Rényi random graph. In Sections 4.4 and 4.5 we study the supercritical regime of the Erdős–Rényi random graph, by proving a law of large numbers for the largest connected component in Section 4.4 and a central limit theorem in Section 4.5. We close this section with notes and discussion in Section 4.6 and exercises in Section 4.7.

4.1 Introduction

In this chapter, we investigate the phase transition in the Erdős–Rényi random graph. The phrase 'phase transition' refers to the fact that there is a sharp transition in the largest connected component. Indeed, all clusters are quite small when the average number of neighbors per vertex is smaller than 1, while there exists a unique giant component containing an asymptotic positive proportion of the vertices when the average number of neighbors per vertex exceeds 1. This phase transition can already be observed in relatively small graphs. For example, Figure 4.1 shows two realizations of Erdős–Rényi random graphs with 100

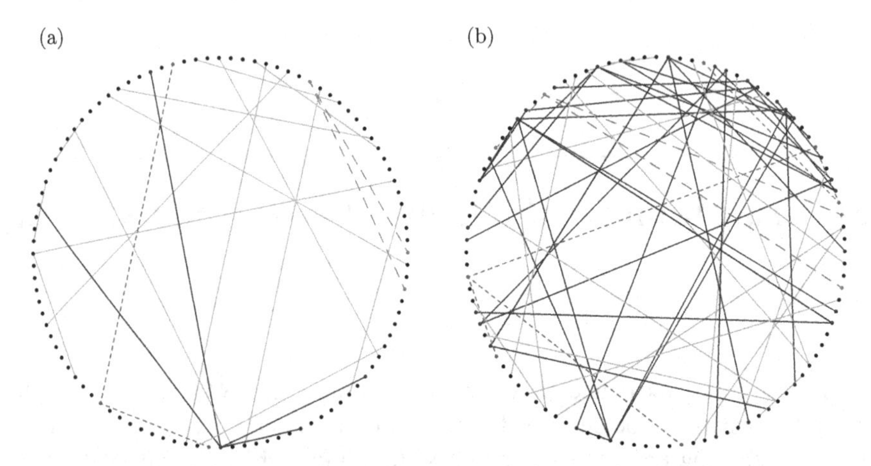

Figure 4.1 Two realizations of Erdős–Rényi random graphs with 100 elements and edge probabilities $1/200$ (see (a)), respectively, $3/200$ (see (b)). The three largest connected components are ordered by the darkness of their edge colors, the remaining connected components have edges with the lightest shade.

elements and expected degree close to $1/2$, respectively, $3/2$. The graph in Figure 4.1(a) is in the subcritical regime, and the connected components are tiny, while the graph in Figure 4.1(b) is in the supercritical regime, and the largest connected component is already substantial.

We next introduce some notation for the Erdős–Rényi random graph, and prove some elementary properties. We recall from Section 1.8 that the Erdős–Rényi random graph has vertex set $[n] = \{1, \ldots, n\}$, and, denoting the edge between vertices $s, t \in [n]$ by st, st is occupied or present with probability p, and vacant or absent otherwise, independently of all the other edges. The parameter p is called the *edge probability*. The above random graph is denoted by $\mathrm{ER}_n(p)$. Exercises 4.1 and 4.2 investigate the number of edges in an Erdős–Rényi random graph.

We now introduce some notation. For two vertices $s, t \in [n]$, we write $s \longleftrightarrow t$ when there exists a path of occupied edges connecting s and t. By convention, we always assume that $v \longleftrightarrow v$. For $v \in [n]$, we denote the *connected component containing* v or *cluster of* v by

$$\mathscr{C}(v) = \{x \in [n] \colon v \longleftrightarrow x\}. \tag{4.1.1}$$

We denote the size of $\mathscr{C}(v)$ by $|\mathscr{C}(v)|$, i.e., $|\mathscr{C}(v)|$ denotes the number of vertices connected to v by an occupied path. The *largest connected component* $\mathscr{C}_{\max}$ is equal to any cluster $\mathscr{C}(v)$ for which $|\mathscr{C}(v)|$ is maximal, so that

$$|\mathscr{C}_{\max}| = \max_{v \in [n]} |\mathscr{C}(v)|. \tag{4.1.2}$$

Note that the above definition does identify $|\mathscr{C}_{\max}|$ uniquely, but it may not identify $\mathscr{C}_{\max}$ uniquely. We can make this definition unique by requiring that $\mathscr{C}_{\max}$ is the cluster of maximal size containing the vertex with the smallest label. As we will see, the typical size of $\mathscr{C}_{\max}$ depends sensitively on the value λ, where $\lambda = np$ is close to the expected number of neighbors per vertex.

We first define a procedure to find the connected component $\mathscr{C}(v)$ containing a given vertex v in a given graph G. This procedure is closely related to the random-walk perspective for branching processes described in Section 3.3, and works as follows. In the course of the exploration, vertices can have three different statuses: active, neutral or inactive. The status of vertices changes in the course of the exploration of the connected component of v, as follows. At time $t = 0$, only v is active and all other vertices are neutral, and we set $S_0 = 1$. At each time $t \geq 1$, we choose an active vertex w in an arbitrary way (for example, by taking the active vertex with smallest label) and explore all the edges ww', where w' runs over all the neutral vertices. If $ww' \in E(G)$, then we set w' active, otherwise it remains neutral. After searching the entire set of neutral vertices, we set w inactive and we let S_t equal the new number of active vertices at time t. When there are no more active vertices, i.e., when $S_t = 0$ for the first time, the process terminates and $\mathscr{C}(v)$ is the set of all inactive vertices, which immediately implies that $|\mathscr{C}(v)| = t$. Note that at any stage of the process, the size of $\mathscr{C}(v)$ is bounded from below by the sum of the number of active and inactive vertices at that time.

Let S_t be the total number of active vertices at time t. Then, similarly as for the branching process in (3.3.1),

$$S_0 = 1, \qquad S_t = S_{t-1} + X_t - 1, \tag{4.1.3}$$

where the variable X_t is the number of vertices that become active due to the exploration of the tth active vertex w_t, and after its exploration, w_t becomes inactive. This explains (4.1.3).

The above description is true for *any* graph G. We now specialize to the Erdős–Rényi random graph $\mathrm{ER}_n(p)$, where each edge can be independently occupied or vacant. As a result, the distribution of X_t depends on the number of active vertices at time $t-1$, i.e., on S_{t-1}, and not in any other way on which vertices are active, inactive or neutral. More precisely, each neutral w' at time $t-1$ in the random graph has probability p to become active at time t. The edges ww' are examined precisely once, so that the conditional probability that ww' is an edge in $\mathrm{ER}_n(p)$ is always equal to p. After $t-1$ explorations of active vertices, we have $t-1$ inactive vertices and S_{t-1} active vertices. This leaves $n-(t-1)-S_{t-1}$ neutral vertices. Therefore, conditionally on S_{t-1},

$$X_t \sim \mathsf{Bin}\big(n - (t-1) - S_{t-1}, p\big). \tag{4.1.4}$$

We note that the recursion in (4.1.3) is identical to the recursion relation (3.3.1). The only difference is the distribution of the process $(X_i)_{i\in[n]}$, as described in (4.1.4). For branching processes, $(X_i)_{i\geq 1}$ is an i.i.d. sequence, while for the exploration of connected components in $\mathrm{ER}_n(p)$, we see that this is not quite true. However, by (4.1.4), it is 'almost' true as long as the number of active vertices is not too large. We see in (4.1.4) that the first parameter of the binomial distribution, describing the number of trials, *decreases* as time progresses. This is due to the fact that after more explorations, fewer neutral vertices remain, and is sometimes called the *depletion-of-points* effect. The depletion-of-points effect makes $\mathrm{ER}_n(p)$ different from a branching process. See Figure 4.2 for a graphical representation of the exploration.

Let T be the first time t for which $S_t = 0$, i.e.,

$$T = \inf\{t \colon S_t = 0\}, \tag{4.1.5}$$

then $|\mathscr{C}(v)| = T$, see also (3.3.2) for a similar result in the branching process setting. This describes the exploration of a single connected component. While of course for the cluster

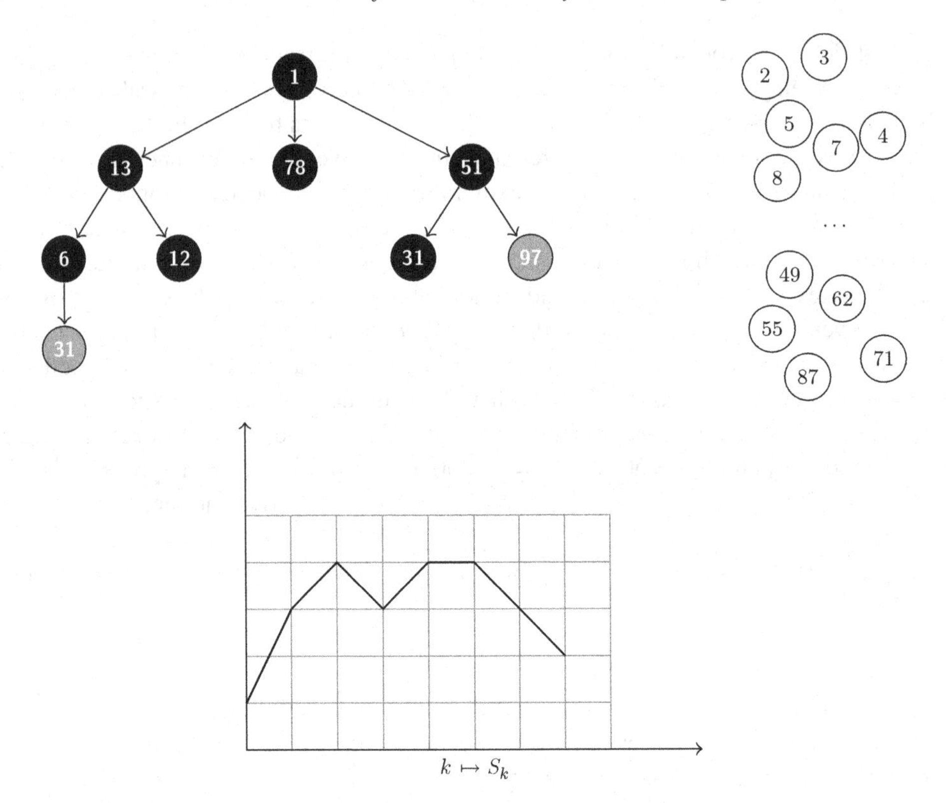

Figure 4.2 The exploration of the cluster of vertex 1 in an Erdős–Rényi random graph, as formulated in (4.1.3) and (4.1.4). We explore the cluster in a breadth-first order. The black vertices have been explored and are inactive, the grey vertices are active, i.e., they have not yet been explored but have been found to be part of $\mathscr{C}(1)$. We have explored 7 vertices, and $X_1 = 3$, $X_2 = 2$, $X_3 = 0$, $X_4 = 2$, $X_5 = 1$, etc. The labels correspond to the vertex labels in the graph. The next vertex to be explored is vertex 97, and there is 1 more vertex to be explored (i.e., $S_7 = 2$). The vertices on the right are the neutral vertices. At each time in the exploration process, the total number of active, inactive and neutral vertices is equal to n. The lowest figure plots the random-walk path arising as the number of active vertices in the exploration of the branching process tree as in (4.1.3) and (4.1.4).

exploration, (4.1.3) and (4.1.4) only make sense when $S_{t-1} \geq 1$, that is, when $t \leq T$, there is no harm in continuing it formally for $t > T$. This will prove to be extremely useful later on. Exercise 4.3 asks you to verify that (4.1.5) is correct when $|\mathscr{C}(v)| = 1, 2$ and 3.

We end this section by introducing some notation. For the Erdős–Rényi random graph, the status of all edges $\{st : 1 \leq s < t \leq n\}$ are i.i.d. random variables taking the value 1 with probability p and the value 0 with probability $1 - p$, 1 denoting that the edge is occupied and 0 that it is vacant. We sometimes call the edge probability p, and sometimes λ/n. We always use the convention that

$$p = \lambda/n. \tag{4.1.6}$$

We write $\mathbb{P}_\lambda$ for the distribution of $\mathrm{ER}_n(p) = \mathrm{ER}_n(\lambda/n)$.

4.1.1 Monotonicity in the Edge Probabilities

In this section, we investigate Erdős–Rényi random graphs with different values of p, and show that the Erdős–Rényi random graph is *monotonically increasing in* p, using a coupling argument. The material in this section makes it clear that components of the Erdős–Rényi random graph are growing with the edge probability p, as one would intuitively expect.

We couple the random graphs $\mathrm{ER}_n(p)$ for *all* $p \in [0, 1]$ on the same probability space. For this, we draw independent uniform random variables U_{st} on $[0, 1]$ for each edge st, and, for fixed p, we declare an edge to be p-occupied if and only if $U_{st} \leq p$. Note that edges are p-occupied independently of each other, and the probability that an edge is p-occupied is $\mathbb{P}(U_{st} \leq p) = p$. Therefore, the resulting graph of p-occupied edges has the same distribution as $\mathrm{ER}_n(p)$. The above coupling shows that the number of occupied edges increases when p increases. Because of the monotone nature of $\mathrm{ER}_n(p)$ one expects that certain events become more likely, and random variables grow larger, when p increases. This is formalized in the following definition:

Definition 4.1 (Increasing events and random variables) We say that an event is *increasing* when, if the event occurs for a given set of occupied edges, it remains to hold when we make some more edges occupied. We say that a random variable X is *increasing* when the events $\{X \geq x\}$ are increasing for each $x \in \mathbb{R}$.

An example of an increasing event is $\{s \longleftrightarrow t\}$. An example of an increasing random variable is $|\mathscr{C}(v)|$ and the size of the maximal cluster $|\mathscr{C}_{\max}|$, where

$$|\mathscr{C}_{\max}| = \max_{v \in [n]} |\mathscr{C}(v)|. \tag{4.1.7}$$

Exercises 4.10 and 4.11 investigate monotonicity properties of $|\mathscr{C}_{\max}|$.

4.1.2 Informal Link to Poisson Branching Processes

We now describe the link between cluster sizes in $\mathrm{ER}_n(\lambda/n)$ and Poisson branching processes in an informal manner. The results in this section are not used in the remainder of the chapter, even though the philosophy forms the core of the proofs given. Fix $\lambda > 0$. Let $S_0^*, S_1^*, \ldots, X_1^*, X_2^*, \ldots, H^*$ refer to the history of a branching process with Poisson offspring distribution with mean λ and $S_0, S_1, \ldots, X_1, X_2, \ldots, H$ refer to the history of the exploration of the connected component of a vertex v in the random graph $\mathrm{ER}_n(\lambda/n)$, where $S_0, S_1, \ldots$ are defined in (4.1.3) above. The event $\{H^* = (x_1, \ldots, x_t)\}$ is the event that the total progeny T^* of the Poisson branching process is equal to t, and the values of $X_1^*, \ldots, X_t^*$ are given by $x_1, \ldots, x_t$. Recall that $\mathbb{P}_\lambda^*$ denotes the law of a Poisson branching process with mean offspring distribution λ. Naturally, by (3.3.2),

$$t = \min\{i : s_i = 0\} = \min\{i : x_1 + \cdots + x_i = i - 1\}, \tag{4.1.8}$$

where

$$s_0 = 1, \qquad s_i = s_{i-1} + x_i - 1. \tag{4.1.9}$$

For any possible history $(x_1, \ldots, x_t)$ (recall (3.3.3)),

$$\mathbb{P}^*_\lambda(H^* = (x_1, \ldots, x_t)) = \prod_{i=1}^{t} \mathbb{P}^*_\lambda(X^*_i = x_i), \tag{4.1.10}$$

where $(X^*_i)_{i \geq 1}$ are i.i.d. Poisson random variables with mean λ, while

$$\mathbb{P}_\lambda(H = (x_1, \ldots, x_t)) = \prod_{i=1}^{t} \mathbb{P}_\lambda(X_i = x_i | X_1 = x_1, \ldots, X_{i-1} = x_{i-1}),$$

where, conditionally on $X_1 = x_1, \ldots, X_{i-1} = x_{i-1}$, the random variable X_i is binomially distributed $\mathsf{Bin}(n - (i-1) - s_{i-1}, \lambda/n)$ (recall (4.1.4) and (4.1.9)).

As shown in Theorem 2.10, the Poisson distribution with parameter λ is the limiting distribution of $\mathsf{Bin}(n, \lambda/n)$. When $m_n = n(1 + o(1))$ and λ, i are fixed, we can extend this to the statement that when X has distribution $\mathsf{Bin}(m_n, \lambda/n)$,

$$\lim_{n\to\infty} \mathbb{P}(X = i) = \mathrm{e}^{-\lambda} \frac{\lambda^i}{i!}. \tag{4.1.11}$$

Therefore, for every $t < \infty$,

$$\lim_{n\to\infty} \mathbb{P}_\lambda\big(H = (x_1, \ldots, x_t)\big) = \mathbb{P}^*_\lambda\big(H^* = (x_1, \ldots, x_t)\big). \tag{4.1.12}$$

Thus, the distribution of finite connected components in the random graph $\mathrm{ER}_n(\lambda/n)$ is closely related to a Poisson branching process with mean λ. This relation is explored further in the remainder of this chapter.

4.2 Comparisons to Branching Processes

In this section, we investigate the relation between connected components in $\mathrm{ER}_n(\lambda/n)$ and binomial branching processes, thus making the informal argument in Section 4.1.2 precise. We start by proving two stochastic domination results for connected components in the Erdős–Rényi random graph. In Theorem 4.2, we give a stochastic upper bound on $|\mathscr{C}(v)|$, and in Theorem 4.3 a lower bound on the cluster tails. These bounds will be used in the following sections to prove results concerning $|\mathscr{C}_{\max}|$.

4.2.1 Stochastic Domination of Connected Components

The following upper bound shows that each connected component is stochastically dominated by the total progeny of a branching process with a binomial offspring distribution:

Theorem 4.2 (Stochastic domination of the cluster size) *For each $k \geq 1$,*

$$\mathbb{P}_{np}(|\mathscr{C}(1)| \geq k) \leq \mathbb{P}_{n,p}(T^{\geq} \geq k), \qquad \textit{i.e.,} \qquad |\mathscr{C}(1)| \preceq T^{\geq}, \tag{4.2.1}$$

where $T^{\geq}$ is the total progeny of a binomial branching process with parameters n and p, and we recall that $\mathbb{P}_\lambda$ denotes the law of $\mathrm{ER}_n(\lambda/n)$.

Proof Let $N_i = n - i - S_i$ denote the number of neutral vertices after i explorations, so that, conditionally on N_{i-1}, $X_i \sim \mathsf{Bin}(N_{i-1}, p)$. Let $Y_i \sim \mathsf{Bin}(n - N_{i-1}, p)$ be, conditionally on N_{i-1}, independent from X_i, and write

$$X_i^{\geq} = X_i + Y_i. \tag{4.2.2}$$

Then, conditionally on $(X_j)_{j=1}^{i-1}$, $X_i^{\geq} \sim \mathsf{Bin}(n, p)$. Since this distribution is independent of $(X_j)_{j=1}^{i-1}$, the sequence $(X_j^{\geq})_{j\geq 1}$ is in fact i.i.d. Also, $X_i^{\geq} \geq X_i$ a.s. since $Y_i \geq 0$ a.s.

Denote

$$S_i^{\geq} = X_1^{\geq} + \cdots + X_i^{\geq} - (i-1). \tag{4.2.3}$$

Then,

$$\begin{aligned}\mathbb{P}_{np}(|\mathscr{C}(1)| \geq k) &= \mathbb{P}(S_t > 0\ \forall t \leq k-1) \leq \mathbb{P}(S_t^{\geq} > 0\ \forall t \leq k-1) \\ &= \mathbb{P}(T^{\geq} \geq k),\end{aligned} \tag{4.2.4}$$

where we recall from Chapter 3 that $T^{\geq} = \min\{t\colon S_t^{\geq} = 0\}$ is the total progeny of a branching process with binomial distribution with parameters n and success probability p. □

In Exercise 4.12, you are asked to use Theorem 4.2 to derive an upper bound on the expected cluster size when $\lambda < 1$.

4.2.2 Lower Bound on the Cluster Tail

The following lower bound shows that the probability that a connected component has size at least k is bounded from below by the probability that the total progeny of a branching process with binomial offspring distribution exceeds k, where now the parameters of the binomial distribution are $n-k$ and p:

Theorem 4.3 (Lower bound on cluster tail) *For every* $k \in [n]$,

$$\mathbb{P}_{np}(|\mathscr{C}(1)| \geq k) \geq \mathbb{P}_{n-k,p}(T^{\leq} \geq k), \tag{4.2.5}$$

where $T^{\leq}$ *is the total progeny of a branching process with binomial distribution with parameters* $n-k$ *and success probability* $p = \lambda/n$.

Note that, since the parameter $n-k$ on the right-hand side of (4.2.5) depends explicitly on k, Theorem 4.3 does *not* imply a stochastic lower bound on $|\mathscr{C}(1)|$.

Proof We again use a coupling approach. Recall that N_i denotes the number of neutral vertices after i explorations. Denote by $\mathcal{T}_k$ the stopping time

$$\mathcal{T}_k = \min\{t\colon N_t \leq n-k\}. \tag{4.2.6}$$

Then, $\mathcal{T}_k \leq k-1$ whenever $|\mathscr{C}(1)| \geq k$, since $N_{k-1} \leq n-(k-1)-1 = n-k$. In terms of $\mathcal{T}_k$, we can trivially rewrite

$$\mathbb{P}_{np}(|\mathscr{C}(1)| \geq k) = \mathbb{P}_{np}(S_t > 0 \forall t \leq \mathcal{T}_k). \tag{4.2.7}$$

We let $(X_i^{\leq})_{i\geq 1}$ denote an i.i.d. sequence of $\mathsf{Bin}(n-k, p)$ random variables. For $i \leq \mathcal{T}_k$, and conditionally on N_{i-1}, let $Y_i \sim \mathsf{Bin}(N_{i-1} - (n-k), p)$ independently of all other random variables involved. Define

$$X_i = X_i^{\leq} + Y_i. \tag{4.2.8}$$

Then, clearly, $X_i \geq X_i^{\leq}$ a.s. for all $i \leq \mathcal{T}_k$, while, conditionally on N_{i-1}, $X_i \sim \mathsf{Bin}(N_{i-1}, p)$ as required for (4.1.4).

Denote

$$S_i^{\leq} = X_1^{\leq} + \cdots + X_i^{\leq} - (i-1). \tag{4.2.9}$$

Then, $S_t^{\leq} \leq S_t$ for all $t \leq \mathcal{T}_k$. Using the above coupling and the fact that $\mathcal{T}_k \leq k-1$, we can therefore bound

$$\begin{aligned} \{S_t > 0\ \forall t \leq \mathcal{T}_k\} &\supseteq \{S_t^{\leq} > 0\ \forall t \leq \mathcal{T}_k\} \supseteq \{S_t^{\leq} > 0\ \forall t \leq k-1\} \\ &= \{T^{\leq} \geq k\}, \end{aligned} \tag{4.2.10}$$

where $T^{\leq} = \min\{t : S_t^{\leq} = 0\}$ is the total progeny of a branching process with binomial distribution with parameters $n-k$ and success probability p. Taking probabilities in (4.2.10) proves Theorem 4.3. □

In the remaining sections we use the results of Theorems 4.2–4.3 to study the behavior of $|\mathscr{C}_{\max}|$ in sub- and supercritical Erdős–Rényi random graphs. The general strategy for the investigation of the largest connected component $|\mathscr{C}_{\max}|$ is as follows. We make use of the bounds on tail probabilities in Theorems 4.2–4.3 in order to compare the cluster sizes to branching processes with a binomial offspring with parameters n and $p = \lambda/n$. By Theorem 3.1, the behavior of branching processes is rather different when the expected offspring is larger than 1 or smaller than or equal to 1. Using Theorem 3.20, we can compare cluster sizes in $\mathrm{ER}_n(\lambda/n)$ to a Poisson branching process with a parameter that is close to the parameter λ in $\mathrm{ER}_n(\lambda/n)$. The results on branching processes in Chapter 3 then allow us to complete the proofs.

In Theorems 4.2–4.3, when $k = o(n)$, the expected offspring of the branching processes is close to $np \approx \lambda$. Therefore, for the Erdős–Rényi random graph, we expect different behavior in the subcritical regime $\lambda < 1$, in the supercritical regime $\lambda > 1$ and in the critical regime $\lambda = 1$. The analysis of the behavior of the largest connected component $|\mathscr{C}_{\max}|$ is substantially different in the subcritical regime where $\lambda < 1$, which is treated in Section 4.3, compared to the supercritical regime $\lambda > 1$, which is treated in Section 4.4. In Section 4.5, we prove a central limit theorem for the giant supercritical component. The critical regime $\lambda = 1$ requires some new ideas, and is treated in Section 5.1 in the next chapter.

4.3 The Subcritical Regime

In this section, we derive bounds for the size of the largest connected component for the Erdős–Rényi random graph in the subcritical regime, i.e., when $\lambda = np < 1$. Let I_λ denote the large deviation rate function for Poisson random variables with mean λ, given by

$$I_\lambda = \lambda - 1 - \log(\lambda). \tag{4.3.1}$$

Recall Exercise 2.20 to see an upper bound on tails of Poisson random variables involving I_λ, as well as the fact that $I_\lambda > 0$ for all $\lambda \neq 1$.

The main results for $\lambda < 1$ are Theorem 4.4, which proves that $|\mathscr{C}_{\max}| \leq a \log n$ whp for any $a > 1/I_\lambda$, and Theorem 4.5, where a matching lower bound on $|\mathscr{C}_{\max}|$ is provided by proving that $|\mathscr{C}_{\max}| \geq a \log n$ whp for any $a < 1/I_\lambda$:

Theorem 4.4 (Upper bound on largest subcritical component) *Fix $\lambda < 1$. Then, for every $a > 1/I_\lambda$, there exists $\delta = \delta(a, \lambda) > 0$ such that*

$$\mathbb{P}_\lambda(|\mathscr{C}_{\max}| \geq a \log n) = O(n^{-\delta}). \tag{4.3.2}$$

Theorem 4.5 (Lower bound on largest subcritical component) *Fix $\lambda < 1$. Then, for every $a < 1/I_\lambda$, there exists $\delta = \delta(a, \lambda) > 0$ such that*

$$\mathbb{P}_\lambda(|\mathscr{C}_{\max}| \leq a \log n) = O(n^{-\delta}). \tag{4.3.3}$$

Theorems 4.4 and 4.5 are proved in Sections 4.3.2 and 4.3.3 below. Together, they prove that $|\mathscr{C}_{\max}|/\log n \xrightarrow{\mathbb{P}} 1/I_\lambda$ as investigated in Exercise 4.13. See Figure 4.3 for an impression of how small subcritical clusters are for $n = 1000$.

4.3.1 Largest Subcritical Cluster: Proof Strategy

We start by describing the strategy of proof. We denote the number of vertices that are contained in connected components of size at least k by

$$Z_{\geq k} = \sum_{v \in [n]} \mathbb{1}_{\{|\mathscr{C}(v)| \geq k\}}. \tag{4.3.4}$$

We can identify $|\mathscr{C}_{\max}|$ as

$$|\mathscr{C}_{\max}| = \max\{k \colon Z_{\geq k} \geq k\}, \tag{4.3.5}$$

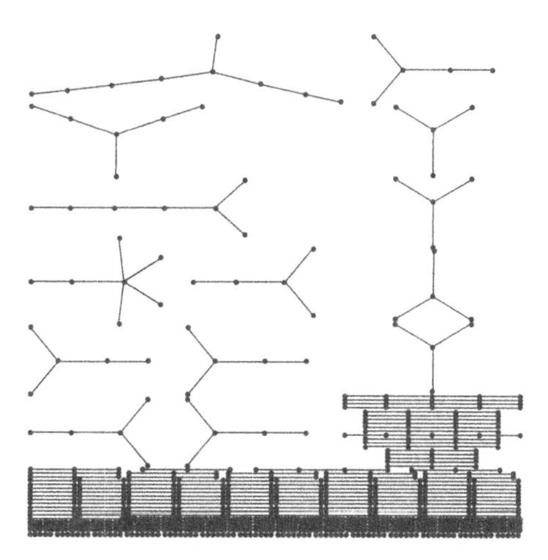

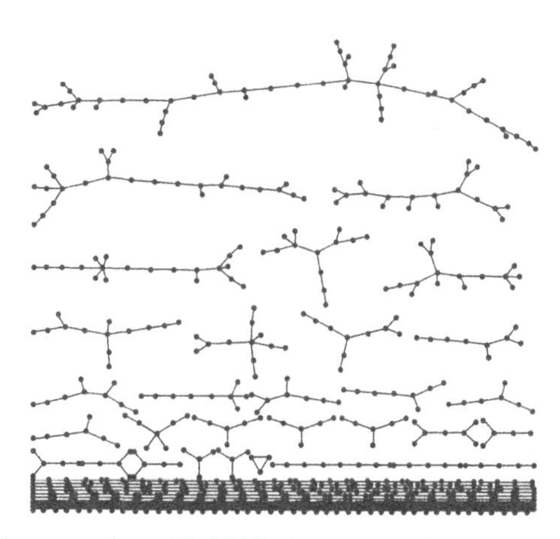

Figure 4.3 Realizations of Erdős–Rényi random graphs with 1000 elements and edge probabilities $\lambda/1000$ with $\lambda = 0.5$ and $\lambda = 0.9$ respectively.

which allows us to prove bounds on $|\mathscr{C}_{\max}|$ by investigating $Z_{\geq k}$ for an appropriately chosen k. In particular, (4.3.5) implies that

$$\{|\mathscr{C}_{\max}| \geq k\} = \{Z_{\geq k} \geq k\}, \tag{4.3.6}$$

see also Exercise 4.14.

To prove Theorem 4.4, we use the union bound to conclude that

$$\mathbb{P}_\lambda(|\mathscr{C}_{\max}| \geq k) \leq n\mathbb{P}_\lambda(|\mathscr{C}(1)| \geq k), \tag{4.3.7}$$

and we use Theorem 4.2 to bound $\mathbb{P}_\lambda(|\mathscr{C}(1)| \geq k_n)$ for $k_n = a \log n$ for any $a > 1/I_\lambda$. This is the strategy of proof of Theorem 4.4. For the details we refer to the formal argument in Section 4.3.2 below.

Alternatively and equivalently, we could use that $\mathbb{E}_\lambda[Z_{\geq k}] = n\mathbb{P}_\lambda(|\mathscr{C}(1)| \geq k)$ and use the first moment method or Markov's inequality (Theorem 2.17) to conclude that whp, $Z_{\geq k_n} = 0$, so that, again whp, $|\mathscr{C}_{\max}| \leq k_n$.

In the proof of Theorem 4.5, we use the second moment method or Chebychev's inequality (Theorem 2.18) on $Z_{\geq k}$ instead. In order to be able to apply this result, we first prove an upper bound on the variance of $Z_{\geq k}$ (see Proposition 4.7 below). We further use Theorem 4.3 to prove a lower bound on $\mathbb{E}_\lambda[Z_{\geq k_n}]$, now for $k_n = a \log n$ for any $a < 1/I_\lambda$. Then, (2.4.5) in Theorem 2.18 proves that $Z_{\geq k_n} > 0$ whp, so that, again whp, $|\mathscr{C}_{\max}| \geq k_n$. We now present the details of the proofs.

4.3.2 Upper Bound Largest Subcritical Cluster

By Theorem 4.2,

$$\mathbb{P}_\lambda(|\mathscr{C}(v)| > t) \leq \mathbb{P}_{n,p}(T > t), \tag{4.3.8}$$

where T is the total progeny of a branching process with a binomial offspring distribution with parameters n and $p = \lambda/n$. To study $\mathbb{P}_{n,p}(T > t)$, we let $(\hat{X}_i)_{i \geq 1}$ be an i.i.d. sequence of $\mathsf{Bin}(n, p)$ random variables, and let

$$\hat{S}_t = \hat{X}_1 + \cdots + \hat{X}_t - (t-1). \tag{4.3.9}$$

Then, by (3.3.2) and (3.3.1),

$$\mathbb{P}_{n,p}(T > t) \leq \mathbb{P}_{n,p}(\hat{S}_t > 0) = \mathbb{P}_{n,p}(\hat{X}_1 + \cdots + \hat{X}_t \geq t) \leq \mathrm{e}^{-tI_\lambda}, \tag{4.3.10}$$

by Corollary 2.20 and the fact that $\hat{X}_1 + \cdots + \hat{X}_t \sim \mathsf{Bin}(nt, \lambda/n)$. We conclude that

$$\mathbb{P}_\lambda(|\mathscr{C}(v)| > t) \leq \mathrm{e}^{-tI_\lambda}. \tag{4.3.11}$$

By the union bound, for $k_n = a \log n$,

$$\mathbb{P}_\lambda(|\mathscr{C}_{\max}| > a \log n) \leq n\mathbb{P}_\lambda(|\mathscr{C}(1)| > a \log n) \leq n^{1-aI_\lambda} = n^{-\delta}, \tag{4.3.12}$$

with $\delta = aI_\lambda - 1 > 0$ whenever $a > 1/I_\lambda$. This proves that the largest connected component is whp bounded by $a \log n$ for every $a > 1/I_\lambda$. □

We now give a second proof of (4.3.11), which is based on a distributional equality of S_t which turns out to be useful in the analysis of the Erdős–Rényi random graph with $\lambda > 1$ as well. The result states that S_t is also binomially distributed, but with a different success probability. In the statement of Proposition 4.6 below, we make essential use of the formal continuation of the recursions in (4.1.3) and (4.1.4) for the breadth-first search, defined right below (4.1.4). Note that, in particular, S_t need not be non-negative.

Proposition 4.6 (The law of S_t) *For all $t \in [n]$, and with $(S_t)_{t\geq 0}$ satisfying the recursion in* (4.1.3)–(4.1.4),

$$S_t + (t-1) \sim \mathsf{Bin}\big(n-1, 1-(1-p)^t\big). \tag{4.3.13}$$

We only use Proposition 4.6 when $|\mathscr{C}(v)| \geq t$, in which case $S_t \geq 0$ *does* hold.

Proof Let N_t represent the number of unexplored vertices at time $t \geq 0$, i.e.,

$$N_t = n - t - S_t. \tag{4.3.14}$$

Note that $X \sim \mathsf{Bin}(m, p)$ precisely when $Y = m - X \sim \mathsf{Bin}(m, 1-p)$. It is more convenient to show the equivalent statement that, for all $t \geq 0$,

$$N_t \sim \mathsf{Bin}\big(n-1, (1-p)^t\big). \tag{4.3.15}$$

Heuristically, (4.3.15) can be understood by noting that each of the vertices $\{2, \ldots, n\}$ has, independently of all other vertices, probability $(1-p)^t$ to stay neutral in the first t explorations. Formally, conditionally on S_{t-1}, $X_t \sim \mathsf{Bin}(n-(t-1)-S_{t-1}, p) = \mathsf{Bin}(N_{t-1}, p)$ by (4.1.4). Thus,

$$\begin{aligned} N_t &= n - t - S_t = n - t - S_{t-1} - X_t + 1 = N_{t-1} - X_t \\ &\sim \mathsf{Bin}(N_{t-1}, 1-p). \end{aligned} \tag{4.3.16}$$

The conclusion follows by recursion on t and $N_0 = n-1$. Exercise 4.15 asks you to complete the proof of (4.3.15). □

To complete the second proof of (4.3.11), we use Proposition 4.6 to see that

$$\mathbb{P}_\lambda(|\mathscr{C}(v)| > t) \leq \mathbb{P}(S_t > 0) \leq \mathbb{P}\big(\mathsf{Bin}(n-1, 1-(1-p)^t) \geq t\big), \tag{4.3.17}$$

where we abuse notation and write $\mathbb{P}(\mathsf{Bin}(m, q) \geq t)$ to denote the probability that a binomial random variable with parameters m and success probability q is at least t. Using Bernoulli's inequality $1-(1-p)^t \leq tp$, we therefore arrive at, for every $s \geq 0$,

$$\mathbb{P}_\lambda(|\mathscr{C}(v)| > t) \leq \mathbb{P}\big(\mathsf{Bin}(n, \frac{t\lambda}{n}) \geq t\big) \leq \mathbb{P}\Big(\mathrm{e}^{s\mathsf{Bin}(n, \frac{t\lambda}{n})} \geq \mathrm{e}^{st}\Big). \tag{4.3.18}$$

By Markov's inequality (Theorem 2.17), and minimizing over $s \geq 0$, we arrive at[1]

$$\begin{aligned} \mathbb{P}_\lambda(|\mathscr{C}(v)| > t) &\leq \min_{s\geq 0} \mathrm{e}^{-st}\mathbb{E}[\mathrm{e}^{s\mathsf{Bin}(n, \frac{t\lambda}{n})}] \\ &= \min_{s\geq 0} \mathrm{e}^{-st}\big[1 + \frac{t\lambda}{n}(\mathrm{e}^s - 1)\big]^n \leq \min_{s\geq 0} \mathrm{e}^{-st}\mathrm{e}^{t\lambda(\mathrm{e}^s-1)}. \end{aligned} \tag{4.3.19}$$

[1] Together, these two bounds give the exponential Markov inequality.

where we have used $1 + x \leq \mathrm{e}^x$ in the last inequality. We arrive at the bound

$$\mathbb{P}_\lambda(|\mathscr{C}(v)| > t) \leq \mathrm{e}^{-I_\lambda t}, \tag{4.3.20}$$

which re-proves (4.3.11). □

4.3.3 Lower Bound Largest Subcritical Cluster

The proof of Theorem 4.5 makes use of a variance estimate on $Z_{\geq k}$. We use the notation

$$\chi_{\geq k}(\lambda) = \mathbb{E}_\lambda\Big[|\mathscr{C}(v)|\mathbb{1}_{\{|\mathscr{C}(v)|\geq k\}}\Big]. \tag{4.3.21}$$

Note that, by exchangeability of the vertices, $\chi_{\geq k}(\lambda)$ does not depend on v. The following proposition gives a variance estimate for $Z_{\geq k}$ in terms of $\chi_{\geq k}(\lambda)$:

Proposition 4.7 (A variance estimate for $Z_{\geq k}$) *For every n and $k \in [n]$ and $\lambda > 0$,*

$$\mathrm{Var}_\lambda(Z_{\geq k}) \leq n\chi_{\geq k}(\lambda). \tag{4.3.22}$$

Proof We use that

$$\mathrm{Var}_\lambda(Z_{\geq k}) = \sum_{i,j\in[n]} \Big[\mathbb{P}_\lambda(|\mathscr{C}(i)| \geq k, |\mathscr{C}(j)| \geq k) - \mathbb{P}_\lambda(|\mathscr{C}(i)| \geq k)^2\Big]. \tag{4.3.23}$$

We split the probability $\mathbb{P}_\lambda(|\mathscr{C}(i)| \geq k, |\mathscr{C}(j)| \geq k)$, depending on whether $i \longleftrightarrow j$ or not, as

$$\begin{aligned}\mathbb{P}_\lambda(|\mathscr{C}(i)| \geq k, |\mathscr{C}(j)| \geq k) = {} & \mathbb{P}_\lambda(|\mathscr{C}(i)| \geq k, i \longleftrightarrow j) \\ & + \mathbb{P}_\lambda(|\mathscr{C}(i)| \geq k, |\mathscr{C}(j)| \geq k, i \not\longleftrightarrow j).\end{aligned} \tag{4.3.24}$$

Clearly,

$$\begin{aligned}&\mathbb{P}_\lambda(|\mathscr{C}(i)| = l, |\mathscr{C}(j)| \geq k, i \not\longleftrightarrow j) \\ &\quad = \mathbb{P}_\lambda(|\mathscr{C}(i)| = l, i \not\longleftrightarrow j)\mathbb{P}_\lambda\big(|\mathscr{C}(j)| \geq k \;\big|\; |\mathscr{C}(i)| = l, i \not\longleftrightarrow j\big).\end{aligned} \tag{4.3.25}$$

When $|\mathscr{C}(i)| = l$ and $i \not\longleftrightarrow j$, all vertices in the components different from the one of i, which includes the component of j, form a random graph $\mathrm{ER}_{n-l}(\lambda/n)$ where the size n is replaced by $n - l$. Since the probability that $|\mathscr{C}(j)| \geq k$ in $\mathrm{ER}_n(p)$ is increasing in n,

$$\mathbb{P}_\lambda(|\mathscr{C}(j)| \geq k \big| |\mathscr{C}(i)| = l, i \not\longleftrightarrow j) \leq \mathbb{P}_\lambda(|\mathscr{C}(j)| \geq k). \tag{4.3.26}$$

Therefore, for every $l \geq 1$,

$$\mathbb{P}_\lambda(|\mathscr{C}(i)| = l, |\mathscr{C}(j)| \geq k, i \not\longleftrightarrow j) - \mathbb{P}_\lambda(|\mathscr{C}(i)| = l)\mathbb{P}_\lambda(|\mathscr{C}(j)| \geq k) \leq 0, \tag{4.3.27}$$

which in turn implies that

$$\begin{aligned}\mathbb{P}_\lambda(|\mathscr{C}(i)| \geq k, |\mathscr{C}(j)| \geq k, i \not\longleftrightarrow j) &= \sum_{l\geq k} \mathbb{P}_\lambda(|\mathscr{C}(i)| = l, |\mathscr{C}(j)| \geq k, i \not\longleftrightarrow j) \\ &\leq \sum_{l\geq k} \mathbb{P}_\lambda(|\mathscr{C}(i)| = l)\mathbb{P}_\lambda(|\mathscr{C}(j)| \geq k) \\ &= \mathbb{P}_\lambda(|\mathscr{C}(i)| \geq k)\mathbb{P}_\lambda(|\mathscr{C}(j)| \geq k).\end{aligned} \tag{4.3.28}$$

We conclude that

$$\mathrm{Var}_\lambda(Z_{\geq k}) \leq \sum_{i,j\in[n]} \mathbb{P}_\lambda(|\mathscr{C}(i)| \geq k, i \longleftrightarrow j), \tag{4.3.29}$$

so that

$$\begin{aligned}\mathrm{Var}_\lambda(Z_{\geq k}) &\leq \sum_{i\in[n]}\sum_{j\in[n]} \mathbb{E}_\lambda\big[\mathbb{1}_{\{|\mathscr{C}(i)|\geq k\}}\mathbb{1}_{\{j\in\mathscr{C}(i)\}}\big] \\ &= \sum_{i\in[n]} \mathbb{E}_\lambda\Big[\mathbb{1}_{\{|\mathscr{C}(i)|\geq k\}}\sum_{j\in[n]}\mathbb{1}_{\{j\in\mathscr{C}(i)\}}\Big].\end{aligned} \tag{4.3.30}$$

Since $\sum_{j\in[n]}\mathbb{1}_{\{j\in\mathscr{C}(i)\}} = |\mathscr{C}(i)|$ and by exchangeability of the vertices,

$$\begin{aligned}\mathrm{Var}_\lambda(Z_{\geq k}) &\leq \sum_{i\in[n]} \mathbb{E}_\lambda[|\mathscr{C}(i)|\mathbb{1}_{\{|\mathscr{C}(i)|\geq k\}}] \\ &= n\mathbb{E}_\lambda[|\mathscr{C}(1)|\mathbb{1}_{\{|\mathscr{C}(1)|\geq k\}}] = n\chi_{\geq k}(\lambda).\end{aligned} \tag{4.3.31}$$

□

Proof of Theorem 4.5 By (4.3.6) (see also Exercise 4.14), $\mathbb{P}_\lambda(|\mathscr{C}_{\max}| < k) = \mathbb{P}_\lambda(Z_{\geq k} = 0)$. Thus, to prove Theorem 4.5, it suffices to prove that $\mathbb{P}_\lambda(Z_{\geq k_n} = 0) = O(n^{-\delta})$, where $k_n = a\log n$ with $a < 1/I_\lambda$. For this, we use Chebychev's inequality (Theorem 2.18). In order to apply Theorem 2.18, we need to derive a lower bound on $\mathbb{E}_\lambda[Z_{\geq k}]$ and an upper bound on $\mathrm{Var}_\lambda(Z_{\geq k})$.

We start by giving a lower bound on $\mathbb{E}_\lambda[Z_{\geq k}]$. We use that

$$\mathbb{E}_\lambda[Z_{\geq k}] = nP_{\geq k}(\lambda), \qquad \text{where} \qquad P_{\geq k}(\lambda) = \mathbb{P}_\lambda(|\mathscr{C}(v)| \geq k). \tag{4.3.32}$$

We take $k_n = a\log n$. We use Theorem 4.3 to see that, with T a binomial branching process with parameters $n - k_n$ and $p = \lambda/n$,

$$P_{\geq k_n}(\lambda) \geq \mathbb{P}_{n-k_n,p}(T \geq k_n). \tag{4.3.33}$$

By Theorem 3.20, with T^* the total progeny of a Poisson branching process with mean $\lambda_n = \lambda(n-k_n)/n$,

$$\mathbb{P}_{n-k_n,p}(T \geq a\log n) = \mathbb{P}^*_{\lambda_n}(T^* \geq a\log n) + O\Big(\frac{a\lambda^2\log n}{n}\Big). \tag{4.3.34}$$

Also, by Theorem 3.16,

$$\mathbb{P}^*_{\lambda_n}(T^* \geq a\log n) = \sum_{k=a\log n}^{\infty} \mathbb{P}^*_{\lambda_n}(T^* = k) = \sum_{k=a\log n}^{\infty} \frac{(\lambda_n k)^{k-1}}{k!}\mathrm{e}^{-\lambda_n k}. \tag{4.3.35}$$

By Stirling's formula,

$$k! = \Big(\frac{k}{\mathrm{e}}\Big)^k\sqrt{2\pi k}\big(1+o(1)\big), \tag{4.3.36}$$

so that, recalling (4.3.1), and using that $I_{\lambda_n} = I_\lambda + o(1)$,

$$\mathbb{P}^*_{\lambda_n}(T^* \geq a \log n) = \lambda^{-1} \sum_{k=a\log n}^{\infty} \frac{1}{\sqrt{2\pi k^3}} \mathrm{e}^{-I_{\lambda_n} k}(1 + o(1)) \geq \mathrm{e}^{-I_\lambda a \log n(1+o(1))}. \tag{4.3.37}$$

As a result, it follows that, with $k_n = a \log n$ and any $0 < \alpha < 1 - I_\lambda a$,

$$\mathbb{E}_\lambda[Z_{\geq k_n}] = n P_{\geq k_n}(\lambda) \geq n^{(1-I_\lambda a)(1+o(1))} \geq n^{\alpha}, \tag{4.3.38}$$

when n is sufficiently large.

We next bound the variance of $Z_{\geq k_n}$ using Proposition 4.7. For an integer-valued random variable X,

$$\begin{aligned}\mathbb{E}[X \mathbb{1}_{\{X\geq k\}}] &= \sum_{t=k}^{\infty} t\mathbb{P}(X = t) = \sum_{t=k}^{\infty}\sum_{s=1}^{t} \mathbb{P}(X = t) \\ &= \sum_{s=1}^{\infty}\sum_{t=k\vee s}^{\infty} \mathbb{P}(X = t) = k\mathbb{P}(X \geq k) + \sum_{s=k+1}^{\infty} \mathbb{P}(X \geq s).\end{aligned} \tag{4.3.39}$$

By (4.3.11),

$$\begin{aligned}\chi_{\geq k_n}(\lambda) &= kP_{\geq k}(\lambda) + \sum_{t=k_n+1}^{n} P_{\geq t}(\lambda) \leq k_n \mathrm{e}^{-(k_n-1)I_\lambda} + \sum_{t=k_n}^{n} \mathrm{e}^{-I_\lambda(t-1)} \\ &\leq k_n \mathrm{e}^{-(k_n-1)I_\lambda} + \frac{\mathrm{e}^{-k_n I_\lambda}}{1 - \mathrm{e}^{-I_\lambda}} = O(k_n n^{-aI_\lambda}).\end{aligned} \tag{4.3.40}$$

We conclude that, by Proposition 4.7 and with $k_n = a \log n$,

$$\mathrm{Var}_\lambda(Z_{\geq k_n}) \leq n\chi_{\geq k_n}(\lambda) \leq O(k_n n^{1-aI_\lambda}), \tag{4.3.41}$$

while, for large enough n,

$$\mathbb{E}_\lambda[Z_{\geq k_n}] \geq n^{\alpha}. \tag{4.3.42}$$

Therefore, by the Chebychev inequality (Theorem 2.17),

$$\mathbb{P}_\lambda(Z_{\geq k_n} = 0) \leq \frac{\mathrm{Var}_\lambda(Z_{\geq k_n})}{\mathbb{E}_\lambda[Z_{\geq k_n}]^2} \leq O(k_n n^{1-aI_\lambda-2\alpha}) = O(n^{-\delta}), \tag{4.3.43}$$

for n large enough and any $\delta < 2\alpha - (1 - I_\lambda a)$, where $2\alpha - (1 - I_\lambda a) > 0$ when $0 < \alpha < 1 - I_\lambda a$. Finally, by Exercise 4.14,

$$\mathbb{P}_\lambda(|\mathscr{C}_{\max}| < k_n) = \mathbb{P}_\lambda(Z_{\geq k_n} = 0), \tag{4.3.44}$$

which completes the proof of Theorem 4.5. □

4.4 The Supercritical Regime

In this section, we fix $\lambda > 1$. The main result is a law of large numbers for the size of the maximal connected component. Below, we write $\zeta_\lambda = 1 - \eta_\lambda$ for the survival probability of a Poisson branching process with mean offspring λ. Note that $\zeta_\lambda > 0$ since $\lambda > 1$, so the following theorem shows that there exists a giant component:

Theorem 4.8 (Law of large numbers for giant component) *Fix $\lambda > 1$. Then, for every $\nu \in (\frac{1}{2}, 1)$, there exists $\delta = \delta(\nu, \lambda) > 0$ such that*

$$\mathbb{P}_\lambda\big(\big||\mathscr{C}_{\max}| - \zeta_\lambda n\big| \geq n^\nu\big) = O(n^{-\delta}). \tag{4.4.1}$$

Theorem 4.8 can be interpreted as follows. A vertex has a large connected component with probability ζ_λ (we are deliberately vague about what 'large' means). Therefore, at least in expectation, there are roughly $\zeta_\lambda n$ vertices with large connected components. Theorem 4.8 implies that almost all these vertices are in fact in the *same* connected component, which is called the *giant component*. We first give an overview of the proof of Theorem 4.8. See Figure 4.4 for an impression of how the giant component emerges for $n = 1000$. When $\lambda = 1.1$, the maximal component is still quite small; when $\lambda = 2$, it is already rather complex in its topology, and, for example, already contains quite a few cycles.

4.4.1 Proof Strategy Law of Large Numbers for the Giant Component

In this section, we give an overview of the proof of Theorem 4.8. We again rely on an analysis of the number of vertices in connected components of size at least k,

$$Z_{\geq k} = \sum_{v\in[n]} \mathbb{1}_{\{|\mathscr{C}(v)|\geq k\}}. \tag{4.4.2}$$

The proof contains 4 main steps. In the **first step**, for $k_n = K \log n$ and K sufficiently large, we compute

$$\mathbb{E}_\lambda[Z_{\geq k_n}] = n\mathbb{P}_\lambda(|\mathscr{C}(v)| \geq k_n). \tag{4.4.3}$$

We evaluate $\mathbb{P}_\lambda(|\mathscr{C}(v)| \geq k_n)$ using the bound in Theorem 4.3. Proposition 4.9 below states that

$$\mathbb{P}_\lambda(|\mathscr{C}(v)| \geq k_n) = \zeta_\lambda(1 + o(1)). \tag{4.4.4}$$

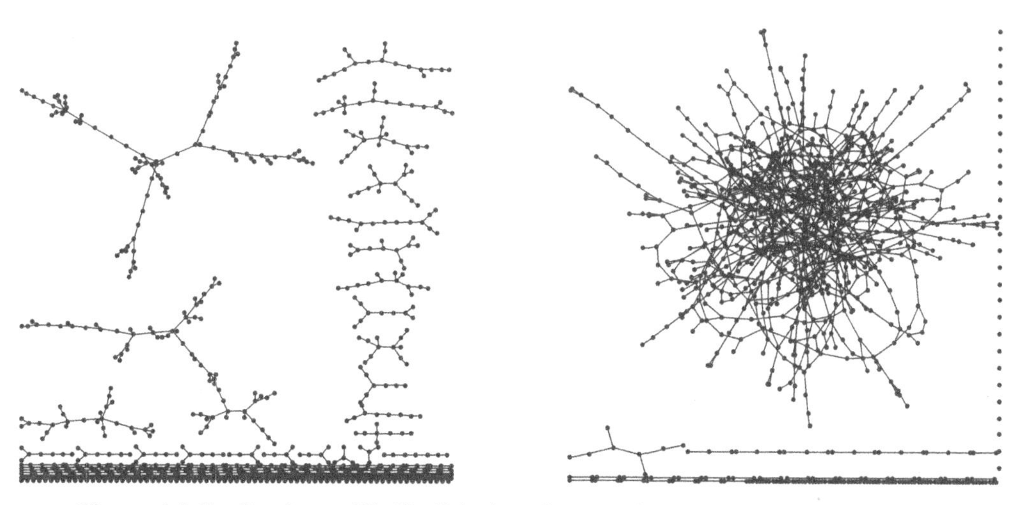

Figure 4.4 Realizations of Erdős–Rényi random graphs with 1000 elements and edge probabilities $\lambda/1000$ with $\lambda = 1.1$ and $\lambda = 2$, respectively.

In the **second step**, we use a variance estimate on $Z_{\geq k}$ in Proposition 4.10, which implies that, whp and for all $\nu \in (\frac{1}{2}, 1)$,

$$|Z_{\geq k_n} - \mathbb{E}_\lambda[Z_{\geq k_n}]| \leq n^\nu. \tag{4.4.5}$$

In the **third step**, we show that, for $k = k_n = K \log n$ for some $K > 0$ sufficiently large and whp, there is no connected component of size in between k_n and αn for any $\alpha < \zeta_\lambda$. This is done by a first moment argument: the expected number of vertices in such connected components is equal to $\mathbb{E}_\lambda[Z_{\geq k_n} - Z_{\geq \alpha n}]$, and we use the bound in Proposition 4.9 described above, as well as Proposition 4.12, which states that, for any $\alpha < \zeta_\lambda$, there exists $J = J(\alpha) > 0$ such that, for all n sufficiently large,

$$\mathbb{P}_\lambda\big(k_n \leq |\mathscr{C}(v)| < \alpha n\big) \leq \mathrm{e}^{-k_n J}. \tag{4.4.6}$$

In the **fourth step**, we prove that for $2\alpha > \zeta_\lambda$, and in the event there are no clusters of size in between k_n and αn and (4.4.5) holds,

$$Z_{\geq k_n} = |\mathscr{C}_{\max}|. \tag{4.4.7}$$

The proof of Theorem 4.8 follows by combining (4.4.3), (4.4.5) and (4.4.7). We now give the details of the proof of Theorem 4.8.

4.4.2 Step 1: Expected Number of Vertices in Large Components

In the first step, we show that the probability that $|\mathscr{C}(v)| \geq k_n$ is, for $k_n \geq a \log n$ with $a > 1/I_\lambda$, close to the survival probability of a Poisson branching process with mean λ. Proposition 4.9 implies (4.4.4).

Proposition 4.9 (Cluster tail is branching process survival probability) *Fix $\lambda > 1$. Then, for $k_n \geq a \log n$ where $a > 1/I_\lambda$ and I_λ is defined in* (4.3.1), *and for n sufficiently large,*

$$\mathbb{P}_\lambda(|\mathscr{C}(v)| \geq k_n) = \zeta_\lambda + O(k_n/n). \tag{4.4.8}$$

Proof For the upper bound on $\mathbb{P}_\lambda(|\mathscr{C}(v)| \geq k_n)$, we first use Theorem 4.2, followed by Theorem 3.20, to deduce

$$\mathbb{P}_\lambda(|\mathscr{C}(v)| \geq k_n) \leq \mathbb{P}_{n,\lambda/n}(T \geq k_n) \leq \mathbb{P}^*_\lambda(T^* \geq k_n) + O(k_n/n), \tag{4.4.9}$$

where T and T^*, respectively, are the total progeny of a binomial branching process with parameters n and λ/n and a Poisson mean λ branching process, respectively. To complete the upper bound, we use Theorem 3.8 to see that

$$\begin{aligned}\mathbb{P}^*_\lambda(T^* \geq k_n) &= \mathbb{P}^*_\lambda(T^* = \infty) + \mathbb{P}^*_\lambda(k_n \leq T^* < \infty)\\ &= \zeta_\lambda + O(\mathrm{e}^{-k_n I_\lambda}) = \zeta_\lambda + O(n^{-aI_\lambda}) = \zeta_\lambda + o(1/n),\end{aligned} \tag{4.4.10}$$

because $k_n \geq a \log n$ with $a > 1/I_\lambda$. Substituting (4.4.10) into (4.4.9) proves the upper bound in Proposition 4.9.

For the lower bound, we use Theorem 4.3 again followed by Theorem 3.20, so that, with $\lambda_n = \lambda(1 - k_n/n)$,

$$\mathbb{P}_\lambda(|\mathscr{C}(v)| \geq k_n) \geq \mathbb{P}_{n-k_n,\lambda/n}(T \geq k_n) \geq \mathbb{P}^*_{\lambda_n}(T^* \geq k_n) + O(k_n/n), \tag{4.4.11}$$

where now T and T^*, respectively, are the total progenies of a binomial branching process with parameters $n - k_n$ and λ/n and a Poisson mean λ_n branching process, respectively.

By Exercise 3.30 for $k_n \geq a \log n$ with $a > 1/I_\lambda$,

$$\mathbb{P}^*_{\lambda_n}(T^* \geq k_n) = \zeta_{\lambda_n} + O(\mathrm{e}^{-k_n I_{\lambda_n}}) = \zeta_{\lambda_n} + o(1/n). \tag{4.4.12}$$

Furthermore, by the mean-value theorem,

$$\eta_{\lambda_n} = \eta_\lambda + (\lambda_n - \lambda)\frac{d}{d\lambda}\eta_\lambda\big|_{\lambda=\lambda_n^*} = \eta_\lambda + O(k_n/n), \tag{4.4.13}$$

for some $\lambda_n^* \in (\lambda_n, \lambda)$, where we use Corollary 3.19 for $\lambda > 1$ and $\lambda_n - \lambda = k_n/n$. Therefore, since $\zeta_\lambda = 1 - \eta_\lambda$, also $\zeta_{\lambda_n} = \zeta_\lambda + O(k_n/n)$. Putting these estimates together proves the lower bound. Together, the upper and lower bound complete the proof of Proposition 4.9. □

4.4.3 Step 2: Concentration of Number of Vertices in Large Clusters

The proof of Theorem 4.8 makes use of a variance estimate on $Z_{\geq k}$. In its statement, we use the notation

$$\chi_{<k}(\lambda) = \mathbb{E}_\lambda[|\mathscr{C}(v)|\mathbb{1}_{\{|\mathscr{C}(v)|<k\}}]. \tag{4.4.14}$$

Proposition 4.10 (A second variance estimate on $Z_{\geq k}$) *For every n and $k \in [n]$,*

$$\mathrm{Var}_\lambda(Z_{\geq k}) \leq (\lambda k + 1)n\chi_{<k}(\lambda). \tag{4.4.15}$$

Note that the variance estimate in Proposition 4.10 is, in the supercritical regime, much better than the variance estimate in Proposition 4.7. Indeed, the bound in Proposition 4.7 reads

$$\mathrm{Var}_\lambda(Z_{\geq k}) \leq n\chi_{\geq k}(\lambda). \tag{4.4.16}$$

However, when $\lambda > 1$, according to Theorem 4.8 (which is currently not yet proved), $|\mathscr{C}(1)| = \Theta(n)$ with positive probability. This suggests that

$$n\chi_{\geq k}(\lambda) = \Theta(n^2), \tag{4.4.17}$$

which leads to a trivial bound on the variance of $Z_{\geq k}$. The bound in Proposition 4.10 is at most $\Theta(k^2 n)$, which is much smaller when k is not too large.

Proof Define

$$Z_{<k} = \sum_{v\in[n]} \mathbb{1}_{\{|\mathscr{C}(v)|<k\}}. \tag{4.4.18}$$

Then, since $Z_{<k} = n - Z_{\geq k}$,

$$\mathrm{Var}_\lambda(Z_{\geq k}) = \mathrm{Var}_\lambda(Z_{<k}). \tag{4.4.19}$$

Therefore, it suffices to prove that $\mathrm{Var}_\lambda(Z_{<k}) \leq (\lambda k + 1)n\chi_{<k}(\lambda)$. For this, we compute

$$\mathrm{Var}_\lambda(Z_{<k}) = \sum_{i,j\in[n]} \big[\mathbb{P}_\lambda(|\mathscr{C}(i)| < k, |\mathscr{C}(j)| < k) - \mathbb{P}_\lambda(|\mathscr{C}(i)| < k)^2\big]. \tag{4.4.20}$$

We again split, depending on whether $i \longleftrightarrow j$ or not:

$$\begin{aligned}\mathrm{Var}_\lambda(Z_{<k}) = &\sum_{i,j\in[n]} \big[\mathbb{P}_\lambda(|\mathscr{C}(i)| < k, |\mathscr{C}(j)| < k, i \not\longleftrightarrow j) - \mathbb{P}_\lambda(|\mathscr{C}(i)| < k)^2\big] \\ &+ \sum_{i,j\in[n]} \mathbb{P}_\lambda(|\mathscr{C}(i)| < k, |\mathscr{C}(j)| < k, i \longleftrightarrow j). \end{aligned} \tag{4.4.21}$$

We compute explicitly, using that $|\mathscr{C}(i)| = |\mathscr{C}(j)|$ when $i \longleftrightarrow j$,

$$\begin{aligned}\sum_{i,j\in[n]} \mathbb{P}_\lambda(|\mathscr{C}(i)| < k, |\mathscr{C}(j)| < k, i \longleftrightarrow j) &= \sum_{i,j\in[n]} \mathbb{E}_\lambda\big[\mathbb{1}_{\{|\mathscr{C}(i)|<k\}}\mathbb{1}_{\{i\longleftrightarrow j\}}\big] \\ &= \sum_{i\in[n]} \mathbb{E}_\lambda\big[\mathbb{1}_{\{|\mathscr{C}(i)|<k\}} \sum_{j\in[n]} \mathbb{1}_{\{i\longleftrightarrow j\}}\big] \\ &= \sum_{i\in[n]} \mathbb{E}_\lambda[|\mathscr{C}(i)|\mathbb{1}_{\{|\mathscr{C}(i)|<k\}}] = n\chi_{<k}(\lambda). \end{aligned} \tag{4.4.22}$$

To compute the first sum on the right-hand side of (4.4.21), we write that, for $l < k$,

$$\begin{aligned}&\mathbb{P}_\lambda(|\mathscr{C}(i)| = l, |\mathscr{C}(j)| < k, i \not\longleftrightarrow j) \\ &\quad= \mathbb{P}_\lambda(|\mathscr{C}(i)| = l)\mathbb{P}_\lambda\big(i \not\longleftrightarrow j \mid |\mathscr{C}(i)| = l\big)\mathbb{P}_\lambda\big(|\mathscr{C}(j)| < k \mid |\mathscr{C}(i)| = l, i \not\longleftrightarrow j\big). \end{aligned} \tag{4.4.23}$$

See Exercise 4.16 for an explicit formula for $\mathbb{P}_\lambda\big(i \not\longleftrightarrow j \big| |\mathscr{C}(i)| = l\big)$. We bound $\mathbb{P}_\lambda\big(i \not\longleftrightarrow j \mid |\mathscr{C}(i)| = l\big) \leq 1$, to obtain

$$\begin{aligned}&\mathbb{P}_\lambda(|\mathscr{C}(i)| = l, |\mathscr{C}(j)| < k, i \not\longleftrightarrow j) \\ &\quad\leq \mathbb{P}_\lambda(|\mathscr{C}(i)| = l)\mathbb{P}_\lambda\big(|\mathscr{C}(j)| < k \big| |\mathscr{C}(i)| = l, i \not\longleftrightarrow j\big). \end{aligned} \tag{4.4.24}$$

When $|\mathscr{C}(i)| = l$ and when $i \not\longleftrightarrow j$, the law of $|\mathscr{C}(j)|$ is identical to the law of $|\mathscr{C}(1)|$ in a random graph with $n - l$ vertices and edge probability $p = \lambda/n$, i.e.,

$$\mathbb{P}_{n,\lambda}(|\mathscr{C}(j)| < k \big| |\mathscr{C}(i)| = l, i \not\longleftrightarrow j) = \mathbb{P}_{n-l,\lambda}(|\mathscr{C}(1)| < k), \tag{4.4.25}$$

where we slightly abuse notation to write $\mathbb{P}_{m,\lambda}$ for the distribution of $\mathrm{ER}_m(\lambda/n)$. Therefore,

$$\begin{aligned}&\mathbb{P}_\lambda(|\mathscr{C}(j)| < k \big| |\mathscr{C}(i)| = l, i \not\longleftrightarrow j) \\ &\quad= \mathbb{P}_{n-l,\lambda}(|\mathscr{C}(1)| < k) \\ &\quad= \mathbb{P}_{n,\lambda}(|\mathscr{C}(1)| < k) + \mathbb{P}_{n-l,\lambda}(|\mathscr{C}(1)| < k) - \mathbb{P}_{n,\lambda}(|\mathscr{C}(1)| < k). \end{aligned} \tag{4.4.26}$$

We can couple $\mathrm{ER}_{n-l}(p)$ and $\mathrm{ER}_n(p)$ by adding the vertices $[n] \setminus [n-l]$, and by letting st, for $s \in [n] \setminus [n-l]$ and $t \in [n]$, be independently occupied with probability p. In this coupling, we note that $\mathbb{P}_{n-l,\lambda}(|\mathscr{C}(1)| < k) - \mathbb{P}_{n,\lambda}(|\mathscr{C}(1)| < k)$ is equal to the probability of the event that $|\mathscr{C}(1)| < k$ in $\mathrm{ER}_{n-l}(p)$, but $|\mathscr{C}(1)| \geq k$ in $\mathrm{ER}_n(p)$.

If $|\mathscr{C}(1)| < k$ in $\mathrm{ER}_{n-l}(p)$, but $|\mathscr{C}(1)| \geq k$ in $\mathrm{ER}_n(p)$, then it follows that at least one of the vertices $[n] \setminus [n-l]$ must be connected to one of the at most k vertices in the connected component of vertex 1 in $\mathrm{ER}_{n-l}(p)$. This has probability at most lkp, so that

$$\mathbb{P}_\lambda(|\mathscr{C}(j)| < k, i \not\longleftrightarrow j \big| |\mathscr{C}(i)| = l) - \mathbb{P}_\lambda(|\mathscr{C}(j)| < k) \leq lk\lambda/n. \tag{4.4.27}$$

Therefore,

$$\begin{aligned}
&\sum_{i,j\in[n]} \big[\mathbb{P}_\lambda(|\mathscr{C}(i)| < k, |\mathscr{C}(j)| < k, i \not\longleftrightarrow j) - \mathbb{P}_\lambda(|\mathscr{C}(i)| < k)\mathbb{P}_\lambda(|\mathscr{C}(j)| < k)\big] \\
&\quad \leq \sum_{l=1}^{k-1} \sum_{i,j\in[n]} \frac{\lambda k l}{n} \mathbb{P}_\lambda(|\mathscr{C}(i)| = l) = \frac{\lambda k}{n} \sum_{i,j\in[n]} \mathbb{E}_\lambda[|\mathscr{C}(i)| \mathbb{1}_{\{|\mathscr{C}(i)|<k\}}] \\
&\quad = nk\lambda \chi_{<k}(\lambda),
\end{aligned} \tag{4.4.28}$$

which, together with (4.4.21)–(4.4.22), completes the proof. □

As a corollary of Proposition 4.10, we obtain that the number of vertices in large components is concentrated:

Corollary 4.11 (Concentration of the number of vertices in large components) *Fix $k_n = K \log n$ and $\nu \in (\frac{1}{2}, 1)$. Then, for K sufficiently large and every $\delta < 2\nu - 1$, as $n \to \infty$,*

$$\mathbb{P}_\lambda(|Z_{\geq k_n} - n\zeta_\lambda| > n^\nu) = O(n^{-\delta}). \tag{4.4.29}$$

Proof By Proposition 4.9,

$$\mathbb{E}_\lambda[Z_{\geq k_n}] = n\mathbb{P}_\lambda(|\mathscr{C}(v)| \geq k_n) = n\zeta_\lambda + O(k_n), \tag{4.4.30}$$

and therefore, since $k_n = o(n^\nu)$ and for n sufficiently large,

$$\{|Z_{\geq k_n} - n\zeta_\lambda| > n^\nu\} \subseteq \{|Z_{\geq k_n} - \mathbb{E}_\lambda[Z_{\geq k_n}]| > n^\nu/2\}. \tag{4.4.31}$$

By Chebychev's inequality (Theorem 2.18), Proposition 4.10 and since $\chi_{<k_n}(\lambda) \leq k_n$,

$$\begin{aligned}
\mathbb{P}_\lambda(|Z_{\geq k_n} - n\zeta_\lambda| > n^\nu) &\leq \mathbb{P}_\lambda(|Z_{\geq k_n} - \mathbb{E}_\lambda[Z_{\geq k_n}]| > n^\nu/2) \leq 4n^{-2\nu}\mathrm{Var}(Z_{\geq k_n}) \\
&\leq 4n^{1-2\nu}\big(\lambda k_n^2 + k_n\big) \leq n^{-\delta},
\end{aligned} \tag{4.4.32}$$

for any $\delta < 2\nu - 1$ and n sufficiently large, since $k_n = K \log n$. □

4.4.4 Step 3: No Middle Ground

We next show that the probability that $k_n \leq |\mathscr{C}(v)| \leq \alpha n$ is exponentially small in k_n when $\alpha < \zeta_\lambda$. In order to state the result, we introduce some notation. Recall the definition of I_λ in (4.3.1). Let

$$J(\alpha; \lambda) = I_{g(\alpha;\lambda)}, \qquad \text{with} \qquad g(\alpha; \lambda) = (1 - \mathrm{e}^{-\lambda\alpha})/\alpha. \tag{4.4.33}$$

Then, $J(\alpha; \lambda) > 0$ whenever $g(\alpha; \lambda) \neq 1$. Since ζ_λ is defined to be the unique solution to $g(\alpha; \lambda) = 1$ and $g(\alpha; \lambda) > 1$ for every $\alpha < \zeta_\lambda$, we arrive at $J(\alpha; \lambda) > 0$ for all $\alpha < \zeta_\lambda$. This implies that the tail bounds in the next proposition really are exponentially small:

Proposition 4.12 (Exponential bound for supercritical clusters smaller than $\zeta_\lambda n$) *Fix $\lambda > 1$ and let k_n be such that $k_n \to \infty$ as $n \to \infty$. Then, for any $\alpha < \zeta_\lambda$,*

$$\mathbb{P}_\lambda(k_n \leq |\mathscr{C}(v)| \leq \alpha n) \leq \mathrm{e}^{-k_n J(\alpha;\lambda)}/[1 - \mathrm{e}^{-J(\alpha;\lambda)}]. \tag{4.4.34}$$

Proof We start by bounding

$$\mathbb{P}_\lambda(k_n \leq |\mathscr{C}(v)| \leq \alpha n) = \sum_{t=k_n}^{\alpha n} \mathbb{P}_\lambda(|\mathscr{C}(v)| = t) \leq \sum_{t=k_n}^{\alpha n} \mathbb{P}_\lambda(S_t = 0), \tag{4.4.35}$$

where $p = \lambda/n$ and we recall (4.1.3). By Proposition 4.6, $S_t \sim \mathsf{Bin}(n-1, 1-(1-p)^t)+1-t$. Therefore, with $p = \lambda/n$,

$$\mathbb{P}_\lambda(S_t = 0) = \mathbb{P}\big(\mathsf{Bin}(n-1, 1-(1-p)^t) = t-1\big), \tag{4.4.36}$$

where we again abuse notation to write $\mathbb{P}(\mathsf{Bin}(m, q) = s)$ for the probability that a binomial random variable with parameters m and q takes the value s. To explain the exponential decay in (4.4.34), we note that, for $p = \lambda/n$ and $t = \lfloor \alpha n \rfloor$,

$$1 - (1-p)^t = 1 - \left(1 - \frac{\lambda}{n}\right)^{\lfloor \alpha n \rfloor} = (1 - \mathrm{e}^{-\lambda\alpha})(1 + o(1)) \qquad \text{for large } n. \tag{4.4.37}$$

The unique solution in $(0, 1]$ to the equation $1 - \mathrm{e}^{-\lambda\alpha} = \alpha$ is $\alpha = \zeta_\lambda$, as shown in Exercise 4.17.

If $\alpha < \zeta_\lambda$, then $\alpha < 1 - \mathrm{e}^{-\lambda\alpha}$, so that $g(\alpha; \lambda) < 1$. Therefore, $J(\alpha; \lambda) = I_{g(\alpha;\lambda)} > 0$, and thus the probability in (4.4.36) is exponentially small. We now fill in the details. First, by (4.4.36) and using that $1 - p \leq \mathrm{e}^{-p}$, so that $1 - (1-p)^t \geq 1 - \mathrm{e}^{-pt}$,

$$\begin{aligned}
\mathbb{P}_\lambda(S_t = 0) &= \mathbb{P}_\lambda\big(\mathsf{Bin}(n-1, 1-(1-p)^t) = t-1\big) \\
&\leq \mathbb{P}_\lambda\big(\mathsf{Bin}(n-1, 1-(1-p)^t) \leq t-1\big) \\
&\leq \mathbb{P}_\lambda\big(\mathsf{Bin}(n, 1-(1-p)^t) \leq t\big) \leq \mathbb{P}_\lambda\big(\mathsf{Bin}(n, 1-\mathrm{e}^{-pt}) \leq t\big) \\
&= \mathbb{P}_\lambda\Big(\mathrm{e}^{-s\mathsf{Bin}(n,1-\mathrm{e}^{-pt})} \geq \mathrm{e}^{-st}\Big),
\end{aligned} \tag{4.4.38}$$

for each $s > 0$. We use Markov's inequality (Theorem 2.17) to bound

$$\begin{aligned}
\mathbb{P}_\lambda(S_t = 0) &\leq \mathrm{e}^{st}\mathbb{E}\Big[\mathrm{e}^{-s\mathsf{Bin}(n,1-\mathrm{e}^{-pt})}\Big] = \mathrm{e}^{st}\Big((1-\mathrm{e}^{-pt})\mathrm{e}^{-s} + \mathrm{e}^{-pt}\Big)^n \\
&= \mathrm{e}^{st}\Big(1 - (1-\mathrm{e}^{-pt})(1-\mathrm{e}^{-s})\Big)^n \leq \exp\big(st - n(1-\mathrm{e}^{-pt})(1-\mathrm{e}^{-s})\big) \\
&= \exp\big(st + n(1-\mathrm{e}^{-\lambda t/n})(1-\mathrm{e}^{-s})\big),
\end{aligned} \tag{4.4.39}$$

where we use that $1 - x \leq \mathrm{e}^{-x}$ in the last inequality. The minimizer of the exponent $s \mapsto st - n(1-\mathrm{e}^{-\lambda t/n})(1-\mathrm{e}^{-s})$ over s is equal to

$$s^* = \log\Big(n(1-\mathrm{e}^{-\lambda t/n})/t\Big), \tag{4.4.40}$$

Write $t = \beta n$ and $g(\beta; \lambda) = (1-\mathrm{e}^{-\lambda\beta})/\beta$. Note that $\lim_{\beta\downarrow 0} g(\beta; \lambda) = \lambda > 1$, and, by Exercise 4.17, $g(\zeta_\lambda; \lambda) = 1$. Further, $\beta \mapsto g(\beta; \lambda)$ is decreasing, since

$$\frac{\partial}{\partial\beta} g(\beta; \lambda) = \mathrm{e}^{-\beta\lambda}\frac{\beta\lambda - (\mathrm{e}^{\beta\lambda} - 1)}{\beta^2} < 0, \tag{4.4.41}$$

by the fact that $\mathrm{e}^x - 1 > x$ for every $x \in \mathbb{R}$. As a result, $s^* \geq 0$ precisely when $t = \lfloor \alpha n \rfloor$ with $\alpha \leq \zeta_\lambda$.

Substitution of $s^* = \log\big(n(1-\mathrm{e}^{-\lambda t/n})/t\big)$ into (4.4.39) yields

$$\mathbb{P}_\lambda(S_t = 0) \le \mathrm{e}^{-t\left(\log g(t/n;\lambda) - 1 - g(t/n;\lambda)\right)} = \mathrm{e}^{-t I_{g(t/n;\lambda)}}. \tag{4.4.42}$$

Since $\alpha \mapsto g(\alpha;\lambda)$ is decreasing with $g(\alpha;\lambda) > 1$ for every $\alpha < \zeta_\lambda$, while $\lambda \mapsto I_\lambda$ is increasing for $\lambda > 1$, we obtain that $\alpha \mapsto I_{g(\alpha;\lambda)}$ is decreasing. Thus, for $t/n \le \alpha < \zeta_\lambda$, it follows that

$$\mathbb{P}_\lambda(S_t = 0) \le \mathrm{e}^{-t I_{g(\alpha;\lambda)}} = \mathrm{e}^{-t J(\alpha;\lambda)}. \tag{4.4.43}$$

We conclude that

$$\begin{aligned}\mathbb{P}_\lambda(k_n \le |\mathscr{C}(v)| \le \alpha n) &\le \sum_{t=k_n}^{\alpha n} \mathbb{P}_\lambda(S_t = 0) \le \sum_{t=k_n}^{\alpha n} \mathrm{e}^{-t J(\alpha;\lambda)} \\ &\le \mathrm{e}^{-k_n J(\alpha;\lambda)}/[1-\mathrm{e}^{-J(\alpha;\lambda)}].\end{aligned} \tag{4.4.44}$$

This completes the proof of Proposition 4.12. □

We finally state a consequence of Proposition 4.12 that shows that whp there is no cluster with intermediate size, i.e., size in between $k_n = K\log n$ and αn. Corollary 4.13 implies (4.4.6):

Corollary 4.13 (No intermediate clusters) *Fix $k_n = K\log n$ and $\alpha < \zeta_\lambda$ and let $\delta = KJ(\alpha;\lambda) - 1$, with $J(\alpha;\lambda)$ as in Proposition 4.12, so that $\delta > 0$ for K sufficiently large. Then, with probability at least $1 - O(n^{-\delta})$, there is no connected component with size in between k_n and αn.*

Proof By the union bound and the exchangeability of the vertices

$$\mathbb{P}_\lambda(\exists v\colon k_n \le |\mathscr{C}(v)| \le \alpha n) \le n\mathbb{P}_\lambda(k_n \le |\mathscr{C}(1)| \le \alpha n) \le Cn\mathrm{e}^{-k_n J(\alpha;\lambda)}, \tag{4.4.45}$$

where we have used Proposition 4.12 and $C = 1/[1-\mathrm{e}^{-J(\alpha;\lambda)}] < \infty$ for the last estimate. When $k_n = K\log n$, the right-hand side is $O(n^{-\delta})$ with $\delta = KJ(\alpha;\lambda) - 1 > 0$ for $K > 0$ sufficiently large. This completes the proof of Corollary 4.13. □

Exercises 4.18, 4.19 and 4.20 investigate the expected cluster size in the supercritical regime.

4.4.5 Step 4: Completion Proof of the LLNs for the Giant Component

In this section, we put the results in the previous three steps together to complete the proof of Theorem 4.8. We fix $\nu \in (\frac{1}{2}, 1)$, $\alpha \in (\zeta_\lambda/2, \zeta_\lambda)$ and take $k_n = K\log n$ with K sufficiently large. Let $\mathcal{E}_n$ be the event that

(1) $|Z_{\ge k_n} - n\zeta_\lambda| \le n^\nu$; and
(2) there does not exist a $v \in [n]$ such that $k_n \le |\mathscr{C}(v)| \le \alpha n$.

In the proof of Theorem 4.8 we use the following lemma:

Lemma 4.14 ($|\mathscr{C}_{\max}|$ equals $Z_{\ge k_n}$ whp) *Fix $\alpha \in (\zeta_\lambda/2, \zeta_\lambda)$. The event $\mathcal{E}_n$ occurs whp, i.e., with δ the minimum of $\delta(K;\alpha)$ in Corollary 4.13 and $\delta(K,\nu)$ in Corollary 4.11, $\mathbb{P}_\lambda(\mathcal{E}_n^c) = O(n^{-\delta})$, and $|\mathscr{C}_{\max}| = Z_{\ge k_n}$ on the event $\mathcal{E}_n$.*

Proof To see that $\mathcal{E}_n$ occurs whp, note that $\mathcal{E}_n^c$ equals the union of complements of the events in (1) and (2) above, so that the union bound implies that

$$\mathbb{P}_\lambda(\mathcal{E}_n^c) \leq \mathbb{P}_\lambda(|Z_{\geq k_n} - n\zeta_\lambda| > n^\nu) + \mathbb{P}_\lambda(\exists v \in [n] \text{ such that } k_n \leq |\mathscr{C}(v)| \leq \alpha n). \tag{4.4.46}$$

We bound these probabilities one by one. By Corollary 4.11, $\mathbb{P}_\lambda(|Z_{\geq k_n} - n\zeta_\lambda| > n^\nu) = O(n^{-\delta})$. By Corollary 4.13,

$$\mathbb{P}_\lambda(\exists v \in [n] \text{ such that } k_n \leq |\mathscr{C}(v)| \leq \alpha n) \leq O(n^{-\delta}). \tag{4.4.47}$$

Together, these estimates imply that $\mathbb{P}_\lambda(\mathcal{E}_n^c) = O(n^{-\delta})$ for K sufficiently large.

To prove that $|\mathscr{C}_{\max}| = Z_{\geq k_n}$ on $\mathcal{E}_n$, note that $\{|Z_{\geq k_n} - \zeta_\lambda n| \leq n^\nu\} \subseteq \{Z_{\geq k_n} \geq 1\}$. Thus, $|\mathscr{C}_{\max}| \leq Z_{\geq k_n}$ when the event $\mathcal{E}_n$ holds. In turn, $|\mathscr{C}_{\max}| < Z_{\geq k_n}$ implies that there are two connected components with size at least k_n. Furthermore, since $\mathcal{E}_n$ occurs, there are no connected components with sizes in between k_n and αn. Therefore, there must be two connected components with size at least αn, which in turn implies that $Z_{\geq k_n} \geq 2\alpha n$. When $2\alpha > \zeta_\lambda$ and n is sufficiently large, this is in contradiction with $Z_{\geq k_n} \leq \zeta_\lambda n + n^\nu$. We conclude that $|\mathscr{C}_{\max}| = Z_{\geq k_n}$ on $\mathcal{E}_n$. □

Proof of Theorem 4.8 By Lemma 4.14 and on the event $\mathcal{E}_n$, $|\mathscr{C}_{\max}| = Z_{\geq k_n}$. Therefore,

$$\mathbb{P}_\lambda\big(\big||\mathscr{C}_{\max}| - \zeta_\lambda n\big| \leq n^\nu\big) \geq \mathbb{P}_\lambda\big(\{\big||\mathscr{C}_{\max}| - \zeta_\lambda n\big| \leq n^\nu\} \cap \mathcal{E}_n\big) = \mathbb{P}_\lambda(\mathcal{E}_n) \geq 1 - O(n^{-\delta}), \tag{4.4.48}$$

since $|Z_{\geq k_n} - n\zeta_\lambda| \leq n^\nu$ on the event $\mathcal{E}_n$. This completes the proof of the law of large numbers of the giant component in Theorem 4.8. □

4.4.6 The Discrete Duality Principle

Using the law of large numbers for the giant component in Theorem 4.8, we can construct a duality principle for Erdős–Rényi random graphs similar to the duality principle for branching processes:

Theorem 4.15 (Discrete duality principle) *Let $\lambda > 1 > \mu_\lambda$ be conjugates as in* (3.6.6). *The vector of connected components in the graph* $\mathrm{ER}_n(\lambda/n)$ *with the giant component removed is close in law to the random graph* $\mathrm{ER}_m(\mu_\lambda/m)$, *where the variable $m = \lceil n\eta_\lambda \rceil$ is the asymptotic number of vertices outside the giant component.*

Theorem 4.15 can be understood from Theorem 4.8, which implies that the giant component has size $n - m = \zeta_\lambda n(1 + o(1))$. In the statement of Theorem 4.15 we make use of the informal notion 'close in law'. This notion can be made precise as follows. Let $\mathrm{ER}_n'(\lambda/n)$ be $\mathrm{ER}_n(\lambda/n)$ with the giant component removed. We write $\mathbb{P}_{n,\lambda}'$ for the law of $\mathrm{ER}_n'(\lambda/n)$, and we let $\mathbb{P}_{m,\mu}$ denote the law of $\mathrm{ER}_m(\mu/m)$. Let $\mathcal{E}$ be an event which is determined by connected components (such as 'the largest connected component exceeds k'). If $\lim_{m\to\infty} \mathbb{P}_{m,\mu_\lambda}(\mathcal{E})$ exists, then

$$\lim_{n\to\infty} \mathbb{P}_{n,\lambda}'(\mathcal{E}) = \lim_{m\to\infty} \mathbb{P}_{m,\mu_\lambda}(\mathcal{E}). \tag{4.4.49}$$

We sketch the proof of Theorem 4.15 only. First of all, all the edges in the complement of the giant component in $\mathrm{ER}_n(p)$ are (conditionally on not being in the giant component) independent. Furthermore, the conditional probability that an edge st is occupied in $\mathrm{ER}_n(p)$ with the giant component removed is, conditionally on $|\mathscr{C}_{\max}| = n - m$, equal to

$$\frac{\lambda}{n} = \frac{\lambda}{m}\frac{m}{n}. \tag{4.4.50}$$

Now, $m \approx \eta_\lambda n$, so that the conditional probability that an edge st is occupied in $\mathrm{ER}_n(p)$ with the giant component removed, conditionally on $|\mathscr{C}_{\max}| \approx \zeta_\lambda n$, is equal to

$$\frac{\lambda}{n} \approx \frac{\lambda\eta_\lambda}{m} = \frac{\mu_\lambda}{m}, \tag{4.4.51}$$

where we have used (3.6.2) and (3.6.4), which implies that $\lambda\eta_\lambda = \mu_\lambda$. Therefore, the conditional probability that an edge st is occupied in $\mathrm{ER}_n(p)$ with the giant component removed, conditionally on $|\mathscr{C}_{\max}| \approx \zeta_\lambda n$, is close to μ_λ/m, and different edges are independent.

We omit further details and refer the reader to Janson et al. (2000, Section 5.6) or Bollobás (2001, Section 6.3) for a complete proof of Theorem 4.15 and a more substantial discussion of the duality principle. The argument in Alon and Spencer (2000, Section 10.5) is similar to the above informal version of the discrete duality principle.

Exercise 4.21 investigates the size of the second largest supercritical cluster using duality.

4.5 CLT for the Giant Component

In this section, we prove a central limit theorem (CLT) for the giant component in the supercritical regime, extending the law of large numbers for the giant component in Theorem 4.8. The main result is as follows:

Theorem 4.16 (Central limit theorem for giant component) *Fix $\lambda > 1$. Then,*

$$\frac{|\mathscr{C}_{\max}| - \zeta_\lambda n}{\sqrt{n}} \xrightarrow{d} Z, \tag{4.5.1}$$

where Z is a normal random variable with mean 0 and variance

$$\sigma_\lambda^2 = \frac{\zeta_\lambda(1-\zeta_\lambda)}{(1-\lambda+\lambda\zeta_\lambda)^2}. \tag{4.5.2}$$

As in the proof of Theorem 4.8, we make use of the exploration of connected components to prove Theorem 4.16. In the proof, we make essential use of Theorem 4.8, as well as from the bounds on the second largest component in the supercritical regime derived in its proof.

We start by sketching the main ideas in the proof of Theorem 4.16. Fix $k = k_n$, which will be chosen later on. We explore the union of the connected components of the vertices $[k] = \{1, \ldots, k\}$. We use Theorem 4.8 to show that, for an appropriate $k \to \infty$ and whp as $n \to \infty$, this union contains the largest connected component $\mathscr{C}_{\max}$, and it cannot be larger than $|\mathscr{C}_{\max}| + kb_n$, where $b_n = K\log n$ is an upper bound on the second largest component in the supercritical regime. Taking $k = o(n^\nu)$ with $\nu < \frac{1}{2}$, this union of components is equal to $|\mathscr{C}_{\max}| + o(\sqrt{n})$. As a result, a CLT for the union of components implies one for $|\mathscr{C}_{\max}|$. We now describe the exploration of the union of the connected components of $[k]$.

Let $S_0 = k$ and, for $t \geq 1$,

$$S_t = S_{t-1} + X_t - 1, \tag{4.5.3}$$

where

$$X_t \sim \mathsf{Bin}\big(n - S_{t-1} - (t-1), p\big). \tag{4.5.4}$$

Equations (4.5.3) and (4.5.4) are similar to the ones in (4.1.3) and (4.1.4). We next derive the distribution of S_t in a similar way as in Proposition 4.6:

Proposition 4.17 (The law of S_t revisited) *For all $t, k \in [n]$,*

$$S_t + (t-k) \sim \mathsf{Bin}(n-k, 1-(1-p)^t). \tag{4.5.5}$$

Moreover, for all $l, t \in [n]$ satisfying $l \geq t$, and conditionally on S_t,

$$S_l + (l-t) - S_t \sim \mathsf{Bin}\big(n - t - S_t, 1 - (1-p)^{l-t}\big). \tag{4.5.6}$$

For $k = 1$, the equality in distribution in (4.5.5) in Proposition 4.17 reduces to Proposition 4.6.

Proof We adapt the proof of Proposition 4.6. For $t \geq 1$, let N_t represent the number of unexplored vertices at time t, i.e., as in (4.3.14),

$$N_t = n - t - S_t. \tag{4.5.7}$$

To prove (4.5.5), it is more convenient to show the equivalent statement that, for all $t \geq 1$,

$$N_t \sim \mathsf{Bin}\big(n-k, (1-p)^t\big). \tag{4.5.8}$$

To see this informally, we note that each of the vertices $[n] \setminus [k]$ has, independently of all other vertices, probability $(1-p)^t$ to stay neutral in the first t explorations. Formally, conditionally on S_{t-1}, $X_t \sim \mathsf{Bin}(n - S_{t-1} - (t-1), p) = \mathsf{Bin}(N_{t-1}, p)$ by (4.5.4). Thus, noting that $N_0 = n - k$ and

$$\begin{aligned} N_t &= n - t - S_t = n - t - S_{t-1} - X_t + 1 \\ &= n - (t-1) - S_{t-1} - X_t \\ &= N_{t-1} - X_t \sim \mathsf{Bin}(N_{t-1}, 1-p), \end{aligned} \tag{4.5.9}$$

the conclusion follows by recursion on $t \geq 1$ and Exercise 4.15. We note that (4.5.9) also implies that for any $l \geq t$,

$$N_l \sim \mathsf{Bin}(N_t, (1-p)^{l-t}). \tag{4.5.10}$$

Substituting $N_t = n - t - S_t$, this implies that

$$\begin{aligned} n - l - S_l &\sim \mathsf{Bin}\big(n - t - S_t, (1-p)^{l-t}\big) \\ &\sim n - t - S_t - \mathsf{Bin}\big(n - t - S_t, 1 - (1-p)^{l-t}\big), \end{aligned} \tag{4.5.11}$$

which, in turn, is equivalent to the statement that, for all $l \geq t$ and conditionally on S_t,

$$S_l + (l-t) - S_t \sim \mathsf{Bin}\big(n - t - S_t, 1 - (1-p)^{l-t}\big). \tag{4.5.12}$$

□

We now state a corollary of Proposition 4.17 which states that $S_{\lfloor nt \rfloor}$ satisfies a central limit theorem. In its statement, we make use of the asymptotic mean

$$\mu_t = 1 - t - \mathrm{e}^{-\lambda t} \tag{4.5.13}$$

and asymptotic variance

$$v_t = \mathrm{e}^{-\lambda t}(1 - \mathrm{e}^{-\lambda t}). \tag{4.5.14}$$

The central limit theorem for $S_{\lfloor nt \rfloor}$ reads as follows:

Corollary 4.18 (CLT for $S_{\lfloor nt \rfloor}$) *Fix $k = k_n = o(\sqrt{n})$, and let $S_0 = k$. Then, for every $t \in (0, 1)$, the random variable $\frac{S_{\lfloor nt \rfloor} - n\mu_t}{\sqrt{nv_t}}$ converges in distribution to a standard normal random variable.*

Proof The statement follows immediately from the central limit theorem for the binomial distribution when we can show that

$$\mathbb{E}[S_{\lfloor nt \rfloor}] = n\mu_t + o(\sqrt{n}), \qquad \mathrm{Var}(S_{\lfloor nt \rfloor}) = nv_t + o(n). \tag{4.5.15}$$

Indeed, by the central limit theorem for the binomial distribution,

$$\frac{S_{\lfloor nt \rfloor} - \mathbb{E}[S_{\lfloor nt \rfloor}]}{\sqrt{\mathrm{Var}(S_{\lfloor nt \rfloor})}} \xrightarrow{d} Z, \tag{4.5.16}$$

where Z is a standard normal random variable (see also Exercise 4.22). Write

$$\frac{S_{\lfloor nt \rfloor} - n\mu_t}{\sqrt{nv_t}} = \sqrt{\frac{\mathrm{Var}(S_{\lfloor nt \rfloor})}{nv_t}} \frac{S_{\lfloor nt \rfloor} - \mathbb{E}[S_{\lfloor nt \rfloor}]}{\sqrt{\mathrm{Var}(S_{\lfloor nt \rfloor})}} + \frac{\mathbb{E}[S_{\lfloor nt \rfloor}] - n\mu_t}{\sqrt{nv_t}}. \tag{4.5.17}$$

The last term converges to zero by (4.5.15), and the factor $\sqrt{\frac{\mathrm{Var}(S_{\lfloor nt \rfloor})}{nv_t}}$ converges to one. Therefore, (4.5.15) implies the central limit theorem.

To prove the asymptotics of the mean in (4.5.15), we use (4.5.5) to obtain that

$$\mathbb{E}[S_{\lfloor nt \rfloor}] = (n-k)\Big(1 - (1 - \frac{\lambda}{n})^{\lfloor nt \rfloor}\Big) - \big(\lfloor nt \rfloor - k\big) = n\mu_t + o(\sqrt{n}), \tag{4.5.18}$$

as long as $k = o(\sqrt{n})$. To prove the asymptotics of the variance in (4.5.15), we note that

$$\mathrm{Var}(S_{\lfloor nt \rfloor}) = (n-k)(1 - \frac{\lambda}{n})^{\lfloor nt \rfloor}\big(1 - (1 - \frac{\lambda}{n})^{\lfloor nt \rfloor}\big) = nv_t + o(n), \tag{4.5.19}$$

as long as $k = o(n)$. This completes the proof of Corollary 4.18. □

With Corollary 4.18 at hand, we are ready to complete the proof of Theorem 4.16:

Proof of Theorem 4.16 Let $|\mathscr{C}([k])|$ be the size of the union of the components of the vertices in $[k]$. Recall the recursion for $(S_t)_{t \geq 0}$ in (4.5.3)–(4.5.4). Then,

$$|\mathscr{C}([k])| = \min\{m \colon S_m = 0\}. \tag{4.5.20}$$

Let $k = k_n \to \infty$. We prove below that $|\mathscr{C}([k])|$ satisfies a CLT with asymptotic mean $n\zeta_\lambda$ and asymptotic variance $n\sigma_\lambda^2$.

To do this, we start by noting that by Corollary 4.13 and Theorem 4.8, whp, the second largest cluster has size at most $K \log n$. Indeed, by Corollary 4.13 and whp, there do not exist any clusters with size in between $K \log n$ and αn for any $\alpha < \zeta_\lambda$. By Theorem 4.8, $|\mathscr{C}_{\max}| = \zeta_\lambda n(1 + o_{\mathbb{P}}(1))$. Finally, by Lemma 4.14, whp $|\mathscr{C}_{\max}| = Z_{\geq k_n}$ and by Corollary 4.11, $Z_{\geq k_n} = \zeta_\lambda n(1 + o_{\mathbb{P}}(1))$. As a result, there is a *unique* connected component of size larger than $K \log n$ which is $\mathscr{C}_{\max}$.

We conclude that, whp,

$$|\mathscr{C}_{\max}| \leq |\mathscr{C}([k])| \leq |\mathscr{C}_{\max}| + (k-1)K \log n, \tag{4.5.21}$$

so that a central limit theorem for $|\mathscr{C}_{\max}|$ follows from that for $|\mathscr{C}([k])|$ for any $k = k_n \to \infty$ with $k_n = o(\sqrt{n})$.

The central limit theorem for $|\mathscr{C}([k])|$ is proved by upper and lower bounds on the probabilities

$$\mathbb{P}_\lambda\Big(\frac{|\mathscr{C}([k])| - \zeta_\lambda n}{\sqrt{n}} > x\Big). \tag{4.5.22}$$

Upper Bound in CLT $|\mathscr{C}([k])|$

For the upper bound, we use that (4.5.20) implies that, for every ℓ,

$$\mathbb{P}_\lambda(|\mathscr{C}([k])| > \ell) = \mathbb{P}_\lambda(S_m > 0 \ \forall m \leq \ell). \tag{4.5.23}$$

Applying (4.5.23) to $\ell = m_x = \lfloor n\zeta_\lambda + x\sqrt{n} \rfloor$, we obtain

$$\mathbb{P}_\lambda\Big(\frac{|\mathscr{C}([k])| - \zeta_\lambda n}{\sqrt{n}} > x\Big) = \mathbb{P}_\lambda(S_m > 0 \ \forall m \leq m_x) \leq \mathbb{P}_\lambda(S_{m_x} > 0). \tag{4.5.24}$$

To analyse $\mathbb{P}_\lambda(S_{m_x} > 0)$, we rely on Corollary 4.18, for which we start by carefully analysing the mean and variance of S_{m_x}. We use (4.5.13), (4.5.15) and $\mu_{\zeta_\lambda} = 0$, and write μ_t' for the derivative of $t \mapsto \mu_t$, to see that

$$\begin{aligned}\mathbb{E}[S_{m_x}] &= n\mu_{\zeta_\lambda} + \sqrt{n}x\mu'_{\zeta_\lambda} + o(\sqrt{n}) = \sqrt{n}x(\lambda e^{-\lambda\zeta_\lambda} - 1) + o(\sqrt{n}) \\ &= \sqrt{n}x(\lambda e^{-\lambda\zeta_\lambda} - 1) + o(\sqrt{n}),\end{aligned} \tag{4.5.25}$$

where we note that $\lambda e^{-\lambda\zeta_\lambda} - 1 < 0$ for $\lambda > 1$, see Exercise 4.23.

The variance of S_{m_x} is, by (4.5.14) and (4.5.15),

$$\mathrm{Var}(S_{m_x}) = nv_{\zeta_\lambda} + o(n), \tag{4.5.26}$$

where $v_{\zeta_\lambda} > 0$. Thus,

$$\mathbb{P}_\lambda(S_{m_x} > 0) = \mathbb{P}_\lambda\Big(\frac{S_{m_x} - \mathbb{E}[S_{m_x}]}{\sqrt{\mathrm{Var}(S_{m_x})}} > \frac{x(1 - \lambda e^{-\lambda\zeta_\lambda})}{\sqrt{v_{\zeta_\lambda}}}\Big) + o(1). \tag{4.5.27}$$

By Corollary 4.18, the right-hand side converges to

$$\mathbb{P}\Big(Z > \frac{x(1 - \lambda e^{-\lambda\zeta_\lambda})}{\sqrt{v_{\zeta_\lambda}}}\Big) = \mathbb{P}(Z' > x), \tag{4.5.28}$$

where Z' has a normal distribution with mean 0 and variance $v_{\zeta_\lambda}(1 - \lambda e^{-\lambda\zeta_\lambda})^{-2}$. We finally note that $1 - \zeta_\lambda = e^{-\lambda\zeta_\lambda}$ by (3.6.2), so that

$$v_{\zeta_\lambda} = e^{-\lambda\zeta_\lambda}(1 - e^{-\lambda\zeta_\lambda}) = \zeta_\lambda(1 - \zeta_\lambda). \tag{4.5.29}$$

By (4.5.29), the variance of the normal distribution appearing in the lower bound can be rewritten as

$$\frac{v_{\zeta_\lambda}}{(1-\lambda \mathrm{e}^{-\lambda\zeta_\lambda})^2} = \frac{\zeta_\lambda(1-\zeta_\lambda)}{(1-\lambda+\lambda\zeta_\lambda)^2} = \sigma_\lambda^2, \tag{4.5.30}$$

where we recall (4.5.2). By (4.5.24), this completes the upper bound. □

Lower Bound in CLT $|\mathscr{C}([k])|$

The lower bound is slightly more involved. We again use the fact that

$$\mathbb{P}_\lambda\big(|\mathscr{C}([k])| - \zeta_\lambda n > x\sqrt{n}\big) = \mathbb{P}_\lambda(S_m > 0 \; \forall m \le m_x), \tag{4.5.31}$$

where we recall that $m_x = \lfloor n\zeta_\lambda + x\sqrt{n}\rfloor$. Then, for any $\varepsilon > 0$, we bound from below

$$\begin{aligned}\mathbb{P}_\lambda(S_m > 0 \; \forall m \le m_x) &\ge \mathbb{P}_\lambda(S_m > 0 \; \forall m \le m_x, S_{m_x} > \varepsilon\sqrt{n}) \\ &= \mathbb{P}_\lambda(S_{m_x} > \varepsilon\sqrt{n}) - \mathbb{P}_\lambda(S_{m_x} > \varepsilon\sqrt{n}, \exists m < m_x \colon S_m = 0).\end{aligned} \tag{4.5.32}$$

The first term can be handled in a similar way as for the upper bound. Indeed, repeating the steps in the upper bound, we obtain that, for every $\varepsilon > 0$,

$$\mathbb{P}_\lambda(S_{m_x} > \varepsilon\sqrt{n}) = \mathbb{P}\Big(Z > \frac{x(1-\lambda\mathrm{e}^{-\lambda\zeta_\lambda})+\varepsilon}{\sqrt{v_{\zeta_\lambda}}}\Big) + o(1). \tag{4.5.33}$$

The quantity in (4.5.33) converges to $\mathbb{P}(Z' > x + \varepsilon/(1-\lambda\mathrm{e}^{-\lambda\zeta_\lambda}))$, where Z' has a normal distribution with mean 0 and variance σ_λ^2. When $\varepsilon \downarrow 0$, in turn, this converges to $\mathbb{P}(Z' > x)$, as required.

We conclude that to prove the lower bound in the CLT for $|\mathscr{C}([k])|$, it suffices to prove that, for every $\varepsilon > 0$ fixed and as $n \to \infty$,

$$\mathbb{P}_\lambda(S_{m_x} > \varepsilon\sqrt{n}, \exists m < m_x \colon S_m = 0) = o(1). \tag{4.5.34}$$

To bound the probability in (4.5.34), we first use Boole's inequality to get

$$\mathbb{P}_\lambda(S_{m_x} > \varepsilon\sqrt{n}, \exists m < m_x \colon S_m = 0) \le \sum_{m=k}^{m_x-1} \mathbb{P}_\lambda(S_m = 0, S_{m_x} > \varepsilon\sqrt{n}), \tag{4.5.35}$$

where we use that $m \ge k$, since $S_0 = k$ and $S_t - S_{t-1} \ge -1$. For $m \le \alpha n$ with $\alpha < \zeta_\lambda$, by Exercise 4.24,

$$\mathbb{P}_\lambda(S_m = 0) \le \mathrm{e}^{-mJ(m/n,\lambda)}. \tag{4.5.36}$$

We conclude that

$$\begin{aligned}&\mathbb{P}_\lambda(S_{m_x} > \varepsilon\sqrt{n}, \exists m < m_x \colon S_m = 0) \\ &\quad\le \sum_{m=\lfloor\alpha n\rfloor+1}^{m_x-1} \mathbb{P}_\lambda(S_m = 0, S_{m_x} > \varepsilon\sqrt{n}) + \sum_{m=k}^{\lfloor\alpha n\rfloor} \mathrm{e}^{-mJ(m/n,\lambda)} \\ &\quad= \sum_{m=\lfloor\alpha n\rfloor+1}^{m_x-1} \mathbb{P}_\lambda(S_m = 0, S_{m_x} > \varepsilon\sqrt{n}) + o(1),\end{aligned} \tag{4.5.37}$$

since $k = k_n \to \infty$.

We continue by proving a similar bound for $m > \alpha n$, where $\alpha < \zeta_\lambda$ can be chosen arbitrarily close to ζ_λ. Here we make use of the fact that, for m close to $\zeta_\lambda n$, $\mathbb{E}_\lambda[X_m] < 1$, so that $m \mapsto S_m$, for $m \geq \alpha n$ is close to a random walk with negative drift. As a result, the probability that $S_m = 0$, yet $S_{m_x} > \varepsilon\sqrt{n}$ is small.

We now present the details of this argument. We bound, with $X \sim \mathsf{Bin}\big(n - m, 1 - (1 - p)^{m_x - m}\big)$,

$$\begin{aligned}\mathbb{P}_\lambda\big(S_m = 0, S_{m_x} > \varepsilon\sqrt{n}\big) &= \mathbb{P}_\lambda\big(S_{m_x} > \varepsilon\sqrt{n} \mid S_m = 0\big)\mathbb{P}_\lambda(S_m = 0) \\ &= \mathbb{P}_\lambda\big(X > (m_x - m) + \varepsilon\sqrt{n}\big)\mathbb{P}_\lambda(S_m = 0),\end{aligned} \tag{4.5.38}$$

since, by (4.5.6) in Proposition 4.17 and conditionally on $S_m = 0$,

$$S_{m_x} + (m_x - m) \sim \mathsf{Bin}\big(n - m, 1 - (1 - p)^{m_x - m}\big). \tag{4.5.39}$$

We fix $\alpha = \zeta_\lambda - \varepsilon$ for some $\varepsilon > 0$ sufficiently small. We use that $1 - (1 - a)^b \leq ab$ for every a, b with $0 < a < 1, b \geq 1$ to arrive at

$$1 - (1 - p)^{m_x - m} = 1 - \big(1 - \frac{\lambda}{n}\big)^{m_x - m} \leq \frac{\lambda(m_x - m)}{n}. \tag{4.5.40}$$

As a result, using that $n - m \leq n(1 - \zeta_\lambda + \varepsilon)$ since $m \geq \alpha n$ with $\alpha = \zeta_\lambda - \varepsilon$ and with $p = \lambda/n$,

$$\mathbb{E}_\lambda[X] = (n - m)[1 - (1 - p)^{m_x - m}] \leq (m_x - m)\lambda(1 - \zeta_\lambda + \varepsilon). \tag{4.5.41}$$

Since $\lambda > 1$, we can use that $\lambda(1 - \zeta_\lambda) = \lambda \mathrm{e}^{-\lambda\zeta_\lambda} < 1$ by Exercise 4.23, so that, taking $\varepsilon > 0$ so small that $\lambda(1 - \zeta_\lambda + \varepsilon) < 1 - \varepsilon$,

$$\mathbb{E}_\lambda[X] \leq (1 - \varepsilon)(m_x - m). \tag{4.5.42}$$

Therefore,

$$\begin{aligned}\mathbb{P}_\lambda\big(S_m = 0, S_{m_x} > \varepsilon\sqrt{n}\big) &\leq \mathbb{P}_\lambda\big(X > (m_x - m) + \varepsilon\sqrt{n}\big) \\ &\leq \mathbb{P}_\lambda\big(X - \mathbb{E}_\lambda[X] > \varepsilon\big((m_x - m) + \sqrt{n}\big)\big).\end{aligned} \tag{4.5.43}$$

By Theorem 2.21, with $t = \varepsilon\big((m_x - m) + \sqrt{n}\big)$ and using (4.5.42), we obtain

$$\mathbb{P}_\lambda\big(S_m = 0, S_{m_x} > \varepsilon\sqrt{n}\big) \leq \exp\Big(-\frac{t^2}{2\big((1 - \varepsilon)(m_x - m) + t/3\big)}\Big). \tag{4.5.44}$$

We split, depending on whether $m_x - m \geq \varepsilon\sqrt{n}$ or $m_x - m \leq \varepsilon\sqrt{n}$. For $m_x - m \geq \varepsilon\sqrt{n}$ and since $t \geq \varepsilon(m_x - m)$ and $(1 - \varepsilon) + \varepsilon/3 \leq 1$,

$$\begin{aligned}\mathbb{P}_\lambda\big(S_m = 0, S_{m_x} > \varepsilon\sqrt{n}\big) &\leq \exp\Big(-\frac{t}{2\big((1 - \varepsilon)(m_x - m)/t + 1/3\big)}\Big) \\ &\leq \exp\Big(-\frac{t}{2\big((1 - \varepsilon)/\varepsilon + 1/3\big)}\Big) \leq \mathrm{e}^{-\varepsilon t/2} \\ &\leq \exp\Big(-\varepsilon^2(m_x - m)/2\Big) \leq \exp\big(-\varepsilon^2\sqrt{n}/2\big).\end{aligned} \tag{4.5.45}$$

For $m_x - m \leq \varepsilon\sqrt{n}$, since $t \geq \varepsilon\sqrt{n}$,

$$\mathbb{P}_\lambda\big(S_m = 0, S_{m_x} > \varepsilon\sqrt{n}\big) \leq \exp\Big(-\frac{t^2}{2\big((1 - \varepsilon)\varepsilon\sqrt{n} + t/3\big)}\Big) \tag{4.5.46}$$

$$\leq \exp\Big(-\frac{t^2}{2\big((1-\varepsilon)t+t/3\big)}\Big) \leq \exp\big(-3t/8\big)$$
$$\leq \exp\big(-3\varepsilon\sqrt{n}/8\big).$$

We conclude from (4.5.37) and (4.5.45)–(4.5.46) that

$$\begin{aligned}
&\mathbb{P}_\lambda(S_{m_x} > \varepsilon\sqrt{n}, \exists m < m_x\colon S_m = 0) \qquad (4.5.47)\\
&\leq \sum_{m=\lfloor \alpha n\rfloor+1}^{m_x-1} \mathbb{P}_\lambda(S_m = 0, S_{m_x} > \varepsilon\sqrt{n}) + o(1)\\
&\leq n[\exp\Big(-\varepsilon^2\sqrt{n}/2\Big) + \exp\big(-3\varepsilon\sqrt{n}/4\big)] + o(1) = o(1).
\end{aligned}$$

The bounds (4.5.36), (4.5.45) and (4.5.46) complete the proof of (4.5.34), and thus that of the lower bound in the CLT for $|\mathscr{C}([k])|$. □

4.6 Notes and Discussion

Notes on Section 4.1

The informal discussion of the exploration of connected components in the Erdős–Rényi random graph is inspired by the *Probabilistic Methods* book by Alon and Spencer, see in particular (Alon and Spencer, 2000, Section 10.5).

There are several related definitions of the Erdős–Rényi random graph. Many of the classical results are proved for $\mathrm{ER}_n^{(e)}(m)$, which is the random graph on the vertices $[n]$ obtained by adding m edges uniformly at random and without replacement. Since the number of edges in the Erdős–Rényi random graph has a binomial distribution with parameters $n(n-1)/2$ and p, we should think of M corresponding roughly to $pn(n-1)/2$. Also, writing $\mathbb{P}_m$ for the distribution of $\mathrm{ER}_n^{(e)}(m)$, $\mathbb{P}_\lambda$ and $\mathbb{P}_m$ are related by

$$\mathbb{P}_\lambda(\mathcal{E}) = \sum_{M=1}^{n(n-1)/2} \mathbb{P}_m(\mathcal{E})\mathbb{P}\big(\mathsf{Bin}(n(n-1)/2, p) = M\big), \tag{4.6.1}$$

where $\mathcal{E}$ is any event. This allows one to deduce results for $\mathrm{ER}_n^{(e)}(m)$ from the ones for $\mathrm{ER}_n(p)$ and vice versa. The model $\mathrm{ER}_n^{(e)}(m)$ was first studied in the seminal work by Erdős and Rényi (1959), the model $\mathrm{ER}_n(p)$ was introduced by Gilbert (1959) and a model with possibly multiple edges between vertices in Austin et al. (1959). An early reference with a correct, yet not entirely rigorous, analysis of the phase transition is Solomonoff and Rapoport (1951).

The random graph $\mathrm{ER}_n^{(e)}(m)$ has the advantage that we can think of the graph as evolving as a process, by adding the edges one at a time, which also allows us to investigate *dynamical properties*, such as when the first cycle appears. This is also possible for $\mathrm{ER}_n(p)$ using the coupling in Section 4.1.1, but is slightly less appealing.

We refer to the books Alon and Spencer (2000); Bollobás (2001); Janson et al. (2000) for more detailed references of the early literature on random graphs.

Notes on Section 4.2

The idea of exploration processes to investigate cluster sizes is introduced by Martin-Löf (1986) and by Karp (1990). See also (Alon and Spencer, 2000, Section 10.5), where these ideas were formulated slightly more intuitively than we do.

Notes on Section 4.3

The strategy in the proof of Theorems 4.4 and 4.5 is close in spirit to the proof by Alon and Spencer (2000), with ingredients taken from Borgs et al. (2005a), which, in turn, was inspired by Borgs et al. (1999, 2001). In particular, the use of the random variable $Z_{\geq k}$ first appeared in these references. The random variable $Z_{\geq k}$ also plays a crucial role in the analysis of $|\mathscr{C}_{\max}|$ both when $\lambda > 1$ and when $\lambda = 1$. Further, Proposition 4.6 can also be found as (Alon and Spencer, 2000, Claim 3 in Section 10.5). Exercise 4.25 investigates the subcritical behavior of $\mathrm{ER}_n^{(e)}(m)$.

Notes on Section 4.4

Again, the strategy in the proof of Theorem 4.8 is close in spirit to the proof by Alon and Spencer (2000), where we provide more mathematical details. Exercise 4.25 investigates the supercritical behavior of $\mathrm{ER}_n^{(e)}(m)$. In particular, Exercises 4.25 and 4.26 show that $\mathrm{ER}_n^{(e)}(m)$ has a phase transition when $M = n\lambda/2$ at $\lambda = 1$.

Notes on Section 4.5

The central limit theorem for the largest supercritical cluster was proved by Martin-Löf (1998), by Pittel (1990) and by Barraez et al. (2000). In the proof by Pittel (1990), the result follows as a corollary of the main result, involving central limit theorems for various random graph quantities, such as the number of tree components of various sizes. Martin-Löf (1998) studies the giant component in the context of epidemics. His proof makes clever use of a connection to asymptotic stochastic differential equations, and is reproduced by Durrett (2007). Since we do not assume familiarity with stochastic differential equations, we have produced an independent proof that only relies on elementary techniques. A local central limit theorem for the giant component is proved by Bender, Canfield and McKay (1990); see also Behrisch et al. (2014) for related results.

Large deviations for the giant component are performed in a beautiful paper by O'Connell (1998). Remarkably, the large deviation rate function is non-convex. This is related to the fact that when the maximal cluster is quite small, then removing it might leave a setting that is *still* supercritical. For a deviation sufficiently close to the typical value this is ruled out by the duality principle (recall Theorem 4.15).

4.7 Exercises for Chapter 4

Exercise 4.1 (Number of edges in $\mathrm{ER}_n(p)$) What is the distribution of the number of edges in the Erdős–Rényi random graph $\mathrm{ER}_n(p)$?

Exercise 4.2 (CLT for number of edges in $\mathrm{ER}_n(p)$) Prove that the number of edges in $\mathrm{ER}_n(p)$ satisfies a central limit theorem and compute its asymptotic mean and variance.

Exercise 4.3 (Verification of cluster size description) Collect evidence that T in (4.1.5) has the same distribution as $|\mathscr{C}(v)|$ by computing the probabilities of the events that $\{|\mathscr{C}(v)| = 1\}$, $\{|\mathscr{C}(v)| = 2\}$ and $\{|\mathscr{C}(v)| = 3\}$ directly, and the probabilities of $\{T = 1\}$, $\{T = 2\}$ and $\{T = 3\}$ by using (4.1.4), (4.1.3) and (4.1.5).

Exercise 4.4 (CLT for number of edges in $\mathrm{ER}_n(\lambda/n)$) Prove that the number of edges in $\mathrm{ER}_n(\lambda/n)$ satisfies a central limit theorem with asymptotic mean and variance equal to $\lambda n/2$.

Exercise 4.5 (Mean number of triangles in $\mathrm{ER}_n(\lambda/n)$) We say that the distinct vertices (i, j, k) form an occupied triangle when the edges ij, jk and ki are all occupied. Note that (i, j, k) is the same triangle as (i, k, j) and as any other permutation. Compute the expected number of occupied triangles in $\mathrm{ER}_n(\lambda/n)$.

Exercise 4.6 (Mean number of 4-cycles in $\mathrm{ER}_n(\lambda/n)$) We say that the distinct vertices (i, j, k, l) form an occupied 4-cycle when the edges ij, jk, kl and li are all occupied. Note that the 4-cycles (i, j, k, l) and (i, k, j, l) are different. Compute the expected number of occupied 4-cycles in $\mathrm{ER}_n(\lambda/n)$.

Exercise 4.7 (Poisson limits for number of triangles and 4-cycles in $\mathrm{ER}_n(\lambda/n)$) Show that the number of occupied triangles in an Erdős–Rényi random graph on n vertices with edge probability $p = \lambda/n$ has an asymptotic Poisson distribution. Do the same for the number of occupied 4-cycles. Hint: use the method of moments in Theorem 2.4.

Exercise 4.8 (Clustering of $\mathrm{ER}_n(\lambda/n)$) Recall from (1.5.4) in Section 1.5 that the clustering coefficient of a random graph $G = (V, E)$ with V the vertex set and E the edge set, is defined to be

$$\mathrm{CC}_G = \frac{\mathbb{E}[\Delta_G]}{\mathbb{E}[W_G]}, \tag{4.7.1}$$

where

$$\Delta_G = \sum_{i,j,k\in V} \mathbb{1}_{\{ij,ik,jk \text{ occupied}\}}, \qquad W_G = \sum_{i,j,k\in V} \mathbb{1}_{\{ij,ik \text{ occupied}\}}. \tag{4.7.2}$$

Thus (since we are not restricting to $i < j < k$ in Δ_G and to $i < k$ in W_G), Δ_G is six times the number of triangles in G, W_G is two times the number of wedges in G and CC_G is the ratio of the number of expected closed triangles to the expected number of wedges. Compute CC_G for $\mathrm{ER}_n(\lambda/n)$.

Exercise 4.9 (Asymptotic clustering of $\mathrm{ER}_n(\lambda/n)$) Take $G = \mathrm{ER}_n(\lambda/n)$. Show that $W_G/n \xrightarrow{\mathbb{P}} \lambda^2$ by using the second moment method. Use Exercise 4.7 to conclude that

$$\frac{n\Delta_G}{W_G} \xrightarrow{d} \frac{6}{\lambda^2} Y, \tag{4.7.3}$$

where $Y \sim \mathsf{Poi}(\lambda^3/6)$.

Exercise 4.10 ($|\mathscr{C}_{\max}|$ increasing) Show that $|\mathscr{C}_{\max}|$ is an increasing random variable.

Exercise 4.11 Is the event $\{v \in \mathscr{C}_{\max}\}$ an increasing event?

Exercise 4.12 (Upper bound for mean cluster size) Show that, for $\lambda < 1$, $\mathbb{E}_\lambda[|\mathscr{C}(v)|] \leq 1/(1-\lambda)$.

Exercise 4.13 (Convergence in probability of largest subcritical cluster) Prove that Theorems 4.4 and 4.5 imply that $|\mathscr{C}_{\max}|/\log n \xrightarrow{\mathbb{P}} 1/I_\lambda$.

Exercise 4.14 (Relation $|\mathscr{C}_{\max}|$ and $Z_{\geq k}$) Prove (4.3.5) and conclude that $\{|\mathscr{C}_{\max}| \geq k\} = \{Z_{\geq k} \geq k\}$.

Exercise 4.15 (A binomial number of binomial trials) Show that if $N \sim \mathsf{Bin}(n, p)$ and, conditionally on N, $M \sim \mathsf{Bin}(N, q)$, then $M \sim \mathsf{Bin}(n, pq)$. Use this to complete the proof that $N_t \sim \mathsf{Bin}(n-1, (1-p)^t)$.

Exercise 4.16 (Connectivity with given expected cluster size) Show that

$$\mathbb{P}_\lambda\big(1 \not\longleftrightarrow 2 \,\big|\, |\mathscr{C}(1)| = l\big) = 1 - \frac{l-1}{n-1}. \tag{4.7.4}$$

Exercise 4.17 (Uniqueness solution of Poisson survival probability equation) Fix $\lambda > 1$. Prove that the unique solution in $(0, 1]$ to the equation $1 - \mathrm{e}^{-\lambda\alpha} = \alpha$ is $\alpha = \zeta_\lambda$, where ζ_λ is the survival probability of a Poisson branching process with parameter λ.

Exercise 4.18 (Connectivity and expected cluster size) Prove that the expected cluster size of a given vertex

$$\chi(\lambda) = \mathbb{E}_\lambda[|\mathscr{C}(1)|], \tag{4.7.5}$$

satisfies

$$\chi(\lambda) = 1 + (n-1)\mathbb{P}_\lambda(1 \longleftrightarrow 2). \tag{4.7.6}$$

Exercise 4.19 (Connectivity function) Prove that (4.4.1) and Corollary 4.13 imply that, for $\lambda > 1$,

$$\mathbb{P}_\lambda(1 \longleftrightarrow 2) = \zeta_\lambda^2[1 + o(1)]. \tag{4.7.7}$$

Exercise 4.20 (Supercritical expected cluster size) Prove that (4.4.1) implies that the expected cluster size satisfies, for $\lambda > 1$,

$$\chi(\lambda) = \zeta_\lambda^2 n(1 + o(1)). \tag{4.7.8}$$

Exercise 4.21 (Second largest supercritical cluster) Use the duality principle to show that the second largest component of a supercritical Erdős–Rényi random graph $\mathscr{C}_{(2)}$ satisfies

$$\frac{|\mathscr{C}_{(2)}|}{\log n} \xrightarrow{\mathbb{P}} 1/I_{\mu_\lambda}. \tag{4.7.9}$$

Exercise 4.22 (CLT for binomials with general parameters) Prove that if $X_n \sim \mathsf{Bin}(m_n, p_n)$, where $\mathrm{Var}(X_n) = m_n p_n(1 - p_n) \to \infty$, then

$$\frac{X_n - m_n p_n}{\sqrt{m_n p_n(1 - p_n)}} \xrightarrow{d} Z, \tag{4.7.10}$$

where Z is a standard normal random variable. Use this to conclude that (4.5.15) implies (4.5.16).

Exercise 4.23 (Asymptotic mean and variance at $t = \zeta_\lambda$) Prove that $\mu_{\zeta_\lambda} = 0$ and $\mu'_{\zeta_\lambda} = \lambda \mathrm{e}^{-\lambda\zeta_\lambda} - 1 < 0$ for $\lambda > 1$.

Exercise 4.24 (The probability that $S_m = 0$ in CLT proof) Prove that (4.5.36) holds uniformly in $m \leq \alpha n$ with $\alpha < \zeta_\lambda$, by using (4.5.5) in Proposition 4.17 and adapting the proof for $k = 1$ in (4.4.43).

Exercise 4.25 (Subcritical clusters for $\mathrm{ER}_n^{(e)}(m)$) Fix $\lambda < 1$. Use (4.6.1) and Theorems 4.4–4.5 to show that $|\mathscr{C}_{\max}|/\log n \xrightarrow{\mathbb{P}} 1/I_\lambda$ for $\mathrm{ER}_n^{(e)}(m)$ when $M = n\lambda/2$.

Exercise 4.26 (Supercritical clusters for $\mathrm{ER}_n^{(e)}(m)$) Fix $\lambda > 1$. Use (4.6.1) and Theorem 4.8 to show that $|\mathscr{C}_{\max}|/n \xrightarrow{\mathbb{P}} \zeta_\lambda$ for $\mathrm{ER}_n^{(e)}(m)$ when $M = n\lambda/2$.

5

Erdős–Rényi Random Graph Revisited

In the previous chapter, we have proved that the largest connected component of the Erdős–Rényi random graph exhibits a phase transition. In this chapter, we investigate several more properties of the Erdős–Rényi random graph. We focus on the critical behavior of Erdős–Rényi random graphs, their connectivity threshold and their degree structure.

Organization of this Chapter

We start by investigating the critical behavior of the size of largest connected component in the Erdős–Rényi random graph by studying values of p close to $1/n$ in Section 5.1. We look deeper into the critical Erdős–Rényi random graph in Section 5.2, where we use martingale properties to investigate the critical window. After this, in Section 5.3, we investigate the phase transition for connectivity of $\mathrm{ER}_n(p)$, and for p inside the *connectivity critical window*, compute the asymptotic probability that the Erdős–Rényi random graph is connected. In Section 5.4, we study the degree sequence of an Erdős–Rényi random graph. We close this chapter with notes and discussion in Section 5.5 and exercises in Section 5.6.

5.1 The Critical Behavior

In this section, we study the behavior of the largest connected component of $\mathrm{ER}_n(\lambda/n)$, for p close to the critical value $1/n$. In this case, it turns out that there is interesting behavior, where the size of the largest connected component is large, yet much smaller than the volume n:

Theorem 5.1 (Largest critical cluster) *Take $\lambda = 1 + \theta n^{-1/3}$, where $\theta \in \mathbb{R}$. There exists a constant $b = b(\theta) > 0$ such that, for all $\omega > 1$,*

$$\mathbb{P}_{1+\theta n^{-1/3}}\Big(\omega^{-1} n^{2/3} \leq |\mathscr{C}_{\max}| \leq \omega n^{2/3}\Big) \geq 1 - \frac{b}{\omega}. \tag{5.1.1}$$

Theorem 5.1 shows that the largest critical cluster obeys non-trivial scaling when p is of the form $p = (1 + \theta n^{-1/3})/n$. While $|\mathscr{C}_{\max}|$ is logarithmically small in the subcritical regime $\lambda < 1$ by Theorem 4.4, and $|\mathscr{C}_{\max}| = \Theta_{\mathbb{P}}(n)$ in the supercritical regime $\lambda > 1$ by Theorem 4.8, at the critical value $\lambda = 1$, we see that the largest cluster is $\Theta_{\mathbb{P}}(n^{2/3})$. The result in Theorem 5.1 shows that the random variable $|\mathscr{C}_{\max}| n^{-2/3}$ is a *tight* sequence of random variables, in the sense that $|\mathscr{C}_{\max}| n^{-2/3} \leq \omega$ whp for ω sufficiently large. Also,

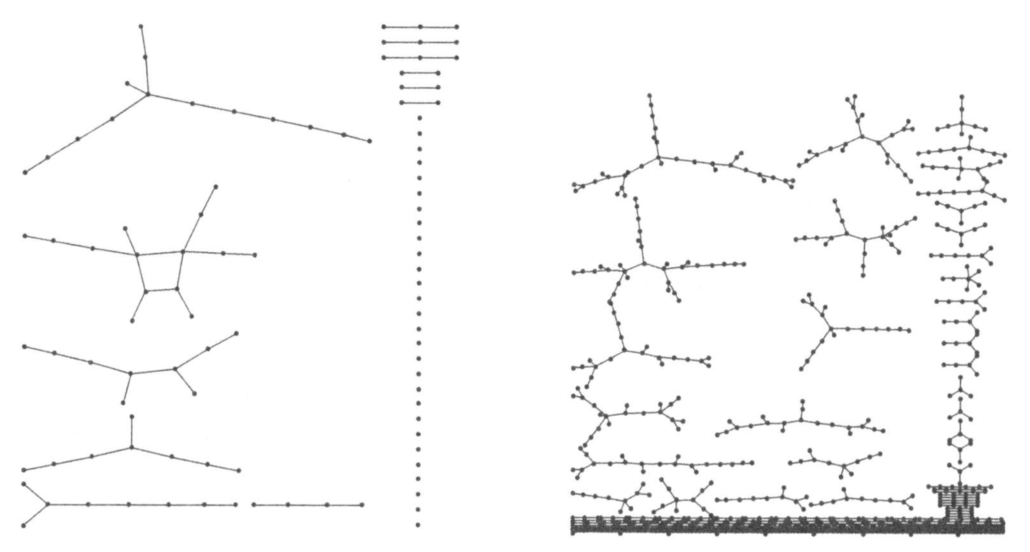

Figure 5.1 Realizations of Erdős–Rényi random graphs with $n = 100$ and $n = 1000$ elements, respectively, and edge probabilities λ/n with $\lambda = 1$.

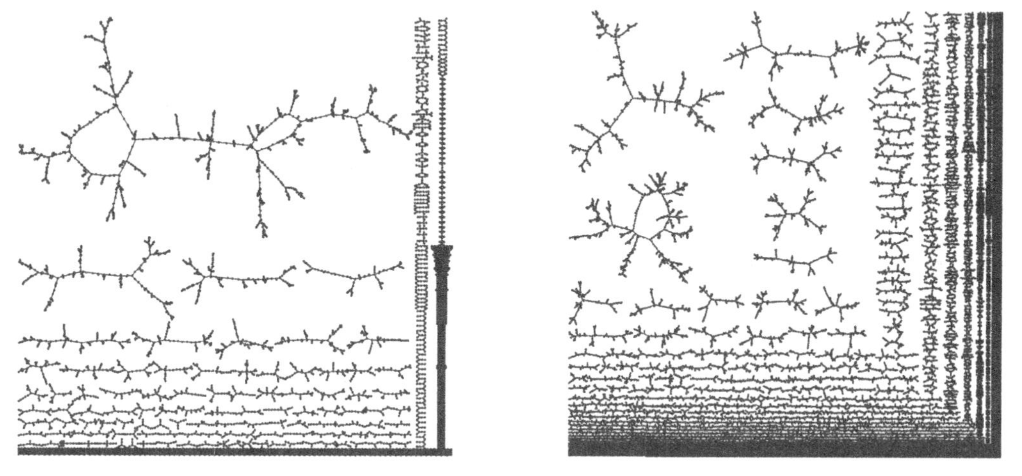

Figure 5.2 Realizations of Erdős–Rényi random graphs with $n = 10000$ and $n = 100000$ elements, respectively, and edge probabilities λ/n with $\lambda = 1$.

$|\mathscr{C}_{\max}|n^{-2/3} \geq \omega^{-1}$ whp, so that also $n^{2/3}/|\mathscr{C}_{\max}|$ is a tight sequence of random variables. See Exercise 5.1 for a proof of these facts. See Figures 5.1–5.2 for plots of critical Erdős–Rényi random graphs with $\lambda = 1$ and $n = 10^l$ with $l = 2, 3, 4, 5$.

5.1.1 Strategy of the Proof

In this section, we formulate two key results for the proof of Theorem 5.1. We start by studying the tail of the distribution of $|\mathscr{C}(v)|$ for the critical case $\lambda = 1 + \theta n^{-1/3}$. We generalize Theorems 4.2–4.3 to $\lambda = 1 + \theta n^{-1/3}$ that are close to the critical value $\lambda = 1$, and prove that bounds as for the Poisson branching process in (3.6.22) in Theorem 3.18 are also valid for $\mathrm{ER}_n(\lambda/n)$ for $\lambda = 1 + \theta n^{-1/3}$.

Proposition 5.2 (Critical cluster tails) *Take $\lambda = 1 + \theta n^{-1/3}$, where $\theta \in \mathbb{R}$, and let $r > 0$. For $k \leq rn^{2/3}$, there exist constants $0 < c_1 < c_2 < \infty$ with $c_1 = c_1(r, \theta)$ such that $\min_{r \leq 1} c_1(r, \theta) > 0$, and c_2 independent of r and θ, such that, for n sufficiently large,*

$$\frac{c_1}{\sqrt{k}} \leq \mathbb{P}_{1+\theta n^{-1/3}}(|\mathscr{C}(1)| \geq k) \leq c_2\Big((\theta \vee 0)n^{-1/3} + \frac{1}{\sqrt{k}}\Big). \tag{5.1.2}$$

Proposition 5.2 implies that the tail of the critical cluster size distribution obeys similar asymptotics as the tail of the total progeny of a critical branching process (see (3.6.22) in Theorem 3.18). The tail in (5.1.2) is only valid for values of k that are not too large. Indeed, when $k > n$, then $\mathbb{P}_\lambda(|\mathscr{C}(v)| \geq k) = 0$. Therefore, there must be a cut-off above which the asymptotics fail to hold. As it turns out, this cut-off is given by $rn^{2/3}$. The upper bound in (5.1.2) holds for a wider range of k; in fact, the proof yields that (5.1.2) is valid for *all* k.

We next study the critical expected cluster size:

Proposition 5.3 (Bound on critical expected cluster size) *Take $\lambda = 1 + \theta n^{-1/3}$ with $\theta < 0$. Then, for all $n \geq 1$,*

$$\mathbb{E}_{1+\theta n^{-1/3}}[|\mathscr{C}(1)|] \leq n^{1/3}/|\theta|. \tag{5.1.3}$$

Proposition 5.3 is intuitively consistent with Theorem 5.1. Indeed, in the critical regime, one can expect the expected cluster size to receive a substantial amount from the contribution where the vertex is in the maximal connected component. This suggests that, for any $v \in [n]$,

$$\begin{aligned}\mathbb{E}_{1+\theta n^{-1/3}}[|\mathscr{C}(1)|] &\sim \mathbb{E}_{1+\theta n^{-1/3}}[|\mathscr{C}(1)|\mathbb{1}_{\{1\in\mathscr{C}_{\max}\}}] = \mathbb{E}_{1+\theta n^{-1/3}}[|\mathscr{C}_{\max}|\mathbb{1}_{\{v\in\mathscr{C}_{\max}\}}] \\ &= \frac{1}{n}\mathbb{E}_{1+\theta n^{-1/3}}[|\mathscr{C}_{\max}|^2],\end{aligned} \tag{5.1.4}$$

where $\sim$ denotes an equality with an uncontrolled error. When $|\mathscr{C}_{\max}|$ is of the order $n^{2/3}$ as Theorem 5.1 suggests,

$$\mathbb{E}_{1+\theta n^{-1/3}}[|\mathscr{C}_{\max}|^2] \sim n^{4/3}. \tag{5.1.5}$$

Therefore, one is intuitively lead to the conclusion

$$\mathbb{E}_{1+\theta n^{-1/3}}[|\mathscr{C}(1)|] \sim n^{1/3}. \tag{5.1.6}$$

The above heuristic is confirmed by Proposition 5.3, at least when $\theta < 0$. With a little more effort, we can show that Proposition 5.3 remains to hold for *all* $\theta \in \mathbb{R}$. We refrain from proving this here, and return to this question in the next section. Exercise 5.2 investigates a lower bound of order $n^{1/3}$ on the expected cluster size for $\lambda = 1 + \theta n^{-1/3}$.

Propositions 5.2 and 5.3 are proved in Section 5.1.2 below. We first prove Theorem 5.1 subject to them.

Proof of Theorem 5.1 subject to Propositions 5.2 and 5.3 The statement in Theorem 5.1 is trivial when $\omega \leq b$, so that, by taking b sufficiently large, we may assume that $\omega \geq 1$ is large. In turn, the statement in Theorem 5.1 is trivial when $\omega^{-1}n^{2/3} \leq 1$ and $\omega n^{2/3} \geq n$, i.e., for $n \leq \omega^{3/2}$, since then $\mathbb{P}_{1+\theta n^{-1/3}}(\omega^{-1}n^{2/3} \leq |\mathscr{C}_{\max}| \leq \omega n^{2/3}) = 1$. Since ω is large, we may also assume that $n \geq N$, where N is large.

We start with the upper bound on $|\mathscr{C}_{\max}|$. We again make use of the fundamental equality $\{|\mathscr{C}_{\max}| \geq k\} = \{Z_{\geq k} \geq k\}$, where we recall that

$$Z_{\geq k} = \sum_{v\in[n]} \mathbb{1}_{\{|\mathscr{C}(v)|\geq k\}}. \tag{5.1.7}$$

By Markov's inequality (Theorem 2.17),

$$\begin{aligned}\mathbb{P}_{1+\theta n^{-1/3}}\big(|\mathscr{C}_{\max}| \geq \omega n^{2/3}\big) &= \mathbb{P}_{1+\theta n^{-1/3}}\big(Z_{\geq \omega n^{2/3}} \geq \omega n^{2/3}\big) \\ &\leq \omega^{-1}n^{-2/3}\mathbb{E}_{1+\theta n^{-1/3}}[Z_{\geq \omega n^{2/3}}].\end{aligned} \tag{5.1.8}$$

By Proposition 5.2,

$$\mathbb{E}_{1+\theta n^{-1/3}}[Z_{\geq \omega n^{2/3}}] = n\mathbb{P}_{1+\theta n^{-1/3}}(|\mathscr{C}(1)| \geq \omega n^{2/3}) \leq c_2 n^{2/3}((\theta \vee 0) + 1/\sqrt{\omega}), \tag{5.1.9}$$

so that

$$\begin{aligned}\mathbb{P}_{1+\theta n^{-1/3}}\big(|\mathscr{C}_{\max}| > \omega n^{2/3}\big) &\leq c_2 n^{2/3}((\theta \vee 0) + 1/\sqrt{\omega})/(\omega n^{2/3}) \\ &\leq \frac{c_2}{\omega}((\theta \vee 0) + 1/\sqrt{\omega}).\end{aligned} \tag{5.1.10}$$

This establishes the upper bound on $|\mathscr{C}_{\max}|$ with $b \leq c_2[(\theta \vee 0) + 1]$.

For the lower bound on $|\mathscr{C}_{\max}|$, we make use of *monotonicity* in λ. The random variable $|\mathscr{C}_{\max}|$ is *increasing* in λ (recall Exercise 4.10). Therefore,

$$\mathbb{P}_{1+\theta n^{-1/3}}\Big(|\mathscr{C}_{\max}| < \omega^{-1}n^{2/3}\Big) \leq \mathbb{P}_{1-\bar{\theta} n^{-1/3}}\Big(|\mathscr{C}_{\max}| < \omega^{-1}n^{2/3}\Big), \tag{5.1.11}$$

where we define $\bar{\theta} = |\theta| \vee 1$. Thus, $-\bar{\theta} \leq -1$. We can thus restrict to prove the result for $\theta \leq -1$, so we assume that $\theta < -1$ from now on. We use Chebychev's inequality (Theorem 2.18), as well as $\{|\mathscr{C}_{\max}| < k\} = \{Z_{\geq k} = 0\}$, to obtain that

$$\begin{aligned}\mathbb{P}_{1+\theta n^{-1/3}}\big(|\mathscr{C}_{\max}| < \omega^{-1}n^{2/3}\big) &= \mathbb{P}_{1+\theta n^{-1/3}}\big(Z_{\geq \omega^{-1}n^{2/3}} = 0\big) \\ &\leq \frac{\mathrm{Var}_{1+\theta n^{-1/3}}(Z_{\geq \omega^{-1}n^{2/3}})}{\mathbb{E}_{1+\theta n^{-1/3}}[Z_{\geq \omega^{-1}n^{2/3}}]^2}.\end{aligned} \tag{5.1.12}$$

By (5.1.2) and with $k = n^{2/3}/\omega$ with $\omega \geq 1$,

$$\mathbb{E}_{1+\theta n^{-1/3}}[Z_{\geq \omega^{-1}n^{2/3}}] = n\mathbb{P}_{1+\theta n^{-1/3}}(|\mathscr{C}(1)| \geq k) \geq c_1\sqrt{\omega} n^{2/3}, \tag{5.1.13}$$

where we have used that $c_1 = c_1(\theta) = \min_{r\leq 1} c_1(r, \theta) > 0$. Also, by Proposition 4.7,

$$\mathrm{Var}_{1+\theta n^{-1/3}}(Z_{\geq \omega^{-1}n^{2/3}}) \leq n\mathbb{E}_{1+\theta n^{-1/3}}[|\mathscr{C}(1)|\mathbb{1}_{\{|\mathscr{C}(1)|\geq \omega^{-1}n^{2/3}\}}]. \tag{5.1.14}$$

By Proposition 5.3, we can further bound, using that $\theta \leq -1$,

$$\mathrm{Var}_{1+\theta n^{-1/3}}(Z_{\geq \omega^{-1}n^{2/3}}) \leq n\mathbb{E}_{1+\theta n^{-1/3}}[|\mathscr{C}(1)|] \leq n^{4/3}. \tag{5.1.15}$$

Substituting (5.1.12)–(5.1.15), we obtain, for n sufficiently large,

$$\mathbb{P}_{1+\theta n^{-1/3}}\big(|\mathscr{C}_{\max}| < \omega^{-1}n^{2/3}\big) \leq \frac{n^{4/3}}{c_1^2\omega n^{4/3}} = \frac{1}{c_1^2\omega}. \tag{5.1.16}$$

This establishes the lower bound on $|\mathscr{C}_{\max}|$.

We conclude that

$$\begin{aligned}\mathbb{P}_{1+\theta n^{-1/3}}\Big(\omega^{-1}n^{2/3} \le |\mathscr{C}_{\max}| \le \omega n^{2/3}\Big) &= 1 - \mathbb{P}_{1+\theta n^{-1/3}}\big(|\mathscr{C}_{\max}| < \omega^{-1}n^{2/3}\big)\\ &\quad - \mathbb{P}_{1+\theta n^{-1/3}}\big(|\mathscr{C}_{\max}| > \omega n^{2/3}\big)\\ &\ge 1 - \frac{1}{c_1^2\omega} - \frac{c_2}{\omega^{3/2}} \ge 1 - \frac{b}{\omega}, \end{aligned} \tag{5.1.17}$$

when $b = c_1^{-2} + c_2$. This completes the proof of Theorem 5.1 subject to Propositions 5.2 and 5.3. □

5.1.2 Critical Cluster Tails and Expected Cluster Size

In this section, we prove Propositions 5.2 and 5.3. We start with the proof of Proposition 5.2.

Proof of Proposition 5.2 Theorem 4.2 gives

$$\mathbb{P}_{1+\theta n^{-1/3}}(|\mathscr{C}(1)| \ge k) \le \mathbb{P}_{n,p}(T \ge k), \tag{5.1.18}$$

where we recall that $\mathbb{P}_{n,p}$ is the law of a binomial branching process with parameters n and $p = \lambda/n = (1 + \theta n^{-1/3})/n$, and T its total progeny. By Theorem 3.20,

$$\mathbb{P}_{1+\theta n^{-1/3}}(|\mathscr{C}(1)| \ge k) \le \mathbb{P}^*_{1+\theta n^{-1/3}}(T^* \ge k) + e_n(k), \tag{5.1.19}$$

where, by (3.7.2),

$$|e_n(k)| \le \frac{1}{n}\sum_{s=1}^{k-1} \mathbb{P}^*_\lambda(T^* \ge s), \tag{5.1.20}$$

and where we recall that $\mathbb{P}^*_\lambda$ is the law of a Poisson branching process with parameter λ, and T^* is its total progeny.

By Theorem 3.18 and Corollary 3.19, it follows that there exists a $C > 0$ independent of θ such that for all $s \ge 1$,

$$\begin{aligned}\mathbb{P}^*_{1+\theta n^{-1/3}}(T^* \ge s) &\le \zeta_\lambda + \sum_{t=s}^{\infty} \mathbb{P}^*_{1+\theta n^{-1/3}}(T^* = t)\\ &\le C\big((\theta \vee 0)n^{-1/3} + 1/\sqrt{s}\big).\end{aligned} \tag{5.1.21}$$

See in particular (3.6.27) in Corollary 3.19 for the first term and (3.6.20) in Theorem 3.18 for the second, for which we also use that $I_\lambda \ge 0$ for every λ. Therefore, for all $k \le n$,

$$\begin{aligned}|e_n(k)| &\le \frac{1}{n}\sum_{s=1}^{k} C\big(\theta_+ n^{-1/3} + 1/\sqrt{s}\big) \le C\Big((\theta \vee 0)kn^{-4/3} + \frac{\sqrt{k}}{n}\Big)\\ &\le C\big((\theta \vee 0)n^{-1/3} + 1/\sqrt{k}\big).\end{aligned} \tag{5.1.22}$$

We conclude that, for all $k \le n$,

$$\mathbb{P}_{1+\theta n^{-1/3}}(|\mathscr{C}(1)| \ge k) \le 2C\big((\theta \vee 0)n^{-1/3} + 1/\sqrt{k}\big). \tag{5.1.23}$$

We proceed with the lower bound in (5.1.2), for which we may assume that $\theta \leq 0$ by monotonicity. We make use of Theorem 4.3 with $k \leq rn^{2/3}$. This gives that

$$\mathbb{P}_{1+\theta n^{-1/3}}(|\mathscr{C}(1)| \geq k) \geq \mathbb{P}_{n-k,p}(T \geq k). \tag{5.1.24}$$

where T is the total progeny of a binomial branching process with parameters $n - k \leq n - rn^{2/3}$ and $p = \lambda/n = (1+\theta n^{-1/3})/n$. We again use Theorem 3.20 for $\lambda_n = 1+(\theta-r)n^{-1/3}$, as in (5.1.19) and (5.1.20). We apply the last bound in (5.1.22) with $\theta \leq 0$, to obtain

$$\begin{aligned}\mathbb{P}_{1+\theta n^{-1/3}}(|\mathscr{C}(1)| \geq k) &\geq \mathbb{P}_{n-k,p}(T \geq k) \\ &\geq \mathbb{P}^*_{\lambda_n}(T^* \geq k) - \frac{C\sqrt{k}}{n} \\ &\geq \mathbb{P}^*_{\lambda_n}(T^* \geq k) - \frac{C\sqrt{r}}{n^{2/3}}.\end{aligned} \tag{5.1.25}$$

Since $\lambda_n \leq 1$,

$$\begin{aligned}\mathbb{P}_{1+\theta n^{-1/3}}(|\mathscr{C}(1)| \geq k) &\geq \sum_{t=k}^{\infty} \mathbb{P}^*_{\lambda_n}(T^* = t) - \frac{C\sqrt{r}}{n^{2/3}} \\ &= \sum_{t=k}^{\infty} \frac{(\lambda_n t)^{t-1}}{t!} \mathrm{e}^{-\lambda_n t} - \frac{C\sqrt{r}}{n^{2/3}} \\ &= \frac{1}{\lambda_n} \sum_{t=k}^{\infty} \mathbb{P}^*_1(T^* = t) \mathrm{e}^{-I_{\lambda_n} t} - \frac{C\sqrt{r}}{n^{2/3}},\end{aligned} \tag{5.1.26}$$

where, for $\lambda_n = 1 + (\theta - r)n^{-1/3}$ and by (3.6.21),

$$I_{\lambda_n} = \lambda_n - 1 - \log \lambda_n = \frac{1}{2}(\lambda_n - 1)^2 + O(|\lambda_n - 1|^3). \tag{5.1.27}$$

It is not hard to see that (see also Exercise 5.4)

$$\frac{(\lambda t)^{t-1}}{t!} \mathrm{e}^{-\lambda t} = \frac{1}{\lambda} \mathrm{e}^{-I_\lambda t} \mathbb{P}^*_1(T^* = t). \tag{5.1.28}$$

We then use (5.1.28), combined with (3.6.22) in Theorem 3.16, to obtain

$$\begin{aligned}\mathbb{P}_{1+\theta n^{-1/3}}(|\mathscr{C}(1)| \geq k) &\geq \sum_{t=k}^{2k} \mathbb{P}^*_1(T^* = t) \mathrm{e}^{-\frac{1}{2}(\lambda_n-1)^2 t(1+o(1))} - \frac{C\sqrt{r}}{n^{2/3}} \\ &\geq \sum_{t=k}^{2k} \frac{C}{\sqrt{t^3}} \mathrm{e}^{-\frac{1}{2}(\lambda_n-1)^2 t(1+o(1))} - \frac{C\sqrt{r}}{n^{2/3}} \\ &\geq \frac{2^{-3/2}C}{\sqrt{k}} \mathrm{e}^{-k(\lambda_n-1)^2(1+o(1))} - \frac{C\sqrt{r}}{n^{2/3}}.\end{aligned} \tag{5.1.29}$$

Here we have used that $\lambda_n - 1 = (\theta - r)n^{-1/3}$, $k(\lambda_n - 1)^2 \leq (\theta - r)^2 r$, which is uniformly bounded. Further, for $n \geq N$,

$$\sqrt{r} n^{-2/3} = \sqrt{rk} n^{-2/3}/\sqrt{k} \leq rn^{-1/3}/\sqrt{k} \leq rN^{-1/3}/\sqrt{k}, \tag{5.1.30}$$

so that

$$\mathbb{P}_{1+\theta n^{-1/3}}(|\mathscr{C}(1)| \geq k) \geq \frac{c_1(r)}{\sqrt{k}}, \tag{5.1.31}$$

with $c_1(r) = (c2^{-3/2}e^{-r(\theta-r)^2} - C\sqrt{r}/N^{-1/3}) > 0$ for $r \leq 1$, whenever N is sufficiently large. This completes the proof of Proposition 5.2. □

Proof of Proposition 5.3 Theorem 4.2 gives that $|\mathscr{C}(1)| \preceq T$, where T is the total progeny of a branching process with a $\mathsf{Bin}(n, \lambda/n)$ offspring distribution, and where $\lambda = 1 + \theta n^{-1/3}$. As a result, for $\theta < 0$, and using Theorem 3.5,

$$\mathbb{E}_{1+\theta n^{-1/3}}[|\mathscr{C}(1)|] \leq \mathbb{E}[T] = 1/(1-\lambda) = n^{1/3}/|\theta|. \tag{5.1.32}$$

This proves the claim. □

5.2 Critical Erdős–Rényi Random Graphs with Martingales

In this section, we use martingales and the exploration of clusters to look deeper into the critical window for the Erdős–Rényi random graph. The main result in this section is the following theorem:

Theorem 5.4 (Cluster tails revisited) *Let $\mathscr{C}_{\max}$ denote the largest component of $\mathrm{ER}_n(1/n)$, and let $\mathscr{C}(v)$ be the component that contains the vertex $v \in [n]$. For any $n > 1000$ and $A > 8$,*

$$\mathbb{P}_1(|\mathscr{C}(v)| > An^{2/3}) \leq 4n^{-1/3}e^{-\frac{A^2(A-4)}{32}}, \tag{5.2.1}$$

and

$$\mathbb{P}_1(|\mathscr{C}_{\max}| > An^{2/3}) \leq \frac{4}{A}e^{-\frac{A^2(A-4)}{32}}. \tag{5.2.2}$$

The proof of Theorem 5.4 uses the exploration process $(S_t)_{t\geq 0}$ defined in (4.1.3), as well as the Optional Stopping Theorem (Theorem 2.28).

5.2.1 The Exploration Process

We recall the exploration process $(S_t)_{t\geq 0}$ defined in (4.1.3). We adapt this process slightly, so as to allow us to explore several clusters after each another. We explain this exploration in more detail in the next section. At each time $t \in [n]$, the number of active vertices is S_t and the number of explored vertices is t. Fix an ordering of the vertices, with v first. At time $t = 0$, the vertex v is active and all other vertices are neutral, so $S_0 = 1$. In step $t \in [n]$, if $S_{t-1} > 0$, then let w_t be the first active vertex; if $S_{t-1} = 0$, then let w_t be the first neutral vertex. Denote by X_t the number of neutral neighbors of w_t in $\mathrm{ER}_n(1/n)$, and change the status of these vertices to *active*. Then, set w_t itself explored.

Write $N_t = n - S_t - t - \mathbb{1}_{\{S_t=0\}}$. Given $S_1, \ldots, S_{t-1}$, the random variable X_t has a $\mathsf{Bin}(N_{t-1}, 1/n)$ distribution, and we have the recursion

$$S_t = \begin{cases} S_{t-1} + X_t - 1 & \text{when } S_{t-1} > 0, \\ X_t & \text{when } S_{t-1} = 0. \end{cases} \tag{5.2.3}$$

At time $T = \min\{t \geq 1 \colon S_t = 0\}$, the set of explored vertices is precisely $|\mathscr{C}(v)|$, so $|\mathscr{C}(v)| = T$.

To prove Theorem 5.4, we couple $(S_t)_{t\geq 0}$ to a random walk with shifted binomial increments. We need the following lemma concerning the *overshoots* of such walks:

Lemma 5.5 (Overshoots of walks) *Let $p \in (0, 1)$ and $(Y_i)_{i\geq 1}$ be i.i.d. random variables with $\mathsf{Bin}(n, p)$ distribution and let $R_t = 1 + \sum_{i=1}^{t}(Y_i - 1)$. Fix an integer $H > 0$, and define*

$$\gamma = \min\{t \geq 1 \colon R_t \geq H \text{ or } S_t = 0\}. \tag{5.2.4}$$

Let $\Xi \subset \mathbb{N}$ be a set of positive integers. Given the event $\{R_\gamma \geq H, \gamma \in \Xi\}$, the conditional distribution of the overshoot $S_\gamma - H$ is stochastically dominated by the binomial distribution $\mathsf{Bin}(n, p)$.

Proof First observe that if Y has a $\mathsf{Bin}(n, p)$ distribution, then the conditional distribution $Y-r$ given $Y \geq r$ is stochastically dominated by $\mathsf{Bin}(n, p)$. To see this, write Y as a sum of n indicator random variables $(I_j)_{j\in[n]}$ and let J be the minimal index such that $\sum_{j=1}^{J} I_j = r$. Given J, the conditional distribution of $Y - r$ is $\mathsf{Bin}(n - J, p)$ which is certainly dominated by $\mathsf{Bin}(n, p)$, independently of J.

For any $l \in \Xi$, conditioned on $\{\gamma = l\} \cap \{R_{l-1} = H - r\} \cap \{S_\gamma \geq H\}$, the overshoot $R_\gamma - H$ equals $Y_l - r$ where Y_l has a $\mathsf{Bin}(n, p)$ distribution conditioned on $Y_l \geq r$. The assertion of the lemma follows by averaging. □

The following corollary, in the spirit of Theorems 2.15 and 2.16, will be useful in the proof:

Corollary 5.6 (Expectations of increasing functions of overshoots) *Let Y have a $\mathsf{Bin}(n, p)$ distribution and let f be an increasing real function. In the notation of the previous lemma,*

$$\mathbb{E}[f(R_\gamma - H) \mid R_\gamma \geq H, \gamma \in \Xi] \leq \mathbb{E}[f(Y)]. \tag{5.2.5}$$

5.2.2 An Easy Upper Bound

Fix a vertex v. To analyze the component of v in $\mathrm{ER}_n(1/n)$, we use the notation established in the previous section. As in the proof of Theorem 4.2, we can couple the sequence $(X_t)_{t\geq 1}$ constructed there, to a sequence $(Y_t)_{t\geq 1}$ of i.i.d. $\mathsf{Bin}(n, 1/n)$ random variables, such that $Y_t \geq X_t$ for all $t \leq n$. The random walk $(R_t)_{t\geq 0}$ defined in Lemma 5.5 satisfies $R_t = R_{t-1} + Y_t - 1$ for all $t \geq 1$ and $R_0 = 1$. Fix an integer $H > 0$ and define γ as in Lemma 5.5. Couple R_t and S_t such that $R_t \geq S_t$ for all $t \leq \gamma$. Since $(R_t)_{t\geq 0}$ is a martingale, optional stopping gives $1 = \mathbb{E}_1[R_\gamma] \geq H\mathbb{P}_1(R_\gamma \geq H)$, whence

$$\mathbb{P}_1(R_\gamma \geq H) \leq 1/H. \tag{5.2.6}$$

Write $R_\gamma^2 = H^2 + 2H(R_\gamma - H) + (R_\gamma - H)^2$ and apply Corollary 5.6 with $f(x) = 2Hx + x^2$ and $\mathbb{E}[Y] = 1$, $\mathbb{E}[Y^2] \leq 2$ to get that, for $H \geq 2$,

$$\mathbb{E}_1[R_\gamma^2 \mid R_\gamma \geq H] \leq \mathbb{E}[f(Y)] = H^2 + 2H + 2 \leq H^2 + 3H. \tag{5.2.7}$$

Now $R_t^2 - (1 - \frac{1}{n})t$ is also a martingale. By the Optional Stopping Theorem (Theorem 2.28), (5.2.6) and (5.2.7),

$$1 + (1 - \frac{1}{n})\mathbb{E}_1[\gamma] = \mathbb{E}_1[R_\gamma^2] = \mathbb{P}_1(R_\gamma \geq H)\mathbb{E}_1[R_\gamma^2 \mid R_\gamma \geq H] \leq H + 3. \tag{5.2.8}$$

Thus, for $2 \leq H \leq n - 3$,

$$\mathbb{E}_1[\gamma] \leq H + 3 \tag{5.2.9}$$

We conclude that, for $2 \leq H \leq n - 3$,

$$\mathbb{P}_1(\gamma \geq H^2) \leq \frac{H+3}{H^2} \leq \frac{2}{H}. \tag{5.2.10}$$

Define $\gamma^* = \gamma \wedge H^2$, so that by the previous inequality and (5.2.6),

$$\mathbb{P}_1(R_{\gamma^*} > 0) \leq \mathbb{P}_1(R_\gamma \geq H) + \mathbb{P}_1(\gamma \geq H^2) \leq \frac{3}{H}. \tag{5.2.11}$$

Let $k = H^2$ and note that if $|\mathscr{C}(v)| > H^2$, we must have $R_{\gamma^*} > 0$ so by (5.2.11) we deduce

$$\mathbb{P}_1(|\mathscr{C}(v)| > k) \leq \frac{3}{\sqrt{k}} \tag{5.2.12}$$

for all $9 \leq k \leq (n-3)^2$. This gives an alternative proof of the upper bound in Proposition 5.2. Recall from (4.3.4) that $Z_{\geq k}$ denotes the number of vertices contained in components larger than k. Then

$$\mathbb{P}_1(|\mathscr{C}_{\max}| > k) \leq \mathbb{P}(Z_{\geq k} > k) \leq \frac{\mathbb{E}_1[Z_{\geq k}]}{k} \leq \frac{n\mathbb{P}_1(|\mathscr{C}(v)| > k)}{k}. \tag{5.2.13}$$

Putting $k = \left(\lfloor \sqrt{An^{2/3}} \rfloor\right)^2$ for any $A > 1$ yields

$$\mathbb{P}_1(|\mathscr{C}_{\max}| > An^{2/3}) \leq \mathbb{P}_1(|\mathscr{C}_{\max}| > k) \leq \frac{3n}{\lfloor \sqrt{An^{2/3}} \rfloor^3} \leq \frac{6}{A^{3/2}}, \tag{5.2.14}$$

since

$$\left(\lfloor \sqrt{An^{2/3}} \rfloor\right)^3 \geq \left(\sqrt{An^{2/3}} - 1\right)^3 \geq nA^{3/2}(1 - 3A^{-1/2}n^{-1/3}) \geq A^{3/2}n/2. \tag{5.2.15}$$

This provides a first easy upper bound. In the next section, we improve upon this estimate, and prove Theorem 5.4.

5.2.3 Proof Critical Cluster Tails Revisited

In this section, we complete the proof of Theorem 5.4. We proceed from (5.2.11). Define the process $(Y_t)_{t\geq 0}$ by

$$Y_t = \sum_{j=1}^{t}(X_{\gamma^*+j} - 1). \tag{5.2.16}$$

The law of X_{γ^*+j} is stochastically dominated by a $\mathsf{Bin}(n - j, \frac{1}{n})$ distribution for $j \leq n$. Hence,

$$\mathbb{E}_1[\mathrm{e}^{c(X_{\gamma^*+j}-1)} \mid \gamma^*] \leq \mathrm{e}^{-c}\Big[1 + \frac{1}{n}(\mathrm{e}^c - 1)\Big]^{n-j} \leq \mathrm{e}^{(c+c^2)\frac{n-j}{n}-c} \leq \mathrm{e}^{c^2-cj/n}, \tag{5.2.17}$$

as $\mathrm{e}^c - 1 \leq c + c^2$ for any $c \in (0, 1)$ and $1 + x \leq \mathrm{e}^x$ for $x \geq 0$. Since this bound is uniform in S_{γ^*} and γ^*, we have

$$\mathbb{E}_1[\mathrm{e}^{cY_t}|R_{\gamma^*}] \leq \mathrm{e}^{tc^2-ct^2/(2n)}. \tag{5.2.18}$$

Write $\mathbb{P}_{1,R}$ for the conditional probability given R_{γ^*}. Then, for any $c \in (0, 1)$,

$$\mathbb{P}_{1,R}(Y_t \geq -R_{\gamma^*}) \leq \mathbb{P}_{1,R}(\mathrm{e}^{cY_t} \geq \mathrm{e}^{-cR_{\gamma^*}}) \leq \mathrm{e}^{tc^2-\frac{ct^2}{2n}}\mathrm{e}^{cR_{\gamma^*}}. \tag{5.2.19}$$

By (5.2.3), if $S_{\gamma^*+j} > 0$ for all $0 \leq j \leq t - 1$, then $Y_j = S_{\gamma^*+j} - S_{\gamma^*}$ for all $j \in [t]$. It follows that

$$\begin{aligned}\mathbb{P}_1(R_{\gamma^*+j} > 0\,\forall j \in [t] \mid R_{\gamma^*} > 0) &\leq \mathbb{E}_1[\mathbb{P}_{1,R}(Y_t \geq -R_{\gamma^*})|R_{\gamma^*} > 0]\\ &\leq \mathrm{e}^{tc^2-ct^2/(2n)}\mathbb{E}_1[\mathrm{e}^{cR_{\gamma^*}} \mid R_{\gamma^*} > 0].\end{aligned} \tag{5.2.20}$$

By Corollary 5.6 with $\Xi = [H^2]$, for $c \in (0, 1)$,

$$\mathbb{E}_1\big[\mathrm{e}^{cR_{\gamma^*}} \mid \gamma \leq H^2, R_\gamma > 0\big] \leq \mathrm{e}^{cH+c+c^2}. \tag{5.2.21}$$

Since $\{R_{\gamma^*} > 0\} = \{\gamma > H^2\} \cup \{\gamma \leq H^2, R_\gamma > 0\}$, which is a disjoint union, the conditional expectation $\mathbb{E}_1[\mathrm{e}^{cR^*_\gamma} \mid R_{\gamma^*} > 0]$ is a weighted average of the conditional expectation in (5.2.21) and of $\mathbb{E}_1[\mathrm{e}^{cR_{\gamma^*}} \mid \gamma > H^2] \leq \mathrm{e}^{cH}$ by (5.2.4). Therefore,

$$\mathbb{E}_1[\mathrm{e}^{cR_{\gamma^*}} \mid R_{\gamma^*} > 0] \leq \mathrm{e}^{cH+c+c^2}, \tag{5.2.22}$$

so that, by (5.2.20),

$$\mathbb{P}_1\left(R_{\gamma^*+j} > 0\,\forall j \in [t] \mid R_{\gamma^*} > 0\right) \leq \mathrm{e}^{tc^2-ct^2/(2n)+cH+c+c^2}. \tag{5.2.23}$$

By our coupling, for any integer $k > H^2$, if $|\mathscr{C}(v)| > k$, then we must have that $R_{\gamma^*} > 0$ and $S_{\gamma^*+j} > 0$ for all $j \in [0, k - H^2]$. Thus, by (5.2.11) and (5.2.23),

$$\begin{aligned}\mathbb{P}_1(|\mathscr{C}(v)| > k) &\leq \mathbb{P}_1(R_{\gamma^*} > 0)\mathbb{P}_1(\forall j \in [0, k - H^2] \quad S_{\gamma^*+j} > 0 \mid R_{\gamma^*} > 0)\\ &\leq \frac{3}{H}\mathrm{e}^{(k-H^2)c^2-\frac{c(k-H^2)^2}{2n}+cH+c+c^2}.\end{aligned} \tag{5.2.24}$$

Take $H = \lfloor n^{1/3} \rfloor$ and $k = \lfloor An^{2/3} \rfloor$ for some $A > 4$; substituting c which attains the minimum of the parabola in the exponent of the right-hand side of (5.2.24) gives

$$\begin{aligned}\mathbb{P}_1(|\mathscr{C}(v)| > An^{2/3}) &\leq 4n^{-1/3} \mathrm{e}^{-\frac{\left((k-H^2)^2/(2n)-H-1\right)^2}{4(k-H^2+1)}} \\ &\leq 4n^{-1/3} \mathrm{e}^{-\frac{\left((A-1-n^{-2/3})^2/2-1-n^{-1/3}\right)^2}{4(A-1+2n^{-1/3}+n^{-2/3})}} \\ &\leq 4n^{-1/3} \mathrm{e}^{-\frac{\left((A-2)^2/2-2\right)^2}{4(A-1/2)}},\end{aligned} \tag{5.2.25}$$

since $H^2 \geq n^{2/3}(1 - 2n^{-1/3})$ and $n > 1000$. As $[(A-2)^2/2 - 2]^2 = A^2(A/2-2)^2$ and $(A/2-2)/(A-1/2) > 1/4$ for $A > 8$, we arrive at

$$\mathbb{P}_1(|\mathscr{C}(v)| > An^{2/3}) \leq 4n^{-1/3} \mathrm{e}^{-A^2(A-4)/(32)}. \tag{5.2.26}$$

Recall that $Z_{\geq k}$ denotes the number of vertices contained in components larger than k. Then

$$\mathbb{P}_1\left(|\mathscr{C}_{\max}| > k\right) \leq \mathbb{P}_1\left(Z_{\geq k} > k\right) \leq \frac{\mathbb{E}_1[Z_{\geq k}]}{k} \leq \frac{n\mathbb{P}_1(|\mathscr{C}(v)| > k)}{k}, \tag{5.2.27}$$

and we conclude that for all $A > 8$ and $n > 1000$,

$$\mathbb{P}_1(|\mathscr{C}_{\max}| > An^{2/3}) \leq \frac{4}{A} \mathrm{e}^{-A^2(A-4)/(32)}. \tag{5.2.28}$$

This completes the proof of Theorem 5.4. □

5.2.4 Connected Components in the Critical Window Revisited

In this section, we discuss the *critical window* of the Erdős–Rényi random graph in more detail, from an intuitive perspective. By Theorem 5.1, we know that, for $p = (1+\theta n^{-1/3})/n$, the largest connected component has size roughly equal to $n^{2/3}$. Therefore, the values of p for which $p = (1+\theta n^{-1/3})/n$ are called the *critical window*. In this section, we discuss the convergence in distribution of the rescaled clusters by Aldous (1997). The point by Aldous (1997) is to prove simultaneous weak convergence of all connected components at once. We start by introducing some notation. Let $|\mathscr{C}_{(j)}(\theta)|$ denote the jth largest cluster of $\mathrm{ER}_n(p)$ for $p = (1+\theta n^{-1/3})/n$. The main result on the behavior within the critical window is the following theorem:

Theorem 5.7 (Weak convergence of largest clusters in critical window) *For $p = (1 + \theta n^{-1/3})/n$, and any $\theta \in \mathbb{R}$, the vector $\mathbf{C}^n(\theta) \equiv \left(n^{-2/3}|\mathscr{C}_{(i)}(\theta)|\right)_{i\geq 1}$ converges in distribution to a random vector $(\gamma_i(\theta))_{i\geq 1}$.*

Theorem 5.7 is stronger than Theorem 5.1 in two ways: (1) Theorem 5.7 proves weak convergence, rather than tightness only, of $|\mathscr{C}_{\max}|/n^{2/3}$; (2) Theorem 5.7 considers *all* connected components, ordered by size, rather than only the maximal one. Aldous gives two explicit descriptions of the distribution of the limiting random variable $(\gamma_i(\theta))_{i\geq 1}$, the first being in terms of lengths of excursions of Brownian motion, the second in terms of the so-called *multiplicative coalescent process*. We intuitively explain these constructions now.

Exploration Process Convergence

We start by explaining the construction in terms of excursions of Brownian motion. Let $(W(s))_{s\geq 0}$ be standard Brownian motion, and let

$$W^{\theta}(s) = W(s) + \theta s - s^2/2 \tag{5.2.29}$$

be Brownian motion with an (inhomogeneous) drift $\theta - s$ at time s. Let

$$B^{\theta}(s) = W^{\theta}(s) - \min_{0\leq s'\leq s} W^{\theta}(s') \tag{5.2.30}$$

correspond to the process $(W^{\theta}(s))_{s\geq 0}$ reflected at 0. We now consider the *excursions* of this process, ordered in their length. Here an excursion γ of $(B^{\theta}(s))_{s\geq 0}$ is a time interval $[l(\gamma), r(\gamma)]$ for which $B^{\theta}(l(\gamma)) = B^{\theta}(r(\gamma)) = 0$, but $B^{\theta}(s) > 0$ for all $s \in (l(\gamma), r(\gamma))$. Let the length $|\gamma|$ of the excursion γ be given by $r(\gamma) - l(\gamma)$. As it turns out (see Aldous (1997, Section 1) for details), the excursions of $(B^{\theta}(s))_{s\geq 0}$ can be ordered by decreasing length, so that $(\gamma_i(\theta))_{i\geq 1}$ are the excursions. Then, the random vector $\mathbf{C}^n(\theta)$ converges in distribution to the ordered excursions $(\gamma_i(\theta))_{i\geq 1}$. The idea behind this is as follows.

We make use of the random-walk representation of the various clusters, which connects the cluster exploration to random walks. However, as for example (4.5.4) shows, the step size distribution is decreasing as we explore more vertices, which means that we arrive at an inhomogeneous and ever-decreasing drift, as in (5.2.29). Since, in general, random walks converge to Brownian motions, this way the connection between these precise processes can be made using functional martingale central limit theorems.

Let us describe this connection informally to explain the idea behind the proof of Theorem 5.7. The proof relies on an *exploration* of the components of the Erdős–Rényi random graph as we have seen several times before (see (4.1.3)–(4.1.4)). To set this exploration up, we now successively explore different clusters, starting from the cluster of a single vertex. The fact that we explore several clusters after each other makes this exploration different from the exploration used so far. Again, this exploration is described in terms of a stochastic process $(S_i)_{i\geq 0}$ that encodes the cluster sizes as well as their structure. To describe the exploration process, we let $S_0 = 0$ and let S_i satisfy the recursion for $i \geq 1$

$$S_i = S_{i-1} + X_i - 1, \tag{5.2.31}$$

where $X_i \sim \mathsf{Bin}(N_{i-1}, p)$. Here N_{i-1} denotes the number of neutral vertices after we have explored $i - 1$ vertices (but note that these vertices could now be part of different connected components). In this exploration, $-\inf_{j\in[i]} S_j$ denotes the number of disjoint clusters that have been fully explored at time i, while, for $i \geq 1$,

$$R_i = S_i - \inf_{j\in[i-1]} S_j + 1 \tag{5.2.32}$$

denotes the number of active vertices at time i in the cluster that we are currently exploring (with $R_0 = 0$). Here we again call a vertex *active* when it has been found to be part of the cluster that is currently explored, but its neighbors in the random graph have not yet been identified. Then, the number of neutral vertices, i.e., the vertices not yet found to be in any of the clusters explored up to time i, equals

$$N_i = n - i - R_i = n - i - S_i + \inf_{j \in [i-1]} S_j - 1. \tag{5.2.33}$$

We can see $(R_i)_{i \geq 0}$ as the *reflection* of the process $(S_i)_{i \geq 0}$. When $R_i = 0$, we have fully explored a cluster, and this occurs precisely when $S_i = \inf_{j \in [i-1]} S_j - 1$, i.e., the process $(S_i)_{i \geq 0}$ reaches a new all-time minimum. In particular, when we start exploring $\mathscr{C}(v_1)$, then

$$|\mathscr{C}(v_1)| = \inf\{i > 0 \colon R_i = 0\}. \tag{5.2.34}$$

After having explored $\mathscr{C}(v_1)$, we explore $\mathscr{C}(v_2)$ for some $v_2 \notin \mathscr{C}(v_1)$, and obtain that

$$|\mathscr{C}(v_2)| = \inf\{i > |\mathscr{C}(v_1)| \colon R_i = 0\} - |\mathscr{C}(v_1)|. \tag{5.2.35}$$

Iterating this procedure, we see that

$$\begin{aligned} |\mathscr{C}(v_j)| = \inf\{i > |\mathscr{C}(v_1)| + \cdots + |\mathscr{C}(v_{j-1})| \colon R_i = 0\} \\ - (|\mathscr{C}(v_1)| + \cdots + |\mathscr{C}(v_2)|). \end{aligned} \tag{5.2.36}$$

We conclude that the total number of clusters that are fully explored up to time i indeed equals $\inf_{j \in [i]} S_j$. Since $n^{-1/3} S_{tn^{2/3}}$ will be seen to converge in distribution, and we are investigating the exploration process at times $tn^{2/3}$, we will simplify our lives considerably, and approximate (5.2.33) to $N_i \approx n - i$. This means that the random variables $(X_i)_{i \geq 1}$ in (5.2.31) are close to being independent, though not identically distributed, with

$$X_i \approx \mathsf{Bin}(n - i, p) = \mathsf{Bin}\big(n - i, (1 + \theta n^{-1/3})/n\big). \tag{5.2.37}$$

Here, and in what follows, $\approx$ denotes an uncontrolled approximation.

It is well known that a binomial random variable with parameters n and success probability μ/n is close to a Poisson random variable with mean μ (recall e.g., Theorem 2.10). In our case,

$$\mu = (n - i)p = (n - i)(1 + \theta n^{-1/3})/n \approx 1 + \theta n^{-1/3} - i/n. \tag{5.2.38}$$

Note that when $i = tn^{2/3}$, both correction terms are of the same order in n. Thus, we will approximate

$$S_i \approx \sum_{j=1}^{i} (Y_j - 1), \tag{5.2.39}$$

where $Y_j \sim \mathsf{Poi}(1 + \theta n^{-1/3} - j/n)$ are independent. Since sums of independent Poisson variables are Poisson again, we thus obtain that

$$S_i \approx S_i^*, \tag{5.2.40}$$

where

$$\begin{aligned} S_i^* &\sim \mathsf{Poi}\Big(\sum_{j=1}^{i} (1 + \theta n^{-1/3} - j/n)\Big) - i \\ &\approx \mathsf{Poi}\Big(i + i\theta n^{-1/3} - \frac{i^2}{2n}\Big) - i, \end{aligned} \tag{5.2.41}$$

and $(S_i^* - S_{i-1}^*)_{i\geq 1}$ are independent. Now we multiply by $n^{-1/3}$ and take $i = tn^{2/3}$ to obtain

$$n^{-1/3} S^*_{tn^{2/3}} \sim n^{-1/3}\Big(\mathsf{Poi}\big(tn^{2/3} + t\theta n^{1/3} - \tfrac{1}{2}t^2 n^{1/3}\big) - tn^{2/3}\Big). \tag{5.2.42}$$

Since a Poisson process is to leading order deterministic, we can approximate

$$\mathsf{Poi}\big(tn^{2/3} + t\theta n^{1/3} - \tfrac{1}{2}t^2 n^{1/3}\big) \approx \mathsf{Poi}(tn^{2/3}) + t\theta n^{1/3} - \tfrac{1}{2}t^2 n^{1/3}. \tag{5.2.43}$$

Then, we can use the CLT for the process $n^{-1/3}(\mathsf{Poi}(tn^{2/3}) - tn^{2/3})$ in the form

$$\Big(n^{-1/3}(\mathsf{Poi}(tn^{2/3}) - tn^{2/3})\Big)_{t\geq 0} \xrightarrow{d} (W(t))_{t\geq 0} \tag{5.2.44}$$

to approximate

$$\begin{aligned} n^{-1/3} S^*_{tn^{2/3}} &\sim n^{-1/3}(\mathsf{Poi}(tn^{2/3}) - tn^{2/3}) + t\theta - \tfrac{1}{2}t^2 \\ &\xrightarrow{d} W(t) + t\theta - \tfrac{1}{2}t^2. \end{aligned} \tag{5.2.45}$$

This informally explains the proof of Theorem 5.7. To make this proof rigorous, one typically resorts to Martingale Functional Central Limit Theorems; see for example how Aldous (1997) does this nicely.

Multiplicative Coalescents

Here we explain the connection between critical clusters and the so-called *multiplicative coalescent.* We will be very brief and thus skim over important details. This is a difficult topic, and we refer to the original paper by Aldous (1997) for more details. To explain the connection to the multiplicative coalescent, we interpret the θ-variable in $p = (1 + \theta n^{-1/3})/n$ as a *time* variable. The aim is to study what happens when θ increases. We can couple all the edge variables in terms of uniform random variables such that edge e is p-occupied precisely when $U_e \leq p$. Increasing p from $p + dp$ then means that we add precisely those edges e for which $U_e \in [p, p + dp]$. These extra edges might cause connected components to merge, which occurs precisely when an edge appears that connects two vertices in the different connected components. Let us now investigate in more detail what this process looks like.

When we have two connected components of sizes $xn^{2/3}$ and $yn^{2/3}$, say, and we increase θ to $\theta + d\theta$, then the probability that these two clusters merge is roughly equal to the number of possible connecting edges, which is $xn^{2/3} \times yn^{2/3} = xyn^{4/3}$ times the probability that an edge turns from vacant to occupied when p increases from $p = (1 + \theta n^{-1/3})/n$ to $(1 + (\theta + d\theta)n^{-1/3})/n$, which is $d\theta n^{-4/3}$. Thus, this probability is, for small $d\theta$, close to

$$xyd\theta. \tag{5.2.46}$$

We see that distinct clusters meet at a rate proportional to the rescaled product of their sizes. The continuous process that does this precisely is called the *multiplicative coalescent.* Using the above ideas, Aldous is able to show that the limit of $\mathbf{C}^n(\theta)$ equals such a multiplicative coalescent process. Of course, it is far from obvious that one can make sense of the stochastic process in continuous time for which masses merge at a rate that is the product of their current masses, and Aldous heavily relies on the connections between multiplicative coalescents and the exploration processes that we have described earlier.

5.3 Connectivity Threshold

In this section, we investigate the connectivity threshold for the Erdős–Rényi random graph. As we can see in Theorem 4.8, for every $1 < \lambda < \infty$, the maximal connected component for the Erdős–Rényi random graph when $p = \lambda/n$ has size $\zeta_\lambda n(1 + o_{\mathbb{P}}(1))$, where $\zeta_\lambda > 0$ is the survival probability of a Poisson branching process with parameter λ. Since extinction is certain when the root has no offspring,

$$\zeta_\lambda \leq 1 - \mathbb{P}^*(Z_1^* = 0) = 1 - \mathrm{e}^{-\lambda} < 1. \tag{5.3.1}$$

Therefore, the Erdős–Rényi random graph with edge probability $p = \lambda/n$ is whp *disconnected* for each fixed $\lambda < \infty$. In this section, we discuss how $\lambda = \lambda_n \to \infty$ should be chosen to make $\mathrm{ER}_n(\lambda/n)$ whp connected.

5.3.1 Sharp Transition Connectivity

We now investigate the threshold for connectivity for an appropriate choice $\lambda = \lambda_n \to \infty$. We see that there is a sharp transition in the connectivity of the Erdős–Rényi random graph:

Theorem 5.8 (Connectivity threshold) *For $\lambda - \log n \to \infty$, the Erdős–Rényi random graph is with high probability connected, while for $\lambda - \log n \to -\infty$, the Erdős–Rényi random graph is with high probability disconnected.*

In the proof, we investigate the number of isolated vertices. Define

$$Y = \sum_{i \in [n]} I_i, \qquad \text{where} \qquad I_i = \mathbb{1}_{\{|\mathscr{C}(i)|=1\}}, \tag{5.3.2}$$

to be the number of isolated vertices. Clearly, there exists at least one isolated vertex when $Y \geq 1$, so that the graph is disconnected when $Y \geq 1$. Remarkably, it turns out that when there is no isolated vertex, i.e., when $Y = 0$, then the random graph is also, with high probability, connected. See Proposition 5.10 below for the precise formulation of this result. Thus, the connectivity threshold is the threshold for the disappearance of isolated vertices.

By Proposition 5.10, we need to investigate the probability that $Y \geq 1$. In the case where $|\lambda - \log n| \to \infty$, we make use of the Markov and Chebychev inequalities (Theorems 2.17 and 2.18) combined with a first and second moment argument using a variance estimate in Proposition 5.9:

Proposition 5.9 (Mean and variance of number of isolated vertices) *For every $\lambda \leq n/2$ with $\lambda \geq 1/2$,*

$$\mathbb{E}_\lambda[Y] = n\mathrm{e}^{-\lambda}(1 + O(\lambda^2/n)), \tag{5.3.3}$$

and, for every $\lambda \leq n$,

$$\mathrm{Var}_\lambda(Y) \leq \mathbb{E}_\lambda[Y] + \frac{\lambda}{n-\lambda}\mathbb{E}_\lambda[Y]^2. \tag{5.3.4}$$

Proof Since $|\mathscr{C}(i)| = 1$ precisely when all edges emanating from i are vacant, and using $1 - x \le \mathrm{e}^{-x}$,

$$\mathbb{E}_\lambda[Y] = n\mathbb{P}_\lambda(|\mathscr{C}(1)| = 1) = n(1 - \frac{\lambda}{n})^{n-1} \le n\mathrm{e}^{-\lambda}\mathrm{e}^{\lambda/n}. \tag{5.3.5}$$

Also, using that $1 - x \ge \mathrm{e}^{-x-x^2}$ for $0 \le x \le \frac{1}{2}$, we obtain

$$\begin{aligned}\mathbb{E}_\lambda[Y] = n\mathbb{P}_\lambda(|\mathscr{C}(1)| = 1) &\ge n\mathrm{e}^{-(n-1)\frac{\lambda}{n}(1+\frac{\lambda}{n})} \\ &\ge n\mathrm{e}^{-\lambda(1+\frac{\lambda}{n})} = n\mathrm{e}^{-\lambda}\mathrm{e}^{-\lambda^2/n}.\end{aligned} \tag{5.3.6}$$

Since $\lambda \ge 1/2$, we have that $O(\lambda/n) = O(\lambda^2/n)$, so this proves (5.3.3).

To prove (5.3.4), if $\lambda = n$, there is nothing to prove. Therefore, we assume that $\lambda < n$. We use the exchangeability of the vertices to compute

$$\mathbb{E}_\lambda[Y^2] = n\mathbb{P}_\lambda(|\mathscr{C}(1)| = 1) + n(n-1)\mathbb{P}_\lambda(|\mathscr{C}(1)| = 1, |\mathscr{C}(2)| = 1). \tag{5.3.7}$$

Therefore, we obtain

$$\begin{aligned}\mathrm{Var}_\lambda(Y) = n\big[&\mathbb{P}_\lambda(|\mathscr{C}(1)| = 1) - \mathbb{P}_\lambda(|\mathscr{C}(1)| = 1, |\mathscr{C}(2)| = 1)\big] \\ &+ n^2\big[\mathbb{P}_\lambda(|\mathscr{C}(1)| = 1, |\mathscr{C}(2)| = 1) - \mathbb{P}_\lambda(|\mathscr{C}(1)| = 1)^2\big].\end{aligned} \tag{5.3.8}$$

The first term is bounded above by $\mathbb{E}_\lambda[Y]$. The second term can be computed by using (5.3.5), together with

$$\mathbb{P}_\lambda(|\mathscr{C}(1)| = 1, |\mathscr{C}(2)| = 1) = (1 - \frac{\lambda}{n})^{2n-3}. \tag{5.3.9}$$

Therefore, by (5.3.5) and (5.3.9), we obtain

$$\begin{aligned}\mathbb{P}_\lambda(&|\mathscr{C}(1)| = 1, |\mathscr{C}(2)| = 1) - \mathbb{P}_\lambda(|\mathscr{C}(1)| = 1)^2 \\ &= \mathbb{P}_\lambda(|\mathscr{C}(1)| = 1)^2\big[(1 - \frac{\lambda}{n})^{-1} - 1\big] \\ &= \frac{\lambda}{n(1-\frac{\lambda}{n})}\mathbb{P}_\lambda(|\mathscr{C}(1)| = 1)^2.\end{aligned} \tag{5.3.10}$$

We conclude that

$$\mathrm{Var}_\lambda(Y) \le \mathbb{E}_\lambda[Y] + \frac{\lambda}{n-\lambda}\mathbb{E}_\lambda[Y]^2. \tag{5.3.11}$$

□

Proposition 5.10 (Connectivity and isolated vertices) *For all $0 \le \lambda \le n$ and $n \ge 2$,*

$$\mathbb{P}_\lambda\big(\mathrm{ER}_n(\lambda/n) \textit{ connected}\big) \le \mathbb{P}_\lambda(Y = 0). \tag{5.3.12}$$

Moreover, if there exists an $a > 1/2$ such that $\lambda \ge a \log n$, then, for $n \to \infty$,

$$\mathbb{P}_\lambda\big(\mathrm{ER}_n(\lambda/n) \textit{ connected}\big) = \mathbb{P}_\lambda(Y = 0) + o(1). \tag{5.3.13}$$

Proof The bound in (5.3.12) is obvious. For (5.3.13), we use that

$$\mathbb{P}_\lambda\big(\mathrm{ER}_n(\lambda/n) \text{ connected}\big) = \mathbb{P}_\lambda(Y=0) - \mathbb{P}_\lambda\big(\mathrm{ER}_n(\lambda/n) \text{ disconnected}, Y=0\big). \tag{5.3.14}$$

□

To prove (5.3.13), we make use of a computation involving *trees*. For $k = 2, \ldots, n$, we denote by X_k the number of occupied trees of size equal to k on $[n]$ that cannot be extended to a tree of larger size. Thus, each tree that is counted in X_k has size precisely equal to k, and when we denote its vertices by $v_1, \ldots, v_k$, all the edges between v_i and $v \notin \{v_1, \ldots, v_k\}$ are vacant. Moreover, there are precisely $k-1$ occupied edges between the v_i that are such that these vertices together with the occupied edges form a tree. Such a tree is sometimes called a *spanning tree*. Note that a connected component of size k can contain more than one tree of size k, since the connected component may contain cycles. Note furthermore that when $\mathrm{ER}_n(\lambda/n)$ is disconnected, but $Y=0$, there must be a $k \in \{2, \ldots, n/2\}$ for which $X_k \geq 1$.

We conclude from Boole's inequality and Markov's inequality (Theorem 2.17) that

$$\begin{aligned}\mathbb{P}_\lambda\big(\mathrm{ER}_n(\lambda/n) \text{ disconnected}, Y=0\big) &\leq \mathbb{P}_\lambda\big(\cup_{k=2}^{n/2}\{X_k \geq 1\}\big)\\ &\leq \sum_{k=2}^{n/2}\mathbb{P}_\lambda(X_k \geq 1) \leq \sum_{k=2}^{n/2}\mathbb{E}_\lambda[X_k].\end{aligned} \tag{5.3.15}$$

Therefore, we need to bound $\mathbb{E}_\lambda[X_k]$. For this, we note that there are $\binom{n}{k}$ ways of choosing k vertices, and, by Cayley's Theorem (Theorem 3.17), there are k^{k-2} labeled trees containing k vertices. Therefore,

$$\mathbb{E}_\lambda[X_k] = \binom{n}{k}k^{k-2}q_k, \tag{5.3.16}$$

where q_k is the probability that any tree of size k is occupied and all the edges from the tree to other vertices are vacant. We compute

$$q_k = \Big(\frac{\lambda}{n}\Big)^{k-1}\Big(1-\frac{\lambda}{n}\Big)^{k(n-k)} \leq \Big(\frac{\lambda}{n}\Big)^{k-1}\mathrm{e}^{-\lambda k(n-k)/n}. \tag{5.3.17}$$

We conclude that

$$\mathbb{E}_\lambda[X_k] \leq n\lambda^{k-1}\frac{k^{k-2}}{k!}\mathrm{e}^{-\frac{\lambda}{n}k(n-k)}. \tag{5.3.18}$$

If we further use that $k! \geq k^k\mathrm{e}^{-k}$ by the Stirling bound, and also use that $\lambda \geq 1$, then we arrive at

$$\mathbb{E}_\lambda[X_k] \leq n(\mathrm{e}\lambda)^k\frac{1}{k^2}\mathrm{e}^{-\frac{\lambda}{n}k(n-k)}. \tag{5.3.19}$$

Let $f(\lambda) = \lambda\mathrm{e}^{-\lambda(1-k/n)}$ and note that, since $k \leq n/2$,

$$f'(\lambda) = \mathrm{e}^{-\lambda(1-k/n)}\big[1-\lambda(1-k/n)\big] \leq \mathrm{e}^{-\lambda(1-k/n)}(1-\lambda/2) \leq 0, \tag{5.3.20}$$

when $\lambda \geq 2$. Thus, it suffices to investigate $\lambda = a\log n$ for some $a > 1/2$. For $k \in \{2,3,4\}$, for $\lambda = a\log n$ for some $a > 1/2$,

$$\mathbb{E}_\lambda[X_k] \leq n(\mathrm{e}\lambda)^4\mathrm{e}^{-\lambda k}\mathrm{e}^{o(1)} = o(1). \tag{5.3.21}$$

For all $k \le n/2$ with $k \ge 5$, we bound $k(n-k) \ge kn/2$, so that

$$\mathbb{E}_\lambda[X_k] \le n(\mathrm{e}\lambda \mathrm{e}^{-\lambda/2})^k. \tag{5.3.22}$$

As a result, for $\lambda = a \log n$ with $a > 1/2$, and all $k \ge 5$, and using that $\lambda \mapsto \lambda \mathrm{e}^{-\lambda/2}$ is decreasing for $\lambda \ge 2$,

$$\mathbb{E}_\lambda[X_k] \le n^{1-k/4}. \tag{5.3.23}$$

We conclude that, using (5.3.21) and (5.3.23),

$$\begin{aligned} &\mathbb{P}_\lambda\big(\mathrm{ER}_n(\lambda/n) \text{ disconnected}, Y = 0\big) \\ &\quad \le \sum_{k=2}^{n/2} \mathbb{E}_\lambda[X_k] \le o(1) + \sum_{k=5}^{n/2} n^{1-k/4} = o(1). \end{aligned} \tag{5.3.24}$$

□

Proof of Theorem 5.8 The proof makes essential use of Proposition 5.10. We start by proving that for $\lambda - \log n \to -\infty$, the Erdős–Rényi random graph is with high probability disconnected. We use (5.3.3) to note that

$$\mathbb{E}_\lambda[Y] = n\mathrm{e}^{-\lambda}(1 + o(1)) = \mathrm{e}^{-\lambda + \log n}(1 + o(1)) \to \infty. \tag{5.3.25}$$

By the Chebychev inequality (Theorem 2.18), and the fact that $\lambda \le \log n$,

$$\mathbb{P}_\lambda(Y = 0) \le \frac{\mathbb{E}_\lambda[Y] + \frac{\lambda}{n-\lambda}\mathbb{E}_\lambda[Y]^2}{\mathbb{E}_\lambda[Y]^2} = \mathbb{E}_\lambda[Y]^{-1} + \frac{\lambda}{n-\lambda} \to 0. \tag{5.3.26}$$

Proposition 5.10 completes the proof that the Erdős–Rényi random graph is whp disconnected for $\lambda - \log n \to -\infty$.

When $\lambda - \log n \to \infty$ with $\lambda \le 2\log n$, then we start by noticing that by (5.3.24), it suffices to show that $\mathbb{P}_\lambda(Y = 0) \to 1$. By the Markov inequality (Theorem 2.17) and (5.3.5),

$$\mathbb{P}_\lambda(Y = 0) = 1 - \mathbb{P}_\lambda(Y \ge 1) \ge 1 - \mathbb{E}_\lambda[Y] \ge 1 - n\mathrm{e}^{-\lambda}O(1) \to 1. \tag{5.3.27}$$

This proves that the Erdős–Rényi random graph is whp connected when $\lambda - \log n \to \infty$ with $\lambda \le 2\log n$. Since connectivity of $\mathrm{ER}_n(\lambda/n)$ is an increasing property in λ, this also proves the claim for $\lambda - \log n \to \infty$ with $\lambda \ge 2\log n$, which completes the proof that $\mathrm{ER}_n(\lambda/n)$ is whp connected when $\lambda - \log n \to \infty$. □

5.3.2 Critical Window for Connectivity*

In this section, we investigate the critical window for connectivity by considering connectivity of $\mathrm{ER}_n(\lambda/n)$ when $\lambda = \log n + t$ for *fixed* $t \in \mathbb{R}$. The main result in this section is as follows:

Theorem 5.11 (Critical window for connectivity) *For $\lambda = \log n + t \to \infty$, the Erdős–Rényi random graph is connected with probability* $\mathrm{e}^{-\mathrm{e}^{-t}}(1 + o(1))$.

Proof In the proof, we again rely on Proposition 5.10. We fix $\lambda = \log n + t$ for some $t \in \mathbb{R}$. We prove a Poisson approximation for Y that reads that $Y \xrightarrow{d} Z$, where Z is a Poisson random variable with parameter

$$\lim_{n\to\infty} \mathbb{E}_\lambda[Y] = \mathrm{e}^{-t}, \tag{5.3.28}$$

where we recall (5.3.3). Therefore, the convergence in distribution of Y to a Poisson random variable with mean e^{-t} implies that

$$\mathbb{P}_\lambda(Y = 0) = \mathrm{e}^{-\mathrm{e}^{-t}} + o(1), \tag{5.3.29}$$

and the result follows by Proposition 5.10.

In order to show that $Y \xrightarrow{d} Z$, we use Theorems 2.4 and 2.5, so that it suffices to prove that for all $r \geq 1$, and recalling that $I_i = \mathbb{1}_{\{|\mathscr{C}(i)|=1\}}$ as defined in (5.3.2),

$$\lim_{n\to\infty} \mathbb{E}[(Y)_r] = \lim_{n\to\infty} \sum_{i_1,\ldots,i_r}^{*} \mathbb{P}_\lambda\big(I_{i_1} = \cdots = I_{i_r} = 1\big) = \mathrm{e}^{-tr}, \tag{5.3.30}$$

where the sum ranges over all $i_1, \ldots, i_r \in [n]$ which are *distinct*. By exchangeability of the vertices, $\mathbb{P}_\lambda\big(I_{i_1} = \cdots = I_{i_r} = 1\big)$ is independent of the precise choice of the indices $i_1, \ldots, i_r$, so that

$$\mathbb{P}_\lambda\big(I_{i_1} = \cdots = I_{i_r} = 1\big) = \mathbb{P}_\lambda\big(I_1 = \cdots = I_r = 1\big). \tag{5.3.31}$$

Using that there are $n(n-1)\cdots(n-r+1)$ distinct choices of $i_1, \ldots, i_r \in [n]$, we arrive at

$$\mathbb{E}[(Y)_r] = \frac{n!}{(n-r)!}\mathbb{P}_\lambda\big(I_1 = \cdots = I_r = 1\big). \tag{5.3.32}$$

The event $\{I_1 = \cdots = I_r = 1\}$ occurs precisely when all edges st with $s \in [r]$ and $t \in [n]$ are vacant. There are $r(r-1)/2 + r(n-r) = r(2n-r-1)/2$ of such edges, and since these edges are all independent, we arrive at

$$\begin{aligned}\mathbb{P}_\lambda\big(I_1 = \cdots = I_r = 1\big) &= (1 - \frac{\lambda}{n})^{r(2n-r-1)/2} = (1 - \frac{\lambda}{n})^{nr}(1 - \frac{\lambda}{n})^{-r(r+1)/2} \\ &= n^{-r}\mathbb{E}_\lambda[Y]^r(1 + o(1)),\end{aligned} \tag{5.3.33}$$

using that $\mathbb{E}_\lambda[Y] = n(1 - \lambda/n)^{n-1}$. Thus,

$$\lim_{n\to\infty} \mathbb{E}[(Y)_r] = \lim_{n\to\infty} \frac{n!}{(n-r)!} n^{-r}\mathbb{E}_\lambda[Y]^r = \mathrm{e}^{-tr}, \tag{5.3.34}$$

where we use (5.3.28). This completes the proof of Theorem 5.11. □

Exercise 5.5 investigates the second moment of the number of isolated vertices, while Exercise 5.6 investigates an upper bound on the probability that $\mathrm{ER}_n(p)$ is connected.

5.4 Degree Sequence of the Erdős–Rényi Random Graph

As described in Chapter 1, the degree sequences of various real-world networks obey power laws. In this section, we investigate the degree sequence of the Erdős–Rényi random graph for fixed $\lambda > 0$. In order to be able to state the result, we first introduce some notation. We write

$$p_k = e^{-\lambda}\frac{\lambda^k}{k!}, \qquad k \geq 0, \tag{5.4.1}$$

for the Poisson probability mass function with parameter λ. Let D_i denote the degree of vertex i and write

$$P_k^{(n)} = \frac{1}{n}\sum_{i\in[n]} \mathbb{1}_{\{D_i=k\}} \tag{5.4.2}$$

for the empirical degree distribution. The main result in this section is as follows:

Theorem 5.12 (Degree sequence of the Erdős–Rényi random graph) *Fix $\lambda > 0$. Then, for every ε_n such that $n\varepsilon_n^2 \to \infty$,*

$$\mathbb{P}_\lambda\big(\max_{k\geq 0} |P_k^{(n)} - p_k| \geq \varepsilon_n\big) \to 0. \tag{5.4.3}$$

Before proving Theorem 5.12, we give some simulation results of the degree distribution in $\mathrm{ER}_n(\lambda/n)$ for various n and λ. In Figure 5.3, a realization of the degree distribution of $\mathrm{ER}_n(\lambda/n)$ is plotted for $n = 100, \lambda = 2$. In Figure 5.4, a realization of the degree distribution of $\mathrm{ER}_n(\lambda/n)$ is plotted for $n = 10{,}000, \lambda = 1$ is compared to the probability mass function of a Poisson random variable with parameter 1. In Figure 5.5, we see that for $n = 1{,}000{,}000$ and $\lambda = 1$, the two distributions are virtually identical.

Proof We note that

$$\mathbb{E}_\lambda[P_k^{(n)}] = \mathbb{P}_\lambda(D_1 = k) = \binom{n-1}{k}\Big(\frac{\lambda}{n}\Big)^k\Big(1-\frac{\lambda}{n}\Big)^{n-k-1}. \tag{5.4.4}$$

Furthermore,

$$\sum_{k\geq 0}\Big|p_k - \mathbb{P}_\lambda(D_1 = k)\Big| = \sum_{k\geq 0}\big|\mathbb{P}(X^* = k) - \mathbb{P}(X_n = k)\big|, \tag{5.4.5}$$

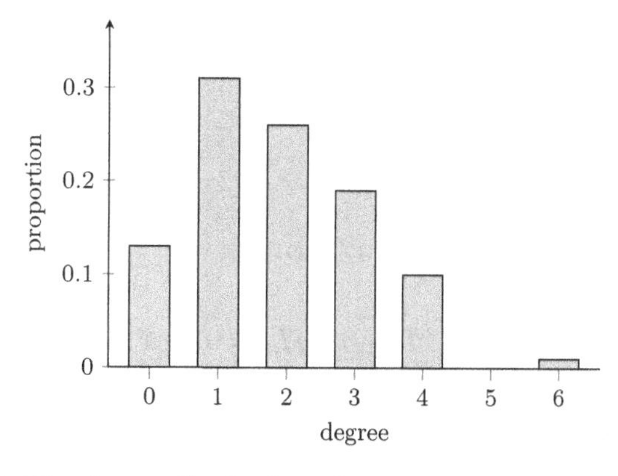

Figure 5.3 The degree distribution of $\mathrm{ER}_n(\lambda/n)$ with $n = 100, \lambda = 2$.

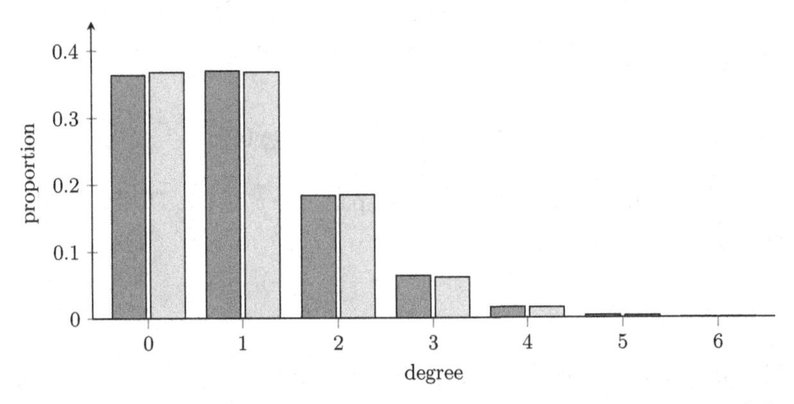

Figure 5.4 The degree distribution of $\mathrm{ER}_n(\lambda/n)$ with $n = 10{,}000$, $\lambda = 1$ (left) and the probability mass function of a Poisson random variable with parameter 1 (right).

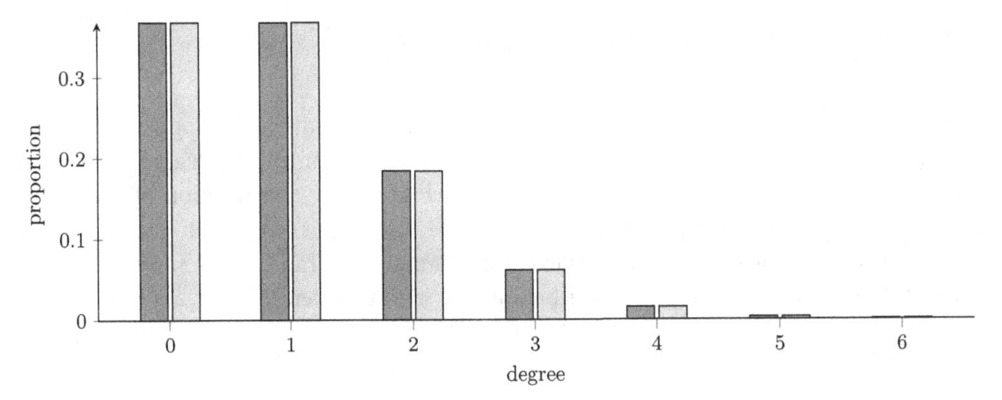

Figure 5.5 The degree distribution of $\mathrm{ER}_n(\lambda/n)$ with $n = 1{,}000{,}000$, $\lambda = 1$ (left) and the probability mass function of a Poisson random variable with parameter 1 (right).

where X^* is a Poisson random variable with mean λ, and X_n is a binomial random variable with parameters $n - 1$ and $p = \lambda/n$. We will use a coupling argument to bound this difference. Indeed, we let $Y_n = X_n + I_n$, where I_n has a Bernoulli distribution with parameter $p = \lambda/n$ independently of X_n. We can couple Y_n and X_n such that the probability that they are different is bounded by $p = \lambda/n$, i.e.,

$$\begin{aligned}
&\sum_{k=0}^{\infty} \big|\mathbb{P}(X_n = k) - \mathbb{P}(Y_n = k)\big| \qquad (5.4.6)\\
&\quad = \sum_{k=0}^{\infty} \big|\mathbb{P}(X_n = k) - \mathbb{P}(X_n = k, I_n = 0) - \mathbb{P}(X_n = k-1, I_n = 1)\big|\\
&\quad = \frac{\lambda}{n} \sum_{k=0}^{\infty} |\mathbb{P}(X_n = k) - \mathbb{P}(X_n = k-1)| \leq \frac{2\lambda}{n}.
\end{aligned}$$

Therefore, for all $k \geq 0$,

$$\sum_{k=0}^{\infty} \left|\mathbb{P}(X^* = k) - \mathbb{P}(X_n = k)\right| \tag{5.4.7}$$
$$\leq \sum_{k=0}^{\infty} \left|\mathbb{P}(X_n = k) - \mathbb{P}(Y_n = k)\right| + \sum_{k=0}^{\infty} \left|\mathbb{P}(X^* = k) - \mathbb{P}(Y_n = k)\right|$$
$$\leq \frac{2\lambda}{n} + \sum_{k=0}^{\infty} \left|\mathbb{P}(X^* = k) - \mathbb{P}(Y_n = k)\right| \leq \frac{2\lambda + \lambda^2}{n},$$

where we have also used Theorem 2.10. Since $\frac{2\lambda+\lambda^2}{n} \leq \frac{\varepsilon_n}{2}$, we have just shown that $\sum_{k\geq 0} |p_k - \mathbb{E}_\lambda[P_k^{(n)}]| \leq \varepsilon_n/2$ for n sufficiently large. Thus, it suffices to prove that

$$\mathbb{P}_\lambda\Big(\max_{k\geq 0} |P_k^{(n)} - \mathbb{E}_\lambda[P_k^{(n)}]| \geq \frac{\varepsilon_n}{2}\Big) = o(1). \tag{5.4.8}$$

For this, we use Boole's inequality to bound

$$\mathbb{P}_\lambda\Big(\max_{k\geq 0} |P_k^{(n)} - \mathbb{E}_\lambda[P_k^{(n)}]| \geq \frac{\varepsilon_n}{2}\Big) \leq \sum_{k=0}^{\infty} \mathbb{P}_\lambda\big(|P_k^{(n)} - \mathbb{E}_\lambda[P_k^{(n)}]| \geq \frac{\varepsilon_n}{2}\big). \tag{5.4.9}$$

By Chebychev's inequality (Theorem 2.18),

$$\mathbb{P}_\lambda\big(|P_k^{(n)} - \mathbb{E}_\lambda[P_k^{(n)}]| \geq \frac{\varepsilon_n}{2}\big) \leq 4\varepsilon_n^{-2} \mathrm{Var}_\lambda(P_k^{(n)}). \tag{5.4.10}$$

We then note that

$$\mathrm{Var}_\lambda(P_k^{(n)}) = \frac{1}{n}\Big[\mathbb{P}_\lambda(D_1 = k) - \mathbb{P}_\lambda(D_1 = k)^2\Big]$$
$$+ \frac{n-1}{n}\Big[\mathbb{P}_\lambda(D_1 = D_2 = k) - \mathbb{P}_\lambda(D_1 = k)^2\Big]. \tag{5.4.11}$$

We again use a coupling argument. We let X_1, X_2 be two independent $\mathsf{Bin}(n-2, \lambda/n)$ random variables, and I_1, I_2 two independent Bernoulli random variables with success probability λ/n. Then, the law of (D_1, D_2) is the same as the one of $(X_1 + I_1, X_2 + I_1)$ while $(X_1 + I_1, X_2 + I_2)$ are two independent copies of D_1. Thus,

$$\mathbb{P}_\lambda(D_1 = D_2 = k) = \mathbb{P}_\lambda\big((X_1 + I_1, X_2 + I_1) = (k, k)\big), \tag{5.4.12}$$
$$\mathbb{P}_\lambda(D_1 = k)^2 = \mathbb{P}_\lambda\big((X_1 + I_1, X_2 + I_2) = (k, k)\big), \tag{5.4.13}$$

so that

$$\mathbb{P}_\lambda(D_1 = D_2 = k) - \mathbb{P}_\lambda(D_1 = k)^2 \tag{5.4.14}$$
$$\leq \mathbb{P}_\lambda\Big((X_1 + I_1, X_2 + I_1) = (k, k), (X_1 + I_1, X_2 + I_2) \neq (k, k)\Big).$$

When $(X_1 + I_1, X_2 + I_1) = (k, k)$, but $(X_1 + I_1, X_2 + I_2) \neq (k, k)$, we must have that $I_1 \neq I_2$. If $I_1 = 1$, then $I_2 = 0$ and $X_2 = k - 1$, while, if $I_1 = 0$, then $I_2 = 1$ and $X_1 = k$. Therefore,

$$\mathbb{P}_\lambda(D_1 = D_2 = k) - \mathbb{P}_\lambda(D_1 = k)^2 \leq \frac{\lambda}{n}\big[\mathbb{P}_\lambda(X_1 = k) + \mathbb{P}_\lambda(X_2 = k - 1)\big]. \tag{5.4.15}$$

We conclude from (5.4.11) and (5.4.15) that

$$\mathrm{Var}_\lambda(P_k^{(n)}) \leq \frac{\lambda}{n}\big[\mathbb{P}_\lambda(X_1 = k) + \mathbb{P}_\lambda(X_2 = k-1)\big] + \frac{1}{n}\mathbb{P}_\lambda(D_1 = k), \tag{5.4.16}$$

so that, by (5.4.9)–(5.4.10),

$$\begin{aligned} &\mathbb{P}_\lambda\big(\max_{k\geq 0} |P_k^{(n)} - \mathbb{E}_\lambda[P_k^{(n)}]| \geq \varepsilon_n/2\big) \qquad (5.4.17)\\ &\leq \frac{4}{\varepsilon_n^2}\sum_{k\geq 0}\Big[\frac{\lambda}{n}\mathbb{P}_\lambda(X_1 = k) + \frac{\lambda}{n}\mathbb{P}_\lambda(X_2 = k-1) + \frac{1}{n}\mathbb{P}_\lambda(D_1 = k)\Big]\\ &= \frac{4(2\lambda+1)}{\varepsilon_n^2 n} \to 0, \end{aligned}$$

since $\varepsilon_n^2 n \to \infty$. This completes the proof of Theorem 5.12. □

In Chapter 6 below, we give an alternative proof of Theorem 5.12, allowing for weaker bounds on ε_n. In that proof, we use that the Erdős–Rényi random graph is a special case of the generalized random graph with equal weights. See Theorem 6.12 below.

5.5 Notes and Discussion

Notes on Section 5.1

We list some more recent results. Janson and Spencer (2007) give a point process description of the sizes and number of components of size $\varepsilon n^{2/3}$. Pittel (2001) derives an explicit, yet involved, description for the distribution of the limit of $|\mathscr{C}_{\max}|n^{-2/3}$. The proof makes use of generating functions, and the relation between the largest connected component and the number of labeled graphs with a given complexity l. Here, the complexity of a graph is its number of edges minus its number of vertices. Relations between Erdős–Rényi random graphs and the problem of counting the number of labeled graphs has received considerable attention, see e.g. Bollobás (1984b); van der Hofstad and Spencer (2006); Łuczak (1990b); Spencer (1997); Wright (1977, 1980) and the references therein. Consequences of the result by Pittel (2001) are for example that the probability that $|\mathscr{C}_{\max}|n^{-2/3}$ exceeds a for large a decays as $\mathrm{e}^{-a^3/8}$ (in fact, the asymptotics are much finer than this!), and for very small $a > 0$, the probability that $|\mathscr{C}_{\max}|n^{-2/3}$ is smaller than a decays as $\mathrm{e}^{-ca^{-3/2}}$ for some explicit constant $c > 0$. The bound on the upper tails of $|\mathscr{C}_{\max}|n^{-2/3}$ is also proved by Nachmias and Peres (2010), and is valid for all n and a, with the help of relatively simple martingale arguments. Nachmias and Peres (2010) also explicitly prove the bound (5.1.10).

Notes on Section 5.2

Theorem 5.4 is by Nachmias and Peres (2010). Many more precise results are proved there, including detailed bounds on the behavior within the critical window. We state an extension of Theorem 5.4 to the complete critical window:

Theorem 5.13 (Critical window revisited) *Set* $\lambda = 1 + \theta n^{-1/3}$ *for some* $\theta \in \mathbb{R}$ *and consider* $\mathrm{ER}(n, \lambda/n)$. *For* $\theta > 0$ *and* $A > 2\theta + 3$, *for large enough* n,

$$\mathbb{P}_\lambda(|\mathscr{C}(v)| \geq An^{2/3}) \leq \left(\frac{4\theta}{1 - \mathrm{e}^{-4\theta}} + 16\right) n^{-1/3} \mathrm{e}^{-((A-1)^2/2-(A-1)\theta-2)^2/(4A)}, \quad (5.5.1)$$

and

$$\mathbb{P}_\lambda(|\mathscr{C}_{\max}| \geq An^{2/3}) \leq \left(\frac{4\theta}{A(1 - \mathrm{e}^{-4\theta})} + \frac{16}{A}\right) \mathrm{e}^{-((A-1)^2/2-(A-1)\theta-2)^2/(4A)}. \quad (5.5.2)$$

Theorem 5.13 gives bounds that are remarkably close to the truth, as derived by Pittel (2001).

Aldous (1997) also investigates the number of edges in the largest connected components in the critical window, showing that the number of extra edges also converges in distribution. Thus, only a finite number of edges needs to be removed to turn these clusters into trees. Recently, much interest has been devoted to the precise properties of critical clusters. The most refined work is by Addario-Berry, Broutin and Goldschmidt (2010), who identify the limit of the largest clusters seen as *graphs*, under an appropriate topology. This work makes crucial use of the results by Aldous (1997), and extends it significantly.

Theorem 5.7 is due to Aldous (1997), following previous work on the critical window in Bollobás (1984a); Janson et al. (1993); Łuczak (1990a); Łuczak et al. (1994). While Aldous (1997) is the first paper where a result as in Theorem 5.7 is stated explicitly, similar results had been around before Aldous (1997), which explains why Aldous calls Theorem 5.7 a 'Folk Theorem'. The beauty of Aldous (1997) is that he gives two explicit descriptions of the distribution of the limiting random variable $(\gamma_i(t))_{i\geq 1}$, the first being in terms of lengths of excursions of Brownian motion, the second in terms of the so-called *multiplicative coalescent process*. Finally, in van der Hofstad et al. (2009), an explicit local limit theorem is proved for all largest connected components inside the critical window.

The relation between the Erdős–Rényi random graph and coalescing processes can also be found in the standard book by Bertoin on coagulation and fragmentation (Bertoin, 2006, Section 5.2) and the references therein. In fact, $\mathrm{ER}_n(p)$ for the entire regime of $p \in [0, 1]$ can be understood using coalescent processes, for which the multiplicative coalescent is most closely related to random graphs.

Notes on Section 5.3

Connectivity of the Erdős–Rényi random graph was investigated in the early papers on the subject. Theorem 5.8 and its extension, Theorem 5.11, were first proved by Rényi (1959), where versions of Theorems 5.8–5.11 were proved for $\mathrm{ER}(n, M)$. Bollobás gives two separate proofs in (Bollobás, 2001, Pages 164-165). Exercise 5.6 is based on a suggestion of Mislav Mišković.

Notes on Section 5.4

The degrees of Erdős–Rényi random graphs have attracted considerable attention. In particular, when ordering the degrees by size as $d_1 \geq d_2 \geq \cdots \geq d_n$, various properties have been shown, such as the fact that there is, with high probability, a *unique* vertex with degree d_1 (see Erdős and Wilson (1977)). See Bollobás (1981) or Bollobás (2001) for more details. The result on the degree sequence proved here is a weak consequence of the result by Janson (2007, Theorem 4.1), where even asymptotic normality was shown for the number of vertices with degree k, for all k simultaneously.

5.6 Exercises for Chapter 5

Exercise 5.1 (Tightness of $|\mathscr{C}_{\max}|/n^{2/3}$ and $n^{2/3}/|\mathscr{C}_{\max}|$ in Theorem 5.1) Fix $\lambda = 1 + \theta n^{-1/3}$ for some $\theta \in \mathbb{R}$ and n sufficiently large. Prove that Theorem 5.1 implies that $|\mathscr{C}_{\max}|/n^{2/3}$ and $n^{2/3}/|\mathscr{C}_{\max}|$ are tight sequences of random variables.

Exercise 5.2 (Critical expected cluster size) Prove that Proposition 5.2 also implies that $\mathbb{E}_{1+\theta n^{-1/3}}[|\mathscr{C}(1)|] \geq cn^{1/3}$ for some $c > 0$. Therefore, for $\lambda = 1 + \theta n^{-1/3}$ with $\theta < 0$, the bound in Proposition 5.3 is asymptotically sharp.

Exercise 5.3 (Derivative of expected cluster size) Let $\chi_n(\lambda) = \mathbb{E}_\lambda[|\mathscr{C}(v)|]$ be the expected cluster size in $\mathrm{ER}_n(\lambda/n)$. Prove that

$$\frac{\partial}{\partial\lambda}\chi_n(\lambda) \leq \chi_n(\lambda)^2, \tag{5.6.1}$$

and use this inequality to deduce that for all $\lambda \leq 1$

$$\chi_n(\lambda) \geq \frac{1}{\chi_n(1)^{-1} + (\lambda - 1)}. \tag{5.6.2}$$

Exercise 5.4 (Equality total progeny probabilities) Prove that

$$\frac{(\lambda t)^{t-1}}{t!}\mathrm{e}^{-\lambda t} = \frac{1}{\lambda}\mathrm{e}^{-I_\lambda t}\mathbb{P}_1^*(T^* = t). \tag{5.6.3}$$

Exercise 5.5 (Second moment of the number of isolated vertices) Prove directly that the second moment of Y converges to the second moment of Z, by using (5.3.10).

Exercise 5.6 (Upper bound on connectivity) Use Proposition 4.6 to prove that

$$\mathbb{P}(\mathrm{ER}_n(p) \text{ connected}) \leq (1 - (1-p)^{n-1})^{n-1}. \tag{5.6.4}$$

Hint: Use that

$$\mathbb{P}(\mathrm{ER}_n(p) \text{ connected}) = \mathbb{P}_{np}(|\mathscr{C}(1)| = n) = \mathbb{P}_{np}(S_t > 0\ \forall t \in [n-1], S_{n-1} = 0). \tag{5.6.5}$$

Exercise 5.7 (Disconnected phase $\mathrm{ER}_n(p)$) Use the previous exercise to give an alternative proof that $\mathbb{P}(\mathrm{ER}_n(p) \text{ connected}) = o(1)$ when $p = \lambda/n$ with $\lambda - \log n \to -\infty$.

Exercise 5.8 (Isolated edges $\mathrm{ER}_n(p)$) Fix $\lambda = a \log n$. Let M denote the number of isolated edges in $\mathrm{ER}_n(p)$ with $p = \lambda/n$, i.e., edges that are occupied but for which the vertices at either end have no other neighbors. Show that $M \xrightarrow{\mathbb{P}} \infty$ when $a > \frac{1}{2}$, while $M \xrightarrow{\mathbb{P}} 0$ for $a < \frac{1}{2}$.

Exercise 5.9 (Isolated edges $\mathrm{ER}_n(p)$ (cont.)) Fix $\lambda = \frac{1}{2}\log n + t$. Let M denote the number of isolated edges in $\mathrm{ER}_n(p)$ with $p = \lambda/n$. Show that there exists a random variable X such that $M \xrightarrow{d} X$, and identify the limiting distribution X.

Exercise 5.10 (Number of connected components $\mathrm{ER}_n(p)$) Fix $\lambda = \log n + t$ with $t \in \mathbb{R}$. Let N denote the number of connected components in $\mathrm{ER}_n(p)$ with $p = \lambda/n$. Use Theorem 5.11, as well as the ideas in its proof, to show that $N \xrightarrow{d} 1 + Z$, where $Z \sim \mathsf{Poi}(\mathrm{e}^{-t})$.

Exercise 5.11 (Vertices of degree 2) Fix $\lambda = 2\log n + t$. Let N_2 denote the number of vertices of degree 2 in $\mathrm{ER}_n(p)$ with $p = \lambda/n$. Show that there exists a random variable X such that $N_2 \xrightarrow{d} X$, and identify the limiting distribution X.

Exercise 5.12 (Minimal degree 2) Fix $\lambda = 2\log n + t$. Let D_{min} denote the minimal degree in $\mathrm{ER}_n(p)$ with $p = \lambda/n$. Use the previous exercise to show that $\mathbb{P}_\lambda(D_{\mathrm{min}} = 2)$ converges, and identify its limit.

Exercise 5.13 (Minimal degree 3) Let D_{min} denote the minimal degree in $\mathrm{ER}_n(p)$ with $p = \lambda/n$. For which λ does $\mathbb{P}_\lambda(D_{\mathrm{min}} = 3)$ converge to a limit in $(0, 1)$? For such λ, identify the limiting value.

Exercise 5.14 (Maximal degree in $\mathrm{ER}_n(\lambda/n)$) Fix $\lambda > 0$. Show that the maximal degree $D_{\mathrm{max}} = \max_{i\in[n]} D_i$ in $\mathrm{ER}_n(\lambda/n)$ satisfies that

$$D_{\mathrm{max}}/\log n \xrightarrow{\mathbb{P}} 0. \tag{5.6.6}$$

Exercise 5.15 (Maximal degree in $\mathrm{ER}_n(\lambda/n)$ (cont.)) Fix $\lambda > 0$. How should we choose a_n such that the maximal degree $D_{\mathrm{max}} = \max_{i\in[n]} D_i$ in $\mathrm{ER}_n(\lambda/n)$ satisfies that

$$D_{\mathrm{max}}/a_n \xrightarrow{\mathbb{P}} 1? \tag{5.6.7}$$

Exercise 5.16 (Tails of Poisson) Fix $\lambda > 0$ and let X^* be a Poisson random variable with parameter λ. Show that there is $C = C(A) > 0$ such that for every $A > 0$

$$\mathbb{P}(X^* \geq x) \leq C\mathrm{e}^{-Ax}. \tag{5.6.8}$$

Part III

Models for Complex Networks

Intermezzo: Back to Real-World Networks...

Theorem 5.12 shows that the degree sequence of the Erdős–Rényi random graph is close to a Poisson distribution with parameter λ. A Poisson distribution has thin tails, for example, its moment generating function is always finite (see e.g. Exercise 5.16). Recall Figure 5.4–5.5 for a comparison of the degree sequence of an Erdős–Rényi random graph with $n = 10{,}000$ and $n = 1{,}000{,}000$ and $\lambda = 1$ to the probability mass function of a Poisson random variable with parameter $\lambda = 1$, which are quite close. See the figure below for a log-log plot of the probability mass function of a Poisson random variable with $\lambda = 1$ and $\lambda = 10$.

This figure is quite different from the log-log plots in real-world networks, such as redrawn in the figure below. As a result, the Erdős–Rényi random graph cannot be used to model real-world networks where power-law degree sequences are observed. Therefore, several alternative models have been proposed. In this intermezzo, we discuss degree sequences of networks and informally introduce three random graph models for complex networks that will be discussed in detail in the following three chapters.

In Chapters 6–8 we will be interested in the properties of the *degree sequence of a graph.* A natural question is which sequences of numbers can occur as the degree sequence of a simple graph. A sequence $(d_i)_{i\in[n]}$ with $d_1 \geq d_2 \geq \cdots \geq d_n$ is called *graphic* if it is the degree sequence of a simple graph. Thus, the question is which degree sequences are graphic? Erdős and Gallai (1960) prove that a degree sequence $(d_i)_{i\in[n]}$ is graphic if and

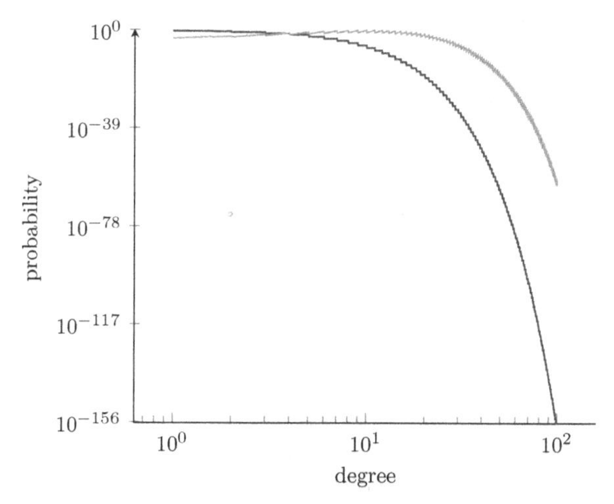

Figure I.1 The probability mass function of a Poisson random variable with $\lambda = 1$ and $\lambda = 10$ in log-log scale.

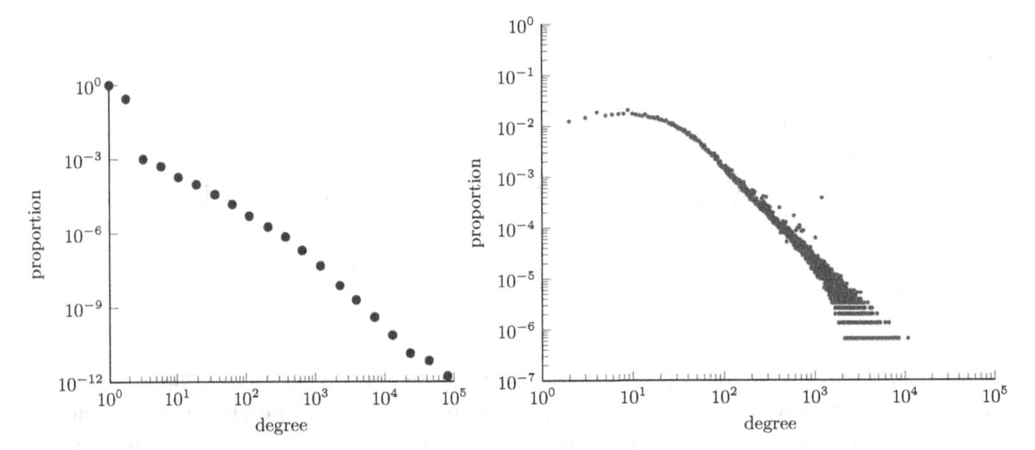

Figure I.2 Log-log plots of degree sequence AS domains as in Figure 1.5, and in the IMDb as in Figure 1.12.

only if $\sum_{i\in[n]} d_i$ is even and

$$\sum_{i=1}^{k} d_i \le k(k-1) + \sum_{i=k+1}^{n} \min(k, d_i), \tag{I.1}$$

for each integer $k \le n-1$. The fact that the total degree of a graph needs to be even is fairly obvious:

Exercise I.1 (Handshake lemma) Show that $\sum_{i\in[n]} d_i$ is even for every graph on the vertices $[n]$, where d_i denotes the degree of vertex $i \in [n]$.

The necessity of (I.1) is relatively easy to see. The left side of (I.1) is the degree of the first k vertices. The first term on the right-hand side of (I.1) is twice the maximal number of edges between the vertices in $[k]$. The second term is a bound on the total degree of the vertices $[k]$ coming from edges that connect to vertices in $[n] \setminus [k]$. The sufficiency is harder to see (see Choudum (1986) for a simple proof of this fact, and Sierksma and Hoogeveen (1991) for seven different proofs). Arratia and Liggett (2005) investigate the asymptotic probability that an i.i.d. sequence of n integer-valued random variables is graphical, the result being in many cases equal to 0 or 1/2, at least when $\mathbb{P}(D \text{ even}) \neq 1$. In the latter case, the limit is equal to 0 when $\lim_{n\to\infty} n\mathbb{P}(D_i \ge n) = \infty$ and 1/2 when $\lim_{n\to\infty} n\mathbb{P}(D_i \ge n) = 0$. Interestingly, when $\lim_{n\to\infty} n\mathbb{P}(D_i \ge n) = c$ for some constant $c > 0$, then the set of limit points of the probability that $(D_i)_{i\in[n]}$ is graphical is a subset of $(0, 1/2)$. The proof is by verifying that (I.1) holds. We will see that in the *sparse* setting, most sequences having an even sum are graphic.

As argued above, the Erdős–Rényi random graph is not a good model for many real-world networks. This can be understood by noting that real-world networks are quite *inhomogeneous*, in the sense that there is much variability in the roles that vertices play. The Erdős–Rényi random graph, on the other hand, is completely *egalitarian*, in the sense that, at least in distribution, every vertex plays an *identical* role. Since the degrees of all the vertices are close to Poissonian, and Poisson distributions have thin tails, this means that the degree distribution of the Erdős–Rényi random graph is completely different from those seen in real-world networks. We conclude that random graph models for complex networks need to incorporate inhomogeneity. In Chapters 6–8, we will discuss three models that incorporate this inhomogeneity in three different ways.

The first model is the so-called *generalized random graph* (GRG), and was first introduced in by Britton, Deijfen and Martin-Löf in Britton et al. (2006). This model stays the closest to the Erdős–Rényi random graph, in the sense that edges in the model remain to be present independently, but steps away from the homogeneous setting by varying the edge probabilities. In the GRG, each vertex $i \in [n]$ receives a *weight* w_i. This weight can be thought of as the propensity of the vertex to have edges, and will turn out to be quite close to the *expected* degree of the vertex. Given the weights, edges are present independently, but the occupation probabilities for different edges are not identical, but moderated by the weights of the vertices. Naturally, this can be done in several different ways. The most general version is presented in the seminal paper by Bollobás, Janson and Riordan (2007), which we explain in detail in [II, Chapter 2]. In the generalized random graph, the edge probability of the edge between vertex i and j with weights $(w_i)_{i\in[n]}$ is equal to

$$p_{ij} = \frac{w_i w_j}{\ell_n + w_i w_j}, \tag{I.2}$$

where ℓ_n is the total weight of all vertices given by

$$\ell_n = \sum_{i\in[n]} w_i. \tag{I.3}$$

Thus, in this model the inhomogeneity is modelled using vertex weights, and the edges incident to vertices with higher weights are more likely to be present, enforcing the degrees of vertices with higher weights to be larger than those of vertices with smaller weights. When the weights are chosen appropriately, this can give rise to random graphs with highly varying vertex degrees. We will focus on the degree distribution of such random graphs when the weights are chosen to be sufficiently regular. We also discuss several related models where the edge probabilities are close, but not exactly equal, to the choice in (I.2). For details, we refer to Chapter 6.

The second model is the *configuration model*, in which the degrees of the vertices are fixed. Indeed, we write d_i for the degree of vertex i, and let, similarly to (I.3), $\ell_n = \sum_{i\in[n]} d_i$ denote the total degree. We assume that ℓ_n is even. We make a graph where vertex i has precisely degree d_i. Thus, in this model, the inhomogeneity is due to the fact that vertices have different degrees. To create a graph with the specified degrees, we think of each vertex having d_i half-edges attached to it. Two half-edges can be paired to each other to form an edge. The configuration model is the model where all half-edges are connected in a uniform fashion, i.e., where the half-edges are uniformly matched. The nice aspect of this model is that we have complete freedom in how we choose the degrees. In the GRG, instead, vertices with degree zero will always be present in the sparse setting, and this might not be realistic for many real-world networks. In the configuration model, we can take the degrees to be like the ones seen in real-world networks, or even take them precisely the same. When we choose the degrees to be precisely equal, then we create a random graph that has precisely the same degrees as the real-world network and this model is sometimes called the *null model*. By comparing the null model to the real network, one is able to judge whether the real-world network can be appropriately modeled by the random graph with the same degree distribution, or whether it has more structure, such as community structure, high clustering, or degree dependencies.

Unfortunately, when we match the half-edges uniformly, we may create self-loops and multiple edges in the configuration model. Thus, the configuration model produces a *multigraph* rather than a simple graph, which is in many real-world networks unwanted. We focus in this chapter on the degree distribution when we erase self-loops and multiple edges, as well as on the probability that the resulting graph is simple. A main conclusion is that conditionally on simplicity, the resulting random graph is *uniform* amongst the collection of all possible graphs with the specified degree sequence. Another main conclusion is that for most regular degree sequences, for example when the average degree ℓ_n/n remains bounded and ℓ_n is even, there *is* a graph with those degrees, so that most 'nice' degree sequences are graphic. The construction of the configuration model also allows us to compute *how many* graphs there are with a specific degree sequence, which is a beautiful example of the *probabilistic method*. The configuration model is studied in detail in Chapter 7.

The third model is the so-called *preferential attachment model*, in which the *growth of the random graph* is modeled by adding edges to the already existing graph in such a way that vertices with large degree are more likely to be connected to the newly added edges. This creates a preference for vertices that have a large degree. In this model, the inhomogeneity is due to the age of the vertices, that is, older vertices have larger degrees than younger vertices. See Chapter 8 for details.

All these models have in common that the degree sequence converges to some limiting distribution that can have various shapes, particularly including power laws. In Chapters 6–8, we focus on degree properties of the random graphs involved, as well as other basic properties of the graphs involved. We further describe extensions and generalizations of these models, and briefly discuss how these generalizations affect the structure of the random graphs involved. We shall study connectivity properties, such as the number and sizes of connected components as well as the graph distances in these models, in the sequel to this book, in [II, Chapters 2–7], respectively.

6

Generalized Random Graphs

In this chapter, we discuss *inhomogeneous random graphs*, in which the equal edge probabilities of the Erdős–Rényi random graph are replaced by edge occupation statuses that are independent but not equally distributed. Indeed, in the models studied here, the edge probabilities are moderated by certain vertex weights. These weights can be taken to be *deterministic* or *random*, and both options will be considered in this chapter. An important example, on which we focus in this chapter, is the so-called *generalized random graph*. We show that this model has a power-law degree distribution when the weights do so. As such, this is one of the simplest adaptions of the Erdős–Rényi random graph having a power-law degree sequence.

Organisation of this Chapter

This chapter is organized as follows. In Section 6.1, we motivate the model and in Section 6.2 we introduce it formally. In Section 6.3, we investigate the degree of a fixed vertex in the generalized random graph, and in Section 6.4, we investigate the degree sequence of the generalized random graph. In Section 6.5, we study the generalized random graph with i.i.d. vertex weights. In Section 6.6, we show that the generalized random graph, conditioned on its degrees, is a uniform random graph with these degrees. In Section 6.7, we study when two inhomogeneous random graphs are asymptotically equivalent, meaning that they have the same asymptotic probabilities. Finally, in Section 6.8, we introduce several more models of inhomogeneous random graphs similar to the generalized random graph that have been studied in the literature, such as the so-called Chung-Lu or random graph with prescribed expected degrees and the Norros–Reittu or Poisson graph process model. We close this chapter with notes and discussion in Section 6.9 and exercises in Section 6.10.

6.1 Motivation of the Model

In the Erdős–Rényi random graph, every vertex plays the same role, leading to a completely *homogeneous* random graph. Of course, vertices do have different degrees, but the amount of variability in degrees in the Erdős–Rényi random graph is much smaller than in many real-world applications. In this chapter, we explain a possible way of extending the Erdős–Rényi random graph model to include a higher amount of inhomogeneity. We start with a simple example:

Example 6.1 (Population of two types) Suppose that we have a complex network in which two distinct types of vertices are present. The first type has on average m_1 neighbors, the second type m_2, where $m_1 \neq m_2$. How can we construct a random graph in which such heterogeneity can be incorporated?

Example 6.2 (Power-law degrees) In Chapter 1, many examples of real-world networks were given where the degrees are quite variable, including hubs having quite high degrees. How can we construct a random graph with a high amount of variability in the vertex degrees, for example in the form of power-law degrees?

Example 6.1 deals with a population having only two types. In many situations, there can be many different types. An important example is the case where the degrees obey a power law as in Example 6.2, for which any finite number of types is insufficient. We think of the different types as giving rise to *inhomogeneity* in the random graph. In the so-called generalized random graph, we model this inhomogeneity by adding *vertex weights*. Vertices with high weights are more likely to have many neighbors than vertices with small weights. Vertices with extremely high weights could act as the hubs observed in many real-world networks. Let us now formally introduce the generalized random graph model.

6.2 Introduction of the Model

In the generalized random graph model, the edge probability of the edge between vertices i and j, for $i \neq j$, is equal to

$$p_{ij} = p_{ij}^{(\mathrm{GRG})} = \frac{w_i w_j}{\ell_n + w_i w_j}, \tag{6.2.1}$$

where $\boldsymbol{w} = (w_i)_{i \in [n]}$ are the *vertex weights* and ℓ_n is the total weight of all vertices given by

$$\ell_n = \sum_{i \in [n]} w_i. \tag{6.2.2}$$

We denote the resulting graph by $\mathrm{GRG}_n(\boldsymbol{w})$. In many cases, the vertex weights actually depend on n, and it would be more appropriate, but also more cumbersome, to write the weights as $\boldsymbol{w}^{(n)} = (w_i^{(n)})_{i \in [n]}$. To keep notation simple, we refrain from making the dependence on n explicit.

A special case of the generalized random graph is when we take $w_i \equiv \frac{n\lambda}{n-\lambda}$, in which case $p_{ij} = \lambda/n$ for all $i, j \in [n]$, so that we retrieve the Erdős–Rényi random graph $\mathrm{ER}_n(\lambda/n)$ (see Exercise 6.1).

Without loss of generality, we assume that $w_i > 0$. Note that vertex $i \in [n]$ will be isolated with probability 1 when $w_i = 0$, so that we can remove i from the graph. The vertex weights moderate the inhomogeneity in the random graph, vertices with high weights have higher edge occupation probabilities than vertices with low weights. Therefore, this suggests that we can create graphs with flexible degree sequences by choosing the weights in an appropriate way. We investigate the degree structure of the generalized random graph in more detail in this chapter. Let us first explain in more detail how we can model the network with two populations in Example 6.1:

Example 6.3 (Population of two types (cont.)) *In Example 6.1, we let the vertices of type 1 have weight m_1 and the vertices of type 2 have weight m_2. Let n_1 and n_2 denote the number of vertices of weight 1 and 2, respectively, so that*

$$\ell_n = n_1 m_1 + n_2 m_2. \tag{6.2.3}$$

Further, the probability that a vertex of type 1 is connected to another vertex of type 1 is equal to $m_1^2/(\ell_n + m_1^2)$, while the probability that it is connected to a vertex of type 2 equals $m_1 m_2/(\ell_n + m_1 m_2)$. Therefore, the expected degree of a vertex of type 1 is equal to

$$\begin{aligned}(n_1 - 1)\frac{m_1^2}{\ell_n + m_1^2} + n_2 \frac{m_1 m_2}{\ell_n + m_1 m_2} &= m_1\Big[\frac{(n_1 - 1)m_1}{\ell_n + m_1^2} + \frac{n_2 m_2}{\ell_n + m_1 m_2}\Big] \\ &= m_1(1 + o(1)),\end{aligned} \tag{6.2.4}$$

whenever $m_1^2 + m_1 m_2 = o(\ell_n)$. Similarly, a vertex of type 2 has expected degree $m_2(1+o(1))$ whenever $m_1 m_2 + m_2^2 = o(\ell_n)$. Thus, our graph is such that vertices of type 1 have on average approximately m_1 neighbors, while vertices of type 2 have average degree approximately m_2 whenever $m_1^2 + m_2^2 = o(\ell_n)$.

Naturally, the topology of the generalized random graph depends sensitively upon the choice of the vertex weights $\boldsymbol{w} = (w_i)_{i \in [n]}$. These vertex weights can be rather general, and we both investigate settings where the weights are *deterministic*, as well as where they are *random*. In order to describe the empirical proporties of the weights, we define their *empirical distribution function* to be

$$F_n(x) = \frac{1}{n} \sum_{i \in [n]} \mathbb{1}_{\{w_i \leq x\}}, \qquad x \geq 0. \tag{6.2.5}$$

We can interpret F_n as the distribution of the weight of a uniformly chosen vertex in $[n]$ (see Exercise 6.2). We denote the weight of a uniformly chosen vertex U in $[n]$ by $W_n = w_U$, so that, by Exercise 6.2, W_n has distribution function F_n.

We aim to identify the asymptotic degree distribution in the generalized random graph, i.e., we aim to prove that the proportion of vertices of degree k approaches a limit when we let the size of the network n tend to infinity. The degree distribution can only converge when the vertex weights are sufficiently regular. We often assume that the vertex weights satisfy the following *regularity conditions,* which turn out to imply convergence of the degree distribution in the generalized random graph:

Condition 6.4 (Regularity conditions for vertex weights) *There exists a distribution function F such that, as $n \to \infty$ the following conditions hold:*
(a) Weak convergence of vertex weight. *As $n \to \infty$,*

$$W_n \xrightarrow{d} W, \tag{6.2.6}$$

where W_n and W have distribution functions F_n and F, respectively. Equivalently, for any x for which $x \mapsto F(x)$ is continuous,

$$\lim_{n \to \infty} F_n(x) = F(x). \tag{6.2.7}$$

(b) Convergence of average vertex weight.

$$\lim_{n\to\infty} \mathbb{E}[W_n] = \mathbb{E}[W], \tag{6.2.8}$$

where W_n and W have distribution functions F_n and F, respectively. Further, we assume that $\mathbb{E}[W] > 0$.

(c) Convergence of second moment vertex weight.

$$\lim_{n\to\infty} \mathbb{E}[W_n^2] = \mathbb{E}[W^2]. \tag{6.2.9}$$

Condition 6.4(a) guarantees that the weight of a 'typical' vertex is close to a random variable W that is independent of n. Condition 6.4(b) implies that the average weight of the vertices in $\mathrm{GRG}_n(\boldsymbol{w})$ converges to the expectation of the limiting weight variable. In turn, this implies that the average degree in $\mathrm{GRG}_n(\boldsymbol{w})$ converges to the expectation of the limit random variable of the vertex weights. Condition 6.4(c) ensures the convergence of the second moment of the weights to the second moment of the limiting weight variable.

In most of our results, we assume that Conditions 6.4(a)–(b) hold, in some we also need Condition 6.4(c). We emphasize that Condition 6.4(b) together with (a) is stronger than the convergence of the average weight. An example where $\mathbb{E}[W_n] = \ell_n/n$ converges, but not to $\mathbb{E}[W]$ can be obtained by taking a weight sequence $\boldsymbol{w} = (w_i)_{i\in[n]}$ that satisfies Conditions 6.4(a)–(b) and replacing w_n by $w_n' = \varepsilon n$ for some $\varepsilon > 0$.

Remark 6.5 (Regularity for random weights) In the sequel, we often deal with cases where the weights of the vertices are *random* themselves. For example, this arises when the weights $\boldsymbol{w} = (w_i)_{i\in[n]}$ are realizations of i.i.d. random variables. When the weights are random variables themselves, also the function F_n is a random distribution function. Indeed, in this case F_n is the *empirical distribution function* of the random weights $(w_i)_{i\in[n]}$. We stress that then $\mathbb{E}[W_n]$ is to be interpreted as $\frac{1}{n}\sum_{i\in[n]} w_i$, which is itself random. Therefore, in Condition 6.4, we require random variables to converge, and there are several notions of convergence that may be used. As it turns out, the most convenient notion of convergence is *convergence in probability*. Thus, we replace Condition 6.4(a) by the condition that

$$\mathbb{P}_n(W_n \leq x) \xrightarrow{\mathbb{P}} \mathbb{P}(W \leq x) = F(x), \tag{6.2.10}$$

for each continuity point x of $x \mapsto F(x)$ and where $\mathbb{P}_n$ denotes the conditional probability given the (random) degrees $(w_i)_{i\in[n]}$. Equation (6.2.10) is equivalent to the statement that, for every continuity point $x \geq 0$ of F and every $\varepsilon > 0$,

$$\mathbb{P}\Big(\big|\mathbb{P}_n(W_n \leq x) - \mathbb{P}(W \leq x)\big| > \varepsilon\Big) \to 0. \tag{6.2.11}$$

Similarly, Condition 6.4(b) and (c) are replaced by

$$\mathbb{E}_n[W_n] \xrightarrow{\mathbb{P}} \mathbb{E}[W], \qquad \mathbb{E}_n[W_n^2] \xrightarrow{\mathbb{P}} \mathbb{E}[W^2], \tag{6.2.12}$$

where $\mathbb{E}_n$ denotes expectation with respect to $\mathbb{P}_n$, so that $\mathbb{E}_n[W_n] = \frac{1}{n}\sum_{i\in[n]} w_i$. Equation (6.2.12) is equivalent to the statement that, for every $\varepsilon > 0$,

$$\mathbb{P}\big(|\mathbb{E}_n[W_n] - \mathbb{E}[W]| > \varepsilon\big) \to 0, \qquad \mathbb{P}\big(|\mathbb{E}_n[W_n^2] - \mathbb{E}[W^2]| > \varepsilon\big) \to 0. \tag{6.2.13}$$

Exercise 6.3 investigates bounds on the maximal weight $\max_{i\in[n]} w_i$ under Condition 6.4, showing that $\max_{i\in[n]} w_i = o(n)$ when Conditions 6.4(a)–(b) hold, while $\max_{i\in[n]} w_i = o(\sqrt{n})$ when Conditions 6.4(a)–(c) hold.

We now discuss two key examples of choices of vertex weights:

Key Example of Generalized Random Graph with Deterministic Weights

Let F be a distribution function for which $F(0) = 0$ and fix

$$w_i = [1 - F]^{-1}(i/n), \tag{6.2.14}$$

where $[1 - F]^{-1}$ is the generalized inverse function of $1 - F$ defined, for $u \in (0, 1)$, by

$$[1 - F]^{-1}(u) = \inf\{x\colon [1 - F](x) \leq u\}. \tag{6.2.15}$$

By convention, we set $[1 - F]^{-1}(1) = 0$. Here the definition of $[1 - F]^{-1}$ is chosen such that

$$[1 - F]^{-1}(1 - u) = F^{-1}(u) = \inf\{x\colon F(x) \geq u\}. \tag{6.2.16}$$

We often make use of (6.2.16), in particular since it implies that $[1 - F]^{-1}(U)$ has distribution function F when U is uniform on $(0, 1)$. For the choice in (6.2.14), we can explicitly compute F_n as

$$\begin{aligned} F_n(x) &= \frac{1}{n}\sum_{i\in[n]} \mathbb{1}_{\{w_i \leq x\}} = \frac{1}{n}\sum_{i\in[n]} \mathbb{1}_{\{[1-F]^{-1}(i/n)\leq x\}} = \frac{1}{n}\sum_{j=0}^{n-1} \mathbb{1}_{\{[1-F]^{-1}(1-\frac{j}{n})\leq x\}} \\ &= \frac{1}{n}\sum_{j=0}^{n-1} \mathbb{1}_{\{F^{-1}(\frac{j}{n})\leq x\}} = \frac{1}{n}\sum_{j=0}^{n-1} \mathbb{1}_{\{\frac{j}{n}\leq F(x)\}} = \frac{1}{n}\big(\lfloor nF(x)\rfloor + 1\big) \wedge 1, \end{aligned} \tag{6.2.17}$$

where we write $j = n - i$ in the third equality and use (6.2.16) in the fourth equality. Exercise 6.4 shows that Condition 6.4(a) holds for $(w_i)_{i\in[n]}$ as in (6.2.14).

Note that $F_n(x) \geq F(x)$ for every $x \geq 0$ by (6.2.17), which shows that W_n is stochastically dominated by W. In particular, this implies that for *increasing* functions $x \mapsto h(x)$,

$$\frac{1}{n}\sum_{i\in[n]} h(w_i) \leq \mathbb{E}[h(W)], \tag{6.2.18}$$

as formalized in Exercises 6.5 and 6.6.

An example of the generalized random graph arises when we take F to have a *power-law distribution*, for which, for some $a \geq 0$ and $\tau > 1$,

$$F(x) = \begin{cases} 0 & \text{for } x \leq a, \\ 1 - (a/x)^{\tau-1} & \text{for } x > a. \end{cases} \tag{6.2.19}$$

Here

$$[1 - F]^{-1}(u) = au^{-1/(\tau-1)}, \tag{6.2.20}$$

so that

$$w_i = a\big(i/n\big)^{-1/(\tau-1)}. \tag{6.2.21}$$

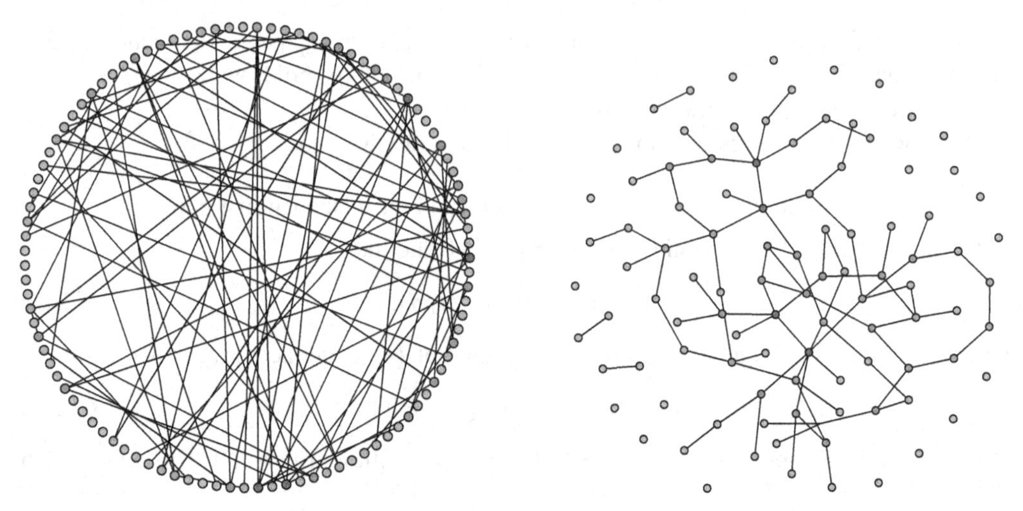

Figure 6.1 A realization of $\mathrm{GRG}_n(\boldsymbol{w})$ for $(w_i)_{i\in[n]}$ as in (6.2.21) with $n = 100$ and $\tau = 3.5$ on a circle (left), and in a way that minimizes a force field (right). Darker vertices have a higher degree than lighter vertices.

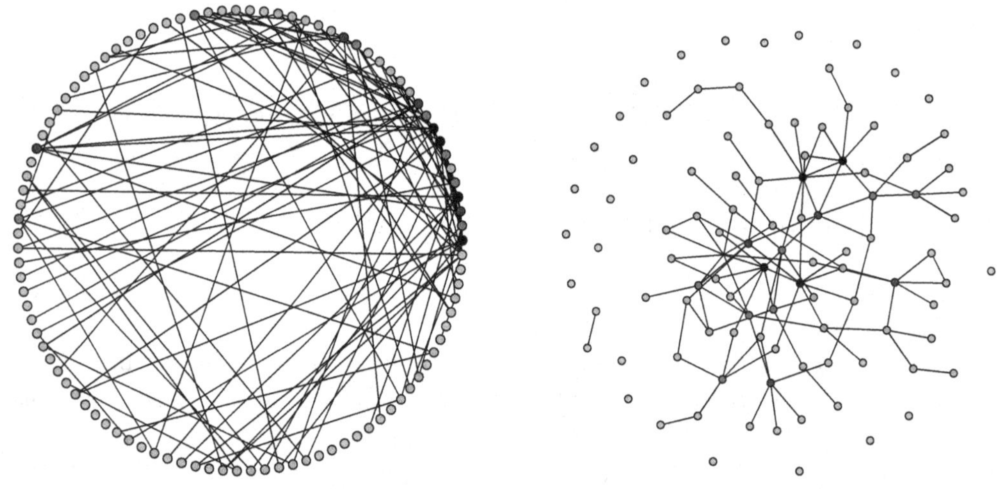

Figure 6.2 A realization of $\mathrm{GRG}_n(\boldsymbol{w})$ for $(w_i)_{i\in[n]}$ as in (6.2.21) with $n = 100$ and $\tau = 2.5$ on a circle (left), and in a way that minimizes a force field (right). Darker vertices have a higher degree than lighter vertices.

See Figures 6.1–6.2 for realizations of $\mathrm{GRG}_n(\boldsymbol{w})$ for $(w_i)_{i\in[n]}$ as in (6.2.21) with $n = 100$ and $\tau = 3.5$ and $\tau = 2.5$, respectively. This predicts that weights as in (6.2.21) correspond to power-law degrees. Exercise 6.7 investigates the maximal weight in the case (6.2.21).

The Generalized Random Graph with I.I.D. Weights

The generalized random graph can be studied both with deterministic weights as well as with independent and identically distributed (i.i.d.) weights. Since we often deal with ratios of the form $w_i w_j/(\sum_{k\in[n]} w_k)$, we assume that $\mathbb{P}(w = 0) = 0$ to avoid situations where all weights are zero.

Both models, i.e., with weights $(w_i)_{i\in[n]}$ as in (6.2.14), and with i.i.d. weights $(w_i)_{i\in[n]}$, have their own merits (see Section 6.9 for more details). The great advantage of i.i.d. weights is that the vertices in the resulting graph are, in distribution, the same. More precisely, the vertices are completely *exchangeable*, like in the Erdős–Rényi random graph $\mathrm{ER}_n(p)$. Unfortunately, when we take the weights to be i.i.d., then in the resulting graph the edges are no longer independent (despite the fact that they are *conditionally* independent given the weights), see Exercise 6.8 for a proof of this fact.

For i.i.d. weights, the empirical distribution function F_n of the weights is given by

$$F_n(x) = \frac{1}{n} \sum_{i\in[n]} \mathbb{1}_{\{w_i \leq x\}}. \tag{6.2.22}$$

When the weights are independent and identically distributed with distribution function F, then it is well known that this empirical distribution function is close to F (this is the Glivenko–Cantelli Theorem). Therefore, Condition 6.4(a) holds. Conditions 6.4(b)–(c) hold by the weak law of large numbers under the appropriate moment conditions on the limiting weight variable W.

The Total Number of Edges in $\mathrm{GRG}_n(\boldsymbol{w})$.

We close this introductory section by investigating the total number of edges $E(\mathrm{GRG}_n(\boldsymbol{w}))$ in $\mathrm{GRG}_n(\boldsymbol{w})$:

Theorem 6.6 (Total number of edges in $\mathrm{GRG}_n(\boldsymbol{w})$) *Assume that Conditions 6.4(a)–(b) hold. Then,*

$$\frac{1}{n} E(\mathrm{GRG}_n(\boldsymbol{w})) \xrightarrow{\mathbb{P}} \tfrac{1}{2}\mathbb{E}[W]. \tag{6.2.23}$$

Proof We apply a second moment method. We start by investigating the first moment of the number of edges, and prove that

$$\mathbb{E}[E(\mathrm{GRG}_n(\boldsymbol{w}))]/n \to \tfrac{1}{2}\mathbb{E}[W]. \tag{6.2.24}$$

For this, we note that

$$\mathbb{E}[E(\mathrm{GRG}_n(\boldsymbol{w}))] = \tfrac{1}{2} \sum_{i,j\in[n]:\, i\neq j} p_{ij} = \tfrac{1}{2} \sum_{i,j\in[n]:\, i\neq j} \frac{w_i w_j}{\ell_n + w_i w_j}. \tag{6.2.25}$$

We bound this from above by

$$\mathbb{E}[E(\mathrm{GRG}_n(\boldsymbol{w}))] \leq \tfrac{1}{2} \sum_{i,j\in[n]} \frac{w_i w_j}{\ell_n} = \ell_n/2, \tag{6.2.26}$$

which gives an upper bound on $\mathbb{E}[E(\mathrm{GRG}_n(\boldsymbol{w}))]$.

We next prove the corresponding lower bound in (6.2.24), for which it is inconvenient that $w_i w_j$ can be quite large, even when Conditions 6.4(a)–(b) hold. For this, we use an appropriate *truncation argument*. Fix a sequence $a_n \to \infty$. Since $x \mapsto x/(\ell_n + x)$ is increasing, we can bound the expected number of edges from below by

$$\mathbb{E}[E(\mathrm{GRG}_n(\boldsymbol{w}))] \geq \tfrac{1}{2} \sum_{i,j\in[n]:\, i\neq j} \frac{(w_i \wedge a_n)(w_j \wedge a_n)}{\ell_n + (w_i \wedge a_n)(w_j \wedge a_n)}. \tag{6.2.27}$$

Therefore, with $\ell_n(a_n) = \sum_{i\in[n]}(w_i \wedge a_n)$,

$$\begin{aligned}
&\ell_n(a_n)^2/\ell_n - 2\mathbb{E}[E(\mathrm{GRG}_n(\boldsymbol{w}))] \qquad (6.2.28)\\
&\quad \le \sum_{i\in[n]} \frac{(w_i \wedge a_n)^2}{\ell_n + (w_i \wedge a_n)^2} + \sum_{i,j\in[n]} (w_i \wedge a_n)(w_j \wedge a_n)\Big[\frac{1}{\ell_n} - \frac{1}{\ell_n + (w_i \wedge a_n)(w_j \wedge a_n)}\Big]\\
&\quad = \sum_{i\in[n]} \frac{(w_i \wedge a_n)^2}{\ell_n + (w_i \wedge a_n)^2} + \sum_{i,j\in[n]} \frac{(w_i \wedge a_n)^2(w_j \wedge a_n)^2}{\ell_n(\ell_n + (w_i \wedge a_n)(w_j \wedge a_n))}\\
&\quad \le \sum_{i\in[n]} \frac{(w_i \wedge a_n)^2}{\ell_n}\Big(1 + \sum_{i\in[n]} \frac{(w_i \wedge a_n)^2}{\ell_n}\Big).
\end{aligned}$$

Since

$$\sum_{i\in[n]} \frac{(w_i \wedge a_n)^2}{\ell_n} \le a_n, \qquad (6.2.29)$$

the right-hand side of (6.2.28) is $o(n)$ when we choose $a_n = o(\sqrt{n})$. By (6.2.26) and (6.2.28), for (6.2.24), it suffices to prove that $\ell_n(a_n)^2/(n\ell_n) \to \mathbb{E}[W]$ and $\ell_n/n \to \mathbb{E}[W]$.

By Condition 6.4(b), $\ell_n/n = \mathbb{E}[W_n] \to \mathbb{E}[W]$. Further, we claim that $\ell_n(a_n)/n = \mathbb{E}[(W_n \wedge a_n)] \to \mathbb{E}[W]$ by Conditions 6.4(a)–(b), which can be seen by establishing upper and lower bounds. Indeed, the upper bound follows easily since $\mathbb{E}[(W_n \wedge a_n)] \le \mathbb{E}[W_n] \to \mathbb{E}[W]$. For the lower bound, and for every $K \ge 0$, $\mathbb{E}[(W_n \wedge a_n)] \ge \mathbb{E}[(W_n \wedge K)] \to \mathbb{E}[(W \wedge K)]$ since $a_n \to \infty$, by Condition 6.4(a) and dominated convergence. Since $\mathbb{E}[(W \wedge K)] \to \mathbb{E}[W]$ as $K \to \infty$ by monotone convergence, we also obtain the lower bound $\liminf_{n\to\infty} \mathbb{E}[(W_n \wedge a_n)] \ge \mathbb{E}[W]$. We conclude that (6.2.24) holds.

We continue by bounding $\mathrm{Var}(E(\mathrm{GRG}_n(\boldsymbol{w})))$, for which we use that the edge statuses $(X_{ij})_{1\le i<j\le n}$ are independent Bernoulli variables with $\mathbb{P}(X_{ij} = 1) = p_{ij}$, to obtain

$$\begin{aligned}
\mathrm{Var}(E(\mathrm{GRG}_n(\boldsymbol{w}))) &= \tfrac{1}{2} \sum_{i,j\in[n]:\, i\neq j} \mathrm{Var}(X_{ij}) = \tfrac{1}{2} \sum_{i,j\in[n]:\, i\neq j} p_{ij}(1 - p_{ij}) \qquad (6.2.30)\\
&\le \tfrac{1}{2} \sum_{i,j\in[n]:\, i\neq j} p_{ij} = \mathbb{E}[E(\mathrm{GRG}_n(\boldsymbol{w}))].
\end{aligned}$$

As a result, $\mathrm{Var}(E(\mathrm{GRG}_n(\boldsymbol{w}))) \le \mathbb{E}[E(\mathrm{GRG}_n(\boldsymbol{w}))]$, which is $o(\mathbb{E}[E(\mathrm{GRG}_n(\boldsymbol{w}))]^2)$ by (6.2.24) and the fact that $\mathbb{E}[W] > 0$ by Condition 6.4(a). We conclude from Chebychev's inequality (Theorem 2.18) that

$$E(\mathrm{GRG}_n(\boldsymbol{w}))/\mathbb{E}[E(\mathrm{GRG}_n(\boldsymbol{w}))] \xrightarrow{\mathbb{P}} 1, \qquad (6.2.31)$$

which, together with (6.2.24), proves the claim. □

6.3 Degrees in the Generalized Random Graph

In this section, we study the degrees of vertices in $\mathrm{GRG}_n(\boldsymbol{w})$. In order to state the main results, we start with some definitions. We write D_i for the degree of vertex i in $\mathrm{GRG}_n(\boldsymbol{w})$. Thus, D_i is given by

$$D_i = \sum_{j \in [n]} X_{ij}, \tag{6.3.1}$$

where X_{ij} is the indicator that the edge ij is occupied. By convention, we set $X_{ij} = X_{ji}$. The random variables $(X_{ij})_{1 \le i < j \le n}$ are independent Bernoulli variables with $\mathbb{P}(X_{ij} = 1) = p_{ij}$ as defined in (6.2.1). Our first main result concerning the vertex degrees is as follows:

Theorem 6.7 (Degree of GRG with deterministic weights) *(a) There exists a coupling $(\hat{D}_i, \hat{Z}_i)$ of the degree D_i of vertex i and a Poisson random variable Z_i with parameter w_i, such that*

$$\mathbb{P}(\hat{D}_i \neq \hat{Z}_i) \le \frac{w_i^2}{\ell_n}\Big(1 + 2\frac{\mathbb{E}[W_n^2]}{\mathbb{E}[W_n]}\Big). \tag{6.3.2}$$

(b) Assume that Conditions 6.4(a)–(b) hold. When p_{ij} given by (6.2.1) are all such that $\lim_{n\to\infty} \max_{1\le i<j\le m} p_{ij} = 0$, the degrees $D_1, \ldots, D_m$ of vertices $1, \ldots, m$ are asymptotically independent.

Before proving Theorem 6.7, we state a consequence of it for the degree sequence when the weights satisfy Condition 6.4. To be able to state this consequence, we need the following definition:

Definition 6.8 (Mixed Poisson distribution) A random variable X has a mixed Poisson distribution with mixing distribution F when, for every $k \in \mathbb{N}_0$,

$$\mathbb{P}(X = k) = \mathbb{E}\big[\mathrm{e}^{-W}\frac{W^k}{k!}\big], \tag{6.3.3}$$

where W is a random variable with distribution function F.

Not every random variable can be obtained as a mixed Poisson distribution. In Exercises 6.9–6.12, aspects of mixed Poisson distributions are further investigated.

By Theorem 6.7, the degree of vertex i is close to Poisson with parameter w_i. Thus, when we choose a vertex *uniformly at random*, and we denote the outcome by U, then the degree of that vertex is close to a Poisson distribution with *random* parameter $w_U = W_n$. Since $W_n \xrightarrow{d} W$ by Condition 6.4, this suggests the following result:

Corollary 6.9 (Degree of uniformly chosen vertex in GRG) *Assume that Conditions 6.4(a)–(b) hold. Then,*

(a) the degree of a uniformly chosen vertex converges in distribution to a mixed Poisson random variable with mixing distribution F;
(b) the degrees of m uniformly chosen vertices in $[n]$ are asymptotically independent.

We now prove Theorem 6.7 and Corollary 6.9:

Proof of Theorem 6.7 We make essential use of Theorem 2.10, in particular, the coupling of a sum of Bernoulli random variables with a Poisson random variable in (2.2.21). We recall that

$$D_i = \sum_{j \in [n]:\, j \neq i} X_{ij}, \tag{6.3.4}$$

where X_{ij} are independent Bernoulli random variables with success probabilities $p_{ij} = \frac{w_i w_j}{\ell_n + w_i w_j}$. By Theorem 2.10, there exists a Poisson random variable $\hat{Y}_i$ with parameter

$$\lambda_i = \sum_{j \neq i} \frac{w_i w_j}{\ell_n + w_i w_j}, \tag{6.3.5}$$

and a random variable $\hat{D}_i$ where $\hat{D}_i$ has the same distribution as D_i, such that

$$\mathbb{P}(\hat{D}_i \neq \hat{Y}_i) \leq \sum_{j \neq i} p_{ij}^2 = \sum_{j \neq i} \frac{w_i^2 w_j^2}{(\ell_n + w_i w_j)^2} \leq w_i^2 \sum_{j \in [n]} \frac{w_j^2}{\ell_n^2} = \frac{w_i^2}{\ell_n} \frac{\mathbb{E}[W_n^2]}{\mathbb{E}[W_n]}. \tag{6.3.6}$$

Thus, in order to prove the claim, it suffices to prove that we can, in turn, couple $(\hat{D}_i, \hat{Y}_i)$ to $(\hat{D}_i, \hat{Z}_i)$, where $\hat{Z}_i$ is a Poisson random variable with parameter w_i, such that

$$\mathbb{P}(\hat{Y}_i \neq \hat{Z}_i) \leq w_i^2 \sum_{j \in [n]} \frac{w_j^2}{\ell_n^2} + \frac{w_i^2}{\ell_n}. \tag{6.3.7}$$

For this, we note that

$$\lambda_i \leq \sum_{j \neq i} \frac{w_i w_j}{\ell_n} \leq \frac{w_i}{\ell_n} \sum_{j \in [n]} w_j = w_i. \tag{6.3.8}$$

Let $\varepsilon_i = w_i - \lambda_i \geq 0$. Then, we let $\hat{V}_i \sim \mathsf{Poi}(\varepsilon_i)$ be independent of $(\hat{D}_i, \hat{Y}_i)$, and write $\hat{Z}_i = \hat{Y}_i + \hat{V}_i$, so that by Markov's inequality (Theorem 2.17)

$$\mathbb{P}(\hat{Y}_i \neq \hat{Z}_i) = \mathbb{P}(\hat{V}_i \neq 0) = \mathbb{P}(\hat{V}_i \geq 1) \leq \mathbb{E}[\hat{V}_i] = \varepsilon_i. \tag{6.3.9}$$

To bound ε_i, we note that

$$\begin{aligned}
\varepsilon_i &= w_i - \sum_{j \neq i} \frac{w_i w_j}{\ell_n + w_i w_j} = \sum_{j \in [n]} w_i w_j \Big(\frac{1}{\ell_n} - \frac{1}{\ell_n + w_i w_j} \Big) + \frac{w_i^2}{\ell_n + w_i^2} \\
&= \sum_{j \in [n]} \frac{w_j^2 w_i^2}{\ell_n(\ell_n + w_i w_j)} + \frac{w_i^2}{\ell_n + w_i^2} \leq \frac{w_i^2}{\ell_n} + \sum_{j \in [n]} \frac{w_j^2 w_i^2}{\ell_n^2} \\
&= w_i^2 \Big(\frac{1}{\ell_n} + \sum_{j \in [n]} \frac{w_j^2}{\ell_n^2} \Big).
\end{aligned} \tag{6.3.10}$$

We conclude that

$$\mathbb{P}(\hat{D}_i \neq \hat{Z}_i) \leq \mathbb{P}(\hat{D}_i \neq \hat{Y}_i) + \mathbb{P}(\hat{Y}_i \neq \hat{Z}_i) \leq \frac{w_i^2}{\ell_n} + 2 w_i^2 \sum_{j \in [n]} \frac{w_j^2}{\ell_n^2}, \tag{6.3.11}$$

as required. This proves Theorem 6.7(a).

To prove Theorem 6.7(b), it suffices to prove that we can couple $(D_i)_{i \in [m]}$ to an *independent* vector $(\hat{D}_i)_{i \in [m]}$ such that

$$\mathbb{P}\Big((D_i)_{i \in [m]} \neq (\hat{D}_i)_{i \in [m]} \Big) = o(1). \tag{6.3.12}$$

To this end, we recall that X_{ij} denotes the indicator that the edge ij is occupied, so that $X_{ij} = X_{ji}$. The random variables $(X_{ij})_{1\leq i<j\leq n}$ are independent Bernoulli random variables with parameters $(p_{ij})_{1\leq i<j\leq n}$ given in (6.2.1). We let $(\hat{X}_{ij})_{1\leq i<j\leq n}$ with $\hat{X}_{ij} = \hat{X}_{ji}$ denote an *independent* copy of $(X_{ij})_{1\leq i<j\leq n}$, and let, for $i \in [n]$,

$$\hat{D}_i = \sum_{j<i} \hat{X}_{ij} + \sum_{j=i+1}^{n} X_{ij}. \tag{6.3.13}$$

Then, we observe the following: (1) Since $(\hat{X}_{ij})_{1\leq i<j\leq n}$ is an independent copy of $(X_{ij})_{1\leq i<j\leq n}$, $\hat{D}_i$ has the same distribution as D_i, for every $i \in [n]$. (2) Set $i < j$. While D_i and D_j are *dependent* since they both contain $X_{ij} = X_{ji}$, $\hat{D}_i$ and $\hat{D}_j$ are independent since $\hat{D}_i$ contains X_{ij}, while $\hat{D}_j$ contains $\hat{X}_{ji} = \hat{X}_{ij}$, which is an independent copy of X_{ij}. It is straightforward to extend this to the statement that $(\hat{D}_i)_{i\in[m]}$ are sums of *mutually independent* Bernoulli random variables, and, therefore, are *independent* random variables. (3) Finally, $(D_i)_{i\in[m]} \neq (\hat{D}_i)_{i\in[m]}$ precisely when there exists at least one edge ij with $i, j \in [m]$ such that $X_{ij} \neq \hat{X}_{ij}$. Since X_{ij} and $\hat{X}_{ij}$ are Bernoulli random variables, $X_{ij} \neq \hat{X}_{ij}$ implies that either $X_{ij} = 0$, $\hat{X}_{ij} = 1$ or $X_{ij} = 1$, $\hat{X}_{ij} = 0$. Thus, by Boole's inequality,

$$\mathbb{P}\Big((D_i)_{i\in[m]} \neq (\hat{D}_i)_{i\in[m]}\Big) \leq 2\sum_{i,j=1}^{m} \mathbb{P}(X_{ij} = 1) = 2\sum_{i,j=1}^{m} p_{ij}. \tag{6.3.14}$$

By assumption, $\lim_{n\to\infty} p_{ij} = 0$, so that (6.3.12) holds for every $m \geq 2$ fixed. This proves Theorem 6.7(b). □

Exercise 6.13 investigates the asymptotic independence of the degrees of a growing number of vertices.

Proof of Corollary 6.9 We start by giving the proof of part (a) when we assume that Conditions 6.4(a)–(c) hold. By (6.3.2) together with the fact that $\max_{i\in[n]} w_i = o(\sqrt{n})$ by Exercise 6.3, the degree of vertex i is close to a Poisson random variable with parameter w_i. Thus, Theorem 6.7 implies that the degree of a uniformly chosen vertex in $[n]$ is close in distribution to a Poisson random variable with parameter w_U, where U is a uniform vertex in $[n]$. This is a mixed Poisson distribution with mixing distribution equal to w_U, and w_U has the same distribution as W_n.

Since a mixed Poisson random variable converges to a limiting mixed Poisson random variable whenever the mixing distribution converges in distribution, it suffices to show that the weight $W_n = w_U$ of a uniform vertex has a limiting distribution given by F. This follows from Condition 6.4(a).

We next extend the proof to the setting where only Conditions 6.4(a)–(b) hold. For this, fix some $a_n \to \infty$ sufficiently slowly. Then, whp, the weight of a uniformly chosen vertex $W_n = w_U$ satisfies $w_U \leq a_n$. Assume that this is the case. Then, we can split

$$D_U = D_U^{(1)} + D_U^{(2)}, \tag{6.3.15}$$

where $D_U^{(1)}$ counts the number of edges of U to vertices of weight at most a_n, and $D_U^{(2)}$ equals the number of edges of U to vertices of weight larger than a_n. Note that

$$\mathbb{E}[D_U^{(2)}] = \frac{1}{n} \sum_{i,j\in[n]} \frac{w_i w_j}{\ell_n + w_i w_j} \mathbb{1}_{\{w_i \leq a_n < w_j\}} \leq \frac{1}{n} \sum_{j\in[n]} w_j \mathbb{1}_{\{w_j > a_n\}} \tag{6.3.16}$$
$$= \mathbb{E}[W_n \mathbb{1}_{\{W_n > a_n\}}] = o(1),$$

so that $\mathbb{P}(D_U^{(2)} \geq 1) = o(1)$. As a result, it suffices to prove that $D_U^{(1)} \xrightarrow{d} D$, where $D \sim \mathsf{Poi}(W)$ has a mixed Poisson distribution with mixing distribution W. This proof is similar to that of Theorem 6.7(a) and is left as Exercise 6.14.

The proof of part (b) is a minor adaptation of the proof of Theorem 6.7(b). We only need to prove the asymptotic independence. Let $(U_i)_{i\in[m]}$ be independent vertices chosen uniformly at random from $[n]$. Then, whp, all vertices $(U_i)_{i\in[m]}$ are *distinct*, since the probability that $U_i = U_j$ for some $i, j \in [m]$ is, by the union bound, bounded by m^2/n. Further, when all vertices $(U_i)_{i\in[m]}$ are distinct, the dependence between the degrees of the vertices $(U_i)_{i\in[m]}$ arises only through the edges between the vertices $(U_i)_{i\in[m]}$. Now, the expected number of occupied edges between the vertices $(U_i)_{i\in[m]}$, conditionally on $(U_i)_{i\in[m]}$, is bounded by

$$\sum_{i,j=1}^{m} \frac{w_{U_i} w_{U_j}}{\ell_n + w_{U_i} w_{U_j}} \leq \sum_{i,j=1}^{m} \frac{w_{U_i} w_{U_j}}{\ell_n} = \frac{1}{\ell_n} \Big(\sum_{i=1}^{m} w_{U_i} \Big)^2. \tag{6.3.17}$$

The random variables $(w_{U_i})_{i\in[m]}$ are i.i.d., so that the expected number of occupied edges between m uniformly chosen vertices is bounded above by

$$\frac{1}{\ell_n} \mathbb{E}\Big[\Big(\sum_{i=1}^{m} w_{U_i} \Big)^2 \Big] = \frac{m}{\ell_n} \mathbb{E}[w_{U_1}^2] + \frac{m(m-1)}{\ell_n} \mathbb{E}[w_{U_1}]^2. \tag{6.3.18}$$

We can bound

$$\mathbb{E}[w_{U_1}^2] \leq (\max_{i\in[n]} w_i) \mathbb{E}[w_{U_1}] = o(n) \mathbb{E}[W_n] = o(n), \tag{6.3.19}$$

by Conditions 6.4(a)–(b) and Exercise 6.3. Therefore, the expected number of edges between the vertices $(U_i)_{i\in[m]}$ is $o(1)$, so that with high probability there are none. We conclude that we can couple the degrees $(D_{U_i})_{i\in[m]}$ of m uniform vertices to m *independent* mixed Poisson random variables $(Y_i)_{i\in[m]}$ with mixing distribution w_U such that $\mathbb{P}\big((D_{U_i})_{i\in[m]} \neq (Y_i)_{i\in[m]}\big) = o(1)$. Since the random variables $(Y_i)_{i\in[m]}$ converge in distribution to independent mixed Poisson random variables with mixing distribution F by part (a), this completes the argument. □

6.4 Degree Sequence of Generalized Random Graph

Theorem 6.7 investigates the degree of a *single vertex* in the generalized random graph. In this section, we extend the result to the convergence of the empirical degree sequence. For $k \geq 0$, we let

$$P_k^{(n)} = \frac{1}{n} \sum_{i\in[n]} \mathbb{1}_{\{D_i = k\}} \tag{6.4.1}$$

denote the degree sequence of $\mathrm{GRG}_n(\boldsymbol{w})$. Due to Theorem 6.7, one would expect that this degree sequence is close to a mixed Poisson distribution. We denote the probability mass function of such a mixed Poisson distribution by p_k, i.e., for $k \geq 0$,

$$p_k = \mathbb{E}\big[\mathrm{e}^{-W}\frac{W^k}{k!}\big], \tag{6.4.2}$$

where W is a random variable having distribution function F from Condition 6.4. Theorem 6.10 shows that indeed the degree sequence $(P_k^{(n)})_{k\geq 0}$ is close to the mixed Poisson distribution with probability mass function $(p_k)_{k\geq 0}$ in (6.4.2):

Theorem 6.10 (Degree sequence of $\mathrm{GRG}_n(\boldsymbol{w})$) *Assume that Conditions 6.4(a)–(b) hold. Then, for every $\varepsilon > 0$,*

$$\mathbb{P}\Big(\sum_{k=0}^{\infty} |P_k^{(n)} - p_k| \geq \varepsilon\Big) \to 0, \tag{6.4.3}$$

where $(p_k)_{k\geq 0}$ is given by (6.4.2).

Proof of Theorem 6.10 By Exercise 2.16 and the fact that $(p_k)_{k\geq 0}$ is a probability mass function,

$$\sum_{k\geq 0} |P_k^{(n)} - p_k| = 2d_{\mathrm{TV}}(P^{(n)}, p) \xrightarrow{\mathbb{P}} 0 \tag{6.4.4}$$

if and only if $\max_{k\geq 0} |P_k^{(n)} - p_k| \xrightarrow{\mathbb{P}} 0$. Thus, we need to show that $\mathbb{P}\big(\max_{k\geq 0} |P_k^{(n)} - p_k| \geq \varepsilon\big)$ vanishes for every $\varepsilon > 0$. We use that

$$\mathbb{P}\big(\max_{k\geq 0} |P_k^{(n)} - p_k| \geq \varepsilon\big) \leq \sum_{k\geq 0} \mathbb{P}\big(|P_k^{(n)} - p_k| \geq \varepsilon\big). \tag{6.4.5}$$

Note that

$$\mathbb{E}[P_k^{(n)}] = \mathbb{P}(D_U = k), \tag{6.4.6}$$

and, by Corollary 6.9(a),

$$\lim_{n\to\infty} \mathbb{P}(D_U = k) = p_k. \tag{6.4.7}$$

Also, it is not hard to see that the convergence is *uniform in k*, that is, for every $\varepsilon > 0$, and for n sufficiently large,

$$\max_k |\mathbb{E}[P_k^{(n)}] - p_k| \leq \frac{\varepsilon}{2}. \tag{6.4.8}$$

Exercise 6.17 proves (6.4.8).

By (6.4.5) and (6.4.8), it follows that, for n sufficiently large,

$$\mathbb{P}\Big(\max_{k\geq 0} |P_k^{(n)} - p_k| \geq \varepsilon\Big) \leq \sum_{k\geq 0} \mathbb{P}\Big(|P_k^{(n)} - \mathbb{E}[P_k^{(n)}]| \geq \varepsilon/2\Big). \tag{6.4.9}$$

By Chebychev's inequality (Theorem 2.18),

$$\mathbb{P}\big(|P_k^{(n)} - \mathbb{E}[P_k^{(n)}]| \geq \varepsilon/2\big) \leq \frac{4}{\varepsilon^2}\mathrm{Var}(P_k^{(n)}), \tag{6.4.10}$$

so that

$$\mathbb{P}\Big(\max_{k\geq 0} |P_k^{(n)} - p_k| \geq \varepsilon\Big) \leq \frac{4}{\varepsilon^2} \sum_{k\geq 0} \mathrm{Var}(P_k^{(n)}). \tag{6.4.11}$$

We use the definition in (6.4.1) to see that

$$\begin{aligned} \mathbb{E}[(P_k^{(n)})^2] &= \frac{1}{n^2} \sum_{i,j\in[n]} \mathbb{P}(D_i = D_j = k) \\ &= \frac{1}{n^2} \sum_{i\in[n]} \mathbb{P}(D_i = k) + \frac{1}{n^2} \sum_{i,j\in[n]:\, i\neq j} \mathbb{P}(D_i = D_j = k). \end{aligned} \tag{6.4.12}$$

Therefore,

$$\begin{aligned} \mathrm{Var}(P_k^{(n)}) \leq &\frac{1}{n^2} \sum_{i\in[n]} [\mathbb{P}(D_i = k) - \mathbb{P}(D_i = k)^2] \\ &+ \frac{1}{n^2} \sum_{i,j\in[n]:\, i\neq j} [\mathbb{P}(D_i = D_j = k) - \mathbb{P}(D_i = k)\mathbb{P}(D_j = k)]. \end{aligned} \tag{6.4.13}$$

We let

$$Y_{i;j} = \sum_{l\in[n]:\, l\neq i,j} X_{il}, \qquad Y_{j;i} = \sum_{l\in[n]:\, l\neq i,j} X_{jl}, \tag{6.4.14}$$

where we recall that $(X_{ij})_{i,j\in[n]}$ are independent $\mathsf{Be}(p_{ij})$ random variables. Then, the law of (D_i, D_j) is the same as the one of $(Y_{i;j} + X_{ij}, Y_{j;i} + X_{ij})$. Further, $(Y_{i;j} + X_{ij}, Y_{j;i} + \hat{X}_{ij})$, where $\hat{X}_{ij}$ has the same distribution as X_{ij} and is independent of $(X_{ij})_{i,j\in[n]}$, are two independent random variables with the same marginals as D_i and D_j. Therefore,

$$\mathbb{P}(D_i = D_j = k) = \mathbb{P}\Big((Y_{i;j} + X_{ij}, Y_{j;i} + X_{ij}) = (k, k)\Big), \tag{6.4.15}$$

$$\mathbb{P}(D_i = k)\mathbb{P}(D_j = k) = \mathbb{P}\Big((Y_{i;j} + X_{ij}, Y_{j;i} + \hat{X}_{ij}) = (k, k)\Big), \tag{6.4.16}$$

so that

$$\begin{aligned} &\mathbb{P}(D_i = D_j = k) - \mathbb{P}(D_i = k)\mathbb{P}(D_j = k) \\ &\leq \mathbb{P}\Big((Y_{i;j} + X_{ij}, Y_{j;i} + X_{ij}) = (k, k), (Y_{i;j} + X_{ij}, Y_{j;i} + \hat{X}_{ij}) \neq (k, k)\Big). \end{aligned} \tag{6.4.17}$$

When $(Y_{i;j} + X_{ij}, Y_{j;i} + X_{ij}) = (k, k)$, but $(Y_{i;j} + X_{ij}, Y_{j;i} + \hat{X}_{ij}) \neq (k, k)$, we must have that $X_{ij} \neq \hat{X}_{ij}$. If $X_{ij} = 1$, then $\hat{X}_{ij} = 0$ and $Y_{j;i} + \hat{X}_{ij} = k$, while if $X_{ij} = 0$, then $\hat{X}_{ij} = 1$ and $Y_{i;j} + X_{ij} = k$. Therefore, by the independence of the events $\{\hat{X}_{ij} = 1\}$ and $\{Y_{i;j} + X_{ij} = k\}$, as well as the independence of the events $\{\hat{X}_{ij} = 0\}$ and $\{Y_{j;i} + X_{ij} = k\}$, and the fact that $Y_{i;j} + X_{ij}$ has the same distribution as D_i, while $Y_{j;i} + \hat{X}_{ij}$ has the same distribution as D_j,

$$\mathbb{P}(D_i = D_j = k) - \mathbb{P}(D_i = k)\mathbb{P}(D_j = k) \leq p_{ij}[\mathbb{P}(D_i = k) + \mathbb{P}(D_j = k)]. \tag{6.4.18}$$

We conclude from (6.4.13) that

$$\sum_{k\geq 0} \mathrm{Var}(P_k^{(n)}) \leq \frac{1}{n} + \frac{1}{n^2} \sum_{i,j\in[n]} p_{ij} \to 0, \tag{6.4.19}$$

since $\sum_{i,j\in[n]} p_{ij} = O(n)$ (recall (6.2.24)). The result follows from (6.4.11), (6.4.19) and the remark at the beginning of the proof. □

6.5 Generalized Random Graph with I.I.D. Weights

In this section, we discuss the special case where $(w_i)_{i\in[n]}$ are independent and identically distributed. To avoid confusion with W_n, which is the weight of a vertex chosen uniformly at random from $[n]$, we continue to write the weights as $(w_i)_{i\in[n]}$, bearing in mind that now these weights are *random*. Note that there now is *double randomness*. Indeed, there is randomness due to the fact that the weights $(w_i)_{i\in[n]}$ are random themselves, and then there is the randomness in the occupation status of the edges conditionally on the weights $(w_i)_{i\in[n]}$. By Exercise 6.8, the edge statuses in $\mathrm{GRG}_n(\boldsymbol{w})$ are *not* independent.

We now investigate the degrees and degree sequence of $\mathrm{GRG}_n(\boldsymbol{w})$:

Corollary 6.11 (Degrees of $\mathrm{GRG}_n(\boldsymbol{w})$) *When $(w_i)_{i\in[n]}$ are i.i.d. random variables with distribution function F with a finite mean,*

(a) the degree D_i of vertex i converges in distribution to a mixed Poisson random variable with mixing distribution F;
(b) the degrees $D_1, \ldots, D_m$ of vertices $1, \ldots, m$ are asymptotically independent.

Proof To see that Corollary 6.11 follows from Theorem 6.7, we note that Conditions 6.4(a)–(b) hold by the convergence of the empirical distribution function as discussed below (6.2.22). We further need to show that $\max_{1\leq i<j\leq m} p_{ij} = o_{\mathbb{P}}(1)$ and $\mathbb{E}[W_n^2] = \frac{1}{n}\sum_{j\in[n]} w_j^2 = o_{\mathbb{P}}(n)$. For this, note that when $(w_i)_{i\in[n]}$ are i.i.d. copies of a random variable W with distribution function F, we have that $\frac{1}{n}\sum_{i\in[n]} w_i^2 = o_{\mathbb{P}}(n)$ since W has a finite mean (see Exercise 6.18). Exercise 6.18 completes the proof of Corollary 6.11. □

We continue to investigate the degree sequence in the case of i.i.d. weights:

Theorem 6.12 (Degree sequence of $\mathrm{GRG}_n(\boldsymbol{w})$) *When $(w_i)_{i\in[n]}$ are i.i.d. random variables with distribution function F having finite mean, then, for every $\varepsilon > 0$,*

$$\mathbb{P}\Big(\sum_{k\geq 0} |P_k^{(n)} - p_k| \geq \varepsilon\Big) \to 0, \tag{6.5.1}$$

where $(p_k)_{k\geq 0}$ is the probability mass function of a mixed Poisson distribution with mixing distribution F.

We leave the proof of Theorem 6.12, which is quite similar to the proof of Theorem 6.10, to the reader as Exercise 6.19.

We next turn our attention to the case where the weights $(w_i)_{i\in[n]}$ are i.i.d. with *infinite mean.* We denote the distribution of w_i by F. Our goal is to obtain a random graph which has a power-law degree sequence with a power-law exponent $\tau \in (1,2)$. We see that this is a non-trivial issue:

Theorem 6.13 (Degrees of $\mathrm{GRG}_n(\boldsymbol{w})$ with i.i.d. conditioned weights) *Let $(w_i)_{i\in[n]}$ be i.i.d. random variables with distribution function F, and let $(w_i^{(n)})_{i\in[n]}$ be i.i.d. copies of the random variable W conditioned on $W \leq a_n$. Then, for every $a_n \to \infty$ such that $a_n = o(n)$,*

(a) the degree $D_k^{(n)}$ of vertex k in the GRG with weights $(w_i^{(n)})_{i\in[n]}$, converges in distribution to a mixed Poisson random variable with mixing distribution F;
(b) the degrees $(D_i^{(n)})_{i\in[m]}$ of vertices $1,\ldots,m$ are asymptotically independent.

Proof Theorem 6.13 follows by a simple adaptation of the proof of Theorem 6.7 and is left as Exercise 6.21. □

We finally show that the conditioning on $w_i \leq a_n$ in Theorem 6.13 is necessary by proving that if we do not condition the weights to be at most a_n, then the degree distribution changes:

Theorem 6.14 (Degrees of $\mathrm{GRG}_n(\boldsymbol{w})$ with i.i.d. infinite mean weights) *Let $(w_i)_{i\in[n]}$ be i.i.d. random variables with distribution function F satisfying that for some $\tau \in (1,2)$ and $c > 0$,*

$$\lim_{x\to\infty} x^{\tau-1}[1 - F(x)] = c. \tag{6.5.2}$$

Let the edge probabilities $(p_{ij})_{1\leq i<j\leq n}$ conditionally on the weights $(w_i)_{i\in[n]}$ be given by

$$p_{ij} = \frac{w_i w_j}{n^{1/(\tau-1)} + w_i w_j}. \tag{6.5.3}$$

Then,

(a) the degree D_k of vertex k converges in distribution to a mixed Poisson random variable with parameter $\gamma W^{\tau-1}$, where

$$\gamma = c\int_0^\infty (1+x)^{-2} x^{-(\tau-1)} dx. \tag{6.5.4}$$

(b) the degrees $(D_i)_{i\in[m]}$ of vertices $1,\ldots,m$ are asymptotically independent.

The proof of Theorem 6.14 is deferred to Section 6.6 below. By Exercise 6.22 a mixed Poisson distribution with mixing distribution $\gamma W^{\tau-1}$ does not obey a power law with exponent τ, but with exponent 2 instead.

We note that the choice of the edge probabilities in (6.5.3) is different from the choice in (6.2.1). Indeed, the term $\ell_n = \sum_{i\in[n]} w_i$ in the denominator in (6.2.1) is replaced by $n^{1/(\tau-1)}$ in (6.5.3). Since (6.5.2) implies that

$$n^{-1/(\tau-1)} \sum_{i\in[n]} w_i \xrightarrow{d} S, \tag{6.5.5}$$

where S is a stable random variable with parameter $\tau - 1 \in (0, 1)$, we expect that the behavior for the choice (6.2.1) is similar (recall Theorem 2.33).

6.6 Generalized Random Graph Conditioned on Its Degrees

In this section, we investigate the distribution of $\mathrm{GRG}_n(\boldsymbol{w})$ in more detail. The main result in this section is that the generalized random graph conditioned on its degree sequence is a uniform simple random graph with that degree sequence (see Theorem 6.15 below).

We start by introducing some notation. We let $X = (X_{ij})_{1\leq i<j\leq n}$, where X_{ij} are independent random variables with

$$\mathbb{P}(X_{ij} = 1) = 1 - \mathbb{P}(X_{ij} = 0) = p_{ij}, \tag{6.6.1}$$

where p_{ij} is given in (6.2.1). We assume that $p_{ij} < 1$ for every $i, j \in [n]$. Then, with $q_{ij} = 1 - p_{ij}$ and for $x = (x_{ij})_{1\leq i<j\leq n}$,

$$\mathbb{P}(X = x) = \prod_{1\leq i<j\leq n} p_{ij}^{x_{ij}} q_{ij}^{1-x_{ij}}. \tag{6.6.2}$$

We define the *odds-ratios* $(r_{ij})_{1\leq i<j\leq n}$ by

$$r_{ij} = \frac{p_{ij}}{q_{ij}}, \tag{6.6.3}$$

which is well defined, since $q_{ij} = 1 - p_{ij} > 0$ due to the fact that $p_{ij} < 1$. Then

$$p_{ij} = \frac{r_{ij}}{1 + r_{ij}}, \qquad q_{ij} = \frac{1}{1 + r_{ij}}, \tag{6.6.4}$$

so that

$$\mathbb{P}(X = x) = \prod_{1\leq i<j\leq n} \frac{1}{1 + r_{ij}} \prod_{1\leq i<j\leq n} r_{ij}^{x_{ij}}. \tag{6.6.5}$$

We now specialize to the setting of the generalized random graph, and choose

$$r_{ij} = u_i u_j, \tag{6.6.6}$$

for some weights $(u_i)_{i\in[n]}$. Later, we will choose

$$u_i = \frac{w_i}{\sqrt{\ell_n}}, \tag{6.6.7}$$

in which case we return to (6.2.1) since

$$p_{ij} = \frac{r_{ij}}{1 + r_{ij}} = \frac{u_i u_j}{1 + u_i u_j} = \frac{w_i w_j}{\ell_n + w_i w_j}. \tag{6.6.8}$$

Then, with

$$G(u) = \prod_{1\leq i<j\leq n} (1 + u_i u_j), \tag{6.6.9}$$

we obtain

$$\mathbb{P}(X = x) = G(u)^{-1} \prod_{1 \le i < j \le n} (u_i u_j)^{x_{ij}} = G(u)^{-1} \prod_{i \in [n]} u_i^{d_i(x)}, \tag{6.6.10}$$

where $(d_i(x))_{i \in [n]}$ is given by

$$d_i(x) = \sum_{j \in [n]} x_{ij}, \tag{6.6.11}$$

i.e., $d_i(x)$ is the degree of vertex i in the generalized random graph configuration $x = (x_{ij})_{1 \le i < j \le n}$, where, by convention, we assume that $x_{ii} = 0$ and $x_{ij} = x_{ji}$. The proof of the last equality in (6.6.10) is left as Exercise 6.23.

From (6.6.10), and using that $\sum_x \mathbb{P}(X = x) = 1$, it follows that

$$\prod_{1 \le i < j \le n} (1 + u_i u_j) = G(u) = \sum_x \prod_{i \in [n]} u_i^{d_i(x)}. \tag{6.6.12}$$

Furthermore, it also follows from (6.6.10) that the distribution of X conditionally on $\{d_i(X) = d_i \ \forall 1 \le i \le n\}$ is *uniform*. That is, all graphs with the same degree sequence have the same probability. This wonderful and surprising result is formulated in the following theorem:

Theorem 6.15 (GRG conditioned on degrees has uniform law) *The GRG with edge probabilities $(p_{ij})_{1 \le i < j \le n}$ given by*

$$p_{ij} = \frac{u_i u_j}{1 + u_i u_j}, \tag{6.6.13}$$

conditioned on $\{d_i(X) = d_i \ \forall i \in [n]\}$, is uniform over all graphs with degree sequence $(d_i)_{i \in [n]}$.

Proof For x satisfying $d_i(x) = d_i$ for all $i \in [n]$, we can write out

$$\begin{aligned}\mathbb{P}(X = x \mid d_i(X) = d_i \forall i \in [n]) &= \frac{\mathbb{P}(X = x)}{\mathbb{P}(d_i(X) = d_i \ \forall i \in [n])} \\ &= \frac{\mathbb{P}(X = x)}{\sum_{y: d_i(y) = d_i \ \forall i \in [n]} \mathbb{P}(X = y)}.\end{aligned} \tag{6.6.14}$$

By (6.6.10), we have that (6.6.14) simplifies to

$$\begin{aligned}\mathbb{P}(X = x \mid d_i(X) = d_i \ \forall i \in [n]) &= \frac{\prod_{i \in [n]} u_i^{d_i(x)}}{\sum_{y: d_i(y) = d_i \ \forall i} \prod_{i \in [n]} u_i^{d_i(y)}} \\ &= \frac{\prod_{i \in [n]} u_i^{d_i}}{\sum_{y: d_i(y) = d_i \ \forall i} \prod_{i \in [n]} u_i^{d_i}} \\ &= \frac{1}{\#\{y : d_i(y) = d_i \ \forall i \in [n]\}},\end{aligned} \tag{6.6.15}$$

that is, the distribution is uniform over all graphs with the prescribed degree sequence. □

We next compute the generating function of all degrees, that is, for $t_1, \dots, t_n \in \mathbb{R}$, we compute, with $D_i = d_i(X)$,

$$\mathbb{E}\Big[\prod_{i\in[n]} t_i^{D_i}\Big] = \sum_x \mathbb{P}(X = x) \prod_{i\in[n]} t_i^{d_i(x)}. \tag{6.6.16}$$

By (6.6.10) and (6.6.12),

$$\mathbb{E}\Big[\prod_{i\in[n]} t_i^{D_i}\Big] = G(u)^{-1} \sum_x \prod_{i\in[n]} (u_i t_i)^{d_i(x)} = \frac{G(tu)}{G(u)}, \tag{6.6.17}$$

where $(tu)_i = t_i u_i$. By (6.6.9),

$$\mathbb{E}\Big[\prod_{i\in[n]} t_i^{D_i}\Big] = \prod_{1\leq i<j\leq n} \frac{1 + u_i t_i u_j t_j}{1 + u_i u_j}. \tag{6.6.18}$$

Therefore, we have proved the following nice property:

Proposition 6.16 (Generating function of degrees of $\mathrm{GRG}_n(\boldsymbol{w})$) *For the edge probabilities given by* (6.2.1),

$$\mathbb{E}\Big[\prod_{i\in[n]} t_i^{D_i}\Big] = \prod_{1\leq i<j\leq n} \frac{\ell_n + w_i t_i w_j t_j}{\ell_n + w_i w_j}. \tag{6.6.19}$$

Proof The proof of Proposition 6.16 follows by (6.6.18), when using (6.6.7) to obtain the edge probabilities in (6.2.1). □

Exercises 6.24, 6.25 and 6.26 investigate consequences of Proposition 6.16. We finally make use of Proposition 6.16 to prove Theorem 6.14:

Proof of Theorem 6.14 We study the generating function of the degree D_k, where the weights $(w_i)_{i\in[n]}$ are i.i.d. random variables with distribution function F in (6.5.2). We recall that, conditionally on the weights $(w_i)_{i\in[n]}$, the probability that edge ij is occupied is equal to $w_i w_j/(n^{1/(\tau-1)} + w_i w_j)$. We note that

$$\mathbb{E}[t^{D_k}] = \mathbb{E}\Big[\prod_{i\neq k} \frac{1 + t w_i w_k n^{-1/(\tau-1)}}{1 + w_i w_k n^{-1/(\tau-1)}}\Big], \tag{6.6.20}$$

where the expectation is over the i.i.d. random variables $(w_i)_{i\in[n]}$. Denote $\phi_w\colon (0,\infty) \mapsto (0,\infty)$ by

$$\phi_w(x) = \frac{1 + twx}{1 + wx}. \tag{6.6.21}$$

Then, by the independence of the weights $(w_i)_{i\in[n]}$,

$$\mathbb{E}[t^{D_k} \mid w_k = w] = \mathbb{E}\Big[\prod_{i\neq k} \phi_w\big(w_i n^{-1/(\tau-1)}\big)\Big] = \psi_n(w)^{n-1}, \tag{6.6.22}$$

where

$$\psi_n(w) = \mathbb{E}\Big[\phi_w\big(w_i n^{-1/(\tau-1)}\big)\Big]. \tag{6.6.23}$$

We claim that

$$\psi_n(w) = 1 + \frac{1}{n}(t-1)\gamma w^{\tau-1} + o(n^{-1}). \tag{6.6.24}$$

This completes the proof since it implies that

$$\mathbb{E}[t^{D_k} \mid w_k = w] = \psi_n(w)^{n-1} = \mathrm{e}^{(t-1)\gamma w^{\tau-1}}(1+o(1)), \tag{6.6.25}$$

which in turn implies that

$$\lim_{n\to\infty} \mathbb{E}[t^{D_k}] = \mathbb{E}[\mathrm{e}^{(t-1)\gamma w_k^{\tau-1}}]. \tag{6.6.26}$$

Since $\mathbb{E}[\mathrm{e}^{(t-1)\gamma w_k^{\tau-1}}]$ is the probability generating function of a mixed Poisson random variable with mixing distribution $\gamma w_k^{\tau-1}$ (see Exercise 6.27), (6.6.24) indeed completes the proof.

We complete the proof of Theorem 6.14 by showing that (6.6.24) holds. For this, we first note, with W having distribution function F,

$$\begin{aligned}\psi_n(w) &= \mathbb{E}\Big[\phi_w\big(Wn^{-1/(\tau-1)}\big)\Big] \\ &= 1 + \mathbb{E}\Big[\phi_w\big(Wn^{-1/(\tau-1)}\big) - 1\Big].\end{aligned} \tag{6.6.27}$$

In Exercise 6.28, it is proved that for every differentiable function $h\colon [0,\infty) \to \mathbb{R}$, with $h(0) = 0$ and every random variable $X \geq 0$ with distribution function F, the following partial integration formula holds:

$$\mathbb{E}[h(X)] = \int_0^\infty h'(x)[1 - F(x)]dx. \tag{6.6.28}$$

Applying (6.6.28) to $h(x) = \phi_w\big(xn^{-1/(\tau-1)}\big) - 1$ and $X = W$ yields

$$\begin{aligned}\psi_n(w) &= 1 + n^{-1/(\tau-1)}\int_0^\infty \phi_w'\big(xn^{-1/(\tau-1)}\big)[1 - F(x)]dx \\ &= 1 + \int_0^\infty \phi_w'(x)[1 - F(xn^{1/(\tau-1)})]dx.\end{aligned} \tag{6.6.29}$$

Thus,

$$n(\psi_n(w) - 1) = \int_0^\infty \frac{\phi_w'(x)}{x^{\tau-1}}(n^{1/(\tau-1)}x)^{\tau-1}[1 - F(xn^{1/(\tau-1)})]dx. \tag{6.6.30}$$

By assumption, $x^{\tau-1}[1 - F(x)]$ is a bounded function that converges to c. As a result, by the Dominated convergence theorem (Theorem A.1) and the fact that $x \mapsto x^{-(\tau-1)}\phi_w'(x)$ is integrable on $(0,\infty)$ for $\tau \in (1,2)$, see also Exercise 6.29,

$$\lim_{n\to\infty}\int_0^\infty \frac{\phi_w'(x)}{x^{\tau-1}}(n^{1/(\tau-1)}x)^{\tau-1}[1 - F(xn^{1/(\tau-1)})]dx = c\int_0^\infty \frac{\phi_w'(x)}{x^{\tau-1}}dx. \tag{6.6.31}$$

We complete the proof of (6.6.24) by noting that

$$\phi_w'(x) = \frac{tw}{1+wx} - \frac{w(1+twx)}{(1+wx)^2} = \frac{w(t-1)}{(1+wx)^2}, \tag{6.6.32}$$

so that

$$c\int_0^\infty \frac{\phi_w'(x)}{x^{\tau-1}}dx = c\int_0^\infty \frac{w(t-1)}{(1+wx)^2x^{\tau-1}}dx = \gamma(t-1)w^{\tau-1}. \tag{6.6.33}$$

The asymptotic independence in Theorem 6.14(b) is similar, now using (6.6.19) in Proposition 6.16 for fixed $t_1, \ldots, t_m$, and $t_{m+1} = 1, \ldots, t_n = 1$, and proving that the limit factorizes. We refrain from giving the details. □

6.7 Asymptotic Equivalence of Inhomogeneous Random Graphs

There are numerous papers that introduce models along the lines of the generalized random graph, in that they have (conditionally) independent edge statuses. The most general model has been introduced by Bollobás, Janson and Riordan (2007). In that paper, the properties of such random graphs (such as diameter, phase transition and average distances) have been studied using comparisons to multitype branching processes. These generalizations are studied in more detail in [II, Chapter 2]. We start by investigating when two inhomogeneous random graph sequences are asymptotically equivalent.

We first introduce the notion of asymptotic equivalence for general random variables. Before we can do so, we say that $(\mathcal{X}, \mathcal{F})$ is a measurable space when $\mathcal{X}$ is the state space, i.e., the space of all possible outcomes, and $\mathcal{F}$ the set of all possible events. We are particularly interested in *finite* measurable spaces, in which case $\mathcal{X}$ is a finite set and $\mathcal{F}$ can be taken to be the set of all subsets of $\mathcal{X}$. However, all notions that will be introduced in this section, can be more generally defined.

Definition 6.17 (Asymptotic equivalence of sequences of random variables) Let $(\mathcal{X}_n, \mathcal{F}_n)$ be a sequence of measurable spaces. Let $\mathbb{P}_n$ and $\mathbb{Q}_n$ be two probability measures on $(\mathcal{X}_n, \mathcal{F}_n)$. Then, we say that the sequences $(\mathbb{P}_n)_{n\geq 1}$ and $(\mathbb{Q}_n)_{n\geq 1}$ are asymptotically equivalent if, for every sequence $\mathcal{E}_n \in \mathcal{F}_n$ of events,

$$\lim_{n\to\infty} \mathbb{P}_n(\mathcal{E}_n) - \mathbb{Q}_n(\mathcal{E}_n) = 0. \tag{6.7.1}$$

Thus, $(\mathbb{P}_n)_{n\geq 1}$ and $(\mathbb{Q}_n)_{n\geq 1}$ are asymptotically equivalent when they have asymptotically equal probabilities.

The main result that we prove in this section is the following theorem that gives a sharp criterion on when two inhomogeneous random graph sequences are asymptotically equivalent. In its statement, we write $\boldsymbol{p} = (p_{ij})_{1\leq i<j\leq n}$ for the edge probabilities in the graph, and $\mathrm{IRG}_n(\boldsymbol{p})$ for the inhomogeneous random graph for which the edges are independent and the probability that the edge ij is present equals p_{ij}.

Theorem 6.18 (Asymptotic equivalence of inhomogeneous random graphs) *Let* $\mathrm{IRG}_n(\boldsymbol{p})$ *and* $\mathrm{IRG}_n(\boldsymbol{q})$ *be two inhomogeneous random graphs with edge probabilities* $\boldsymbol{p} = (p_{ij})_{1\leq i<j\leq n}$ *and* $\boldsymbol{q} = (q_{ij})_{1\leq i<j\leq n}$ *respectively. Assume that there exists* $\varepsilon > 0$ *such that* $\max_{1\leq i<j\leq n} p_{ij} \leq 1-\varepsilon$. *Then* $\mathrm{IRG}_n(\boldsymbol{p})$ *and* $\mathrm{IRG}_n(\boldsymbol{q})$ *are asymptotically equivalent when*

$$\lim_{n\to\infty} \sum_{1\leq i<j\leq n} \frac{(p_{ij}-q_{ij})^2}{p_{ij}} = 0. \tag{6.7.2}$$

When the edge probabilities $\boldsymbol{p} = (p_{ij})_{1\leq i<j\leq n}$ *and* $\boldsymbol{q} = (q_{ij})_{1\leq i<j\leq n}$ *are themselves random variables, with* $\max_{1\leq i<j\leq n} p_{ij} \leq 1-\varepsilon$ *a.s., then* $\mathrm{IRG}_n(\boldsymbol{p})$ *and* $\mathrm{IRG}_n(\boldsymbol{q})$ *are asymptotically equivalent when*

$$\sum_{1\leq i<j\leq n} \frac{(p_{ij}-q_{ij})^2}{p_{ij}} \xrightarrow{\mathbb{P}} 0. \tag{6.7.3}$$

We note that, in particular, $\mathrm{IRG}_n(\boldsymbol{p})$ and $\mathrm{IRG}_n(\boldsymbol{q})$ are asymptotically equivalent when they can be coupled in such a way that $\mathbb{P}(\mathrm{IRG}_n(\boldsymbol{p}) \neq \mathrm{IRG}_n(\boldsymbol{q})) = o(1)$, which is what our proof will yield. This is quite a strong result. The remainder of this section is devoted to the proof of Theorem 6.18. We start by introducing the necessary ingredients.

There is a strong relation between asymptotic equivalence of random variables and coupling, in the sense that two sequences of random variables are asymptotically equivalent precisely when they can be coupled such that they agree with high probability. Recall the results in Section 2.2 that we use and extend in this section. Let $p = (p_x)_{x\in\mathcal{X}}$ and $q = (q_x)_{x\in\mathcal{X}}$ be two discrete probability measures on the space $\mathcal{X}$, and recall that the total variation distance between p and q is given by

$$d_{\mathrm{TV}}(p,q) = \frac{1}{2}\sum_x |p_x - q_x|. \tag{6.7.4}$$

By (2.2.19)–(2.2.20), we see that two sequences of discrete probability measures $p^{(n)} = (p_x^{(n)})_{x\in\mathcal{X}}$ and $q^{(n)} = (q_x^{(n)})_{x\in\mathcal{X}}$ are asymptotically equivalent when

$$d_{\mathrm{TV}}(p^{(n)}, q^{(n)}) \to 0. \tag{6.7.5}$$

In fact, this turns out to be an equivalent definition, as proved in Exercise 6.30.

When p and q correspond to $\mathsf{Be}(p)$ and $\mathsf{Be}(q)$ distributions, then it is rather simple to show that

$$d_{\mathrm{TV}}(p,q) = |p-q|. \tag{6.7.6}$$

Now, for $\mathrm{IRG}_n(\boldsymbol{p})$ and $\mathrm{IRG}_n(\boldsymbol{q})$, the edge occupation variables are all independent $\mathsf{Be}(p_{ij})$ and $\mathsf{Be}(q_{ij})$ random variables. Thus, we can couple each of the edges in such a way that the probability that a particular edge is distinct is equal to

$$d_{\mathrm{TV}}(p_{ij}, q_{ij}) = |p_{ij} - q_{ij}|, \tag{6.7.7}$$

so that we are led to the naive bound

$$d_{\mathrm{TV}}(\mathrm{IRG}_n(\boldsymbol{p}), \mathrm{IRG}_n(\boldsymbol{q})) \leq \sum_{1\leq i<j\leq n} |p_{ij} - q_{ij}|, \tag{6.7.8}$$

which is far worse than (6.7.2). As we will see later on, there are many examples for which

$$\sum_{1\leq i<j\leq n} \frac{(p_{ij}-q_{ij})^2}{p_{ij}} = o(1), \qquad \text{but} \qquad \sum_{1\leq i<j\leq n} |p_{ij} - q_{ij}| \neq o(1). \tag{6.7.9}$$

Thus, the coupling used in the proof of Theorem 6.18 is substantially stronger.

To explain this seeming contradiction, it is useful to investigate the setting of the Erdős–Rényi random graph $\mathrm{ER}_n(p)$. Fix p and q, assume that $q \leq p$ and that $p \leq 1-\varepsilon$. Then, by Theorem 6.18, $\mathrm{ER}_n(p)$ and $\mathrm{ER}_n(q)$ are asymptotically equivalent when

$$\sum_{1\leq i<j\leq n} \frac{(p_{ij}-q_{ij})^2}{p_{ij}} \leq n^2(p-q)^2/p = O(n^3(p-q)^2), \tag{6.7.10}$$

when we assume that $p \geq \varepsilon/n$. Thus, it suffices that $p-q=o(n^{-3/2})$ to obtain asymptotic equivalence. On the other hand, the right-hand side of (6.7.8) is $o(1)$ when $p-q=o(n^{-2})$, which is rather stronger. This can be understood by noting that if we condition on the number of edges M, then the conditional distribution of $\mathrm{ER}_n(p)$ *conditionally on* $M=m$ does not depend on the precise value of p involved. As a result, we obtain that the asymptotic equivalence of $\mathrm{ER}_n(p)$ and $\mathrm{ER}_n(q)$ follows precisely when we have asymptotic equivalence of the number of edges in $\mathrm{ER}_n(p)$ and $\mathrm{ER}_n(q)$. For this, we note that $M \sim \mathsf{Bin}(n(n-1)/2, p)$ for $\mathrm{ER}_n(p)$, while the number of edges M' for $\mathrm{ER}_n(q)$ satisfies $M' \sim \mathsf{Bin}(n(n-1)/2, q)$. By Exercise 4.2 as well as Exercise 4.22, binomial distributions with a variance that tends to infinity satisfy a central limit theorem. When M and M' both satisfy central limit theorems with equal asymptotic variances, it turns out that the asymptotic equivalence of M and M' follows when the asymptotic means are equal, as investigated in Exercise 6.31.

We apply Exercise 6.31 with $m=n(n-1)/2$ to obtain that $\mathrm{ER}_n(p)$ and $\mathrm{ER}_n(q)$ are asymptotically equivalent precisely when $n^2(p-q)^2/p=o(1)$, and, assuming that $p=\lambda/n$, this is equivalent to $p-q=o(n^{-3/2})$. This explains the result in Theorem 6.18 for the Erdős–Rényi random graph, and also shows that the result is optimal in this case.

We now proceed by proving Theorem 6.18. In this section, rather than working with the total variation distance between two measures, it is more convenient to use the so-called *Hellinger distance*, which is defined, for discrete measures $p=(p_x)_{x\in\mathcal{X}}$ and $q=(q_x)_{x\in\mathcal{X}}$ by

$$d_{\mathrm{H}}(p,q)=\sqrt{\frac{1}{2}\sum_{x\in\mathcal{X}}(\sqrt{p_x}-\sqrt{q_x})^2}. \tag{6.7.11}$$

It is readily seen that d_{H} and d_{TV} are intimately related, since (see Exercise 6.32)

$$d_{\mathrm{H}}(p,q)^2 \leq d_{\mathrm{TV}}(p,q) \leq 2^{1/2} d_{\mathrm{H}}(p,q). \tag{6.7.12}$$

We define

$$\rho(p,q)=2d_{\mathrm{H}}(\mathsf{Be}(p),\mathsf{Be}(q))^2=\left(\sqrt{p}-\sqrt{q}\right)^2+\left(\sqrt{1-p}-\sqrt{1-q}\right)^2, \tag{6.7.13}$$

and note that (see Exercise 6.34)

$$\rho(p,q) \leq (p-q)^2\left(p^{-1}+(1-p)^{-1}\right). \tag{6.7.14}$$

In particular, Exercise 6.34 implies that when $p \leq 1-\varepsilon$, then

$$\rho(p,q) \leq C(p-q)^2/p \tag{6.7.15}$$

for some $C=C(\varepsilon)>0$. Now we are ready to complete the proof of Theorem 6.18:

Proof of Theorem 6.18 Let $\mathrm{IRG}_n(\boldsymbol{p})$ and $\mathrm{IRG}_n(\boldsymbol{q})$ with $\boldsymbol{p} = (p_{ij})_{1\leq i<j\leq n}$ and $\boldsymbol{q} = (q_{ij})_{1\leq i<j\leq n}$ be two inhomogeneous random graphs. Asymptotic equivalence of $\mathrm{IRG}_n(\boldsymbol{p})$ and $\mathrm{IRG}_n(\boldsymbol{q})$ is equivalent to asymptotic equivalence of the edge variables, which are independent Bernoulli random variables with success probabilities $\boldsymbol{p} = (p_{ij})_{1\leq i<j\leq n}$ and $\boldsymbol{q} = (q_{ij})_{1\leq i<j\leq n}$. In turn, asymptotic equivalence of the edge variables is equivalent to the fact that $d_{\mathrm{H}}(\boldsymbol{p},\boldsymbol{q}) = o(1)$, which is what we prove now.

For two discrete probability measures $p = (p_x)_{x\in\mathcal{X}}$ and $q = (q_x)_{x\in\mathcal{X}}$, we denote

$$H(p,q) = 1 - d_{\mathrm{H}}(p,q)^2 = \sum_{x\in\mathcal{X}} \sqrt{p_x}\sqrt{q_x}, \tag{6.7.16}$$

so that

$$d_{\mathrm{H}}(\boldsymbol{p},\boldsymbol{q}) = \sqrt{1 - H(\boldsymbol{p},\boldsymbol{q})}. \tag{6.7.17}$$

For $\mathrm{IRG}_n(\boldsymbol{p})$ and $\mathrm{IRG}_n(\boldsymbol{q})$ with $\boldsymbol{p} = (p_{ij})_{1\leq i<j\leq n}$ and $\boldsymbol{q} = (q_{ij})_{1\leq i<j\leq n}$, the edges are independent, so that, with ρ defined in (6.7.13),

$$H(\boldsymbol{p},\boldsymbol{q}) = \prod_{1\leq i<j\leq n} \left(1 - \frac{1}{2}\rho(p_{ij},q_{ij})\right). \tag{6.7.18}$$

As a result, $d_{\mathrm{H}}(\boldsymbol{p},\boldsymbol{q}) = o(1)$ precisely when $H(\boldsymbol{p},\boldsymbol{q}) = 1 + o(1)$. By (6.7.18) and using that $(1-x)(1-y) \geq 1 - x - y$ and $1 - x \leq \mathrm{e}^{-x}$ for $x, y \geq 0$,

$$1 - \frac{1}{2}\sum_{1\leq i<j\leq n} \rho(p_{ij},q_{ij}) \leq H(\boldsymbol{p},\boldsymbol{q}) \leq \mathrm{e}^{-\frac{1}{2}\sum_{1\leq i<j\leq n}\rho(p_{ij},q_{ij})}, \tag{6.7.19}$$

so that $H(\boldsymbol{p},\boldsymbol{q}) = 1 - o(1)$ precisely when $\sum_{1\leq i<j\leq n}\rho(p_{ij},q_{ij}) = o(1)$. By (6.7.15), we further obtain that when $\max_{1\leq i<j\leq n} p_{ij} \leq 1 - \varepsilon$ for some $\varepsilon > 0$, then

$$\sum_{1\leq i<j\leq n} \rho(p_{ij},q_{ij}) \leq C \sum_{1\leq i<j\leq n} \frac{(p_{ij}-q_{ij})^2}{p_{ij}} = o(1), \tag{6.7.20}$$

by (6.7.2). This completes the proof of the first part of Theorem 6.18. For the second part, we note that if (6.7.3) holds, then we can find a sequence ε_n such that

$$\mathbb{P}\Big(\sum_{1\leq i<j\leq n} \frac{(p_{ij}-q_{ij})^2}{p_{ij}} \leq \varepsilon_n\Big) = 1 - o(1). \tag{6.7.21}$$

Then, the asymptotic equivalence of $\mathrm{IRG}_n(\boldsymbol{p})$ and $\mathrm{IRG}_n(\boldsymbol{q})$ is, in turn, equivalent to the asymptotic equivalence of $\mathrm{IRG}_n(\boldsymbol{p})$ and $\mathrm{IRG}_n(\boldsymbol{q})$ *conditionally on*

$$\sum_{1\leq i<j\leq n} \frac{(p_{ij}-q_{ij})^2}{p_{ij}} \leq \varepsilon_n.$$

For the latter, we can use the first part of Theorem 6.18. □

In fact, tracing back the above proof, we see that under the assumptions of Theorem 6.18, we also obtain that $\rho(p,q) \geq c(p-q)^2/p$ for some $c = c(\varepsilon) \geq 0$. Thus, we can strengthen Theorem 6.18 to the fact that $\mathrm{IRG}_n(\boldsymbol{p})$ and $\mathrm{IRG}_n(\boldsymbol{q})$ are asymptotically equivalent if and only if (6.7.2) holds.

6.8 Related Inhomogeneous Random Graph Models

We now discuss two examples of inhomogeneous random graphs which have appeared in the literature, and are related to the generalized random graph, and investigate when they are asymptotically equivalent. We start with the random graph with prescribed expected degrees.

6.8.1 Chung-Lu Model: Random Graph with Prescribed Expected Degrees

In this section, we prove a coupling result for the Chung-Lu random graph, where the edge probabilities are given by

$$p_{ij}^{(\mathrm{CL})} = \frac{w_i w_j}{\ell_n} \wedge 1, \tag{6.8.1}$$

where again

$$\ell_n = \sum_{i\in[n]} w_i. \tag{6.8.2}$$

We denote the resulting graph by $\mathrm{CL}_n(\boldsymbol{w})$. When

$$\max_{i\in[n]} w_i^2 \leq \ell_n, \tag{6.8.3}$$

we may forget about the minimum with 1 in (6.8.1). We will assume that the bound $\max_{i\in[n]} w_i^2 \leq \ell_n$ holds throughout this section.

Naturally, when $w_i/\sqrt{\ell_n}$ is quite small, there is hardly any difference between edge weights $p_{ij} = w_i w_j/(\ell_n + w_i w_j)$ and $p_{ij} = w_i w_j/\ell_n$. Therefore, one expects that $\mathrm{CL}_n(\boldsymbol{w})$ and $\mathrm{GRG}_n(\boldsymbol{w})$ behave rather similarly. We make use of Theorem 6.18, and investigate the asymptotic equivalence of $\mathrm{CL}_n(\boldsymbol{w})$ and $\mathrm{GRG}_n(\boldsymbol{w})$:

Theorem 6.19 (Asymptotic equivalence of CL and GRG with deterministic weights) *The random graphs* $\mathrm{CL}_n(\boldsymbol{w})$ *and* $\mathrm{GRG}_n(\boldsymbol{w})$ *are asymptotically equivalent when*

$$\sum_{i\in[n]} w_i^3 = o(n^{3/2}). \tag{6.8.4}$$

Proof We make use of Theorem 6.18. For this, we compute, for fixed ij, and using the fact that $1 - 1/(1+x) \leq x$,

$$p_{ij}^{(\mathrm{CL})} - p_{ij} = \frac{w_i w_j}{\ell_n} - \frac{w_i w_j}{\ell_n + w_i w_j} = \frac{w_i w_j}{\ell_n}\Big[1 - \frac{1}{1 + \frac{w_i w_j}{\ell_n}}\Big] \leq \frac{w_i^2 w_j^2}{\ell_n^2}. \tag{6.8.5}$$

Moreover, $\max_{i\in[n]} w_i^2 = o(n)$ by (6.8.4), so that, for n sufficiently large

$$p_{ij} = \frac{w_i w_j}{\ell_n + w_i w_j} \geq w_i w_j/(2\ell_n), \tag{6.8.6}$$

we arrive at

$$\sum_{1\leq i<j\leq n} \frac{(p_{ij} - p_{ij}^{(\mathrm{CL})})^2}{p_{ij}} \leq 2\ell_n^{-3} \sum_{1\leq i<j\leq n} w_i^3 w_j^3 \leq \ell_n^{-3}\Big(\sum_{i\in[n]} w_i^3\Big)^2 = o(1), \tag{6.8.7}$$

again by (6.8.4). □

When Conditions 6.4(a)–(c) hold, Exercise 6.3 implies that $\max_{i\in[n]} w_i = o(\sqrt{n})$, so that

$$\sum_{i\in[n]} w_i^3 = o(\sqrt{n}) \sum_{i\in[n]} w_i^2 = o(n^{3/2})\mathbb{E}[W_n^2] = o(n^{3/2}). \tag{6.8.8}$$

Thus, we have proved the following corollary:

Corollary 6.20 (Asymptotic equivalence of CL and GRG) *Assume that Conditions 6.4(a)–(c) hold. Then, the random graphs $\mathrm{CL}_n(\boldsymbol{w})$ and $\mathrm{GRG}_n(\boldsymbol{w})$ are asymptotically equivalent.*

We can prove stronger results linking the degree sequences of $\mathrm{CL}_n(\boldsymbol{w})$ and $\mathrm{GRG}_n(\boldsymbol{w})$ for deterministic weights given by (6.2.14) when $\mathbb{E}[W] < \infty$, by splitting between vertices with small and high weights, but we refrain from doing so.

6.8.2 Norros–Reittu Model or the Poisson Graph Process

Norros and Reittu (2006) introduce a random *multigraph* with a Poisson number of edges in between any two vertices i and j, with parameter equal to $w_i w_j/\ell_n$. The graph is defined as a *graph process*, where at each time t, a new vertex is born with an associated weight w_t. The number of edges between i and t is $\mathsf{Poi}(w_i w_t/\ell_t)$ distributed. Furthermore, at each time each of the older edges is erased with probability equal to w_t/ℓ_t. We claim that the number of edges between vertices i and j at time t is a Poisson random variable with mean $w_i w_j/\ell_t$, and that the number of edges between the various pairs of vertices are *independent*. To see this, we start by observing that, by Exercise 6.35, a Poisson number of Bernoulli variables is Poisson with a different mean.

By Exercise 6.35, the number of edges between vertices i and j at time t is a Poisson random variable with mean $w_i w_j/\ell_t$, and the number of edges between different pairs are independent. Indeed, making repeated use of Exercise 6.35 shows that the number of edges at time t between vertices i and j, for $i < j$, is Poisson with parameter

$$\frac{w_i w_j}{\ell_j} \prod_{s=j+1}^{t} \left(1 - \frac{w_s}{\ell_s}\right) = \frac{w_i w_j}{\ell_j} \prod_{s=j+1}^{t} \left(\frac{\ell_{s-1}}{\ell_s}\right) = \frac{w_i w_j}{\ell_t}, \tag{6.8.9}$$

as required. The independence of the number of edges between different pairs of vertices follows by the independence in the construction of the graph.

The Norros–Reittu graph process produces a multigraph. However, when the weights are sufficiently bounded, it can be seen that the resulting graph is with positive probability simple (see Exercise 6.36). Exercise 6.37 investigates the degree of a fixed vertex in the Norros–Reittu model.

We now discuss the Norros–Reittu model at time n, ignoring the dynamic formulation given above. The Norros–Reittu model is a *multigraph*, where the number of edges between vertices i and j is $\mathsf{Poi}(w_i w_j/\ell_n)$. The Norros–Reittu random graph is obtained by erasing all self-loops and merging all multiple edges. We denote this graph by $\mathrm{NR}_n(\boldsymbol{w})$. For $\mathrm{NR}_n(\boldsymbol{w})$, the probability that there is at least one edge between vertices i and j exists is, conditionally on the weights $(w_i)_{i\in[n]}$, given by

$$p_{ij}^{(\mathrm{NR})} = \mathbb{P}(\mathsf{Poi}(w_i w_j/\ell_n) \geq 1) = 1 - \mathrm{e}^{-w_i w_j/\ell_n}, \tag{6.8.10}$$

and the occupation statuses of different edges are independent random variables. We next return to the relation between the various random graph models discussed in this section.

We say that a random graph G_n is *stochastically dominated* by the random graph G'_n when, with $(X_{ij})_{1\leq i<j\leq n}$ and $(X'_{ij})_{1\leq i<j\leq n}$ denoting the occupation statuses of the edges in G_n and G'_n respectively, there exists a *coupling* $\big((\hat{X}_{ij})_{1\leq i<j\leq n}, (\hat{X}'_{ij})_{1\leq i<j\leq n}\big)$ of $(X_{ij})_{1\leq i<j\leq n}$ and $(X'_{ij})_{1\leq i<j\leq n}$ such that

$$\mathbb{P}\big(\hat{X}_{ij} \leq \hat{X}'_{ij}\ \forall i, j \in [n]\big) = 1. \tag{6.8.11}$$

We write $G_n \preceq G'_n$ when the random graph G_n is stochastically dominated by the random graph G'_n.

When the statuses of the edges are *independent*, then (6.8.11) is equivalent to the bound that, for all $i, j \in [n]$,

$$p_{ij} = \mathbb{P}(X_{ij} = 1) \leq p'_{ij} = \mathbb{P}(X'_{ij} = 1). \tag{6.8.12}$$

By (6.8.12) and the fact that $\frac{x}{1+x} \leq 1 - \mathrm{e}^{-x} \leq (x \wedge 1)$ for every $x \geq 0$,

$$\mathrm{GRG}_n(\boldsymbol{w}) \preceq \mathrm{NR}_n(\boldsymbol{w}) \preceq \mathrm{CL}_n(\boldsymbol{w}). \tag{6.8.13}$$

This provides a good way of comparing the various inhomogeneous random graph models discussed in this chapter. See for example Exercise 6.39, which investigates when $\mathrm{NR}_n(\boldsymbol{w})$ is asymptotically equivalent to $\mathrm{GRG}_n(\boldsymbol{w})$.

6.9 Notes and Discussion

Notes on Section 6.2

In the generalized random graph studied by Britton, Deijfen and Martin-Löf (2006), the situation where the vertex weights are i.i.d. is investigated, and ℓ_n in the denominator of the edge probabilities in (6.2.1) is replaced by n, which leads to a minor change. Indeed, when the weights have finite mean, then $\ell_n = \mathbb{E}[W]n(1 + o_{\mathbb{P}}(1))$, by the law of large numbers. If we would replace ℓ_n by $\mathbb{E}[W]n$ in (6.2.1), then the edge occupation probabilities become

$$\frac{w_i w_j}{\mathbb{E}[W]n + w_i w_j}, \tag{6.9.1}$$

so that this change amounts to replacing w_i by $w_i/\sqrt{\mathbb{E}[W]}$. Therefore, at least at a heuristic level, there is hardly any difference between the definition of p_{ij} in (6.2.1) and the choice $p_{ij} = w_i w_j/(n + w_i w_j)$ in Britton et al. (2006).

In the literature, both the cases with i.i.d. weights as well as the one with deterministic weights have been studied. Chung and Lu (2002a,b, 2003, 2006c); Lu (2002), study the Chung-Lu model, as defined in Section 6.8, with deterministic weights. Van den Esker, van der Hofstad and Hooghiemstra (2008) study more general settings, including the one with deterministic weights as in (6.2.14). Britton et al. (2006), on the other hand, study the generalized random graph where the weights are i.i.d. Van den Esker, van der Hofstad and Hooghiemstra (2008) investigate several cases including i.i.d. degrees, in the case where the degrees have finite-variance degrees, for the Chung-Lu model, the Norros–Reittu model, as well as the generalized random graph.

The advantage of deterministic weights is that there is no double randomness, which makes the model easier to analyze. The results are often also more general, since the results for random weights are a simple consequence of the ones for deterministic weights. On the other hand, the advantage of working with i.i.d. weights is that the vertices are exchangeable, and, in contrast to the deterministic weights case, not many assumptions need to be made. For deterministic weights, one often has to make detailed assumptions concerning the precise structure of the weights.

Notes on Section 6.3

The results in this section are novel, and are inspired by the ones by Britton, Deijfen and Martin-Löf (2006).

Notes on Section 6.4

The results in this section are novel, and are inspired by the ones in Britton et al. (2006).

Notes on Section 6.5

Theorem 6.14 is Britton et al. (2006, Proof of Theorem 3.2), whose proof we follow. Exercise 6.22 is novel.

Notes on Section 6.6

The proof in Section 6.6 follows the argument in Britton et al. (2006, Section 3). Theorem 6.7 is an extension of Britton et al. (2006, Theorem 3.1), in which Corollary 6.11 was proved under the extra assumption that w_i have a finite $(1+\varepsilon)-$moment.

Notes on Section 6.7

Section 6.7 follows the results by Janson (2010). Theorem 6.18 is Janson (2010, Corollary 2.12). Janson (2010) studies many more examples and results, also investigating the notion of *asymptotic contiguity* of random graphs, which is a slightly weaker notion than asymptotic equivalence, and holds when events that have vanishing probability under one measure also have vanishing probabilities under the other. There are deep relations between convergence in probability and in distribution and asymptotic equivalence and contiguity; see Janson (2010, Remark 1.4).

Notes on Section 6.8

The random graph with prescribed expected degrees, or Chung-Lu model, has been studied extensively by Chung and Lu (2002a,b, 2003, 2006c); Lu (2002). See in particular the recent book Chung and Lu (2006a), in which many of these results are summarized.

The Norros–Reittu or Poissonian random graph process first appeared in Norros and Reittu (2006). We return to this model in [II, Chapter 3], where we prove a beautiful relation to branching processes for the vertex neighborhoods in this model.

6.10 Exercises for Chapter 6

Exercise 6.1 (The Erdős–Rényi random graph) Prove that $p_{ij} = \lambda/n$ when $w_i = n\lambda/(n-\lambda)$ for all $i \in [n]$.

Exercise 6.2 (Weight of uniformly chosen vertex) Let U be a vertex chosen uniformly at random from $[n]$. Show that the weight w_U of U has distribution function F_n.

Exercise 6.3 (Bound on the maximal weight assuming Conditions 6.4(b)–(c)) Prove that Conditions 6.4(a)–(b) imply that $\max_{i\in[n]} w_i = o(n)$, while Conditions 6.4(a)–(c) imply that $\max_{i\in[n]} w_i = o(\sqrt{n})$. When the degrees $\boldsymbol{w} = (w_i)_{i\in[n]}$ are *random*, these bounds hold in probability. (Bear in mind that $\boldsymbol{w}^{(n)} = (w_i^{(n)})_{i\in[n]}$ may depend on n!)

Exercise 6.4 (Condition 6.4(a)) Prove that Condition 6.4(a) holds for $(w_i)_{i\in[n]}$ as in (6.2.14).

Exercise 6.5 (Moments of $\boldsymbol{w}$ and F (van den Esker et al. (2008))) Prove that $u \mapsto [1 - F]^{-1}(u)$ is non-increasing, and conclude that, for every non-decreasing function $x \mapsto h(x)$ and for $(w_i)_{i\in[n]}$ as in (6.2.14),

$$\frac{1}{n}\sum_{i\in[n]} h(w_i) \leq \mathbb{E}[h(W)], \tag{6.10.1}$$

where W is a random variable with distribution function F.

Exercise 6.6 (Moments of $\boldsymbol{w}$ and F (van den Esker et al. (2008)) (cont.)) Set $\alpha > 0$, assume that $\mathbb{E}[W^\alpha] < \infty$ where W is a random variable with distribution function F. Use Lebesgue's dominated convergence theorem (Theorem A.1) to prove that for $(w_i)_{i\in[n]}$ as in (6.2.14),

$$\frac{1}{n}\sum_{i\in[n]} w_i^\alpha \to \mathbb{E}[W^\alpha]. \tag{6.10.2}$$

Conclude that Condition 6.4(b) holds when $\mathbb{E}[W] < \infty$, and Condition 6.4(c) when $\mathbb{E}[W^2] < \infty$.

Exercise 6.7 (Bounds on $\boldsymbol{w}$) Fix $(w_i)_{i\in[n]}$ as in (6.2.14). Prove there exists a $c' > 0$ such that $w_j \leq w_1 \leq c' n^{1/(\tau-1)}$ for all $j \in [n]$ and all large enough n, when the distribution function F in (6.2.14) satisfies

$$1 - F(x) \leq c x^{-(\tau-1)}. \tag{6.10.3}$$

Exercise 6.8 (Dependence edges in $\mathrm{GRG}_n(\boldsymbol{w})$ with i.i.d. weights) Let $(w_i)_{i\in[n]}$ be an i.i.d. sequence of weights, where $(w_i)_{i\in[n]}$ are copies of a random variable W for which $\mathbb{E}[W^2] < \infty$. Assume further that there exists $\varepsilon > 0$ such that $\mathbb{P}(W \leq \varepsilon) = 0$. Prove that

$$n\mathbb{P}(12 \text{ present}) = n\mathbb{P}(23 \text{ present}) \to \mathbb{E}[W], \tag{6.10.4}$$

while

$$n^2\mathbb{P}(12 \text{ and } 23 \text{ present}) \to \mathbb{E}[W^2]. \tag{6.10.5}$$

Conclude that the statuses of different edges that share a vertex are *dependent* whenever $\mathrm{Var}(W) > 0$.

Exercise 6.9 (Not every random variable is mixed Poisson) Give an example of an integer-valued random variable that cannot be represented as a mixed Poisson distribution.

Exercise 6.10 (Characteristic function of mixed Poisson distribution) Let X have a mixed Poisson distribution with mixing distribution F and moment generating function M_W, i.e., for $t \in \mathbb{C}$,

$$M_W(t) = \mathbb{E}[\mathrm{e}^{tW}], \tag{6.10.6}$$

where W has distribution function F. Show that the characteristic function of X is given by

$$\phi_X(t) = \mathbb{E}[\mathrm{e}^{itX}] = M_W(\mathrm{e}^{it} - 1). \tag{6.10.7}$$

Exercise 6.11 (Mean and variance mixed Poisson distribution) Let X have a mixed Poisson distribution with mixing distribution F. Express the mean and variance of X into the moments of W, where W has distribution function F.

Exercise 6.12 (Tail behavior mixed Poisson) Suppose that there exist constants $0 < c_1 < c_2 < \infty$ such that, for $x \geq 1$,

$$c_1 x^{1-\tau} \leq 1 - F(x) \leq c_2 x^{1-\tau}. \tag{6.10.8}$$

Show that there exist $0 < c_1' < c_2' < \infty$ such that the distribution function G of a mixed Poisson distribution with mixing distribution F satisfies that, for $x \geq 1$,

$$c_1' x^{1-\tau} \leq 1 - G(x) \leq c_2' x^{1-\tau}. \tag{6.10.9}$$

Exercise 6.13 (Independence of a growing number of degrees for bounded weights) Assume that the conditions in Corollary 6.9 hold, and further suppose that there exists a $\varepsilon > 0$ such that $\varepsilon \leq w_i \leq \varepsilon^{-1}$ for every i, so that the weights are uniformly bounded from above and below. Then, prove that we can couple $(D_i)_{i\in[m]}$ to an independent vector $(\hat{D}_i)_{i\in[m]}$ such that (6.3.12) holds whenever $m = o(\sqrt{n})$. As a result, even the degrees of a growing number of vertices can be coupled to independent degrees.

Exercise 6.14 (Convergence of degrees between vertices of truncated weight) Assume that Conditions 6.4(a)–(b) hold and take $a_n = o(n^{1/3})$. Recall from (6.3.15) that $D_U^{(1)}$ counts the number of edges of U to vertices of weight at most a_n. Prove that, conditionally on $w_U \leq a_n$, $D_U^{(1)} \xrightarrow{d} D$, where $D \sim \mathsf{Poi}(W)$ has a mixed Poisson distribution with mixing distribution W.

Exercise 6.15 (Isolated vertices) Prove that Theorem 6.10 implies that $P_0^{(n)} \xrightarrow{\mathbb{P}} p_0 > 0$. Conclude that $\mathrm{GRG}_n(\boldsymbol{w})$ is not connected when Conditions 6.4(a)–(b) hold.

Exercise 6.16 (Vertices of all degrees exist in $\mathrm{GRG}_n(\boldsymbol{w})$) Prove that Theorem 6.10 implies that $P_k^{(n)} \xrightarrow{\mathbb{P}} p_k > 0$ for every $k \geq 0$. Conclude that $\mathrm{GRG}_n(\boldsymbol{w})$ has a positive proportion of vertices of degree k for *every* $k \geq 0$.

Exercise 6.17 (Uniform convergence of mean degree sequence) Prove (6.4.8).

Exercise 6.18 (Bound on sum of squares of i.i.d. random variables) Show that when $(w_i)_{i\in[n]}$ are i.i.d. random variables with distribution function F with a finite mean, then

$$\frac{1}{n}\max_{i\in[n]} w_i \xrightarrow{\mathbb{P}} 0. \tag{6.10.10}$$

Conclude that

$$\frac{1}{n^2}\sum_{i\in[n]} w_i^2 \xrightarrow{\mathbb{P}} 0. \tag{6.10.11}$$

Hint: Use that

$$\begin{aligned}\mathbb{P}(\max_{i\in[n]} w_i \geq \varepsilon n) &\leq \sum_{i\in[n]} \mathbb{P}(w_i \geq \varepsilon n)\\ &= n\mathbb{P}(W \geq \varepsilon n).\end{aligned} \tag{6.10.12}$$

Then use a variant of the Markov inequality (Theorem 2.17) to show that $\mathbb{P}(W \geq \varepsilon n) = o(1/n)$.

Exercise 6.19 (Proof of Theorem 6.12) Complete the proof of Theorem 6.12, now using Corollary 6.9, as well as the equality

$$\begin{aligned}\mathbb{E}[(P_k^{(n)})^2] &= \frac{1}{n^2}\sum_{1\leq i,j\leq n} \mathbb{P}(D_i = D_j = k)\\ &= \frac{1}{n}\mathbb{P}(D_1 = k) + \frac{2}{n^2}\sum_{1\leq i<j\leq n} \mathbb{P}(D_i = D_j = k).\end{aligned} \tag{6.10.13}$$

Exercise 6.20 (Condition for infinite mean) Show that the mean of W is infinite precisely when the distribution function F of W satisfies

$$\int_0^\infty [1 - F(x)]dx = \infty. \tag{6.10.14}$$

Exercise 6.21 (Proof of Theorem 6.13) Prove Theorem 6.13 by adapting the proof of Theorem 6.7. In particular, show that $\sum_{j\in[n]} \frac{(w_j^{(n)})^2}{\ell_n^2} = o_{\mathbb{P}}(1)$ in the setting of Theorem 6.13.

Exercise 6.22 (Tail of degree law for $\tau \in (1, 2)$) Let the distribution function F satisfy (6.5.2), and let Y be a mixed Poisson random variable with parameter $W^{\tau-1}$, where W has distribution function F. Show that Y is such that there exists a constant $c > 0$ such that

$$\mathbb{P}(Y \geq y) = cy^{-1}(1 + o(1)). \tag{6.10.15}$$

Exercise 6.23 (Equality for probability mass function GRG) Prove the last equality in (6.6.10).

Exercise 6.24 (Alternative proof Theorem 6.7) Use Proposition 6.16 to give an alternative proof of Theorem 6.7.

Exercise 6.25 (Degree of vertex 1 in $\mathrm{ER}_n(\lambda/n)$) Show that for the Erdős–Rényi random graph with $p = \lambda/n$, the degree of vertex 1 is close to a Poisson random variable with mean λ by using (6.6.19). Hint: Use that the Erdős–Rényi random graph is obtained by taking $w_i \equiv \frac{\lambda}{1-\frac{\lambda}{n}}$.

Exercise 6.26 (Asymptotic independence of vertex degrees in $\mathrm{ER}_n(\lambda/n)$) Show that for the Erdős–Rényi random graph with $p = \lambda/n$, the degrees of vertices $1, \ldots, m$ are asymptotically independent.

Exercise 6.27 (Identification of limiting vertex degree) Prove that $\mathbb{E}[\mathrm{e}^{(t-1)\gamma W^{\tau-1}}]$ is the probability generating function of a mixed Poisson random variable with mixing distribution $\gamma W^{\tau-1}$.

Exercise 6.28 (A partial integration formula) Prove that for every differentiable function $h\colon [0, \infty) \to \mathbb{R}$, with $h(0) = 0$ and every random variable $X \geq 0$ with distribution function F_X, the following partial integration formula holds:

$$\mathbb{E}[h(X)] = \int_0^\infty h'(x)[1 - F_X(x)]dx, \tag{6.10.16}$$

provided that $\int_0^\infty |h'(x)|[1 - F_X(x)]dx < \infty$.

Exercise 6.29 (Conditions for dominated convergence) Verify the conditions for dominated convergence for the integral on the left-hand side of (6.6.31).

Exercise 6.30 (Asymptotic equivalence and total variation distance) Use (2.2.4) and Definition 6.17 to prove that $p^{(n)} = (p_x^{(n)})_{x\in\mathcal{X}}$ and $q^{(n)} = (q_x^{(n)})_{x\in\mathcal{X}}$ are asymptotically equivalent if and only if $d_{\mathrm{TV}}(p^{(n)}, q^{(n)}) \to 0$.

Exercise 6.31 (Asymptotic equivalence of binomials with increasing variances Janson (2010)) Let M and M' be two binomial random variables with $M \sim \mathsf{Bin}(m, p_m)$ and $M' \sim \mathsf{Bin}(m, q_m)$ for some m. Assume that $m \to \infty$. Show that M and M' are asymptotically equivalent when $m(p_m - q_m)/\sqrt{mp_m} = o(1)$. When $p_m \to 0$, show that this is equivalent to $mp_m \to \infty$ and $q_m/p_m \to 1$.

Exercise 6.32 (Total variation and Hellinger distance) Prove that, for discrete probability measures $p = (p_x)_{x\in\mathcal{X}}$ and $q = (q_x)_{x\in\mathcal{X}}$,

$$d_{\mathrm{H}}(p, q)^2 \leq d_{\mathrm{TV}}(p, q) \leq 2^{1/2} d_{\mathrm{H}}(p, q). \tag{6.10.17}$$

Exercise 6.33 (Asymptotic equivalence and Hellinger distance) Use Exercises 6.30 and 6.32 to prove that $p^{(n)} = (p_x^{(n)})_{x\in\mathcal{X}}$ and $q^{(n)} = (q_x^{(n)})_{x\in\mathcal{X}}$ are asymptotically equivalent if and only if $d_{\mathrm{H}}(p^{(n)}, q^{(n)}) \to 0$.

Exercise 6.34 (Bound on Hellinger distance Bernoulli variables) Prove (6.7.14).

Exercise 6.35 (Poisson number of Bernoulli variables is Poisson) Let X be a Poisson random variable with mean λ, and let $(I_i)_{i=1}^{\infty}$ be an independent and identically distributed sequence of $\mathsf{Be}(p)$ random variables. Prove that

$$Y = \sum_{i=1}^{X} I_i \tag{6.10.18}$$

has a Poisson distribution with mean λp.

Exercise 6.36 (Simplicity of the Norros–Reittu random graph) Compute the probability that the Norros–Reittu random graph is simple at time n, meaning that it has no self-loops nor multiple edges.

Exercise 6.37 (The degree of a fixed vertex) Assume that Conditions 6.4(a)–(b) hold. Prove that the degree of a uniform vertex $U_n \in [n]$ in the Norros–Reittu graph at time n has an asymptotic mixed Poisson distribution with mixing distribution F, the asymptotic distribution function of W_n.

Exercise 6.38 (Stochastic domination of increasing random variables) Let $G_n \preceq G'_n$ as defined in (6.8.11). Let the random variable $X(G)$ be an increasing random variable of the edge occupation random variables of the graph G. Let $X_n = X(G_n)$ and $X'_n = X(G'_n)$. Show that $X_n \preceq X'_n$.

Exercise 6.39 (Asymptotic equivalence of IRGs) Assume that Conditions 6.4(a)–(c) hold. Show that $\mathrm{NR}_n(\boldsymbol{w})$ is asymptotically equivalent to $\mathrm{GRG}_n(\boldsymbol{w})$.

7

Configuration Model

In this chapter, we investigate graphs with fixed degrees. Ideally, we would like to investigate uniform random graphs having a prescribed degree sequence, i.e, a degree sequence which is given to us beforehand. An example of such a situation could arise from a real-world network of which we know the degree sequence, and we would be interested in generating a random, or even uniform, graph with *precisely* the same degrees.

Such a random graph can be created by uniformly matching half-edges, at the expense of possibly creating self-loops and multiple edges, and the result is called the *configuration model*. The main results in this chapter describe the degree structure in such random graphs, as well as the probability for the random graph to be simple. We also draw connections between the configuration model and the generalized random graphs defined in Chapter 6.

7.1 Motivation for the Configuration Model

The configuration model is a model in which the degrees of vertices are fixed beforehand. Such a model is more flexible than the generalized random graph. For example, the generalized random graph always has a positive proportion of vertices of degree 0, 1, 2, etc. (see Exercises 6.15–6.16). In some real-world networks, however, it is natural to investigate graphs where every vertex has at least one or two neighbors. We start by discussing a few examples where random graphs with prescribed degrees appear naturally:

Example 7.1 (Facebook wall posts) In Facebook wall posts, the vertices of the network are Facebook users, and each directed edge represents one post, linking the user writing the post to the user on whose wall the post is written. This gives rise to a (directed) network of a small subset of posts to the walls of other users on Facebook. Since users may write multiple posts on a wall, the network allows multiple edges between pairs of vertices. Since users may write on their own wall, the network contains loops. The degree distribution in log-log scale is displayed in Figure 7.1, and we see that the degree distribution resembles a power law.[1] We see that the degrees are, by definition, all at least 1. Thus, this data set is poorly modeled by the generalized random graph, which always has many vertices of degree 0.

[1] This data set is from Viswanath et al. (2009) and can be obtained from `http://konect.uni-koblenz.de/networks/facebook-wosn-wall`.

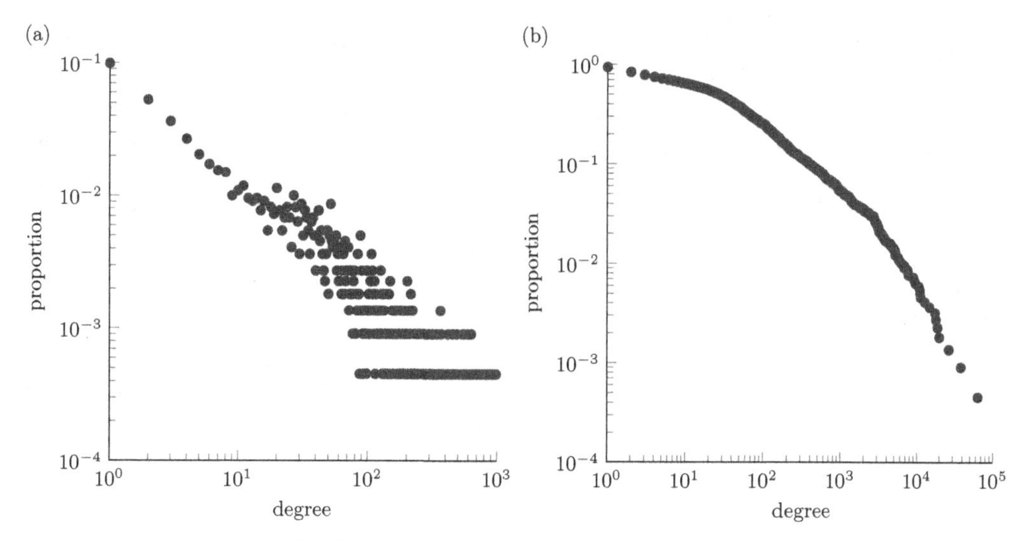

Figure 7.1 Degree distributions for networks of Facebook wall posts (a) Log-log plot of the probability mass function. (b) Log-log plot of the complementary cumulative distribution function.

Example 7.2 (Population of two types revisited) Suppose that we have a complex network in which two distinct types of vertices are present. The first type has precisely m_1 neighbors, the second type precisely m_2. How can we obtain a uniform graph from the collection of graphs satisfying these restrictions?

Example 7.3 (Regular graphs) How many simple graphs are there in which every vertex has degree precisely r? How can we generate a random instance of such a graph, such that each such simple graph has precisely equal probability?

Example 7.4 (Real-world network and its degrees) Suppose that we have a complex network of size n in which vertex $i \in [n]$ has degree d_i. How can we decide whether this network resembles a *uniform* random graph with the same degree sequence, or whether it inherently has more structure? For this, we would need to be able to draw a uniform random graph from the collection of all graphs having the specified degree sequence.

As it turns out, it is not an easy task to generate graphs having prescribed degrees, in particular because there may not exist any (recall (I.3) on page 180). We shall therefore introduce a model that produces a *multigraph* with the prescribed degrees, and which, when conditioned on simplicity, is uniform over all simple graphs with the prescribed degree sequence. This random multigraph is called the *configuration model*. We discuss the connections between the configuration model and a uniform simple random graph having the same degree sequence, and give an asymptotic formula for the number of simple graphs with a given degree sequence.

Organization of this Chapter

This chapter is organized as follows. In Section 7.2, we introduce the configuration model. In Section 7.3, we investigate its properties, given that the degrees satisfy some regularity conditions. We investigate two ways of turning the configuration model into a simple graph, namely, by *erasing* the self-loops and *merging* multiple edges as studied in Section 7.3, or by *conditioning* on obtaining a simple graph as studied in Section 7.4. For the latter, we compute the asymptotic probability of it to be simple. This also allows us to compute the asymptotic number of graphs with a given degree sequence in the case where the maximal degree is not too large. In Section 7.5, we discuss the tight relations that exist between the configuration model conditioned on being simple, and the generalized random graph conditioned on its degrees. This relation will prove to be quite useful when deducing results for the generalized random graph from those for the configuration model. In Section 7.6, we treat the special case of i.i.d. degrees. In Section 7.7, we discuss some further results on the configuration model that we present without proofs. In Section 7.8, we discuss some related models. We close this chapter in Section 7.9 with notes and discussion and in Section 7.10 with exercises.

7.2 Introduction to the Model

Fix an integer n that will denote the number of vertices in the random graph. Consider a sequence of degrees $\boldsymbol{d} = (d_i)_{i\in[n]}$. The aim is to construct an undirected (multi)graph with n vertices, where vertex j has degree d_j. Without loss of generality, we assume throughout this chapter that $d_j \geq 1$ for all $j \in [n]$, since when $d_j = 0$, vertex j is isolated and can be removed from the graph. One possible random graph model is then to take the uniform measure over such undirected and simple graphs. Here, we call a multigraph *simple* when it has no self-loops, and no multiple edges between any pair of vertices. However, the set of undirected simple graphs with n vertices where vertex j has degree d_j may be empty. For example, in order for such a graph to exist, we must assume that the total degree

$$\ell_n = \sum_{j\in[n]} d_j \tag{7.2.1}$$

is even. We wish to construct a simple graph such that $\boldsymbol{d} = (d_i)_{i\in[n]}$ are the degrees of the n vertices. However, even when $\ell_n = \sum_{j\in[n]} d_j$ *is* even, this is not always possible, as explained in more detail in (I.3) on page 180. See also Exercise 7.1.

Since it is not always possible to construct a simple graph with a given degree sequence, instead, we construct a *multigraph*, that is, a graph *possibly* having self-loops and multiple edges between pairs of vertices. One way of obtaining such a multigraph with the given degree sequence is to pair the half-edges attached to the different vertices in a uniform way. Two half-edges together form an edge, thus creating the edges in the graph. Let us explain this in more detail.

To construct the multigraph where vertex j has degree d_j for all $j \in [n]$, we have n separate vertices and incident to vertex j, we have d_j half-edges. Every half-edge needs to be connected to another half-edge to form an edge, and by forming all edges we build the graph. For this, the half-edges are numbered in an arbitrary order from 1 to ℓ_n. We start by randomly connecting the first half-edge with one of the $\ell_n - 1$ remaining half-edges. Once paired, two half-edges form a single edge of the multigraph, and the half-edges are removed from the list of half-edges that need to be paired. Hence, a half-edge can be seen as the left or the right half of an edge. We continue the procedure of randomly choosing and pairing the

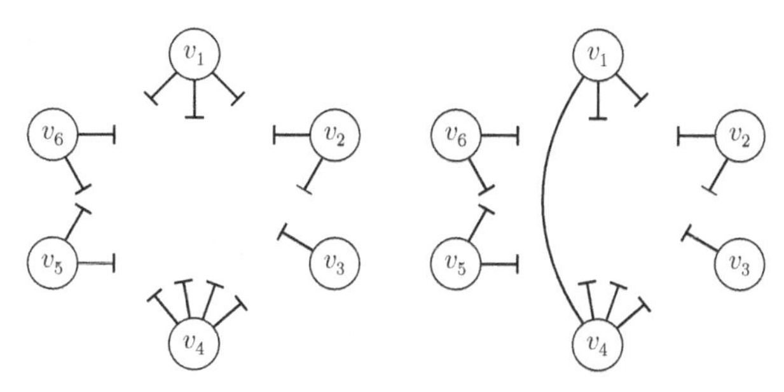

Figure 7.2 The pairing of half-edges in the configuration model. In this figure, $n = 6$, and $d_1 = 3, d_2 = 2, d_3 = 1, d_4 = 4, d_5 = d_6 = 2$. The first half-edge incident to vertex 1 is paired to the first half-edge incident to vertex 4.

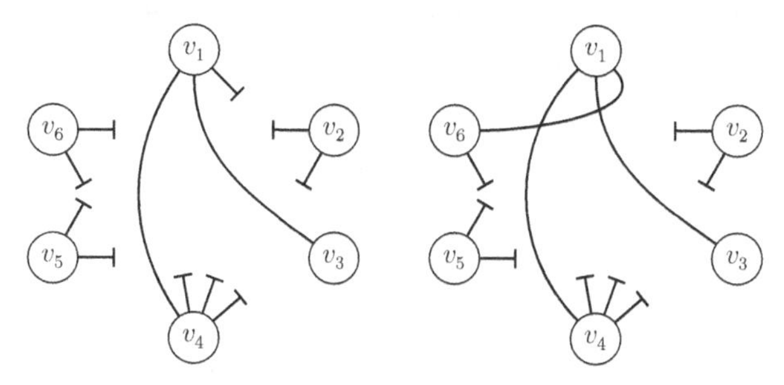

Figure 7.3 The pairing of half-edges in the configuration model. In this figure, $n = 6$, and $d_1 = 3, d_2 = 2, d_3 = 1, d_4 = 4, d_5 = d_6 = 2$. The second half-edge incident to vertex 1 is paired to the half-edge incident to vertex 3, while the third half-edge incident to vertex 1 is paired to the first half-edge incident to vertex 6.

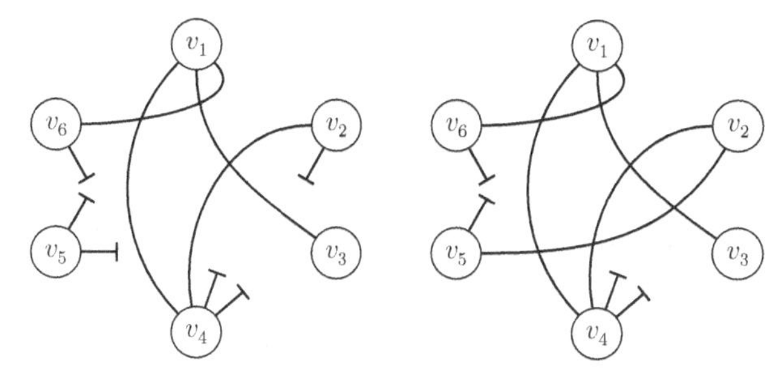

Figure 7.4 The pairing of half-edges in the configuration model. In this figure, $n = 6$, and $d_1 = 3, d_2 = 2, d_3 = 1, d_4 = 4, d_5 = d_6 = 2$. The first half-edge incident to vertex 2 is paired to the second half-edge incident to vertex 4, while the second half-edge incident to vertex 2 is paired to the second half-edge incident to vertex 5.

half-edges until all half-edges are connected, and call the resulting graph the *configuration model with degree sequence* $\boldsymbol{d}$, abbreviated as $\mathrm{CM}_n(\boldsymbol{d})$. See Figures 7.2–7.6 for an example of a realization of the pairing of half-edges and the resulting random graph.

Unfortunately, vertices having self-loops, as well as multiple edges between pairs of vertices, may occur. However, we will see that self-loops and multiple edges are relatively scarce when $n \to \infty$ and the degrees are sufficiently nice. Clearly, when the total degree $\ell_n = \sum_{j \in [n]} d_j$ is even, then the above procedure does produce a multigraph with the right degree sequence. Here, in the degree sequence of the multigraph, a self-loop contributes two

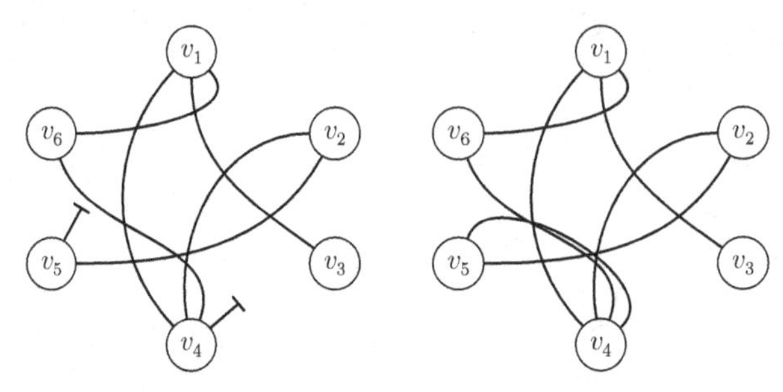

Figure 7.5 The pairing of half-edges in the configuration model. In this figure, $n = 6$, and $d_1 = 3, d_2 = 2, d_3 = 1, d_4 = 4, d_5 = d_6 = 2$. The third half-edge incident to vertex 4 is paired to the half-edge incident to vertex 3, while the third half-edge incident to vertex 1 is paired to the last half-edge incident to vertex 6 and the fourth half-edge incident to vertex 4 is paired to the last half-edge incident to vertex 5.

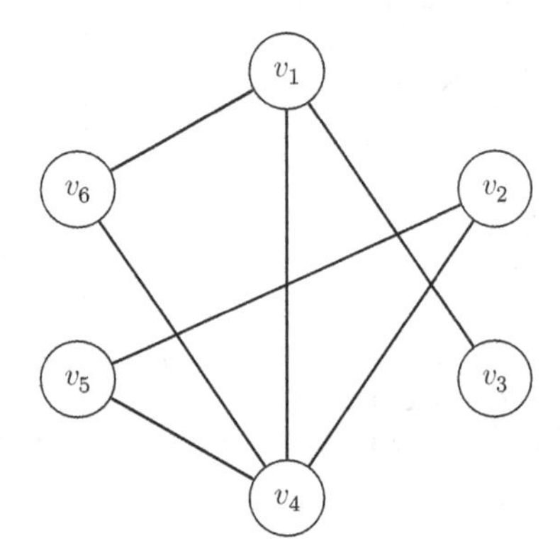

Figure 7.6 The realization of the configuration model. The pairing can create self-loops, for example, when the first half-edge incident to vertex 1 would pair to one of the other half-edges incident to vertex 1. The pairing can also create multiple edges, for example when also the second half-edge incident to vertex 1 would pair to a half-edge incident to vertex 4.

to the degree of the vertex incident to it, while each of the multiple edges contributes one to the degree of each of the two vertices incident to it.

One may wonder, or even worry, about the influence of the arbitrary ordering of the half-edges in the pairing procedure. As it turns out, the process of pairing half-edges is *exchangeable*, meaning that the order does not matter for the distribution of the final outcome. In fact, this is true more generally as long as, conditionally on the paired half-edges so far, the next half-edge is paired to any of the other half-edges with equal probability. This also allows one to pair half-edges in a *random* order, which is useful, for example when exploring the neighborhood of a vertex. See Figure 7.7 for realizations of the configuration model where all the degrees are 3.

To explain the term configuration model, we now present an equivalent way of defining the configuration model in terms of *uniform matchings*. For this, we construct a *second graph*, with vertices $1, \dots, \ell_n$. The vertices in the new graph correspond to the half-edges of the random multigraph in the configuration model. We pair the vertices in the new graph

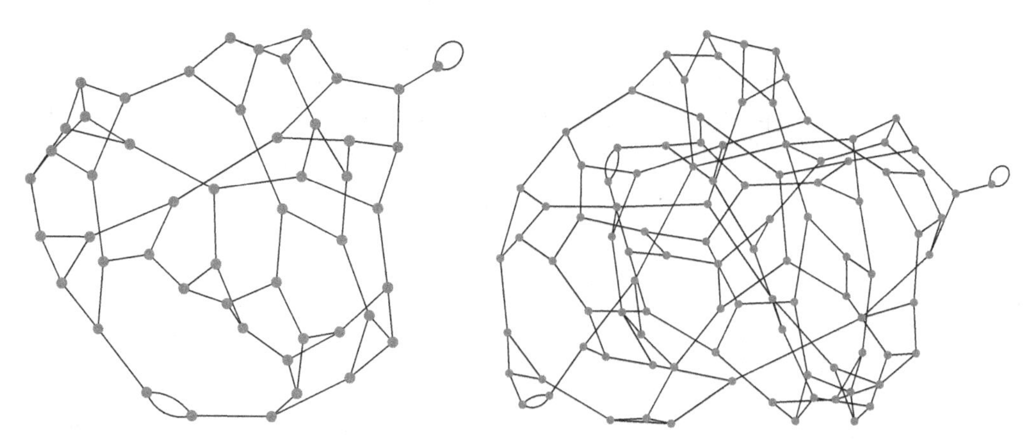

Figure 7.7 Realizations of the configuration model with all degrees equal to 3 and $n = 50$ and $n = 100$, respectively.

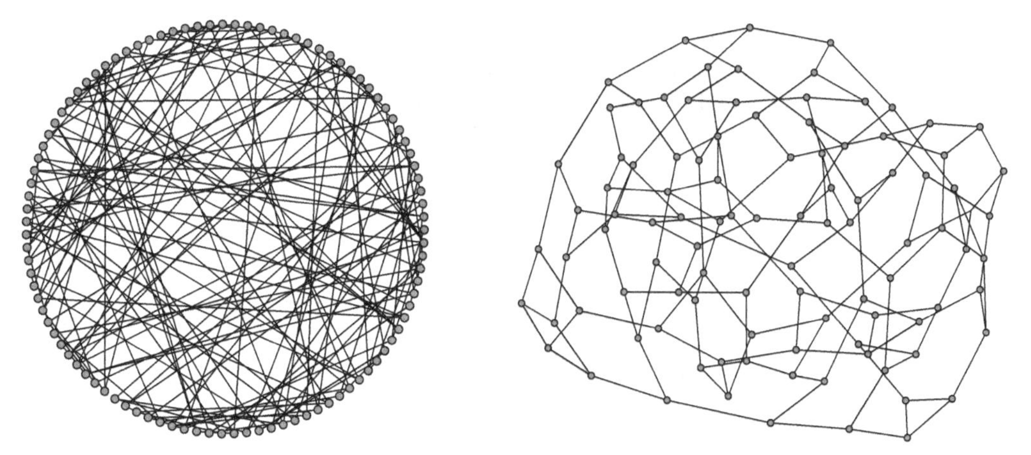

Figure 7.8 A realization of the configuration model with all degrees equal to 3 and $n = 100$ on a circle (left), and in a way that minimizes a force field (right).

in a uniform way to produce a *uniform matching*. For this, we pair vertex 1 with a uniform other vertex. After this, we pair the first not yet paired vertex to a uniform vertex which is not yet paired. The procedure stops when all vertices are paired to another (unique) vertex. We denote the resulting graph by $\mathrm{Conf}_n(\boldsymbol{d})$. Thus, $\mathrm{Conf}_n(\boldsymbol{d})$ can be written as $\mathrm{Conf}_n(\boldsymbol{d}) = \{i\sigma(i)\colon i \in [\ell_n]\}$, where $\sigma(i)$ is the label of the vertex to which vertex $i \in [\ell_n]$ is paired. The pairing of the vertices $1, \ldots, \ell_n$ is called a *configuration*, and each configuration has the same probability. Exercise 7.2 investigates the number of pairings on $2m$ vertices.

To construct the graph of the configuration model from the above configuration, we *identify* vertices $1, \ldots, d_1$ in $\mathrm{Conf}_n(\boldsymbol{d})$ to form vertex 1 in $\mathrm{CM}_n(\boldsymbol{d})$, and vertices $d_1 + 1, \ldots, d_1 + d_2$ in $\mathrm{Conf}_n(\boldsymbol{d})$ to form vertex 2 in $\mathrm{CM}_n(\boldsymbol{d})$, etc. Therefore, precisely d_j vertices in $\mathrm{Conf}_n(\boldsymbol{d})$ are identified with vertex j in $\mathrm{CM}_n(\boldsymbol{d})$.

In the above identification, the number of edges in $\mathrm{CM}_n(\boldsymbol{d})$ between vertices $i, j \in [n]$ is the number of vertices in $\mathrm{Conf}_n(\boldsymbol{d})$ that are identified with $i \in [n]$ and are paired to a vertex in $\mathrm{Conf}_n(\boldsymbol{d})$ that is identified with vertex $j \in [n]$. As a consequence, the degree of vertex j in $\mathrm{CM}_n(\boldsymbol{d})$ is *precisely* equal to d_j. The resulting graph is a *multigraph*, since both self-loops and multiple edges between vertices are possible. We can identify the graph as $\mathrm{CM}_n(\boldsymbol{d}) = (X_{ij})_{i,j\in[n]}$, where X_{ij} is the number of edges between vertices $i, j \in [n]$ and X_{ii} is the number of self-loops of vertex $i \in [n]$, so that, for all $i \in [n]$,

$$d_i = X_{ii} + \sum_{j\in[n]} X_{ij}. \tag{7.2.2}$$

Here, the number of self-loops of vertex i, X_{ii}, appears twice, so that a self-loop contributes 2 to the degree. By convention, we again set $X_{ij} = X_{ji}$. Since the uniform matching of the ℓ_n vertices in $\mathrm{Conf}_n(\boldsymbol{d})$ is sometimes referred to as the *configuration*, the resulting graph $\mathrm{CM}_n(\boldsymbol{d})$ is called *the configuration model.*

Exercise 7.2 also gives us an easy way to see that the order of pairings does not matter for the final outcome. Indeed, as long as each configuration has probability $1/(\ell_n - 1)!!$, the precise order in which the pairings take place is irrelevant. This is formalized in the following definition:

Definition 7.5 (Adaptable pairing schemes) A *pairing scheme* is a sequence $(x_i)_{i\in[\ell_n/2]}$, where x_i denotes the ith half-edge to be paired. We call a pairing scheme *adaptable* when the choice of x_m only depends on $(x_j, y_j)_{j=1}^{m-1}$, where y_j denotes the half-edge to which x_j is paired. We call an adaptable pairing scheme *uniform* when

$$\mathbb{P}\big(x_m \text{ is paired to } y_m \mid x_m, (x_j, y_j)_{j=1}^{m-1}\big) = \frac{1}{\ell_n - 2m + 1}, \tag{7.2.3}$$

for every $y_m \notin \{x_1, \ldots, x_m\} \cup \{y_1, \ldots, y_{m-1}\}$.

When we apply a uniform adaptable pairing scheme, the resulting graph has the same law as the configuration model:

Lemma 7.6 (Adaptable pairing schemes) *For every uniform adaptable pairing, every configuration σ in $\mathrm{Conf}_n(\boldsymbol{d}) = \{i\sigma(i) \colon i \in [\ell_n]\}$ has probability $1/(\ell_n - 1)!!$. Consequently, the resulting multigraph has the same distribution as the configuration model* $\mathrm{CM}_n(\boldsymbol{d})$.

Proof A configuration σ occurs precisely when $x_j = i$ implies that $y_j = \sigma(i)$ for every $j \in [\ell_n/2]$. Here, we choose $(x_j)_{j\in[\ell_n/2]}$ such that $(x_j)_{j\in[\ell_n/2]}$ together with $(y_j)_{j\in[\ell_n/2]}$ partition the half-edges. Thus, the configuration specifies precisely which y_j should be paired to each x_j. Let $\mathbb{P}(\sigma)$ denote the probability that our uniform adaptable pairing gives rise to the configuration σ. Then

$$\begin{aligned}\mathbb{P}(\sigma) &= \mathbb{P}(x_j \text{ is paired to } y_j \ \forall j \in [\ell_n/2]) \\ &= \prod_{m=1}^{\ell_n/2} \mathbb{P}\big(x_m \text{ is paired to } y_m \mid x_m, (x_j, y_j)_{j=1}^{m-1}\big) \\ &= \prod_{m=1}^{\ell_n/2} \frac{1}{\ell_n - 2m + 1} = \frac{1}{(\ell_n - 1)!!},\end{aligned} \tag{7.2.4}$$

as required. □

We note (see, e.g., Janson et al. (1993, Section 1)) that not all multigraphs have the same probability, i.e., not every multigraph is equally likely and the measure obtained is not the uniform measure on all multigraphs with the prescribed degree sequence. Indeed, there is a weight $1/j!$ for every edge of multiplicity j, and a factor $1/2$ for every self-loop:

Proposition 7.7 (The law of $\mathrm{CM}_n(\boldsymbol{d})$) *Let $G = (x_{ij})_{i,j\in[n]}$ be a multigraph on the vertices $[n]$ which is such that*

$$d_i = x_{ii} + \sum_{j\in[n]} x_{ij}. \tag{7.2.5}$$

Then,

$$\mathbb{P}(\mathrm{CM}_n(\boldsymbol{d}) = G) = \frac{1}{(\ell_n - 1)!!} \frac{\prod_{i\in[n]} d_i!}{\prod_{i\in[n]} 2^{x_{ii}} \prod_{1\leq i\leq j\leq n} x_{ij}!}. \tag{7.2.6}$$

Proof By Exercise 7.2, the number of configurations is equal to $(\ell_n - 1)!!$. Each configuration has the same probability, so that

$$\mathbb{P}(\mathrm{CM}_n(\boldsymbol{d}) = G) = \frac{1}{(\ell_n - 1)!!} N(G), \tag{7.2.7}$$

where $N(G)$ is the number of configurations that, after identifying the vertices, give rise to the multigraph G. We note that if we permute the half-edges incident to a vertex, then the resulting multigraph remains unchanged while the configuration is different, and there are precisely $\prod_{i\in[n]} d_i!$ ways to permute the half-edges incident to all vertices. Some of these permutations, however, give rise to the same configuration. The factor $x_{ij}!$ compensates for the multiple edges between vertices $i, j \in [n]$, and the factor $2^{x_{ii}}$ compensates for the fact that the pairing kl and lk in $\mathrm{Conf}_n(\boldsymbol{d})$ give rise to the same configuration. □

Exercises 7.3 and 7.4 investigate the distribution of the configuration model for simple choices of the degrees and $n = 2$.

The flexibility in choosing the degree sequence $\boldsymbol{d}$ gives us a similar flexibility as in choosing the vertex weights $\boldsymbol{w}$ in Chapter 6. However, in this case, the choice of the vertex degrees gives a much more direct control over the topology of the graph. For example, for $\mathrm{CM}_n(\boldsymbol{d})$, it is possible to build graphs with fixed degrees as in Figure 7.7, or for which all degrees are at least a certain value. In many applications, such flexibility is rather convenient. For example, it allows us to generate a (multi)graph with *precisely* the same degrees as a real-world network, so that we can investigate whether the real-world network is similar to it or not.

We slightly abuse notation and consistently denote our degree sequence by $\boldsymbol{d}$. In fact, since we have a degree sequence for the configuration model for each fixed n, we are actually dealing with sequences of finite sequences, and it would be more appropriate to instead write the degree sequence as $\boldsymbol{d}^{(n)} = (d_1^{(n)}, \ldots, d_n^{(n)})$, but we refrain from doing so to keep notation simple. In some cases, the sequence $\boldsymbol{d}^{(n)}$ really can be seen as the first n entries of an infinite sequence (for example in the case where the degrees are i.i.d. random variables).

As in Chapter 6, we again impose *regularity conditions* on the degree sequence $\boldsymbol{d}$. In order to state these assumptions, we introduce some notation. We denote the degree of a uniformly chosen vertex U in $[n]$ by $D_n = d_U$. The random variable D_n has distribution function F_n given by

$$F_n(x) = \frac{1}{n} \sum_{j \in [n]} \mathbb{1}_{\{d_j \leq x\}}, \tag{7.2.8}$$

which is the *empirical distribution of the degrees.* We assume that the vertex degrees satisfy the following *regularity conditions:*

Condition 7.8 (Regularity conditions for vertex degrees)
(a) Weak convergence of vertex weight. *There exists a distribution function F such that*

$$D_n \xrightarrow{d} D, \tag{7.2.9}$$

where D_n and D have distribution functions F_n and F, respectively.
Equivalently (see also Exercise 7.5 below), for any x,

$$\lim_{n\to\infty} F_n(x) = F(x). \tag{7.2.10}$$

Further, we assume that $F(0) = 0$, i.e., $\mathbb{P}(D \geq 1) = 1$.
(b) Convergence of average vertex degrees.

$$\lim_{n\to\infty} \mathbb{E}[D_n] = \mathbb{E}[D], \tag{7.2.11}$$

where D_n and D have distribution functions F_n and F from part (a), respectively.
(c) Convergence of second moment vertex degrees.

$$\lim_{n\to\infty} \mathbb{E}[D_n^2] = \mathbb{E}[D^2], \tag{7.2.12}$$

where again D_n and D have distribution functions F_n and F from part (a), respectively.

Similarly to Condition 6.4 in Chapter 6, we almost always assume that Conditions 7.8(a)–(b) hold, and only sometimes assume Condition 7.8(c). We note that, since the degrees d_i only take values in the integers, so does D_n, and therefore so must the limiting random variable D. As a result, the limiting distribution function F is constant between integers, and makes a jump $\mathbb{P}(D = x)$ at $x \in \mathbb{N}$. This implies that the distribution function F does have discontinuity points, and the weak convergence in (7.2.9) usually implies (7.2.10) only at continuity points. However, since F_n is constant in between integers, we do obtain this implication, as proved in Exercise 7.5.

Remark 7.9 (Regularity for random degrees) In the sequel, and similarly to the discussion of random weights for $\mathrm{GRG}_n(\boldsymbol{w})$ in Remark 6.5, we will often deal with cases where the degrees of the vertices are *random* themselves. For example, this arises when the degrees $\boldsymbol{d} = (d_i)_{i\in[n]}$ are realizations of i.i.d. random variables. This also arises when the degrees are those obtained from the degree sequence of the generalized random graph $\mathrm{GRG}_n(\boldsymbol{w})$. These two examples will be quite prominent in this chapter, and thus we expand on how to interpret Condition 7.8 in the case of random degrees. We note that, when the degrees are random variables themselves, also the function F_n is a random distribution function. Therefore, in Condition 7.8(a), we require random variables to converge, and there are several notions of convergence that may be used in Condition 7.8. As it turns out, the most convenient notion of convergence is convergence in probability. Thus, bearing Exercise 7.5 in mind, we replace Condition 7.8(a) by the condition that, for every $k \in \mathbb{N}$,

$$\mathbb{P}_n(D_n = k) \xrightarrow{\mathbb{P}} \mathbb{P}(D = k), \tag{7.2.13}$$

where $\mathbb{P}_n$ denote the conditional probability given the (random) degrees $(d_i)_{i\in[n]}$. Equation (7.2.13) is equivalent to the statement that, for every $k \in \mathbb{N}$ and every $\varepsilon > 0$,

$$\mathbb{P}\Big(\big|\mathbb{P}_n(D_n = k) - \mathbb{P}(D = k)\big| > \varepsilon\Big) \to 0. \tag{7.2.14}$$

Similarly, Condition 7.8(b) and (c) are replaced by

$$\mathbb{E}_n[D_n] \xrightarrow{\mathbb{P}} \mathbb{E}[D], \qquad \mathbb{E}_n[D_n^2] \xrightarrow{\mathbb{P}} \mathbb{E}[D^2], \tag{7.2.15}$$

where $\mathbb{E}_n$ denotes expectation with respect to $\mathbb{P}_n$. Equation (7.2.15) is equivalent to the statement that, for every $\varepsilon > 0$,

$$\mathbb{P}\big(\big|\mathbb{E}_n[D_n] - \mathbb{E}[D]\big| > \varepsilon\big) \to 0, \qquad \mathbb{P}\big(\big|\mathbb{E}_n[D_n^2] - \mathbb{E}[D^2]\big| > \varepsilon\big) \to 0. \tag{7.2.16}$$

Instead of defining $\mathrm{CM}_n(\boldsymbol{d})$ in terms of the degrees, we could have defined it in terms of the number of vertices with fixed degrees. Indeed, let

$$n_k = \sum_{i\in[n]} \mathbb{1}_{\{d_i=k\}} \tag{7.2.17}$$

denote the number of vertices with degree k. Then, clearly, apart from the vertex labels, the degree sequence $\boldsymbol{d}$ is uniquely determined by the sequence $(n_k)_{k\geq 0}$. Condition 7.8(a) is equivalent to the statement that $\lim_{n\to\infty} n_k/n = \mathbb{P}(D = k)$, while Condition 7.8(b) is equivalent to the statement that $\lim_{n\to\infty} \sum_{k\geq 0} kn_k/n = \mathbb{E}[D]$.

We next describe two canonical ways of obtaining a degree sequence $\boldsymbol{d}$ such that Condition 7.8 holds:

The Configuration Model with Fixed Degrees Moderated by F

Fix a distribution function F of an integer-valued random variable D. We could take $d_i = [1 - F]^{-1}(i/n)$, as we did in for the weights in $\mathrm{GRG}_n(\boldsymbol{w})$ in (6.2.14). In this case, indeed, the degrees are integer due to the fact that F is a distribution function of an integer-valued random variable. Many of the arguments in Chapter 6 can be repeated for this setting.

We now focus on a slightly different construction. We take the number of vertices with degree k to be equal to

$$n_k = \lceil nF(k)\rceil - \lceil nF(k-1)\rceil, \tag{7.2.18}$$

and take the corresponding degree sequence $\boldsymbol{d} = (d_i)_{i\in[n]}$ the unique ordered degree sequence compatible with $(n_k)_{k\geq 0}$. For this sequence, Conditions 7.8(a)–(b) are satisfied (see Exercises 7.6 and 7.7).

The nice thing about our example is that

$$F_n(k) = \frac{1}{n}\lceil nF(k)\rceil, \tag{7.2.19}$$

which is slightly nicer than for the setting $d_i = [1 - F]^{-1}(i/n)$ as described in (6.2.17). In particular, $D_n \preceq D$, since $F_n(x) \geq F(x)$ for every x. As a result, Condition 7.8(b) holds whenever $\mathbb{E}[D] < \infty$, and Condition 7.8(c) whenever $\mathbb{E}[D^2] < \infty$; see Exercise 7.7.

The Configuration Model with I.I.D. Degrees

The next canonical example arises by assuming that the degrees $\boldsymbol{d} = (d_i)_{i\in[n]}$ are an i.i.d. sequence of random variables. When we extend the construction of the configuration model to i.i.d. degrees $\boldsymbol{d}$, we should bear in mind that the total degree

$$\ell_n = \sum_{i\in[n]} d_i \tag{7.2.20}$$

is in many cases odd with probability *exponentially* close to $1/2$; see Exercise 7.8.

There are different possible solutions to overcome the problem of an odd total degree ℓ_n, each producing a graph with similar characteristics. We make use of the following solution: If ℓ_n is odd, then we add a half-edge to the nth vertex, so that d_n is increased by 1, i.e., $d_n = d_n' + \mathbb{1}_{\{\ell_{n-1}+d_n' \text{ odd}, i=n\}}$, where d_n' is an independent copy of d_1 and independent of the i.i.d. sequence $(d_i)_{i\in[n-1]}$. This single half-edge hardly makes any difference in what follows, and we will ignore this effect. Also, we warn the reader that now d_i is *random*, while the use of the small letter suggests that d_i is deterministic. This notation is to avoid confusion between D_n, which is the degree of a random vertex, and d_n, which is the degree of vertex n.

It is not hard to see that Condition 7.8 follows from the Law of Large Numbers (see Exercise 7.9).

In Figure 7.9, we plot the degree-sequences of $\mathrm{CM}_n(\boldsymbol{d})$ for $n = 1{,}000{,}000$ and $\tau = 2.5$ and $\tau = 3.5$. These plots indeed look very like the log-log plots of degree sequences in real-world networks shown in Chapter 1. In Figures 7.10–7.11 for realizations of $\mathrm{CM}_n(\boldsymbol{d})$ for $(d_i)_{i\in[n]}$ such that $d_i = [1 - F]^{-1}(i/n)$ and F satisfying (6.2.20) with $n = 100$ and $\tau = 3.5$ and $\tau = 2.5$, respectively.

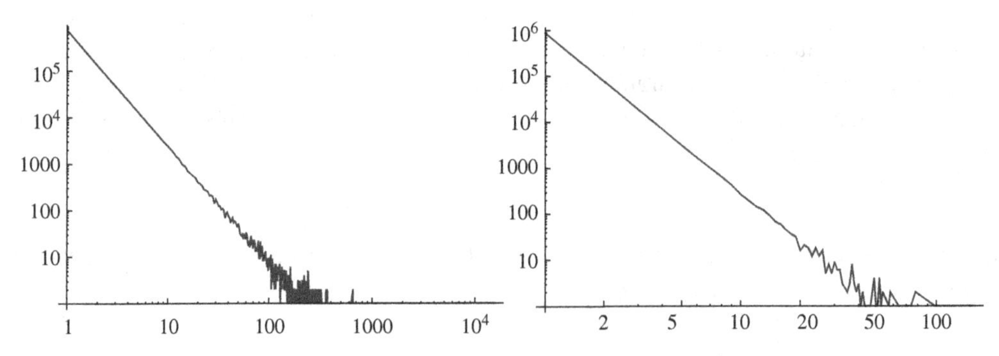

Figure 7.9 The degree sequences of a configuration model with a power-law degree distribution with i.i.d. degrees, $n = 1{,}000{,}000$ and $\tau = 2.5$ and $\tau = 3.5$, respectively, in log-log scale.

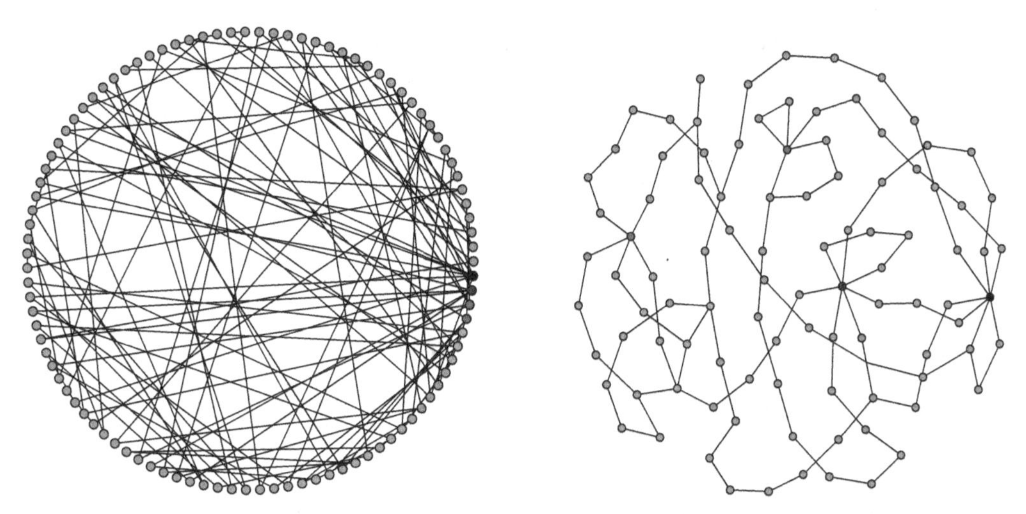

Figure 7.10 A realization of $\mathrm{CM}_n(\boldsymbol{d})$ for $d_i = [1 - F]^{-1}(i/n)$ and F satisfying (6.2.20) with $n = 100$ and $\tau = 3.5$ on a circle (left), and in a way that minimizes a force field (right). Darker vertices have a higher degree than lighter vertices.

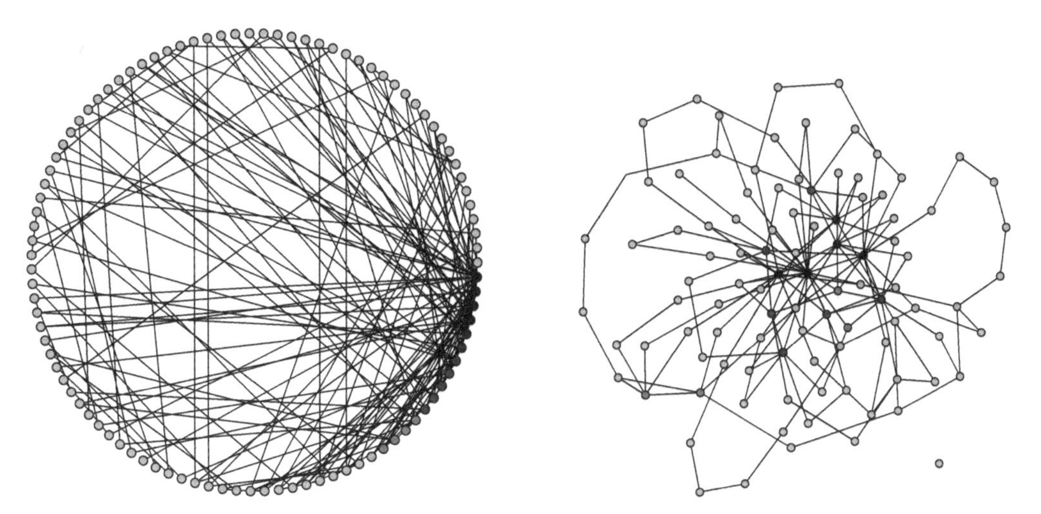

Figure 7.11 A realization of $\mathrm{CM}_n(\boldsymbol{d})$ for $d_i = [1 - F]^{-1}(i/n)$ and F satisfying (6.2.20) with $n = 100$ and $\tau = 2.5$ on a circle (left), and in a way that minimizes a force field (right). Darker vertices have a higher degree than lighter vertices.

7.3 Erased Configuration Model

We start by defining the erased configuration model. We fix the degrees $\boldsymbol{d}$. We start with the multigraph $\mathrm{CM}_n(\boldsymbol{d})$ and erase all self-loops, if any exist. After this, we merge all multiple edges into single edges. Thus, the erased configuration model is a simple random graph, where two vertices are connected by an edge if and only if there is (at least) one edge connecting them in the original multigraph $\mathrm{CM}_n(\boldsymbol{d})$ definition of the configuration model.

We denote the degrees in the erased configuration model by $\boldsymbol{D}^{(\mathrm{er})} = (D_i^{(\mathrm{er})})_{i\in[n]}$, so that

$$D_i^{(\mathrm{er})} = d_i - 2s_i - m_i, \tag{7.3.1}$$

where $(d_i)_{i\in[n]}$ are the degrees in $\mathrm{CM}_n(\boldsymbol{d})$, $s_i = x_{ii}$ is the number of self-loops of vertex i in $\mathrm{CM}_n(\boldsymbol{d})$ and

$$m_i = \sum_{j\neq i}(x_{ij} - 1)\mathbb{1}_{\{x_{ij}\geq 2\}} \tag{7.3.2}$$

is the number of multiple edges removed from i.

Denote the empirical degree sequence $(p_k^{(n)})_{k\geq 1}$ in $\mathrm{CM}_n(\boldsymbol{d})$ by

$$p_k^{(n)} = \mathbb{P}(D_n = k) = \frac{1}{n}\sum_{i\in[n]}\mathbb{1}_{\{d_i=k\}}, \tag{7.3.3}$$

and denote the related degree sequence in the erased configuration model $(P_k^{(\mathrm{er})})_{k\geq 1}$ by

$$P_k^{(\mathrm{er})} = \frac{1}{n}\sum_{i\in[n]}\mathbb{1}_{\{D_i^{(\mathrm{er})}=k\}}. \tag{7.3.4}$$

From the notation it is clear that $(p_k^{(n)})_{k\geq 1}$ is a *deterministic* sequence when $\boldsymbol{d} = (d_i)_{i\in[n]}$ is deterministic, while $(P_k^{(\mathrm{er})})_{k\geq 1}$ is a *random* sequence, since the erased degrees $(D_i^{(\mathrm{er})})_{i\in[n]}$ is a random vector even when $\boldsymbol{d} = (d_i)_{i\in[n]}$ is deterministic.

Now we are ready to state the main result concerning the degree sequence of the erased configuration model:

Theorem 7.10 (Degree sequence of erased configuration model with fixed degrees) *For fixed degrees* $\boldsymbol{d}$ *satisfying Conditions 7.8(a)–(b), the degree sequence of the erased configuration model* $(P_k^{(\mathrm{er})})_{k\geq 1}$ *converges in probability to* $(p_k)_{k\geq 1}$*. More precisely, for every* $\varepsilon > 0$,

$$\mathbb{P}\Big(\sum_{k=1}^{\infty}|P_k^{(\mathrm{er})} - p_k| \geq \varepsilon\Big) \to 0, \tag{7.3.5}$$

where $p_k = \mathbb{P}(D = k)$ *as in Condition 7.8(a).*

Proof By Condition 7.8(a) and the fact that pointwise convergence of a probability mass function is equivalent to convergence in total variation distance (recall Exercise 2.16),

$$\lim_{n\to\infty}\sum_{k=1}^{\infty}|p_k^{(n)} - p_k| = 0. \tag{7.3.6}$$

Therefore, we can take n so large that

$$\sum_{k=1}^{\infty}|p_k^{(n)} - p_k| \leq \varepsilon/2. \tag{7.3.7}$$

We now aim to show that, whp, $\sum_{k=1}^{\infty}|P_k^{(\mathrm{er})} - p_k^{(n)}| < \varepsilon/2$.

We start by proving the result under the extra assumption that

$$\max_{i\in[n]} d_i = o(\sqrt{n}). \tag{7.3.8}$$

For this, we bound $\mathbb{P}(\sum_{k=1}^{\infty} |P_k^{(\mathrm{er})} - p_k^{(n)}| \geq \varepsilon/2)$ by using (7.3.1), which implies that $D_i^{(\mathrm{er})} \neq d_i$ if and only if $2s_i + m_i \geq 1$. Thus,

$$\sum_{k=1}^{\infty} |P_k^{(\mathrm{er})} - p_k^{(n)}| \leq \frac{1}{n}\sum_{k=1}^{\infty}\sum_i |\mathbb{1}_{\{D_i^{(\mathrm{er})}=k\}} - \mathbb{1}_{\{d_i=k\}}|, \tag{7.3.9}$$

and write out that

$$\begin{aligned}\mathbb{1}_{\{D_i^{(\mathrm{er})}=k\}} - \mathbb{1}_{\{d_i=k\}} &= \mathbb{1}_{\{D_i^{(\mathrm{er})}=k, d_i>k\}} - \mathbb{1}_{\{D_i^{(\mathrm{er})}<k, d_i=k\}} \\ &= \mathbb{1}_{\{s_i+m_i>0\}}\big(\mathbb{1}_{\{D_i^{(\mathrm{er})}=k\}} - \mathbb{1}_{\{d_i=k\}}\big).\end{aligned} \tag{7.3.10}$$

Therefore,

$$|\mathbb{1}_{\{D_i^{(\mathrm{er})}=k\}} - \mathbb{1}_{\{d_i=k\}}| \leq \mathbb{1}_{\{s_i+m_i>0\}}\big(\mathbb{1}_{\{D_i^{(\mathrm{er})}=k\}} + \mathbb{1}_{\{d_i=k\}}\big), \tag{7.3.11}$$

so that

$$\begin{aligned}\sum_{k=1}^{\infty} |P_k^{(\mathrm{er})} - p_k^{(n)}| &\leq \frac{1}{n}\sum_{k=1}^{\infty}\sum_{i\in[n]} |\mathbb{1}_{\{D_i^{(\mathrm{er})}=k\}} - \mathbb{1}_{\{d_i=k\}}| \\ &\leq \frac{1}{n}\sum_{i\in[n]} \mathbb{1}_{\{s_i+m_i>0\}}\sum_{k=1}^{\infty}\big(\mathbb{1}_{\{D_i^{(\mathrm{er})}=k\}} + \mathbb{1}_{\{d_i=k\}}\big) \\ &= \frac{2}{n}\sum_{i\in[n]} \mathbb{1}_{\{s_i+m_i>0\}} \leq \frac{2}{n}\sum_{i\in[n]}(s_i+m_i).\end{aligned} \tag{7.3.12}$$

We denote the number of self-loops by S_n and the number of multiple edges by M_n, i.e.,

$$S_n = \sum_{i\in[n]} s_i, \qquad M_n = \frac{1}{2}\sum_{i\in[n]} m_i. \tag{7.3.13}$$

Then, by (7.3.12),

$$\mathbb{P}\Big(\sum_{k=1}^{\infty} |P_k^{(\mathrm{er})} - p_k^{(n)}| \geq \varepsilon/2\Big) \leq \mathbb{P}\big(2S_n + 4M_n \geq \varepsilon n/2\big), \tag{7.3.14}$$

so that Theorem 7.10 follows if

$$\mathbb{P}(2S_n + 4M_n \geq \varepsilon n/2) \to 0. \tag{7.3.15}$$

By the Markov inequality (Theorem 2.17), we obtain

$$\mathbb{P}(2S_n + 4M_n \geq \varepsilon n/2) \leq \frac{4}{\varepsilon n}\big(\mathbb{E}[S_n] + 2\mathbb{E}[M_n]\big). \tag{7.3.16}$$

Bounds on $\mathbb{E}[S_n]$ and $\mathbb{E}[M_n]$ are provided in the following proposition:

Proposition 7.11 (Bounds on the expected number of self-loops and multiple edges) *The expected number of self-loops S_n in the configuration model $\mathrm{CM}_n(\boldsymbol{d})$ satisfies*

$$\mathbb{E}[S_n] = \frac{1}{2}\sum_{i\in[n]} \frac{d_i(d_i-1)}{\ell_n-1} \leq \frac{1}{2}\sum_{i\in[n]} \frac{d_i^2}{\ell_n}, \tag{7.3.17}$$

while the expected number of multiple edges M_n satisfies

$$\mathbb{E}[M_n] \leq \frac{\ell_n^2}{(\ell_n-1)(\ell_n-3)}\Big(\sum_{i\in[n]} \frac{d_i(d_i-1)}{2\ell_n}\Big)^2 \leq 2\Big(\sum_{i\in[n]} \frac{d_i(d_i-1)}{\ell_n}\Big)^2. \tag{7.3.18}$$

Proof For a vertex i, and for $1 \leq s < t \leq d_i$, we define $I_{st,i}$ to be the indicator of the event that half-edge s is paired to half-edge t, where s and t are half-edges incident to vertex i. Here we label the half-edges as (i, s), where $i \in [n]$ is the vertex to which half-edge s is incident and $s \in [d_i]$, and we number the half-edges incident to a vertex in an arbitrary way. Then

$$S_n = \sum_{i\in[n]} \sum_{1\leq s<t\leq d_i} I_{st,i}. \tag{7.3.19}$$

Therefore,

$$\mathbb{E}[S_n] = \sum_{i\in[n]} \sum_{1\leq s<t\leq d_i} \mathbb{E}[I_{st,i}] = \sum_{i\in[n]} \frac{1}{2} d_i(d_i-1)\mathbb{E}[I_{12,i}], \tag{7.3.20}$$

since the probability of producing a self-loop by pairing the half-edges (i, s) and (i, t) does not depend on s and t. Now, $\mathbb{E}[I_{12,i}]$ is equal to the probability that half-edges $(i, 1)$ and $(i, 2)$ are paired to each other, which is equal to $1/(\ell_n-1)$. Therefore, using that $(d_i-1)/(\ell_n-1) \leq d_i/\ell_n$ since $d_i \leq \ell_n$,

$$\mathbb{E}[S_n] = \frac{1}{2}\sum_{i\in[n]} \frac{d_i(d_i-1)}{\ell_n-1} \leq \frac{1}{2}\sum_{i\in[n]} \frac{d_i^2}{\ell_n}. \tag{7.3.21}$$

This completes the proof of (7.3.17).

We continue by proving (7.3.18). Note that when $\ell_n = 3$, there cannot be any multiple edges, so from now on, we assume that $\ell_n \geq 4$. Then, for vertices i and j, and for $1 \leq s_1 < s_2 \leq d_i$ and $1 \leq t_1 \neq t_2 \leq d_j$, we define $I_{s_1t_1,s_2t_2,ij}$ to be the indicator of the event that half-edge (i, s_1) is paired to half-edge (j, t_1) and half-edge (i, s_2) is paired to half-edge (j, t_2). If there are multiple edges between vertices i and j, then $I_{s_1t_1,s_2t_2,ij} = 1$ for some s_1, t_1 and s_2, t_2. We bound

$$\begin{aligned} m_i &= \sum_{j\in[n]:\, j\neq i} (x_{ij}-1)\mathbb{1}_{\{x_{ij}\geq 2\}} \leq \sum_{j\in[n]:\, j\neq i} \binom{x_{ij}}{2} \\ &= \sum_{j\in[n]:\, j\neq i} \sum_{1\leq s_1<s_2\leq d_i} \sum_{1\leq t_1\neq t_2\leq d_j} I_{s_1t_1,s_2t_2,ij}. \end{aligned} \tag{7.3.22}$$

It follows that

$$M_n \leq \frac{1}{2} \sum_{1\leq i\neq j\leq n} \sum_{1\leq s_1<s_2\leq d_i} \sum_{1\leq t_1\neq t_2\leq d_j} I_{s_1t_1,s_2t_2,ij}, \tag{7.3.23}$$

so that

$$\mathbb{E}[M_n] \leq \frac{1}{2} \sum_{1\leq i\neq j\leq n} \sum_{1\leq s_1<s_2\leq d_i} \sum_{1\leq t_1\neq t_2\leq d_j} \mathbb{E}[I_{s_1t_1,s_2t_2,ij}]$$

$$= \frac{1}{4} \sum_{1\leq i\neq j\leq n} d_i(d_i-1)d_j(d_j-1)\mathbb{E}[I_{11,22,ij}]. \tag{7.3.24}$$

Now, since $I_{11,22,ij}$ is an indicator, $\mathbb{E}[I_{11,22,ij}]$ is the probability that $I_{11,22,ij} = 1$, which is equal to the probability that half-edge 1 of vertex i is paired to half-edge 1 of vertex j, and half-edge 2 of vertex i is paired to half-edge 2 of vertex j, which is equal to

$$\mathbb{E}[I_{11,22,ij}] = \frac{1}{(\ell_n-1)(\ell_n-3)}. \tag{7.3.25}$$

Therefore,

$$\mathbb{E}[M_n] \leq \sum_{i,j\in[n]} \frac{d_i(d_i-1)d_j(d_j-1)}{4(\ell_n-1)(\ell_n-3)} \leq \frac{2\Big(\sum_{i\in[n]} d_i(d_i-1)\Big)^2}{\ell_n^2}, \tag{7.3.26}$$

where we use that $8(\ell_n-1)(\ell_n-3) \geq \ell_n^2$ since $\ell_n \geq 4$. This completes the proof of (7.3.18). □

To complete the proof of Theorem 7.10 in the case that $\max_{i\in[n]} d_i = o(\sqrt{n})$ (recall (7.3.8)), we apply Proposition 7.11 to obtain

$$\mathbb{E}[S_n] \leq \sum_{i\in[n]} \frac{d_i^2}{\ell_n} \leq \max_{i\in[n]} d_i = o(\sqrt{n}). \tag{7.3.27}$$

The bound on $\mathbb{E}[M_n]$ is similar. By (7.3.16), this proves the claim.

To prove the result assuming only Conditions 7.8(a)–(b), we start by noting that Conditions 7.8(a)–(b) imply that $\max_{i\in[n]} d_i = o(n)$ (recall Exercise 6.3). We further note that $\sum_{k=1}^{\infty} |P_k^{(\mathrm{er})} - p_k^{(n)}| \geq \varepsilon/2$ implies that the degrees of at least $\varepsilon n/2$ vertices are changed by the erasure procedure. Take $a_n \to \infty$ to be determined later on, and take n large enough that there are at most $\varepsilon n/4$ vertices $i \in [n]$ of degree $d_i \geq a_n$, which is possible by Condition 7.8(a). Then, $\sum_{k=1}^{\infty} |P_k^{(\mathrm{er})} - p_k^{(n)}| \geq \varepsilon$ implies that there are at least $\varepsilon n/4$ vertices of degree at most a_n whose degrees are changed by the erasure procedure. Let

$$S_n(a_n) = \sum_{i\in[n]} s_i \mathbb{1}_{\{d_i\leq a_n\}}, \qquad M_n(a_n) = \frac{1}{2}\sum_{i\in[n]} m_i \mathbb{1}_{\{d_i\leq a_n\}} \tag{7.3.28}$$

denote the number of self-loops and multiple edges incident to vertices of degree at most a_n. Then, it is straightforward to adapt Proposition 7.11 to show that

$$\mathbb{E}[S_n(a_n)] \leq \sum_{i\in[n]} \frac{d_i^2 \mathbb{1}_{\{d_i\leq a_n\}}}{\ell_n}, \qquad \mathbb{E}[M_n(a_n)] \leq 2\sum_{i\in[n]} \frac{d_i^2 \mathbb{1}_{\{d_i\leq a_n\}}}{\ell_n} \sum_{j\in[n]} \frac{d_j^2}{\ell_n}. \tag{7.3.29}$$

Therefore, $\mathbb{E}[S_n(a_n)] \leq a_n, \mathbb{E}[M_n(a_n)] \leq a_n \max_{j\in[n]} d_j$. Take $a_n \to \infty$ so slowly that $a_n \max_{j\in[n]} d_j = o(n)$ (which is possible since $\max_{j\in[n]} d_j = o(n)$), then

$$\mathbb{P}(2S_n(a_n) + 4M_n(a_n) \geq \varepsilon n/4) \leq \frac{8}{\varepsilon n}\big(\mathbb{E}[S_n(a_n)] + 2\mathbb{E}[M_n(a_n)]\big) = o(1), \tag{7.3.30}$$

as required. □

7.4 Repeated Configuration Model: Simplicity Probability

In this section, we investigate the probability that the configuration model yields a simple graph, i.e., the probability that the graph produced in the configuration model has no self-loops nor multiple edges. The asymptotics of this probability is derived in the following theorem:

Theorem 7.12 (Probability of simplicity of $\mathrm{CM}_n(\boldsymbol{d})$) *Assume that $\boldsymbol{d} = (d_i)_{i\in[n]}$ satisfies Conditions 7.8(a)–(c). Then, the probability that* $\mathrm{CM}_n(\boldsymbol{d})$ *is a simple graph is asymptotically equal to* $\mathrm{e}^{-\nu/2-\nu^2/4}$, *where*

$$\nu = \mathbb{E}[D(D-1)]/\mathbb{E}[D]. \tag{7.4.1}$$

Throughout the proof, it turns out to be simpler to work with

$$\widetilde{M}_n = \sum_{1\leq i<j\leq n} \sum_{1\leq s_1<s_2\leq d_i} \sum_{1\leq t_1\neq t_2\leq d_j} I_{s_1t_1,s_2t_2,ij}, \tag{7.4.2}$$

so that $M_n \leq \widetilde{M}_n$ by (7.3.22)–(7.3.23). As a result, $M_n = 0$ when $\widetilde{M}_n = 0$. In Exercise 7.11, you are asked to show that actually $M_n = \widetilde{M}_n$ occurs whp. The distinction between M_n and $\widetilde{M}_n$ is much more prominent in the case where $\mathbb{E}[D_n^2] \to \infty$; see also Section 7.7.

Theorem 7.12 is a consequence of the following result:

Proposition 7.13 (Poisson limit of self-loops and multiple edges) *Assume that $\boldsymbol{d} = (d_i)_{i\in[n]}$ satisfies Conditions 7.8(a)–(c). Then (S_n, M_n) converges in distribution to (S, M), where S and M are two* independent *Poisson random variables with means $\nu/2$ and $\nu^2/4$.*

Indeed, Theorem 7.12 is a simple consequence of Proposition 7.13, since $\mathrm{CM}_n(\boldsymbol{d})$ is simple precisely when $S_n = M_n = 0$. By the weak convergence result stated in Proposition 7.13 and the independence of S and M, the probability that $S_n = M_n = 0$ converges to $\mathrm{e}^{-\mu_S-\mu_M}$, where μ_S and μ_M are the means of the limiting Poisson random variables S and M. Using the identification of the means of S and M in Proposition 7.13, this completes the proof of Theorem 7.12. We are left to prove Proposition 7.13.

Proof of Proposition 7.13 Throughout the proof, we assume that S and M are two independent Poisson random variables with means $\nu/2$ and $\nu^2/4$. We make use of Theorem 2.6 which implies that it suffices to prove that the (joint) factorial moments of $(S_n, \widetilde{M}_n)$ converge to those of (S, M). Also, S_n and $\widetilde{M}_n$ are sums of indicators, so that we can use Theorem 2.7 to identify their joint factorial moments.

To prove that $(S_n, \widetilde{M}_n)$ converges in distribution to (S, M), we use Theorem 2.6 to see that we are left to prove that, for every $s, r \geq 0$,

$$\lim_{n\to\infty} \mathbb{E}[(S_n)_s(\widetilde{M}_n)_r] = \Big(\frac{\nu}{2}\Big)^s\Big(\frac{\nu^2}{4}\Big)^r. \tag{7.4.3}$$

We define the sets of indices

$$\mathcal{I}_1 = \{(st, i)\colon i \in [n], 1 \leq s < t \leq d_i\}, \tag{7.4.4}$$

and

$$\mathcal{I}_2 = \{(s_1t_1, s_2t_2, i, j)\colon 1 \leq i < j \leq n, 1 \leq s_1 < s_2 \leq d_i, 1 \leq t_1 \neq t_2 \leq d_j\}, \tag{7.4.5}$$

for the potential self-loops and pairs of multiple edges, respectively. Then

$$S_n = \sum_{(st,i)\in\mathcal{I}_1} I_{st,i}, \qquad \widetilde{M}_n = \sum_{(s_1t_1,s_2t_2,i,j)\in\mathcal{I}_2} I_{s_1t_1,s_2t_2,ij}. \tag{7.4.6}$$

Without loss of generality, we will assume that the sizes $|\mathcal{I}_1|$ and $|\mathcal{I}_2|$ of the index sets $\mathcal{I}_1$ and $\mathcal{I}_2$ grow to infinity with n, since $n_1 = n(1 - o(1))$ when this is not the case, and in this case $(S_n, \widetilde{M}_n) = 0$ whp and also $\nu = 0$. Indeed, note that only vertices with degree at least 2 contribute to $\mathcal{I}_1$ and $\mathcal{I}_2$ in (7.4.4)–(7.4.5).

Further, for $m^{(1)} = (st, i) \in \mathcal{I}_1$ and $m^{(2)} = (s_1t_1, s_2t_2, i, j) \in \mathcal{I}_2$, we write

$$I^{(1)}_{m^{(1)}} = I_{st,i}, \qquad I^{(2)}_{m^{(2)}} = I_{s_1t_1,s_2t_2,ij}. \tag{7.4.7}$$

Recall the notation $\sum^*$ for the sum over distinct indices introduced in Theorem 2.5. By Theorem 2.7,

$$\mathbb{E}[(S_n)_s(\widetilde{M}_n)_r] = \sum^*_{\substack{m_1^{(1)},\ldots,m_s^{(1)}\in\mathcal{I}_1\\ m_1^{(2)},\ldots,m_r^{(2)}\in\mathcal{I}_2}} \mathbb{P}\big(I^{(1)}_{m_1^{(1)}} = \cdots = I^{(1)}_{m_s^{(1)}} = I^{(2)}_{m_1^{(2)}} = \cdots = I^{(2)}_{m_r^{(2)}} = 1\big). \tag{7.4.8}$$

Since all half-edges are uniformly paired,

$$\mathbb{P}\big(I^{(1)}_{m_1^{(1)}} = \cdots = I^{(1)}_{m_s^{(1)}} = I^{(2)}_{m_1^{(2)}} = \cdots = I^{(2)}_{m_r^{(2)}} = 1\big) = \frac{1}{\prod_{i=0}^{s+2r-1}(\ell_n - 1 - 2i)}, \tag{7.4.9}$$

unless there is a conflict in the required pairings, in which case

$$\mathbb{P}\big(I^{(1)}_{m_1^{(1)}} = \cdots = I^{(1)}_{m_s^{(1)}} = I^{(2)}_{m_1^{(2)}} = \cdots = I^{(2)}_{m_r^{(2)}} = 1\big) = 0. \tag{7.4.10}$$

Such a conflict arises precisely when a half-edge is required to be paired to *two different* other half-edges. Since the upper bound in (7.4.9) always holds, we arrive at

$$\begin{aligned}
&\mathbb{E}[(S_n)_s(\widetilde{M}_n)_r] \qquad (7.4.11)\\
&\leq \sum^*_{m_1^{(1)},\ldots,m_s^{(1)}\in\mathcal{I}_1} \sum^*_{m_1^{(2)},\ldots,m_r^{(2)}\in\mathcal{I}_2} \frac{1}{(\ell_n-1)(\ell_n-3)\cdots(\ell_n-2s-4r+1)}\\
&= \frac{|\mathcal{I}_1|(|\mathcal{I}_1|-1)\cdots(|\mathcal{I}_1|-s+1)\cdot|\mathcal{I}_2|(|\mathcal{I}_2|-1)\cdots(|\mathcal{I}_2|-r+1)}{(\ell_n-1)(\ell_n-3)\cdots(\ell_n-2s-4r+1)}.
\end{aligned}$$

Since $|\mathcal{I}_1|, |\mathcal{I}_2|, \ell_n$ all tend to infinity, and s, r remain fixed,

$$\limsup_{n\to\infty} \mathbb{E}[(S_n)_s(\widetilde{M}_n)_r] \leq \Big(\lim_{n\to\infty} \frac{|\mathcal{I}_1|}{\ell_n}\Big)^s \Big(\lim_{n\to\infty} \frac{|\mathcal{I}_2|}{\ell_n^2}\Big)^r. \tag{7.4.12}$$

By Conditions 7.8(b)–(c),

$$\lim_{n\to\infty} \frac{|\mathcal{I}_1|}{\ell_n} = \lim_{n\to\infty} \frac{1}{\ell_n} \sum_{i\in[n]} \frac{d_i(d_i-1)}{2} = \nu/2. \tag{7.4.13}$$

Further, again by Conditions 7.8(b)–(c) and also using that $d_i = o(\sqrt{n})$ by Exercise 6.3 applied to $\boldsymbol{w} = \boldsymbol{d}$, as well as $\ell_n \geq n$,

$$\begin{aligned}\lim_{n\to\infty} \frac{|\mathcal{I}_2|}{\ell_n^2} &= \lim_{n\to\infty} \frac{1}{\ell_n^2} \sum_{1\leq i<j\leq n} \frac{d_i(d_i-1)}{2} d_j(d_j-1) \\ &= \Big(\lim_{n\to\infty} \frac{1}{\ell_n} \sum_{i\in[n]} \frac{d_i(d_i-1)}{2}\Big)^2 - \lim_{n\to\infty} \sum_{i\in[n]} \frac{d_i^2(d_i-1)^2}{2\ell_n^2} = (\nu/2)^2. \end{aligned} \tag{7.4.14}$$

This provides the required upper bound, and completes the proof when $\nu = 0$.

To prove the matching lower bound for $\nu > 0$, we note that, by (7.4.10),

$$\begin{aligned}&\sum^*_{m_1^{(1)},\ldots,m_s^{(1)}\in\mathcal{I}_1} \sum^*_{m_1^{(2)},\ldots,m_r^{(2)}\in\mathcal{I}_2} \frac{1}{\prod_{i=0}^{s+2r-1}(\ell_n - 1 - 2i)} - \mathbb{E}[(S_n)_s(\widetilde{M}_n)_r] \\ &= \sum^*_{m_1^{(1)},\ldots,m_s^{(1)}\in\mathcal{I}_1} \sum^*_{m_1^{(2)},\ldots,m_r^{(2)}\in\mathcal{I}_2} \frac{J_{m_1^{(1)},\ldots,m_s^{(1)},m_1^{(2)},\ldots,m_r^{(2)}}}{(\ell_n-1)(\ell_n-3)\cdots(\ell_n-2s-4r+1)}, \end{aligned} \tag{7.4.15}$$

where $J_{m_1^{(1)},\ldots,m_s^{(1)},m_1^{(2)},\ldots,m_r^{(2)}} = 1$ precisely when there is a conflict in the indices $m_1^{(1)},\ldots,m_s^{(1)},m_1^{(2)},\ldots,m_r^{(2)}$ and $J_{m_1^{(1)},\ldots,m_s^{(1)},m_1^{(2)},\ldots,m_r^{(2)}} = 0$ otherwise. Our aim is to show that the right-hand side of (7.4.15) vanishes.

There is a conflict precisely when there exists a half-edge that must be paired to two *different* half-edges. For this, there has to be a pair of indices in $m_1^{(1)},\ldots,m_s^{(1)},m_1^{(2)},\ldots,m_r^{(2)}$ that create the conflict. There are three such possibilities: (a) the conflict is created by $m_{i_a}^{(1)}, m_{i_b}^{(1)}$ for some a, b; (b) the conflict is created by $m_{i_a}^{(1)}, m_{i_b}^{(2)}$ for some a, b; and (c) the conflict is created by $m_{i_a}^{(2)}, m_{i_b}^{(2)}$ for some a, b. We bound each of these possibilities separately.

In case (a), the number of $m_{i_l}^{(1)}$, for $l \in [s] \setminus \{a, b\}$ and $m_{i_k}^{(2)}$, for $k \in [r]$, is bounded by $|\mathcal{I}_1|^{s-2}|\mathcal{I}_2|^r$. Thus, to show that this contribution to the right-hand side of (7.4.15) vanishes and comparing (7.4.15) with (7.4.11), we see that it suffices to prove that the number of conflicting $m_{i_a}^{(1)}, m_{i_b}^{(1)}$ is $o(|\mathcal{I}_1|^2)$. We note that $m_{i_a}^{(1)}$ creates a conflict with $m_{i_b}^{(1)}$ when $i_a = (st, i)$ and $i_b = (s't', i)$ where at least two indices in s, t, s', t' agree. The number of conflicting pairs of the form $m_{i_a}^{(1)}, m_{i_b}^{(1)}$ is thus bounded by

$$6 \sum_{i\in[n]} d_i^3 = o\Big(\sum_{i\in[n]} d_i(d_i-1)\Big)^2 = o(|\mathcal{I}_1|^2), \tag{7.4.16}$$

as required, where we use that $\max_{i\in[n]} d_i = o(n)$ and $\nu/2 = \lim_{n\to\infty} |\mathcal{I}_1|/\ell_n > 0$, so that $|\mathcal{I}_1| \geq \varepsilon n$ for some $\varepsilon > 0$.

In case (b), the number of $m^{(1)}_{i_l}$, for $l \in [s] \setminus \{a\}$ and $m^{(2)}_{i_k}$, for $k \in [r] \setminus \{b\}$, is bounded by $|\mathcal{I}_1|^{s-1}|\mathcal{I}_2|^{r-1}$. Thus, to show that this contribution to the right-hand side of (7.4.15) vanishes and comparing (7.4.15) with (7.4.11), we now see that it suffices to prove that the number of conflicting $m^{(1)}_{i_a}, m^{(2)}_{i_b}$ is $o(|\mathcal{I}_1||\mathcal{I}_2|)$. For a conflict between $m^{(1)}_{i_a}$ and $m^{(2)}_{i_b}$ to occur, we must have that $i_a = (st, i)$ and $i_b = (s_1t_1, s_2t_2, ij)$ where at least two indices in s, t, s_1, s_2 agree. The number of conflicting pairs of the form $m^{(1)}_{i_a}, m^{(2)}_{i_b}$ is thus bounded by

$$6 \sum_{i\in[n]} d_i^3 \sum_{j\in[n]} d_j^2 = o\Big(\sum_{i\in[n]} d_i(d_i-1)\Big)^3 = o(|\mathcal{I}_1||\mathcal{I}_2|), \tag{7.4.17}$$

as required, where we again use that $\max_{i\in[n]} d_i = o(n)$ and

$$\nu^2/4 = \lim_{n\to\infty} |\mathcal{I}_2|/\ell_n^2 > 0, \tag{7.4.18}$$

so that $|\mathcal{I}_2| \geq \varepsilon n^2$ for some $\varepsilon > 0$.

In case (c), the number of $m^{(2)}_{i_l}$, for $l \in [s]$ and $m^{(2)}_{i_k}$, for $k \in [r] \setminus \{a, b\}$, is bounded by $|\mathcal{I}_1|^s|\mathcal{I}_2|^{r-2}$. Thus, to show that this contribution to the right-hand side of (7.4.15) vanishes and comparing (7.4.15) with (7.4.11), we now see that it suffices to prove that the number of conflicting $m^{(2)}_{i_a}, m^{(2)}_{i_b}$ is $o(|\mathcal{I}_2|^2)$. For a conflict between $m^{(2)}_{i_a}$ and $m^{(2)}_{i_b}$ to occur, we must have that $i_a = (s_1t_1, s_2t_2, ij)$ and $i_b = (s_1't_1', s_2't_2', ik)$ where at least two indices in s_1, s_2, s_1', s_2' agree. The number of conflicting pairs of the form $m^{(2)}_{i_a}, m^{(2)}_{i_b}$ is thus bounded by

$$6 \sum_{i\in[n]} d_i^3 \sum_{j\in[n]} d_j^2 \sum_{k\in[n]} d_k^2 = o\Big(\sum_{i\in[n]} d_i(d_i-1)\Big)^4 = o(|\mathcal{I}_2|^2), \tag{7.4.19}$$

as before and as required. This completes the proof. □

We close this section by showing that the probability of simplicity converges to zero when $\mathbb{E}[D^2] = \infty$:

Proposition 7.14 (Number of self-loops and multiple edges for infinite variance degrees) *Assume that $\boldsymbol{d} = (d_i)_{i\in[n]}$ satisfies Conditions 7.8(a)–(b), and assume that $\mathbb{E}[D^2] = \infty$. Then S_n and $\widetilde{M}_n$ both converge to infinity in probability. Consequently, $\mathbb{P}(\mathrm{CM}_n(\boldsymbol{d})$ simple$) = o(1)$ when $\mathbb{E}[D^2] = \infty$.*

Proof We adapt the proof of Proposition 7.13. We fix $K \geq 1$ large and define the truncated sets of indices

$$\begin{aligned}\mathcal{I}_1(K) &= \{(st, i) \colon i \in [n], 1 \leq s < t \leq d_i \wedge K\},\\ \mathcal{I}_2(K) &= \{(s_1t_1, s_2t_2, i, j) \colon 1 \leq i < j \leq n, 1 \leq s_1 < s_2 \leq d_i \wedge K,\\ &\qquad 1 \leq t_1 \neq t_2 \leq d_j \wedge K\},\end{aligned}$$

for the potential set of self-loops and pairs of multiple edges between the first K half-edges of all vertices, respectively. Again we assume, without loss of generality, that $|\mathcal{I}_1|$ and $|\mathcal{I}_2|$ grow to infinity with n. Clearly, $S_n \succeq S_n(K)$, $\widetilde{M}_n \succeq \widetilde{M}_n(K)$, where now

$$S_n(K) = \sum_{(st,i)\in\mathcal{I}_1(K)} I_{st,i}, \qquad \widetilde{M}_n(K) = \sum_{(s_1t_1,s_2t_2,i,j)\in\mathcal{I}_2(K)} I_{s_1t_1,s_2t_2,ij}. \tag{7.4.20}$$

Following the proof of Proposition 7.13, we obtain that

$$(S_n(K), \widetilde{M}_n(K)) \xrightarrow{d} (S(K), M(K)), \tag{7.4.21}$$

where $(S(K), M(K))$ are independent Poisson variables with means $\nu(K)/2$ and $\nu(K)^2/4$, respectively, where

$$\nu(K) = \frac{\mathbb{E}[(D \wedge K)((D \wedge K) - 1)]}{\mathbb{E}[D]}. \tag{7.4.22}$$

Since $\nu(K) \nearrow \infty$ as $K \to \infty$, it follows that, for every $A \geq 1$,

$$\liminf_{n\to\infty} \mathbb{P}(S_n \geq A) \geq \liminf_{n\to\infty} \mathbb{P}(S_n(K) \geq A) = \mathbb{P}(S(K) \geq A) \to 1, \tag{7.4.23}$$

as $K \to \infty$. This implies that $S_n \xrightarrow{\mathbb{P}} \infty$. A similar argument applies to $\widetilde{M}_n$. □

Exercises 7.12–7.13 extend Proposition 7.14 to $M_n \xrightarrow{\mathbb{P}} \infty$.

7.5 Uniform Simple Graphs and Generalized Random Graphs

In this section, we investigate the relations between the configuration model, uniform simple random graphs with given degrees and the generalized random graph with given weights. These results are 'folklore' in the random graph community, and allow us to use the configuration model to prove results for several other models. We crucially rely on the fact that $\mathrm{CM}_n(\boldsymbol{d})$ is simple with asymptotically positive probability when the degree sequence $(d_i)_{i\in[n]}$ satisfies Conditions 7.8(a)–(c).

Proposition 7.15 (Uniform graphs with given degree sequence) *For any degree sequence $(d_i)_{i\in[n]}$, and conditionally on the event $\{\mathrm{CM}_n(\boldsymbol{d})$ is a simple graph$\}$, $\mathrm{CM}_n(\boldsymbol{d})$ is a uniform simple graph with degree sequence $\boldsymbol{d} = (d_i)_{i\in[n]}$.*

Proof By (7.2.6) in Proposition 7.7, $\mathbb{P}(\mathrm{CM}_n(\boldsymbol{d}) = G)$ is the same for every simple graph G. Therefore, also $\mathbb{P}(\mathrm{CM}_n(\boldsymbol{d}) = G \mid \mathrm{CM}_n(\boldsymbol{d})$ simple) is the same for every simple graph G. □

Alternatively, by Exercise 7.18, conditionally on the configuration, obtained by the uniform matching of half-edges, corresponding to a simple graph, the conditional distribution of the configuration is uniform over all configurations which are such that the corresponding graph is simple.

An important consequence of Theorem 7.12 is that it allows us to compute the asymptotic number of simple graphs with a given degree sequence:

Corollary 7.16 (Number of graphs with given degree sequence) *Assume that the degree sequence $\boldsymbol{d} = (d_i)_{i\in[n]}$ satisfies Conditions 7.8(a)–(c), and that $\ell_n = \sum_{i\in[n]} d_i$ is even. Then, the number of simple graphs with degree sequence $\boldsymbol{d} = (d_i)_{i\in[n]}$ is equal to*

$$\mathrm{e}^{-\nu/2-\nu^2/4} \frac{(\ell_n - 1)!!}{\prod_{i\in[n]} d_i!}(1 + o(1)). \tag{7.5.1}$$

Proof By (7.2.6) in Proposition 7.7, for any simple G,

$$\mathbb{P}(\mathrm{CM}_n(\boldsymbol{d}) = G) = \frac{\prod_{i\in[n]} d_i!}{(\ell_n - 1)!!}. \tag{7.5.2}$$

Therefore,

$$\begin{aligned}\mathbb{P}(\mathrm{CM}_n(\boldsymbol{d}) \text{ simple}) \\ = \frac{\prod_{i\in[n]} d_i!}{(\ell_n - 1)!!} \#\{\text{simple graphs with degree sequence } \boldsymbol{d}\}.\end{aligned} \tag{7.5.3}$$

We multiply by $(\ell_n - 1)!!$, divide by $\prod_{i\in[n]} d_i!$ and use Theorem 7.12 to arrive at the claim. □

A special case of the configuration model is when all degrees are equal to some r. In this case, when we condition on the fact that the resulting graph in the configuration model is simple, we obtain a *uniform random regular graph*. See Figure 7.7 for two examples of configuration models where every vertex has degree 3. Uniform regular random graphs can be seen as a finite approximation of a regular tree. In particular, Corollary 7.16 implies that, when nr is even, the number of regular graphs of degree r graphs is equal to (see also Exercise 7.20)

$$\mathrm{e}^{-(r-1)/2-(r-1)^2/4} \frac{(rn - 1)!!}{(r!)^n} (1 + o(1)). \tag{7.5.4}$$

A further consequence of Theorem 7.12 is that it allows one to prove a property for uniform graphs with a given degree sequence by proving it for the configuration model with that degree sequence:

Corollary 7.17 (Uniform graphs with given degree sequence and $\mathrm{CM}_n(\boldsymbol{d})$) *Assume that $\boldsymbol{d} = (d_i)_{i\in[n]}$ satisfies Conditions 7.8(a)–(c), and that $\ell_n = \sum_{i\in[n]} d_i$ is even. Then, an event $\mathcal{E}_n$ occurs with high probability for a uniform simple random graph with degrees $(d_i)_{i\in[n]}$ when it occurs with high probability for* $\mathrm{CM}_n(\boldsymbol{d})$.

Corollary 7.17 yields a simple strategy to study proporties of uniform simple random graphs with a prescribed degree sequence. Indeed, when we can prove that a statement holds whp for $\mathrm{CM}_n(\boldsymbol{d})$, then it follows whp for a uniform random graph with the same degree sequence. Since $\mathrm{CM}_n(\boldsymbol{d})$ can be constructed in a rather simple manner, this makes it easier to prove properties for $\mathrm{CM}_n(\boldsymbol{d})$ than it is for a uniform random graph with degrees $\boldsymbol{d}$. For completeness, we now prove Corollary 7.17:

Proof Let $\mathrm{UG}_n(\boldsymbol{d})$ denote a uniform simple random graph with degrees $\boldsymbol{d}$. Let $\mathcal{E}_n$ be a subset of simple graphs. We need to prove that if $\lim_{n\to\infty} \mathbb{P}(\mathrm{CM}_n(\boldsymbol{d}) \in \mathcal{E}_n^c) = 0$, then also $\lim_{n\to\infty} \mathbb{P}(\mathrm{UG}_n(\boldsymbol{d}) \in \mathcal{E}_n^c) = 0$. By Proposition 7.15,

$$\begin{aligned}\mathbb{P}(\mathrm{UG}_n(\boldsymbol{d}) \in \mathcal{E}_n^c) &= \mathbb{P}(\mathrm{CM}_n(\boldsymbol{d}) \in \mathcal{E}_n^c \mid \mathrm{CM}_n(\boldsymbol{d}) \text{ simple}) \\ &= \frac{\mathbb{P}(\mathrm{CM}_n(\boldsymbol{d}) \in \mathcal{E}_n^c, \mathrm{CM}_n(\boldsymbol{d}) \text{ simple})}{\mathbb{P}(\mathrm{CM}_n(\boldsymbol{d}) \text{ simple})} \\ &\leq \frac{\mathbb{P}(\mathrm{CM}_n(\boldsymbol{d}) \in \mathcal{E}_n^c)}{\mathbb{P}(\mathrm{CM}_n(\boldsymbol{d}) \text{ simple})}.\end{aligned} \tag{7.5.5}$$

By Theorem 7.12, for which the assumptions are satisfied by the hypotheses in Corollary 7.17, $\liminf_{n\to\infty} \mathbb{P}(\mathrm{CM}_n(\boldsymbol{d}) \text{ simple}) > 0$. Moreover, $\lim_{n\to\infty} \mathbb{P}(\mathrm{CM}_n(\boldsymbol{d}) \in \mathcal{E}_n^c) = 0$, so that $\mathbb{P}(\mathrm{UG}_n(\boldsymbol{d}) \in \mathcal{E}_n^c) \to 0$ as well, as required. □

Unfortunately, the above strategy does not work when $\mathbb{E}[D^2] = \infty$. This is, for example, the case when the degree distribution satisfies a power-law with exponent $\tau \in (2,3)$. Indeed, in this case, the probability that $\mathrm{CM}_n(\boldsymbol{d})$ is simple vanishes by Proposition 7.14. As a result, it is a big open problem how to effectively generate uniform random graphs with degree sequences having infinite variance. We will return to this issue in Section 7.7.

As a consequence of Proposition 7.15 and Theorem 6.15, we see that $\mathrm{GRG}_n(\boldsymbol{w})$ conditionally on its degrees, and $\mathrm{CM}_n(\boldsymbol{d})$ with the same degrees conditioned on producing a simple graph, are equal in distribution. This also partially explains the popularity of the configuration model: Some results for $\mathrm{GRG}_n(\boldsymbol{w})$, and sometimes even the Erdős–Rényi random graph, are more easily proved by proving the result for the configuration model and conditioning on the degree sequence, and using that the degree distribution of the Erdős–Rényi random graph is very close to a sequence of independent Poisson random variables. See [II, Chapters 2–5] for more results in this direction. We formalize this 'folklore' result in the following theorem:

Theorem 7.18 (Relation between $\mathrm{GRG}_n(\boldsymbol{w})$ and $\mathrm{CM}_n(\boldsymbol{d})$) *Let D_i be the degree of vertex i in $\mathrm{GRG}_n(\boldsymbol{w})$ defined in* (6.3.1), *and let $\boldsymbol{D} = (D_i)_{i\in[n]}$. Then,*

$$\mathbb{P}(\mathrm{GRG}_n(\boldsymbol{w}) = G \mid \boldsymbol{D} = \boldsymbol{d}) = \mathbb{P}(\mathrm{CM}_n(\boldsymbol{d}) = G \mid \mathrm{CM}_n(\boldsymbol{d}) \textit{ simple}). \tag{7.5.6}$$

Let $\mathcal{E}_n$ be a subset of simple graphs for which $\mathbb{P}(\mathrm{CM}_n(\boldsymbol{d}) \in \mathcal{E}_n) \xrightarrow{\mathbb{P}} 1$ when $\boldsymbol{d}$ satisfies Conditions 7.8(a)–(c). Assume that the degree sequence $\boldsymbol{D} = (D_i)_{i\in[n]}$ of $\mathrm{GRG}_n(\boldsymbol{w})$ satisfies Conditions 7.8(a)–(c) in probability as explained in Remark 7.9. Then also $\mathbb{P}(\mathrm{GRG}_n(\boldsymbol{w}) \in \mathcal{E}_n) \to 1$.[2]

Conditions 7.8(a)–(c) are often easier to verify than proving that the event $\mathcal{E}_n$ occurs whp. We remark that related versions of Theorem 7.18 can be stated with different hypotheses on the degrees. Then, the statement becomes that if an event $\mathcal{E}_n$ occurs whp for $\mathrm{CM}_n(\boldsymbol{d})$ under the assumptions on the degrees, then $\mathcal{E}_n$ also occurs whp for $\mathrm{GRG}_n(\boldsymbol{w})$ provided we can show that the degrees of $\mathrm{GRG}_n(\boldsymbol{w})$ satisfy these assumptions whp.

Proof Equation (7.5.6) follows from Theorem 6.15 and Corollary 7.17, for every simple graph G with degree sequence $\boldsymbol{d}$. Indeed, these results imply that both $\mathrm{GRG}_n(\boldsymbol{w})$ conditionally on $\boldsymbol{D} = \boldsymbol{d}$ and $\mathrm{CM}_n(\boldsymbol{d})$ conditionally on being simple are uniform simple random graphs with degree sequence $\boldsymbol{d}$. By (7.5.6), for every event $\mathcal{E}_n$,

$$\mathbb{P}(\mathrm{GRG}_n(\boldsymbol{w}) \in \mathcal{E}_n \mid \boldsymbol{D} = \boldsymbol{d}) = \mathbb{P}(\mathrm{CM}_n(\boldsymbol{d}) \in \mathcal{E}_n \mid \mathrm{CM}_n(\boldsymbol{d}) \text{ simple}). \tag{7.5.7}$$

[2] Be aware of the potential confusion in the notation for the degrees. We use $\boldsymbol{D} = (D_i)_{i\in[n]}$ for the degrees in $\mathrm{GRG}_n(\boldsymbol{w})$, but in Conditions 7.8(a)–(c), we use $\boldsymbol{d} = (d_i)_{i\in[n]}$ instead to denote the degrees in $\mathrm{CM}_n(\boldsymbol{d})$. We strive to minimalise this confusion by being explicit about which degree sequence we speak of at all times.

We rewrite

$$\mathbb{P}(\mathrm{GRG}_n(\boldsymbol{w}) \in \mathcal{E}_n^c) = \mathbb{E}\Big[\mathbb{P}(\mathrm{GRG}_n(\boldsymbol{w}) \in \mathcal{E}_n^c \mid \boldsymbol{D})\Big] \tag{7.5.8}$$

$$= \mathbb{E}\Big[\mathbb{P}_n(\mathrm{CM}_n(\boldsymbol{D}) \in \mathcal{E}_n^c \mid \mathrm{CM}_n(\boldsymbol{D}) \text{ simple})\Big]$$

$$\leq \mathbb{E}\Big[\Big(\frac{\mathbb{P}_n(\mathrm{CM}_n(\boldsymbol{D}) \in \mathcal{E}_n^c)}{\mathbb{P}_n(\mathrm{CM}_n(\boldsymbol{D}) \text{ simple})}\Big) \wedge 1\Big],$$

where we write $\mathbb{P}_n$ for the conditional law given the degrees $\boldsymbol{D}$. We note that $\mathbb{E}$ is the expectation with respect to $\boldsymbol{D}$. By assumption, $\mathbb{P}_n(\mathrm{CM}_n(\boldsymbol{D}) \in \mathcal{E}_n^c) \xrightarrow{\mathbb{P}} 0$. Further, since the degree sequence $\boldsymbol{D}$ satisfies Conditions 7.8(a)–(c) in probability (recall Remark 7.9),

$$\mathbb{P}_n(\mathrm{CM}_n(\boldsymbol{D}) \text{ simple}) \xrightarrow{\mathbb{P}} \mathrm{e}^{-\nu/2-\nu^2/4} > 0. \tag{7.5.9}$$

Therefore, by Dominated Convergence (Theorem A.2),

$$\lim_{n\to\infty} \mathbb{E}\Big[\Big(\frac{\mathbb{P}_n(\mathrm{CM}_n(\boldsymbol{D}) \in \mathcal{E}_n^c)}{\mathbb{P}_n(\mathrm{CM}_n(\boldsymbol{D}) \text{ simple})}\Big) \wedge 1\Big] = 0,$$

so that we conclude that $\lim_{n\to\infty} \mathbb{P}(\mathrm{GRG}_n(\boldsymbol{w}) \in \mathcal{E}_n^c) = 0$, as required. □

In the following theorem, we show that Condition 6.4 on the weight sequence $\boldsymbol{w}$ implies that the degrees $\boldsymbol{D}$ in $\mathrm{GRG}_n(\boldsymbol{w})$ satisfy Condition 7.8. To avoid confusion, we now write d_i for the degree of vertex i in $\mathrm{GRG}_n(\boldsymbol{w})$, which before we called D_i and is a *random variable*. The reason is that Condition 7.8 is phrased in terms of D_n, which is the degree of a uniformly chosen vertex, and which causes confusion with the degree of vertex n, which we now call d_n:

Theorem 7.19 (Regularity conditions weights and degrees) *Let d_i be the degree of vertex i in* $\mathrm{GRG}_n(\boldsymbol{w})$ *defined in* (6.3.1), *and let* $\boldsymbol{d} = (d_i)_{i\in[n]}$. *Then,* $\boldsymbol{d}$ *satisfies Conditions 7.8(a)–(c) in probability when* $\boldsymbol{w}$ *satisfies Conditions 6.4(a)–(c), where*

$$\mathbb{P}(D = k) = \mathbb{E}\Big[\frac{W^k}{k!}\mathrm{e}^{-W}\Big] \tag{7.5.10}$$

denotes the mixed-Poisson distribution with mixing distribution W having distribution function F in Condition 6.4(a).

Proof Let $P_k^{(n)} = \frac{1}{n}\sum_{i\in[n]} \mathbb{1}_{\{d_i=k\}}$. Then Theorem 6.10 implies that $P_k^{(n)} \xrightarrow{\mathbb{P}} \mathbb{P}(D = k)$, which implies that Condition 7.8(a) holds in probability when Conditions 6.4(a)–(b) hold (recall Remark 7.9).

Theorem 6.6, together with the fact that $\frac{1}{n}\sum_{i\in[n]} d_i = 2E(\mathrm{GRG}_n(\boldsymbol{w}))/n$, implies that Condition 7.8(b) holds when Condition 6.4(b) holds.

To prove that $\frac{1}{n}\sum_{i\in[n]} d_i^2 \xrightarrow{\mathbb{P}} \mathbb{E}[D^2]$ for $\mathrm{GRG}_n(\boldsymbol{w})$, we first use that Corollary 6.20 implies that it suffices to prove this for $\mathrm{CL}_n(\boldsymbol{w})$. Further, since Condition 7.8(c) holds, Exercise 6.3 implies that $w_i = o(\sqrt{n})$, so that $p_{ij}^{(\mathrm{CL})} = (w_i w_j/\ell_n \wedge 1) = w_i w_j/\ell_n$. We note that it suffices to prove that Condition 6.4(c) implies that

$$\frac{1}{n}\sum_{i\in[n]} d_i(d_i-1) \xrightarrow{\mathbb{P}} \mathbb{E}[D(D-1)] = \mathbb{E}[W^2], \tag{7.5.11}$$

where the last equality follows from the fact that $\mathbb{E}[X(X-1)] = \lambda^2$ when X has a Poisson distribution with parameter λ. For this, we rewrite

$$\sum_{i\in[n]} d_i(d_i-1) = \sum_{i,j,k\in[n]:\ j\neq k} I_{ij}I_{jk} = 2\sum_{i,j,k\in[n]:\ j<k} I_{ij}I_{jk}, \tag{7.5.12}$$

where $(I_{ij})_{1\leq i<j\leq n}$ are independent Bernoulli variables with success probability $p_{ij} = \mathbb{P}(I_{ij}=1) = w_iw_j/\ell_n$. We use a second moment method, and start by computing the first moment as

$$\begin{aligned}\mathbb{E}\Big[\frac{1}{n}\sum_{i\in[n]} d_i(d_i-1)\Big] &= \frac{2}{n}\sum_{i,j,k\in[n]:\ j<k} \frac{w_iw_j}{\ell_n}\frac{w_iw_k}{\ell_n} \\ &= \frac{1}{n}\sum_{i\in[n]} w_i^2 \sum_{j\in[n]\setminus\{i\}} \frac{w_j}{\ell_n} \sum_{k\in[n]\setminus\{i,j\}} \frac{w_k}{\ell_n} \\ &= \frac{1}{n}\sum_{i\in[n]} w_i^2 - \frac{1}{n}\sum_{i\in[n]} \frac{w_i^3}{\ell_n} \sum_{k\in[n]\setminus\{i,j\}} \frac{w_k}{\ell_n} \\ &\quad - \frac{1}{n}\sum_{i,j\in[n]} \frac{w_i^2w_j}{\ell_n}\frac{(w_i+w_j)}{\ell_n} \\ &\to \mathbb{E}[W^2],\end{aligned} \tag{7.5.13}$$

where the last convergence is Condition 6.4(c) and we have used that $\max_{i\in[n]} w_i = o(n)$. This identifies the first moment.

We next bound the variance of $\frac{1}{n}\sum_{i\in[n]} d_i(d_i-1)$, with the aim of showing that it is $o(1)$. For this, we use (7.5.12) to write

$$\mathrm{Var}\Big(\frac{1}{n}\sum_{i\in[n]} d_i(d_i-1)\Big) = \frac{4}{n^2}\sum_{(i,j,k),(i',j',k')} \mathrm{Cov}(I_{ij}I_{jk}, I_{i'j'}I_{j'k'}), \tag{7.5.14}$$

where $\mathrm{Cov}(X,Y)$ denotes the covariance between two random variables X and Y, and where the sum is over (i,j,k) and (i',j',k') such that $i,j,k\in[n]$ with $j<k$ and $i',j',k'\in[n]$ with $j'<k'$. When i,j,k,i',j',k' are all distinct, $\mathrm{Cov}(I_{ij}I_{jk}, I_{i'j'}I_{j'k'}) = 0$. The same applies *except* when two edges in $\{i,j\}, \{i,k\}, \{i',j'\}, \{i',k'\}$ are equal. Recall that $j<k$ and $j'<k'$. This implies that when two edges are equal, either these three edges are incident to the same vertex, or they form a path of length 3. Furthermore, $\mathrm{Cov}(I_{ij}I_{jk}, I_{i'j'}I_{j'k'}) \leq \mathbb{E}[I_{ij}I_{jk}]$ when $(i,j,k) = (i',j',k')$. This leads us to

$$\begin{aligned}\mathrm{Var}\Big(\frac{1}{n}\sum_{i\in[n]} d_i(d_i-1)\Big) &\leq \frac{2}{n^2}\mathbb{E}\Big[\sum_{i\in[n]} d_i(d_i-1)\Big] \\ &\quad + \frac{8}{n^2}\sum_{i,j,k,l\in[n]} p_{ij}p_{ik}p_{il} + \frac{8}{n^2}\sum_{i,j,k\in[n]} p_{ij}p_{ik}p_{jk}.\end{aligned} \tag{7.5.15}$$

Since $p_{ij} \leq w_i w_j/\ell_n$, we arrive at

$$\mathrm{Var}\Big(\frac{1}{n}\sum_{i\in[n]} d_i(d_i-1)\Big) \leq \frac{2}{n^2}\mathbb{E}\Big[\sum_{i\in[n]} d_i(d_i-1)\Big] \tag{7.5.16}$$

$$+\frac{8}{n^2}\sum_{i,j,k,l\in[n]}\frac{w_i^3 w_j w_k w_l}{\ell_n^3}+\frac{8}{n^2}\sum_{i,j,k,l\in[n]}\frac{w_i w_j^2 w_k^2 w_l}{\ell_n^3}$$

$$= O(n^{-1})\mathbb{E}[W_n^2] + O(n^{-2})\sum_{i\in[n]} w_i^3 + O(n^{-1})\mathbb{E}[W_n^2]^2.$$

By Condition 6.4(c), $\mathbb{E}[W_n^2]$ converges, and also $\max_{i\in[n]} w_i = o(n)$. Therefore,

$$\mathrm{Var}\Big(\frac{1}{n}\sum_{i\in[n]} d_i(d_i-1)\Big) = o(1). \tag{7.5.17}$$

Take n large enough so that $\Big|\mathbb{E}\Big[\frac{1}{n}\sum_{i\in[n]} d_i(d_i-1)\Big] - \mathbb{E}[W^2]\Big| \leq \varepsilon/2$. Then

$$\mathbb{P}\Big(\Big|\frac{1}{n}\sum_{i\in[n]} d_i(d_i-1) - \mathbb{E}[W^2]\Big| > \varepsilon\Big) \tag{7.5.18}$$

$$\leq \mathbb{P}\Big(\Big|\frac{1}{n}\sum_{i\in[n]} d_i(d_i-1) - \mathbb{E}\Big[\frac{1}{n}\sum_{i\in[n]} d_i(d_i-1)\Big]\Big| > \varepsilon/2\Big)$$

$$\leq \frac{4}{\varepsilon^2}\mathrm{Var}\Big(\frac{1}{n}\sum_{i\in[n]} d_i(d_i-1)\Big) = o(1).$$

Since this is true for every $\varepsilon > 0$, we obtain that $\frac{1}{n}\sum_{i\in[n]} d_i(d_i-1) \xrightarrow{\mathbb{P}} \mathbb{E}[W^2]$, as required. □

7.6 Configuration Model with I.I.D. Degrees

In this section, we apply the results of the previous sections to the configuration model with i.i.d. degrees. We recall that this setting is obtained when $(d_i)_{i\in[n-1]}, d_n'$ are n i.i.d. random variables, and $d_n = d_n' + \mathbb{1}_{\{d_n'+\sum_{j\in[n-1]} d_j \text{ odd}\}}$. The latter solves the problem that the total degree is odd with probability close to $1/2$ (recall Exercise 7.8). We ignore the effect of the added indicator in the definition of d_n, since it hardly makes any difference.

We note that, similarly to the generalized random graph with i.i.d. weights, the introduction of randomness in the degrees introduces a *double randomness* in the model: firstly the randomness of the degrees, and secondly, the randomness of the pairing of the edges *given* the degrees. We investigate the degree sequence $(P_k^{(n)})_{k\geq 1}$ defined by

$$P_k^{(n)} = \frac{1}{n}\sum_{i\in[n]} \mathbb{1}_{\{d_i=k\}}. \tag{7.6.1}$$

Apart from the dependence on n of d_n, we see that $(P_k^{(n)})_{k\geq 1}$ is precisely equal to the empirical distribution of the degrees $(d_i)_{i\in[n-1]}, d_n'$, which is an i.i.d. sample. Since in $(d_i)_{i\in[n]}$ only one value is different from the i.i.d. sample $(d_i)_{i\in[n-1]}, d_n'$, $P_k^{(n)}$ differs at most $1/n$

from the empirical distribution of n i.i.d. random variables. As a result, by the Strong Law of Large Numbers, for every $k \geq 1$,

$$P_k^{(n)} \xrightarrow{a.s.} p_k \equiv \mathbb{P}(D_1 = k), \tag{7.6.2}$$

so that the empirical distribution of i.i.d. degrees is almost surely close to the probability distribution of each of the degrees. By Exercise 2.16, the above convergence also implies that $d_{\mathrm{TV}}(P^{(n)}, p) \xrightarrow{a.s.} 0$, where $p = (p_k)_{k\geq 1}$ and $P^{(n)} = (P_k^{(n)})_{k\geq 1}$.

Our first main result describes the degree sequence of the erased configuration model:

Theorem 7.20 (Degree sequence of erased configuration model with i.i.d. degrees) *Let $(d_i)_{i\in[n-1]}, d_n'$ be an i.i.d. sequence of finite mean random variables with $\mathbb{P}(d \geq 1) = 1$. The degree sequence of the erased configuration model $(P_k^{(\mathrm{er})})_{k\geq 1}$ with degrees $(d_i)_{i\in[n]}$ converges to $(p_k)_{k\geq 1}$. More precisely,*

$$\mathbb{P}\Big(\sum_{k\geq 1} |P_k^{(\mathrm{er})} - p_k| \geq \varepsilon\Big) \to 0. \tag{7.6.3}$$

Proof By Exercise 7.9, when $\mathbb{E}[d] < \infty$, Conditions 7.8(a)–(b) hold, where the convergence is in probability as explained in Remark 7.9. As a result, Theorem 7.20 follows directly from Theorem 7.10. □

We next investigate the probability of obtaining a simple graph in $\mathrm{CM}_n(\boldsymbol{d})$ with i.i.d. degrees:

Theorem 7.21 (Probability of simplicity in $\mathrm{CM}_n(\boldsymbol{d})$ with i.i.d. degrees) *Let $(d_i)_{i\geq 1}$ be the degrees obtained from an i.i.d. sequence of random variables $(d_i)_{i\in[n-1]}$ and d_n' with $d_n = d_n' + \mathbb{1}_{\{\sum_{i\in[n-1]} d_i + d_n' \text{ odd}\}}$, where the degrees satisfy* $\mathrm{Var}(d) < \infty$ *and* $\mathbb{P}(d \geq 1) = 1$. *Then, the probability that* $\mathrm{CM}_n(\boldsymbol{d})$ *is simple converges to* $\mathrm{e}^{-\nu/2-\nu^2/4}$, *where* $\nu = \mathbb{E}[D(D-1)]/\mathbb{E}[D]$.

Proof By Exercise 7.9, Conditions 7.8(a)–(c) hold when $\mathrm{Var}(d) < \infty$, where the convergence is in probability as explained in Remark 7.9. As a result, Theorem 7.21 follows directly from Theorem 7.12. □

We finally investigate the case where the mean is *infinite*, with the aim to produce a random graph with power-law degrees with an exponent $\tau \in (1, 2)$. In this case, the graph topology is rather different, as the majority of edges is in fact multiple, and self-loops from vertices with high degrees are abundant. As a result, the erased configuration model has rather different degrees compared to those in the multigraph. Therefore, in order to produce a more realistic graph, we need to perform some sort of truncation procedure. We start by investigating the case where we condition the degrees to be bounded above by some $a_n = o(n)$, which in effect, reduces the number of self-loops significantly.

Theorem 7.22 (Degree sequence of erased configuration model with i.i.d. conditioned infinite mean degrees) *Let $(d_i)_{i\in[n-1]}, d_n'$ be i.i.d. copies of a random variable D conditioned on $D \leq a_n$, and define $d_n = d_n' + \mathbb{1}_{\{d_n' + \sum_{i\in[n-1]} d_i \text{ odd}\}}$. Then, for every $a_n = o(n)$, the empirical degree distribution of the erased configuration model $(P_k^{(\mathrm{er})})_{k=1}^{\infty}$ with degrees $(d_i)_{i\in[n]}$*

converges in probability to $(p_k)_{k\geq 1}$, *where* $p_k = \mathbb{P}(D = k)$. *More precisely, for every* $\varepsilon > 0$,

$$\mathbb{P}(\sum_{k\geq 1} |P_k^{(\mathrm{er})} - p_k| \geq \varepsilon) \to 0. \tag{7.6.4}$$

Proof Theorem 7.22 is similar in spirit to Theorem 6.13 for the generalized random graph, and its proof is left as Exercise 7.22. □

We continue by studying the erased configuration model with infinite-mean degrees in the unconditioned case. We assume that there exists a slowly varying function $x \mapsto L(x)$ such that

$$1 - F(x) = x^{1-\tau} L(x), \tag{7.6.5}$$

where $F(x) = \mathbb{P}(D \leq x)$ and where $\tau \in (1, 2)$. We now investigate the degree sequence in the configuration model with infinite-mean degrees, where we do not condition the degrees to be at most a_n. We make substantial use of Theorem 2.33. In order to describe the result, we need a few definitions.

We define the (random) probability distribution $Q = (Q_i)_{i\geq 1}$ as follows. Let, as in Theorem 2.33, $(E_i)_{i\geq 1}$ be i.i.d. exponential random variables with parameter 1, and define $\Gamma_i = \sum_{j=1}^{i} E_j$. Let $(d_i)_{i\geq 1}$ be an i.i.d. sequence of random variables with distribution function F in (7.6.5), and let $d_{(n:n)} \geq d_{(n-1:n)} \geq \cdots \geq d_{(1:n)}$ be the order statistics of $(D_i)_{i\in[n]}$. We recall from Theorem 2.33 that there exists a sequence u_n, with $u_n n^{-1/(\tau-1)}$ slowly varying, such that

$$u_n^{-1} \left(\ell_n, (d_{(n-i+1:n)})_{i=1}^{n}\right) \xrightarrow{d} \left(\sum_{j\geq 1} \Gamma_j^{-1/(\tau-1)}, (\Gamma_i^{-1/(\tau-1)})_{i\geq 1}\right). \tag{7.6.6}$$

We abbreviate $\eta = \sum_{j\geq 1} \Gamma_j^{-1/(\tau-1)}$ and $\xi_i = \Gamma_i^{-1/(\tau-1)}$, and let

$$Q_i = \xi_i / \eta, \tag{7.6.7}$$

so that, by (7.6.6),

$$\sum_{i\geq 1} Q_i = 1. \tag{7.6.8}$$

However, the Q_i are all random variables, so that $Q = (Q_i)_{i\geq 1}$ is a random probability distribution. We further write $M_{Q,k}$ for a multinomial distribution with parameters k and probabilities $Q = (Q_i)_{i\geq 1}$, and U_{Q,D_k} is the number of non-zero components of the random variable M_{Q,D_k}, where D_k is independent of $Q = (Q_i)_{i\geq 1}$ and the multinomial trials. Recall that $D_i^{(\mathrm{er})}$ denotes the degree of vertex i in the erased configuration model.

Theorem 7.23 (Degree sequence of erased configuration model with i.i.d. infinite-mean degrees) *Let* $(d_i)_{i\in[n-1]}, d_n'$ *be i.i.d. copies of a random variable* D *having distribution function* F *satisfying* (7.6.5). *Then*

$$D_1^{(\mathrm{er})} \xrightarrow{d} U_{Q,D}, \tag{7.6.9}$$

where $Q = (Q_i)_{i\geq 1}$ *is given by* (7.6.7), *and the random variables* D *and* $Q = (Q_i)_{i\geq 1}$ *are independent.*

Theorem 7.23 is similar in spirit to Theorem 6.14 for the generalized random graph.

Proof The degree of vertex 1 is given by d_1, which is a realization of the random variable D. We fix $d_1 = D$ and look at the order statistics of the degrees $(d_i)_{i\in[n]\setminus\{1\}}$. These still satisfy the convergence in (7.6.6), and obviously d_1 is asymptotically independent of $(\eta, \xi_1, \xi_2, \ldots)$. Vertex 1 has $d_1 \stackrel{d}{=} D$ half-edges, which need to be paired to other half-edges. The probability that any half-edge is connected to the vertex with the jth largest degree is asymptotically equal to

$$Q_j^{(n)} = d_{(n-j+1:n)}/\ell_n \tag{7.6.10}$$

(here we ignore the possibility of self-loops), where, by Theorem 2.33 (see also (7.6.6)),

$$(Q_j^{(n)})_{j\geq 1} \xrightarrow{d} (\xi_j/\eta)_{j\geq 1}. \tag{7.6.11}$$

Moreover, the vertices to which the D half-edges are connected are close to being independent. As a result, the D half-edges of vertex 1 are paired to D vertices, and the number of edges of vertex 1 that are paired to the vertex with the ith largest degree are asymptotically given by the ith coordinate of $M_{Q^{(n)},D}$. By (7.6.6), the random variable $M_{Q^{(n)},D}$ converges in distribution to $M_{Q,D}$. We note that in the erased configuration model, the degree of the vertex 1 is equal to the number of *distinct* vertices to which 1 is connected, which is therefore equal to the number of distinct outcomes of the random variable $M_{Q,D}$. By definition, this is equal to $U_{Q,D}$. □

We next investigate the properties of the degree distribution, to obtain an equivalent result as in Exercise 6.22:

Theorem 7.24 (Power law with exponent 2 for erased CM with infinite-mean degrees) *Let the distribution function* F *of* D *satisfy* (7.6.5) *with* $L(x) = 1$ *and assume that* $\tau \in (1, 2)$. *Then, there exists a constant* $c > 0$ *such that the asymptotic degree of vertex 1 in the erased configuration model* $U_{Q,D}$ *satisfies that*

$$\mathbb{P}(U_{Q,D} \geq x) \leq c/x. \tag{7.6.12}$$

The result in Theorem 7.24 is similar in spirit to Exercise 6.22. It would be of interest to prove a precise identity here as well.

Proof In order that $U_{Q,D} \geq x$, in $M_{Q,D}$ at least $x/2$ times an i larger than $x/2$ needs to be chosen, where the probability of chosing i is $Q_i = \xi_i/\eta$ in (7.6.7). By (2.6.17), for i large

$$Q_i = \xi_i/\eta = (1 + o_{\mathbb{P}}(1))i^{-1/(\tau-1)}/\eta. \tag{7.6.13}$$

Therefore, whp,

$$\begin{aligned}\sum_{j\geq i} Q_j &= (1 + o_{\mathbb{P}}(1)) \sum_{j\geq i} j^{-1/(\tau-1)}/\eta \\ &= c(1 + o_{\mathbb{P}}(1))i^{(\tau-2)/(\tau-1)}/\eta \leq 2ci^{(\tau-2)/(\tau-1)}/\eta.\end{aligned} \tag{7.6.14}$$

Moreover, conditionally on $D = k$, the number of values larger than $x/2$ that are chosen is equal to a binomial random variable with k trials and success probability

$$q_x = cx^{(\tau-2)/(\tau-1)}/\eta. \tag{7.6.15}$$

Therefore, conditionally on $D = \lceil dx^b \rceil$, and using Theorem 2.21, the probability that at least $x/2$ values larger than $x/2$ are chosen is $o(x^{-a})$ for some $a > 1$ chosen appropriately, when, for some sufficiently large $C > 0$,

$$|dx^b q_x - \frac{x}{2}| \geq C \log x \sqrt{x^b q_x}. \tag{7.6.16}$$

Take $b = 1 - (\tau - 2)/(\tau - 1) = 1/(\tau - 1)$ and d such that $dx^b q_x = x/4$. Then, (7.6.15) and (7.6.16) above imply that, for some $a > 1$,

$$\mathbb{P}(U_{Q,D} \geq x, D \leq \lceil dx^b \rceil) = o(x^{-a}). \tag{7.6.17}$$

As a result,

$$\mathbb{P}(U_{Q,D} \geq x) \leq \mathbb{P}(D_k \geq \lceil dx^b \rceil) + o(x^{-a}) \leq cx^{-b(\tau-1)} = c/x. \tag{7.6.18}$$

□

7.7 Related Results on the Configuration Model

In this section, we discuss some related results on the configuration model. We start by discussing the number of friends of random friends.

Related Degree Structure $\mathrm{CM}_n(\boldsymbol{d})$ and $\mathrm{GRG}_n(\boldsymbol{w})$

Let $B_n^\star$ denote the degree of a random friend of a uniform vertex. More precisely, let $V_1 \in [n]$ be a uniform vertex, and let V_2 be any of the neighbors of V_1 chosen uniformly at random. Thus, V_2 is the vertex incident to the half-edge to which a uniform half-edge incident to V_1 is paired, and $B_n^\star = d_{V_2}$. Further, we let $e = (x, y)$ be a uniform edge, let i be the vertex incident to the half-edge x and j the vertex incident to half-edge y. Then define $(D_{n,1}^\star, D_{n,2}^\star) = (d_i, d_j)$ to be the degrees of the vertices i, j at either end of the edge. The law of e is the same as the law of a random half-edge x, together with the half-edge y to which x is paired. The simple construction of the configuration model allows us to identify the limiting distributions of $B_n^\star$ and $(D_x^\star, D_y^\star)$:

Theorem 7.25 (Related degrees configuration model) *Assume that $\boldsymbol{d} = (d_i)_{i\in[n]}$ satisfies Condition 7.8(a)–(b). Then, $\mathrm{CM}_n(\boldsymbol{d})$ satisfies that*

$$B_n^\star \xrightarrow{d} D^\star, \tag{7.7.1}$$

and

$$(D_{n,1}^\star, D_{n,2}^\star) \xrightarrow{d} (D_1^\star, D_2^\star), \tag{7.7.2}$$

where $\mathbb{P}(D^\star = k) = k\mathbb{P}(D = k)/\mathbb{E}[D]$ denotes the size-biased distribution of D and $(D_1^\star, D_2^\star)$ are two i.i.d. copies of $D^\star$.

Proof By construction of $\mathrm{CM}_n(\boldsymbol{d})$, $B_n^\star$ has the same distribution as the degree of the half-edge to which a half-edge is paired. Thus,

$$\mathbb{P}(B_n^\star = k) = \frac{k-1}{n(\ell_n - 1)} \sum_{i\in[n]} \mathbb{1}_{\{d_i=k\}} + \frac{1}{n} \sum_{i,j\in[n]:\, i\neq j} \frac{k}{\ell_n - 1} \mathbb{1}_{\{d_j=k\}}. \tag{7.7.3}$$

Here the first contribution is due to a self-loop, i.e., $V_1 = V_2$, and the second from $V_1 \neq V_2$. By Conditions 7.8(a)–(b), for every $k \geq 1$,

$$\begin{aligned}\lim_{n\to\infty} \mathbb{P}(B_n^\star = k) &= \lim_{n\to\infty} \sum_{j\in[n]:\, i\neq j} \frac{k}{\ell_n - 1} \mathbb{1}_{\{d_j=k\}} \\ &= \frac{k}{\mathbb{E}[D]} \mathbb{P}(D = k) = \mathbb{P}(D^\star = k).\end{aligned} \tag{7.7.4}$$

The proof (7.7.2) is similar. We note that (x, y) have the same law as two half-edges drawn uniformly at random and without replacement, and $D_{n,1}^\star$, $D_{n,2}^\star$ are the degrees of the vertices incident to them. Thus,

$$\begin{aligned}\mathbb{P}\big((D_{n,1}^\star, D_{n,2}^\star) = (k_1, k_2)\big) &= \frac{k_1(k_1-1)}{n\ell_n(\ell_n - 1)} \mathbb{1}_{\{k_1=k_2\}} \sum_{i\in[n]} \mathbb{1}_{\{d_i=k_1\}} \\ &+ \sum_{i,j\in[n]:\, i\neq j} \frac{k_1k_2}{\ell_n(\ell_n-1)} \mathbb{1}_{\{d_i=k_1, d_j=k_2\}},\end{aligned} \tag{7.7.5}$$

where again the first contribution is due to a self-loop, for which $D_{n,1}^\star = D_{n,2}^\star$. Again by Conditions 7.8(a)–(b), for every $k_1, k_2 \geq 1$,

$$\begin{aligned}\lim_{n\to\infty} \mathbb{P}\big((D_{n,1}^\star, D_{n,2}^\star) = (k_1, k_2)\big) &= \frac{k_1}{\mathbb{E}[D]} \mathbb{P}(D = k_1) \frac{k_2}{\mathbb{E}[D]} \mathbb{P}(D = k_2) \\ &= \mathbb{P}(D^\star = k_1)\mathbb{P}(D^\star = k_2),\end{aligned} \tag{7.7.6}$$

as required. □

Theorem 7.25 has important consequences for the degree structure of neighborhoods of vertices. For example, in Section 1.2 we have shown that in general graphs G, the average number of friends of a random friend of a random individual is larger than that of the friend himself. Recall Theorem 1.2. Theorem 7.25 quantifies this effect in $\mathrm{CM}_n(\boldsymbol{d})$, since $D^\star$ is stochastically larger than D (recall Proposition 2.13).

Further, it can be seen that, assuming that Conditions 7.8(a)–(b) hold, the number of friends of friends X_n of a random individual satisfies

$$X_n \xrightarrow{d} X^\star \equiv \sum_{i=1}^{D} (D_i^\star - 1), \tag{7.7.7}$$

where D has distribution function F, while $(D_i^\star)_{i\geq 1}$ are i.i.d. random variables with the size-biased distribution in (7.7.1) that are independent from D. In particular in the power-law setting, i.e., when D has a power-law distribution as in (7.6.5), then the distribution of $D^\star$, and therefore also that of $X^\star$, satisfies

$$\mathbb{P}(D^\star > x) = x^{2-\tau} L^\star(x), \tag{7.7.8}$$

for some slowly varying function $x \mapsto L^\star(x)$. In particular, the degree tail-exponent $\tau - 1$ is replaced by $\tau - 2$. Therefore, infinite variance for D implies infinite mean for $D^\star$. The ideas leading to (7.7.7) play an important role also in studying distances in $\mathrm{CM}_n(\boldsymbol{d})$ in [II, Chapter 4], and its proof is left as Exercise 7.26.

The general set-up assuming Conditions 7.8(a)–(c), together with the degree asympotics stated in Theorem 7.19, allow us to easily extend Theorem 7.25 to $\mathrm{GRG}_n(\boldsymbol{w})$:

Corollary 7.26 (Related degrees $\mathrm{GRG}_n(\boldsymbol{w})$) *Let d_i be the degree of vertex i in* $\mathrm{GRG}_n(\boldsymbol{w})$ *defined in* (6.3.1), *and let $\boldsymbol{d} = (d_i)_{i\in[n]}$. Then,*

$$B_n^\star \xrightarrow{d} D^\star, \tag{7.7.9}$$

and

$$(D_{n,1}^\star, D_{n,2}^\star) \xrightarrow{d} (D_1^\star, D_2^\star), \tag{7.7.10}$$

where $\mathbb{P}(D^\star = k) = k\mathbb{P}(D = k)/\mathbb{E}[D]$ denotes the size-biased distribution of D, and $(D_1^\star, D_2^\star)$ are two i.i.d. copies of $D^\star$, and

$$\mathbb{P}(D^\star = k) = \mathbb{P}\big(\mathsf{Poi}(W^\star) = k\big), \tag{7.7.11}$$

where $W^\star$ is the size-biased distribution of the random variable W, i.e., $\mathbb{P}(W^\star \leq x) = \mathbb{E}[W\mathbb{1}_{\{W\leq x\}}]/\mathbb{E}[W]$.

Proof By Theorem 7.19, the degrees $\boldsymbol{d} = (d_i)_{i\in[n]}$ satisfy Conditions 7.8(a)–(c). Therefore, the result immediately follows from Theorems 7.18 and 7.25. □

We remark that a *direct* proof of Corollary 7.26 seems difficult, since the degrees of the vertices in a random edge may have an intricate dependency structure.

Simple Random Graphs with Infinite Variance Degrees and $\mathrm{CM}_n(\boldsymbol{d})$

Theorem 7.12 and its consequences such as in Corollary 7.16 give us a way to count the number of simple graphs with a prescribed degree sequence and to derive properties for them. However, these results only apply when Conditions 7.8(a)–(c) hold, so that in particular $\mathbb{E}[D_n^2] \to \mathbb{E}[D^2] < \infty$. When D has a power-law distribution as in (7.6.5), this means that $\tau > 3$. However, as described in more detail in Section 1.7 in Chapter 1, in many real-world networks, power-law exponents having values $\tau \in (2, 3)$ are reported. This setting is investigated in more detail in the following theorem:

Theorem 7.27 (Number of simple graphs with infinite-variance degrees) *Assume that $\boldsymbol{d} = (d_i)_{i\in[n]}$ satisfies Conditions 7.8(a)–(b). Further, assume that $d_{\min} \geq 1$ and $\mathbb{E}[D_n^2] = o(n^{1/8})$. Then, the number of simple graphs with degree sequence $\boldsymbol{d}$ equals*

$$(1 + o(1))\frac{(\ell_n - 1)!!}{\prod_{i\in[n]} d_i!} \tag{7.7.12}$$

$$\times \exp\Big\{-n\nu_n/2 - \beta_n/(3n) + 3/4 + \sum_{1\leq i<j\leq n} \log(1 + d_i d_j/\ell_n)\Big\},$$

where

$$\beta_n = \mathbb{E}[D_n(D_n-1)(D_n-2)]/\mathbb{E}[D_n]. \tag{7.7.13}$$

In particular, Theorem 7.27 implies that when a property holds for $\mathrm{CM}_n(\boldsymbol{d})$ with probability that is $O(\mathrm{e}^{-a_n})$, where $a_n \gg \nu_n^2$ and $a_n \gg \beta_n/n$, then it also holds whp for a uniform simple graph with the same degree sequence. Unfortunately, Theorem 7.27 does not immediately give us a convenient way to *simulate* uniform random graphs with infinite-variance degrees.

The Number of Erased Edges in Erased CM with Infinite-Variance Degrees

Proposition 7.13 implies that the number of erased edges $E_n = S_n + M_n$ in $\mathrm{CM}_n(\boldsymbol{d})$ converges to a Poisson random variable with parameter $\nu/2 + \nu^2/4$ when Conditions 7.8(a)–(c) hold. When $\mathbb{E}[D^2] = \infty$, on the other hand, $E_n \xrightarrow{\mathbb{P}} \infty$ by Proposition 7.14 (note that the proof in particular yields that $S_n \xrightarrow{\mathbb{P}} \infty$). However, it is not so clear just *how fast* $E_n \xrightarrow{\mathbb{P}} \infty$. Theorem 7.22 implies that $E_n = o_{\mathbb{P}}(n)$ when Conditions 7.8(a)–(b) hold (see also Exercise 7.23). Van der Hoorn and Litvak (2015) show that $E_n = o_{\mathbb{P}}(n^{1/(\tau-1)+\delta})$ and $E_n = o_{\mathbb{P}}(n^{4/(\tau-1)-2+\delta})$ for every $\delta > 0$ when the degrees are i.i.d. with a power-law degree distribution with power-law exponent $\tau \in (2,3)$. This upper bound shows a phase transition for $\tau = 2.5$. It would be of interest to investigate in more detail how $E_n \xrightarrow{\mathbb{P}} \infty$.

7.8 Related Random Graph Models

In this section, we describe a few related models that have been investigated in the literature.

The Hierarchical Configuration Model

The configuration model has low clustering, which often makes it inappropriate in applied contexts. Indeed, in Chapter 1, we have seen that many real-world networks, in particular social networks, have a high amount of clustering instead. A possible solution to overcome this low clustering is by introducing a *community* or *household* structure. Consider the configuration model $\mathrm{CM}_n(\boldsymbol{d})$ with a degree sequence $\boldsymbol{d} = (d_i)_{i\in[n]}$ satisfying Condition 7.8(a)–(b). Now we replace each of the vertices by a small graph. Thus, vertex i is replaced by a local graph G_i. We assign each of the d_i half-edges incident to vertex i to a uniform vertex in G_i. As a result, we obtain a graph with two levels of hierarchy, whose local structure is described by the local graphs G_i, whereas its global structure is described by the configuration model $\mathrm{CM}_n(\boldsymbol{d})$. This model is called the hierarchical configuration model. In Exercise 7.27, you are asked to compute the clustering coefficient in the case where all the G_i are complete graphs.

Configuration Model with Clustering

The low clustering of $\mathrm{CM}_n(\boldsymbol{d})$ can be resolved by introducing communities as described above. Alternatively, we can also introduce clustering directly. In the configuration model with clustering, we assign two numbers to a vertex $i \in [n]$. We let $d_i^{(\mathrm{si})}$ denote the number of simple half-edges incident to vertex i, and we let $d_i^{(\mathrm{tr})}$ denote the number of triangles that vertex i is part of. We say that there are $d_i^{(\mathrm{si})}$ half-edges incident to vertex i, and $d_i^{(\mathrm{tr})}$ third-triangles.

The graph is built by (a) recursively choosing two half-edges uniformly at random without replacement, and pairing them into edges (as for $\mathrm{CM}_n(\boldsymbol{d})$); and (b) choosing triples of third-triangles uniformly at random and without replacement, and drawing edges between the three vertices incident to the third-triangles that are chosen.

Let $(D_n^{(\mathrm{si})}, D_n^{(\mathrm{tr})})$ denote the number of simple edges and triangles incident to a uniform vertex in $[n]$, and assume that $(D_n^{(\mathrm{si})}, D_n^{(\mathrm{tr})}) \xrightarrow{d} (D^{(\mathrm{si})}, D^{(\mathrm{tr})})$ for some limiting distribution $(D^{(\mathrm{si})}, D^{(\mathrm{tr})})$. Further, assume that $\mathbb{E}[D_n^{(\mathrm{si})}] \to \mathbb{E}[D^{(\mathrm{si})}]$ and $\mathbb{E}[(D_n^{(\mathrm{si})})^2] \to \mathbb{E}[(D^{(\mathrm{si})})^2]$. Then, it can be shown in a similar way as in the proof of Theorem 7.12 that with a strictly positive probability, there are no self-loops, no multiple edges and no other triangles except for the ones imposed by $(d_i^{(\mathrm{tr})})_{i\in[n]}$. If we further assume that $\mathbb{E}[D_n^{(\mathrm{tr})}] \to \mathbb{E}[D^{(\mathrm{tr})}]$ and $\mathbb{E}[(D_n^{(\mathrm{tr})})^2] \to \mathbb{E}[(D^{(\mathrm{tr})})^2]$, then we can also show that with positive probability, all triangles formed in the model contain precisely 3 distinct vertices. Thus, under these assumptions and with positive probability, this model yields a random graph with prescribed degrees and number of triangles per vertex. In Exercise 7.28, you are asked to compute the clustering coefficient of this model.

The Directed Configuration Model

Many real-world networks are *directed*, in the sense that edges are oriented from their starting vertex to their end vertex. For example, in the World-Wide Web, the vertices are web pages, and the edges are the hyperlinks between them, which are clearly oriented. One could naturally forget about these directions, but that would discard a wealth of information. For example, in citation networks, it makes a substantial difference whether my paper cites your paper, or your paper cites mine.

One way to obtain a directed version of $\mathrm{CM}_n(\boldsymbol{d})$ is to give each edge a direction, chosen with probability $1/2$, independently of all other edges. In this model, however, the correlation coefficient between the in- and out-degree of vertices is close to one, particularly when the degrees are large. In real-world applications, correlations between in- and out-degrees can be positive or negative, depending on the precise application. Therefore, we formulate a general model of directed graphs, where we can prescribe both the in- and out-degrees of vertices.

Fix $\boldsymbol{d}^{(\mathrm{in})} = (d_i^{(\mathrm{in})})_{i\in[n]}$ to be a sequence of in-degrees, where $d_i^{(\mathrm{in})}$ denotes the in-degree of vertex i. Similarly, we let $\boldsymbol{d}^{(\mathrm{out})} = (d_i^{(\mathrm{out})})_{i\in[n]}$ be a sequence of out-degrees. Naturally, we need that

$$\sum_{i\in[n]} d_i^{(\mathrm{in})} = \sum_{i\in[n]} d_i^{(\mathrm{out})} \tag{7.8.1}$$

in order for a graph with in- and out-degree sequence $\boldsymbol{d} = (\boldsymbol{d}^{(\mathrm{in})}, \boldsymbol{d}^{(\mathrm{out})})$ to exist. We think of $d_i^{(\mathrm{in})}$ as the number of in-half-edges incident to vertex i and $d_i^{(\mathrm{out})}$ as the number of out-half-edges incident to vertex i. The directed configuration model $\mathrm{DCM}_n(\boldsymbol{d})$ is obtained by pairing each in-half-edge to a uniformly chosen out-half-edge. The resulting graph is a random multigraph, where each vertex i has in-degree $d_i^{(\mathrm{in})}$ and out-degree $d_i^{(\mathrm{out})}$. Similarly to $\mathrm{CM}_n(\boldsymbol{d})$, $\mathrm{DCM}_n(\boldsymbol{d})$ can have self-loops as well as multiple edges. A self-loop arises at vertex i when one of its in-half-edges pairs to one of its out-half-edges. Let $(D_n^{(\mathrm{in})}, D_n^{(\mathrm{out})})$

denote the in- and out-degree of a vertex chosen uniformly at random from $[n]$. Exercise 7.29 investigates the limiting distribution of the number of self-loops in $\mathrm{DCM}_n(\boldsymbol{d})$.

Assume, similarly to Conditions 7.8(a)–(b), that $(D_n^{(\mathrm{in})}, D_n^{(\mathrm{out})}) \xrightarrow{d} (D^{(\mathrm{in})}, D^{(\mathrm{out})})$, and that $\mathbb{E}[D_n^{(\mathrm{in})}] \to \mathbb{E}[D^{(\mathrm{in})}]$ and $\mathbb{E}[D_n^{(\mathrm{out})}] \to \mathbb{E}[D^{(\mathrm{in})}]$. Naturally, by (7.8.1), this implies that $\mathbb{E}[D^{(\mathrm{out})}] = \mathbb{E}[D^{(\mathrm{in})}]$. The details of this argument are left as Exercise 7.30.

Let

$$p_{k,l} = \mathbb{P}(D^{(\mathrm{in})} = k, D^{(\mathrm{out})} = l) \tag{7.8.2}$$

denote the asymptotic joint in- and out-degree distribution. We refer to $(p_{k,l})_{k,l\geq 0}$ simply as the asymptotic degree distribution of $\mathrm{DCM}_n(\boldsymbol{d})$. The distribution $(p_{k,l})_{k,l\geq 0}$ plays a similar role for $\mathrm{DCM}_n(\boldsymbol{d})$ as $(p_k)_{k\geq 0}$ does for $\mathrm{CM}_n(\boldsymbol{d})$.

7.9 Notes and Discussion

Notes on Section 7.2

The configuration model has a long history. It was introduced by Bollobás (1980) to study uniform random regular graphs (see also Bollobás (2001, Section 2.4)). The introduction was inspired by, and generalized the results in, the work of Bender and Canfield (1978). The original work allowed for a careful computation of the number of regular graphs, using a probabilistic argument. This is the probabilistic method at its best, and also explains the emphasis on the study of the probability for the graph to be simple. The configuration model, as well as uniform random graphs with a prescribed degree sequence, were further studied in greater generality by Molloy and Reed (1995, 1998). Molloy and Reed were the first to investigate the setting where the degrees are different, and they focused on conditions for the resulting graph to have a giant component.

Regularity conditions on the degree sequence have appeared in many places. Condition 7.8 is closest in spirit to the regularity condition by Janson (2009) or in Bhamidi et al. (n.d.). Often, the conditions as stated here are strengthened and include bounds on the maximal degree. We have tried to avoid these as much as possible.

Notes on Section 7.3

The term erased configuration model is first used by Britton, Deijfen and Martin-Löf in Britton et al. (2006, Section 2.1).

Notes on Section 7.4

The term repeated configuration model is first used by Britton, Deijfen and Martin-Löf in Britton et al. (2006, Section 2.2). The result in Theorem 7.10 was proved by Janson (2009). A version of it with stronger assumptions on the degrees was already present in Bollobás classical random graphs book Bollobás (2001, Section 2.4). Janson revisits this problem in Janson (2014). In joint work with Angel and Holmgren (Preprint (2016)), we investigate the total variation distance between $S_n + M_n$ and the Poisson random variable with parameter $\nu_n/2 + \nu_n^2/4$. The main result is that this vanishes when Conditions 7.8(a)–(c) hold and we quantify the convergence. In particular, when $\mathbb{E}[D_n^3] \to \mathbb{E}[D^3] < \infty$, then this total variation distance is $O(1/n)$. Further, they prove that when $\mathbb{E}[D^2] = \infty$, S_n satisfies a central limit theorem with asymptotic mean and variance $\nu_n/2$. Similar results are proved for M_n, under the additional condition that the maximal degree $d_{\max}$ satisfies $d_{\max} = o(\sqrt{n})$. The proof relies on a Chen-Stein approximation for the Poisson distribution.

Notes on Section 7.5

Corollary 7.17 implies that the uniform simple random graph model is *contiguous* to the configuration model, in the sense that events with vanishing probability for the configuration model also have vanishing probability for the uniform simple random graph model with the same degree sequence. See Janson (2010) for an extended discussion of contiguity of random graphs. Theorem 7.18 implies that the generalized random graph conditioned on having degree sequence $\boldsymbol{d}$ is contiguous to the configuration model with that degree sequence, whenever the degree sequence satisfies Conditions 7.8(a)–(c).

The relations between different models as discussed in detail in Section 7.5 are folklore in the random graphs literature, though their proofs are scattered around the literature. The proof that the generalized random graph conditioned on its degrees yields a uniform simple graph with that degree sequence can be found in Britton et al. (2006, Section 2.1).

Notes on Section 7.6

A version of Theorem 7.20 can be found in Britton et al. (2006). Results on the erased configuration model as in Theorems 7.23–7.24 have appeared in Bhamidi et al. (2010), where first passage percolation on $\mathrm{CM}_n(\boldsymbol{d})$ was studied with infinite-mean degrees, both for the erased as well as for the original configuration model, and it is shown that the behavior in the two models is completely different.

Theorem 7.23 is closely related to what is sometimes called the *Bernoulli sieve*. Fix a probability distribution $(p_j)_{j\geq 1}$ on the integers. Draw n i.i.d. samples, and consider the number of different realizations. This can also be seen as drawing box j with probability p_j, and counting the number of boxes with at least one element after n independent trials. In Theorem 7.23, this is done with a *random* probability distribution denoted by $(Q_i)_{i\geq 1}$ as given in (7.6.7), and with a *random* number of draws given by the degree distribution D. The number of occupied boxes in n draws has received quite some attention, with Karlin (1967) proving a central limit theorem in the power-law case. See the survey Gnedin et al. (2007) for a more recent overview.

Notes on Section 7.7

Theorem 7.25 is folklore, even though it has, as far as we know, never appeared explicitly. Parts of it can be found in van der Hofstad and Litvak (2014). Theorem 7.27 is due to Gao and Wormald (2016), where many related results have been obtained.

Notes on Section 7.8

The configuration model with household structure was investigated by Trapman (2007). See also the works by Ball, Sirl and Trapman (2009; 2010) in the context of epidemics on social networks. Particularly when studying epidemics on networks, clustering is highly relevant, as clustering might slow down the spread of infectious diseases. See also the related models by Coupechoux and Lelarge (Preprint (2012)). The hierarchical configuration model where the communities can take other forms compared to complete graphs has been investigated in work with van Leeuwaarden and Stegehuis (Preprint (2015)).

The configuration model with clustering was introduced by Newman (2009) and independently by Miller (2009). Newman performs a generating function analysis to determine when a giant component is expected to exist, while Miller also investigates the effect of clustering on the epidemic threshold. The giant component in the directed configuration model was investigated by Cooper and Frieze (2004). The degree structure of the directed configuration model, as well as ways to simulate it, are discussed by Chen and Olvera-Cravioto (2013).

7.10 Exercises for Chapter 7

Exercise 7.1 (Non-graphical degree sequence) Find a simple example of a $(d_i)_{i\in[n]}$ satisfying that $\ell_n = \sum_{j\in[n]} d_j$ is even and for which no simple graph satisfying that vertex i has degree d_i exists.

Exercise 7.2 (The number of configurations) Prove that there are $(2m-1)!! = (2m-1)(2m-3)\cdots 3\cdot 1$ different ways of pairing vertices $1,\ldots,2m$.

Exercise 7.3 (Example of multigraph) Let $n = 2$, $d_1 = 2$ and $d_2 = 4$. Use the direct connection probabilities to show that the probability that $\mathrm{CM}_n(\boldsymbol{d})$ consists of 3 self-loops equals $1/5$. Hint: Note that when $d_1 = 2$ and $d_2 = 4$, the graph $\mathrm{CM}_n(\boldsymbol{d})$ consists only of self-loops *precisely* when the first half-edge of vertex 1 connects to the second half-edge of vertex 1.

Exercise 7.4 (Example of multigraph (cont.)) Let $n = 2$, $d_1 = 2$ and $d_2 = 4$. Use Proposition 7.7 to show that the probability that $\mathrm{CM}_n(\boldsymbol{d})$ consists of 3 self-loops equals $1/5$.

Exercise 7.5 (Weak convergence integer random variables) Let (D_n) be a sequence of integer random variables such that $D_n \xrightarrow{d} D$. Show that, for all $x \in \mathbb{R}$,

$$\lim_{n\to\infty} F_n(x) = F(x), \tag{7.10.1}$$

and that also $\lim_{n\to\infty} \mathbb{P}(D_n = x) = \mathbb{P}(D = x)$ for every $x \in \mathbb{N}$.

Exercise 7.6 (Regularity condition for configuration model moderated by F) Fix $\mathrm{CM}_n(\boldsymbol{d})$ be such that there are precisely $n_k = \lceil nF(k)\rceil - \lceil nF(k-1)\rceil$ vertices with degree k. Show that Condition 7.8(a) holds.

Exercise 7.7 (Regularity condition for configuration model moderated by F (cont.)) Fix $\mathrm{CM}_n(\boldsymbol{d})$ be such that there are precisely $n_k = \lceil nF(k)\rceil - \lceil nF(k-1)\rceil$ vertices with degree k. Show that Condition 7.8(b) holds whenever $\mathbb{E}[D] < \infty$.

Exercise 7.8 (Probability of i.i.d. sum to be odd) Assume that $(d_i)_{i\geq 1}$ is an i.i.d. sequence of random variables for which $\mathbb{P}(d_i \text{ even}) \neq 1$. Prove that $\ell_n = \sum_{i\in[n]} d_i$ is odd with probability close to $1/2$. For this, note that

$$\mathbb{P}(\ell_n \text{ is odd}) = \frac{1}{2}\Big[1 - \mathbb{E}[(-1)^{\ell_n}]\Big]. \tag{7.10.2}$$

Then compute

$$\mathbb{E}[(-1)^{\ell_n}] = \phi_{d_1}(\pi)^n, \tag{7.10.3}$$

where

$$\phi_{d_1}(t) = \mathbb{E}[\mathrm{e}^{itd_1}] \tag{7.10.4}$$

is the characteristic function of the degree d_1. Prove that, when $\mathbb{P}(d \text{ is even}) \in (0, 1)$, $|\phi_{d_1}(\pi)| < 1$, so that $\mathbb{P}(\ell_n \text{ is odd})$ is exponentially close to $\frac{1}{2}$.

Exercise 7.9 (Regularity condition for configuration model with i.i.d. degrees) Fix $\mathrm{CM}_n(\boldsymbol{d})$ with degrees $\boldsymbol{d}$ given by $(d_i)_{i\in[n]}$, where $(d_i)_{i\in[n-1]}$ is an i.i.d. sequence of integer-valued random variables and $d_n = d_n' + \mathbb{1}_{\{\ell_{n-1}+d_n' \text{ odd}\}}$, where d_n' has the same distribution as d_1 and is independent of $(d_i)_{i\in[n-1]}$. Show that Condition 7.8(a) holds, whereas Conditions 7.8(b) and (c) hold when $\mathbb{E}[D] < \infty$ and $\mathbb{E}[D^2] < \infty$, respectively. Here the convergence is replaced with convergence in probability as explained in Remark 7.9.

Exercise 7.10 (Mean degree sequence equals average degree) Prove that $p_k^{(n)}$ in (7.3.3) satisfies

$$\sum_{k=1}^{\infty} k p_k^{(n)} = \frac{1}{n}\sum_{i\in[n]} d_i = \frac{\ell_n}{n}. \tag{7.10.5}$$

Exercise 7.11 (M_n versus $\widetilde{M}_n$) Show that $\mathbb{P}(M_n < \widetilde{M}_n) = o(1)$ when Conditions 7.8(a)–(c) hold and conclude that Proposition 7.13 also holds for (S_n, M_n).

Exercise 7.12 (Multiple edges between truncated degrees in infinite-variance setting) Assume that Conditions 7.8(a)–(c) hold and let $\mathbb{E}[D^2] = \infty$. Let

$$M_n^{(K)} = \sum_{1\leq i<j\leq n} (X_{ij} - 1)\mathbb{1}_{\{X_{ij}\geq 2\}}\mathbb{1}_{\{d_i,d_j\leq K\}}. \tag{7.10.6}$$

Extend the proof of Proposition 7.14 and prove that $M_n^{(K)} \xrightarrow{d} M^{(K)}$, where $M^{(K)}$ is Poisson with parameter $\nu^{(K)} = \mathbb{E}[D(D-1)\mathbb{1}_{\{D\leq K\}}]/\mathbb{E}[D]$.

Exercise 7.13 (M_n tends to infinity in infinite-variance setting) Assume that Conditions 7.8(a)–(c) hold and let $\mathbb{E}[D^2] = \infty$. Prove that $\mathbb{P}(M_n^{(K)} \neq \widetilde{M}_n^{(K)}) = o(1)$, where $M_n^{(K)}$ is defined in (7.10.6), while

$$\widetilde{M}_n^{(K)} = \sum_{1\leq i<j\leq n} \binom{X_{ij}}{2}\mathbb{1}_{\{d_i,d_j\leq K\}}. \tag{7.10.7}$$

Conclude that $M_n \xrightarrow{\mathbb{P}} \infty$.

Exercise 7.14 (Characterization moments independent Poisson variables) Show that the moments of (X, Y), where (X, Y) are independent Poisson random variables with parameters μ_X and μ_Y, are identified by the relations, for $r \geq 1$,

$$\mathbb{E}[X^r] = \mu_X \mathbb{E}[(X+1)^{r-1}], \tag{7.10.8}$$

and, for $r, s \geq 1$,

$$\mathbb{E}[X^r Y^s] = \mu_Y \mathbb{E}[X^r (Y+1)^{s-1}]. \tag{7.10.9}$$

Exercise 7.15 (Alternative proof of Proposition 7.13) Assume that Conditions 7.8(a)–(c) hold. Show that all joint moments of $(S_n, \widetilde{M}_n)$ converge to those of (S, M), where S and M are two independent Poisson random variables with means $\nu/2$ and $\nu^2/4$. Use this to give an alternative proof of Proposition 7.13 by using Theorem 2.3(e) together with Exercise 7.14.

Exercise 7.16 (Average number of triangles CM) Compute the average number of occupied triangles T_n in $\mathrm{CM}_n(\boldsymbol{d})$, where

$$T_n = \sum_{1 \leq i < j < k \leq n} X_{ij} X_{jk} X_{ki}, \tag{7.10.10}$$

where X_{ij} is the number of edges between i and j.

Exercise 7.17 (Poisson limit triangles CM) Show that the number of occupied triangles in $\mathrm{CM}_n(\boldsymbol{d})$ converges to a Poisson random variable when Conditions 7.8(a)–(c) hold.

Exercise 7.18 (A conditioned uniform variable is again uniform) Let $(\Omega, \mathcal{F}, \mathbb{P})$ be a probability space. Let U be a random variable on this space being uniformly distributed on the finite set $\mathcal{X}$. Let $\varnothing \neq \mathcal{Y} \subseteq \mathcal{X}$. Show that the conditional probability $\mathbb{P}(U = z \mid U \in \mathcal{Y}) = 1/|\mathcal{Y}|$ for all $z \in \mathcal{Y}$.

Exercise 7.19 (Poisson limits for self-loops, multiple edges and triangles) Assume that the degree sequence $(d_i)_{i \in [n]}$ satisfies Conditions 7.8(a)–(c). Let T_n denote the number of triangles in $\mathrm{CM}_n(\boldsymbol{d})$, i.e., the number of $(i, s_i, t_i), (j, s_j, t_j), (k, s_k, t_k)$ such that $i < j < k$ and such that s_i is paired to t_j, s_j is paired to t_k and s_k is paired to t_i. Show that $(S_n, \widetilde{M}_n, T_n)$ converges to three independent Poisson random variables and compute their asymptotic parameters.

Exercise 7.20 (The number of r-regular graphs) Prove (7.5.4).

Exercise 7.21 (The number of simple graphs without triangles) Assume that the fixed degree sequence $(d_i)_{i \in [n]}$ satisfies Conditions 7.8(a)–(c). Compute the number of simple graphs with degree sequence $(d_i)_{i \in [n]}$ not containing any triangle. Hint: use Exercise 7.19.

Exercise 7.22 (Proof of Theorem 7.22) Adapt the proof of Theorem 7.20 to prove Theorem 7.22.

Exercise 7.23 (Proof that the number of erased edges vanishes compared to total) Prove that Theorem 7.10 implies that $E_n = o_{\mathbb{P}}(n)$, where E_n denotes the number of erased edges.

Exercise 7.24 (Proof tightness of number of erased edges) Assume that the degree sequence $(d_i)_{i\in[n]}$ satisfies Conditions 7.8(a)–(c). Let E_n denote the number of erased edges. Show that $E_n \xrightarrow{d} Y$, where $Y \sim \mathsf{Poi}(\nu/2 + \nu^2/4)$.

Exercise 7.25 (Clustering coefficient in CM with infinite variance degrees) Assume that the degree sequence $(d_i)_{i\in[n]}$ satisfies Conditions 7.8(a)–(b), and assume that $d_1, d_2, d_3 \geq \varepsilon n^{1/(\tau-1)}$ for some $\tau \in (2,3)$. Let X_{ij} denote the number of edges in $\mathrm{CM}_n(\boldsymbol{d})$ between $i, j \in [n]$. Let $X_{12}X_{23}X_{13}$ denote the number of triangles between vertices 1, 2, 3. Show that

$$\frac{1}{n} X_{12}X_{23}X_{13} \xrightarrow{\mathbb{P}} \infty. \tag{7.10.11}$$

Conclude that the clustering coefficient as defined in (1.5.4) converges in probability to infinity. This seems unphysical. Argue what goes wrong here.

Exercise 7.26 (Friends-of-friends in $\mathrm{CM}_n(\boldsymbol{d})$) Assume that Conditions 7.8(a)–(b) hold. Prove that the number of friends-of-friends of a random vertex in $[n]$ converges in distribution as stated in (7.7.7).

Exercise 7.27 (Clustering in hierarchical configuration model with complete graphs) Assume that Conditions 7.8(a)–(c) hold. Consider the hierarchical configuration model where each subgraph G_i is a complete graph K_{d_i} and each of its vertices has precisely one half-edge. The half-edges are paired as in $\mathrm{CM}_n(\boldsymbol{d})$. Compute the clustering coefficient for this graph.

Exercise 7.28 (Clustering in configuration model with clustering) Consider the configuration model with clustering in the case where $(D_n^{(\mathrm{si})}, D_n^{(\mathrm{tr})}) \xrightarrow{d} (D^{(\mathrm{si})}, D^{(\mathrm{tr})})$ and $\mathbb{E}[(D_n^{(\mathrm{si})})^2] \to \mathbb{E}[(D^{(\mathrm{si})})^2]$, $\mathbb{E}[(D_n^{(\mathrm{tr})})^2] \to \mathbb{E}[(D^{(\mathrm{tr})})^2]$. Compute the clustering coefficient for this graph.

Exercise 7.29 (Self-loops and multiple edges in $\mathrm{DCM}_n(\boldsymbol{d})$) Adapt the proof of Proposition 7.13 to the directed configuration model. More precisely, show that the number of self-loops in $\mathrm{DCM}_n(\boldsymbol{d})$ converges to a Poisson random variable with parameter $\mathbb{E}[D^{(\mathrm{in})}D^{(\mathrm{out})}]/\mathbb{E}[D^{(\mathrm{in})}]$ when $(D_n^{(\mathrm{in})}, D_n^{(\mathrm{out})}) \xrightarrow{d} (D^{(\mathrm{in})}, D^{(\mathrm{out})})$ and

$$\mathbb{E}[D_n^{(\mathrm{in})}D_n^{(\mathrm{out})}] \to \mathbb{E}[D^{(\mathrm{in})}D^{(\mathrm{out})}]. \tag{7.10.12}$$

What can you say about the number of multiple edges in $\mathrm{DCM}_n(\boldsymbol{d})$?

Exercise 7.30 (Equivalence of convergence in- and out-degree in $\mathrm{DCM}_n(\boldsymbol{d})$) Show that (7.8.1) implies that $\mathbb{E}[D^{(\mathrm{out})}] = \mathbb{E}[D^{(\mathrm{in})}]$ when $(D_n^{(\mathrm{in})}, D_n^{(\mathrm{out})}) \xrightarrow{d} (D^{(\mathrm{in})}, D^{(\mathrm{out})})$, $\mathbb{E}[D_n^{(\mathrm{in})}] \to \mathbb{E}[D^{(\mathrm{in})}]$ and $\mathbb{E}[D_n^{(\mathrm{out})}] \to \mathbb{E}[D^{(\mathrm{in})}]$.

8

Preferential Attachment Models

Most networks grow in time. Preferential attachment models describe growing networks, where the numbers of edges and vertices grow linearly with time.

In preferential attachment models, vertices having a fixed number of edges are sequentially added to the network. Given the graph at time t, the edges incident to the vertex with label $t+1$ are attached to older vertices that are chosen according to a probability distribution that is an affine function of the degrees of the older vertices. This way, vertices that already have a high degree are more likely to attract edges of later vertices, which explains why such models are called 'rich-get-richer' models. In this chapter, we introduce and investigate such models, focusing on the degree structure of preferential attachment models.

We show how the degrees of the old vertices grow with time, and how the proportion of vertices with a fixed degree behaves as time grows. Since the vertices with the largest degrees turn out to be the earliest vertices in the network, preferential attachment models could also be called 'old-get-rich' models.

8.1 Motivation for the Preferential Attachment Model

The generalized random graph model and the configuration model described in Chapters 6 and 7, respectively, are *static* models, i.e., the size of the graph is *fixed*, and we have not modeled the *growth* of the graph. There is a large body of work investigating *dynamic* models for complex networks, often in the context of the World-Wide Web, but also for citation or biological networks. In various forms, such models have been shown to lead to power-law degree sequences. Thus, they offer a possible *explanation* for the occurrence of power-law degree sequences in real-world networks. The existence of power-law degree sequences in various real-world networks is quite striking, and models offering a convincing explanation can teach us about the mechanisms that give rise to their scale-free nature. Let us illustrate this with the example of citation networks:

Example 8.1 (Citation networks) In citation networks, the vertices of the network are scientific papers, and a directed edge between two papers represents a reference of the first paper to the second. Thus, the in-degree of a paper is its number of citations, while the out-degree is its number of references. Figure 8.1 displays the in-degree distribution, in log-log scale, of the citation network of probability and statistics papers in the period 1980 to May 2015 from Web of Science. There are 178,474 papers in this data base, and 992,833 citations, so that the average number of citations equals 5.5. The average number of real references

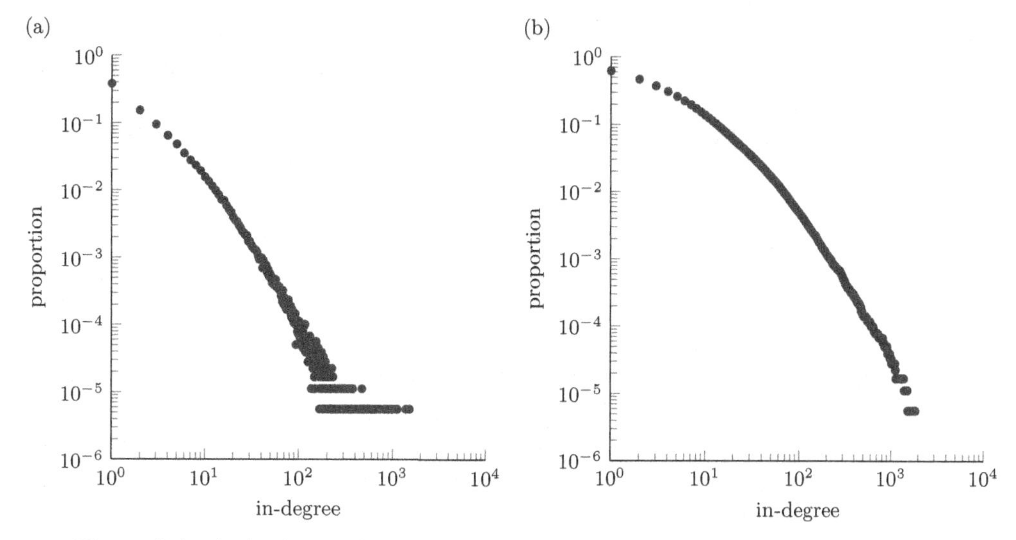

Figure 8.1 The in-degree distribution for the citation network of papers in probability and statistics in Web of Science. (a) Log-log plot of probability mass function. (b) Log-log plot of complementary cumulative distribution function. Data courtesy of Alessandro Garavaglia.

per paper is 15, which is larger since many references are to papers that are not part of the Probability and Statistics Web of Science data base.

We see that the in-degree distribution resembles a power law. In citation networks, papers have a time stamp, which is their time of publication. Citation networks are growing networks, since references remain present forever once they appear. Further, citation networks are to a large extent acyclic, meaning that there are few directed cycles, since references tend to go back in time making it, e.g., less likely that pairs of papers refer to each other. In this chapter, we are interested in possible explanations for the occurrence of power laws, and we aim to explain those by formulating simple rules for the growth of the network.

A possible explanation for the occurrence of power-law degree sequences is offered by the *preferential attachment paradigm*. In preferential attachment models, vertices are added sequentially with a number of edges connected to them. These edges are attached to a receiving vertex with a probability proportional to the degree of the receiving vertex at that time, thus favoring vertices with large degrees. For this model, it is shown that the number of vertices with degree k decays proportionally to k^{-3} (see Bollobás et al. (2001)), and this result is a special case of the more general result on the asymptotic degree sequence of preferential attachment models that we prove in this chapter.

The idea behind preferential attachment models is simple. In a graph that evolves in time, the newly added vertices are connected to the already existing ones. In an Erdős–Rényi random graph, which can also be formulated as an evolving graph where edges are added and removed, these edges would be connected to each individual with equal probability, and we would have to remove edges so as to make the edge probabilities equal (see Exercise 8.1).

Now think of the newly added vertex as a new individual in a social population, which we model as a graph by letting the individuals be the vertices and the edges be the acquaintance relations between pairs of individuals. Is it then realistic that the edges connect to each already-present individual with equal probability, or is the newcomer more likely to get to know socially active individuals, who already know many people? If the latter is true, then we should forget about equal probabilities for the receiving ends of the edges of the newcomer, and introduce a bias in his/her connections towards more social individuals. Phrased in a mathematical way, it should be more likely that the edges are connected to vertices that already have a high degree. A possible model for such a growing graph was proposed by Barabási and Albert (1999), and has incited an enormous research effort since.

Strictly speaking, Barabási and Albert (1999) were not the first to propose such a model, and we start by referring to the old literature on the subject. Yule (1925) was the first to propose a growing model where preferential attachment is present, in the context of the evolution of species. He derives the power-law distribution that we also find in this chapter. Simon (1955) provides a more modern version of the preferential attachment model, as he puts it,

> Because Yule's paper predates the modern theory of stochastic processes, his derivation was necessarily more involved than the one we shall employ here.

The stochastic model of Simon is formulated in the context of the occurrence of words in large pieces of text (as in Zipf's law (Zipf, 1929)). It is based on two assumptions, namely (i) that the probability that the $(k+1)$st word is a word that has already appeared exactly i times is proportional to the number of occurrences of words that have occurred exactly i times, and (ii) that there is a constant probability that the $(k+1)$st word is a new word. Together, these two assumptions give rise to frequency distributions of words that obey a power law, with a power-law exponent that is a simple function of the probability of adding a new word. We shall see a similar effect occurring in this chapter. A second place where the model studied by Simon and Yule can be found is in work by Champernowne (1953), in the context of income distributions in populations.

Barabási and Albert (1999) describe the preferential attachment graph informally as follows:

> To incorporate the growing character of the network, starting with a small number (m_0) of vertices, at every time step we add a new vertex with $m(\leq m_0)$ edges that link the new vertex to m different vertices already present in the system. To incorporate preferential attachment, we assume that the probability Π that a new vertex will be connected to a vertex i depends on the connectivity k_i of that vertex, so that $\Pi(k_i) = k_i / \sum_j k_j$. After t time steps, the model leads to a random network with $t + m_0$ vertices and mt edges.

This description of the model is informal, but Barabási and Albert (1999) must have given it a precise meaning (since, in particular, Barabási and Albert present simulations of the model predicting a power-law degree sequence with exponent close to $\tau = 3$). The model description does not explain how the first edge is connected (note that at time $t = 1$, there are no edges, so the first edge cannot be attached according to the degrees of the existing vertices), and does not give the dependencies between the m edges added at time t. We are left wondering whether these edges are independent, whether we allow for self-loops, whether we should update the degrees after each attachment of a single edge, etc. In fact, each of these choices has, by now, been considered in the literature. The results, in particular the occurrence of power laws and the power-law exponent, do not depend sensitively on the respective choices. See Section 8.9 for an extensive overview of the literature on preferential attachment models.

The first to investigate the model rigorously were Bollobás, Riordan, Spencer and Tusnády (2001). They complain heavily about the lack of a formal definition of Barabási and Albert (1999), arguing that

> The description of the random graph process quoted above (i.e, in Barabási and Albert (1999), edt.) is rather imprecise. First, as the degrees are initially zero, it is not clear how the process is started. More seriously, the expected number of edges linking a new vertex v to earlier vertices is $\sum_i \Pi(k_i) = 1$, rather than m. Also, when choosing in one go a set S of m earlier vertices as the neighbors of v, the distribution of S is *not* specified by giving the marginal probability that each vertex lies in S.

One could say that these differences in formulations form the heart of much confusion between mathematicians and theoretical physicists. To resolve these problems, choices had to be made, and these choices were, according to Bollobás et al. (2001), made first by Bollobás and Riordan (2004). Bollobás and Riordan (2004) specify the initial graph to consist of a vertex with m self-loops, while the degrees are updated in the process of attaching the m edges. This model will be described in full detail in Section 8.2 below.

Organization of this Chapter

This chapter is organized as follows. In Section 8.2, we give a formal introduction of the model. In Section 8.3, we investigate how the degrees of fixed vertices evolve as the graph grows. In Section 8.4, we investigate the degree sequences in preferential attachment models. The main result is Theorem 8.3, which states that the asymptotic degree distribution in the preferential attachment model is a power-law distribution. The proof of Theorem 8.3 consists of two key steps, which are formulated and proved in Sections 8.5 and 8.6, respectively. In Section 8.7 we investigate the maximal degree in a preferential attachment model. In Section 8.8 we discuss some further results on preferential attachment models proved in the literature, and in Section 8.9 we discuss many related preferential attachment models. We close this chapter with notes and discussion in Section 8.10 and exercises in Section 8.11.

8.2 Introduction of the Model

We start by introducing the model. The model we investigate produces a *graph sequence* that we denote by $(\mathrm{PA}_t^{(m,\delta)})_{t\geq 1}$ and which, for every time t, yields a graph of t vertices and mt edges for some $m = 1, 2, \ldots$ We start by defining the model for $m = 1$ when the graph consists of a collection of trees. In this case, $\mathrm{PA}_1^{(1,\delta)}$ consists of a single vertex with a single self-loop. We denote the vertices of $\mathrm{PA}_t^{(1,\delta)}$ by $v_1^{(1)}, \ldots, v_t^{(1)}$. We denote the degree of vertex $v_i^{(1)}$ in $\mathrm{PA}_t^{(1,\delta)}$ by $D_i(t)$, where, by convention, a self-loop increases the degree by 2.

We next describe the evolution of the graph. Conditionally on $\mathrm{PA}_t^{(1,\delta)}$, the growth rule to obtain $\mathrm{PA}_{t+1}^{(1,\delta)}$ is as follows. We add a single vertex $v_{t+1}^{(1)}$ having a single edge. This edge is connected to a second end point, which is equal to $v_{t+1}^{(1)}$ with probability $(1+\delta)/(t(2+\delta)+$

$(1+\delta))$, and to vertex $v_i^{(1)} \in \mathrm{PA}_t^{(1,\delta)}$ with probability $(D_i(t)+\delta)/(t(2+\delta)+(1+\delta))$ for each $i \in [t]$, where $\delta \geq -1$ is a parameter of the model. Thus,

$$\mathbb{P}\big(v_{t+1}^{(1)} \to v_i^{(1)} \mid \mathrm{PA}_t^{(1,\delta)}\big) = \begin{cases} \dfrac{1+\delta}{t(2+\delta)+(1+\delta)} & \text{for } i = t+1, \\ \dfrac{D_i(t)+\delta}{t(2+\delta)+(1+\delta)} & \text{for } i \in [t]. \end{cases} \tag{8.2.1}$$

The above preferential attachment mechanism is called *affine*, since the attachment probabilities in (8.2.1) depend in an affine way on the degrees of the random graph $\mathrm{PA}_t^{(1,\delta)}$. Exercises 8.2 and 8.3 show that (8.2.1) is a probability distribution.

The model with $m > 1$ is defined in terms of the model for $m = 1$ as follows. Fix $\delta \geq -m$. We start with $\mathrm{PA}_{mt}^{(1,\delta/m)}$, and denote the vertices in $\mathrm{PA}_{mt}^{(1,\delta/m)}$ by $v_1^{(1)}, \ldots, v_{mt}^{(1)}$. Then we identify or collapse the m vertices $v_1^{(1)}, \ldots, v_m^{(1)}$ in $\mathrm{PA}_{mt}^{(1,\delta/m)}$ to become the vertex $v_1^{(m)}$ in $\mathrm{PA}_t^{(m,\delta)}$. In doing so, we let all the edges that are incident to any of the vertices in $v_1^{(1)}, \ldots, v_m^{(1)}$ be incident to the new vertex $v_1^{(m)}$ in $\mathrm{PA}_t^{(m,\delta)}$. Then, we collapse the m vertices $v_{m+1}^{(1)}, \ldots, v_{2m}^{(1)}$ in $\mathrm{PA}_{mt}^{(1,\delta/m)}$ to become vertex $v_2^{(m)}$ in $\mathrm{PA}_t^{(m,\delta)}$, etc. More generally, we collapse the m vertices $v_{(j-1)m+1}^{(1)}, \ldots, v_{jm}^{(1)}$ in $\mathrm{PA}_{mt}^{(1,\delta/m)}$ to become vertex $v_j^{(m)}$ in $\mathrm{PA}_t^{(m,\delta)}$. This defines the model for general $m \geq 1$. The resulting graph $\mathrm{PA}_t^{(m,\delta)}$ is a multigraph with precisely t vertices and mt edges, so that the total degree is equal to $2mt$ (see Exercise 8.4).

To explain the description of $\mathrm{PA}_t^{(m,\delta)}$ in terms of $\mathrm{PA}_{mt}^{(1,\delta/m)}$ by collapsing vertices, we note that an edge in $\mathrm{PA}_{mt}^{(1,\delta/m)}$ is attached to vertex $v_k^{(1)}$ with probability proportional to the weight of vertex $v_k^{(1)}$. Here the weight of $v_k^{(1)}$ is equal to the degree of vertex $v_k^{(1)}$ plus δ/m. Now, the vertices $v_{(j-1)m+1}^{(1)}, \ldots, v_{jm}^{(1)}$ in $\mathrm{PA}_{mt}^{(1,\delta/m)}$ are collapsed to form vertex $v_j^{(m)}$ in $\mathrm{PA}_t^{(m,\delta)}$. Thus, an edge in $\mathrm{PA}_t^{(m,\delta)}$ is attached to vertex $v_j^{(m)}$ with probability proportional to the *total weight* of the m vertices $v_{(j-1)m+1}^{(1)}, \ldots, v_{jm}^{(1)}$. Since the sum of the degrees of the m vertices $v_{(j-1)m+1}^{(1)}, \ldots, v_{jm}^{(1)}$ is equal to the degree of vertex $v_j^{(m)}$, this probability is proportional to the degree of vertex $v_j^{(m)}$ in $\mathrm{PA}_t^{(m,\delta)}$ plus δ. We note that in the above construction and for $m \geq 2$, the degrees are updated after each edge is attached. This is what we refer to as *intermediate updating of the degrees.*

The important feature of the model is that newly appearing edges are more likely to connect to vertices with large degrees, thus making these degrees even larger. This effect is called *preferential attachment*. Preferential attachment may explain the *existence* of vertices with quite large degrees. Therefore, the preferential attachment model is sometimes called the *Rich-get-Richer model.* The preferential attachment mechanism is quite reasonable in many real-world networks. For example, one is more likely to get to know a person who already knows many people. However, the precise form of preferential attachment in (8.2.1) is only one of the many possible examples. Similarly, a paper is more likely to be cited by other papers when it has already received many citations. Having said this, it is not obvious *why* the preferential attachment rule should be affine. This turns out to be related to the degree-structure of the resulting random graphs $\mathrm{PA}_t^{(m,\delta)}$ that we investigate in detail in this chapter.

The definition of $(\mathrm{PA}_t^{(m,\delta)})_{t\geq 1}$ in terms of $(\mathrm{PA}_t^{(1,\delta/m)})_{t\geq 1}$ is quite convenient. However, we can also equivalently define the model for $m \geq 2$ directly. We start with $\mathrm{PA}_t^{(1,\delta)}$ consisting of a single vertex with m self-loops. To construct $\mathrm{PA}_{t+1}^{(m,\delta)}$ from $\mathrm{PA}_{t+1}^{(m,\delta)}$, we add a single vertex with m edges attached to it. These edges are attached *sequentially* with intermediate updating of the degrees as follows. The eth edge is connected to vertex $v_i^{(m)}$, for $i \in [n]$, with probability proportional to $D_i(e-1,t)+\delta$, where, for $e = 1,\ldots,m$, $D_i(e,t)$ is the degree of vertex i after the eth edge incident to vertex $v_{t+1}^{(m)}$ is attached, and to vertex $v_{t+1}^{(m)}$ with probability proportional to $D_{t+1}(e-1,t)+1+e\delta/m$. Here we make the convention that $D_{t+1}(0,t) = 0$. This alternative definition makes it perfectly clear how the choices missing in Barabási and Albert (1999) are made. Indeed, the degrees are updated during the process of attaching the edges, and the initial graph at time 1 consists of a single vertex with m self-loops. Naturally, the edges could also be attached sequentially by a different rule, for example by attaching the edges independently according to the distribution for the first edge. Also, one has the choice to allow for self-loops or not. See Figure 8.2 for a realization of $(\mathrm{PA}_t^{(m,\delta)})_{t\geq 1}$ for $m = 2$ and $\delta = 0$, and Figure 8.3 for a realization of $(\mathrm{PA}_t^{(m,\delta)})_{t\geq 1}$ for $m = 2$ and $\delta = -1$. Exercise 8.5 investigates related growth rules.

The above model is a slight variation of models that have appeared in the literature. The model with $\delta = 0$ is the *Barabási–Albert model*, which has received substantial attention in the literature and was first formally defined by Bollobás and Riordan (2004). We have added the extra parameter δ to make the model more general. In the literature, also other slight variations on the above model have been considered. We discuss two of those. In the

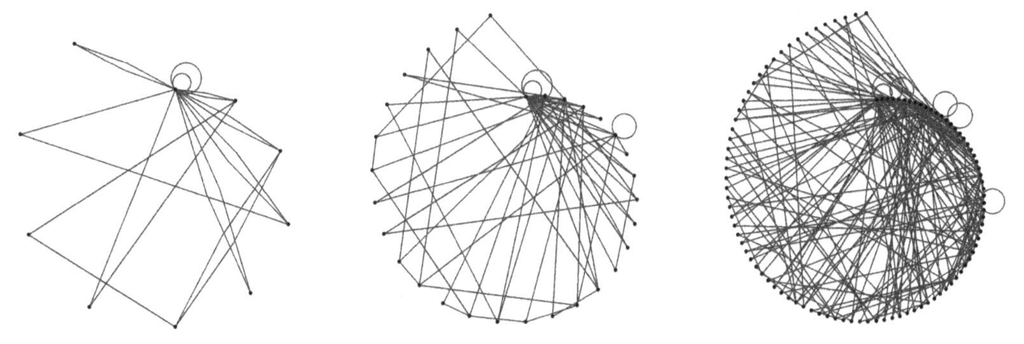

Figure 8.2 Preferential attachment random graph with $m = 2$ and $\delta = 0$ of sizes 10, 30 and 100.

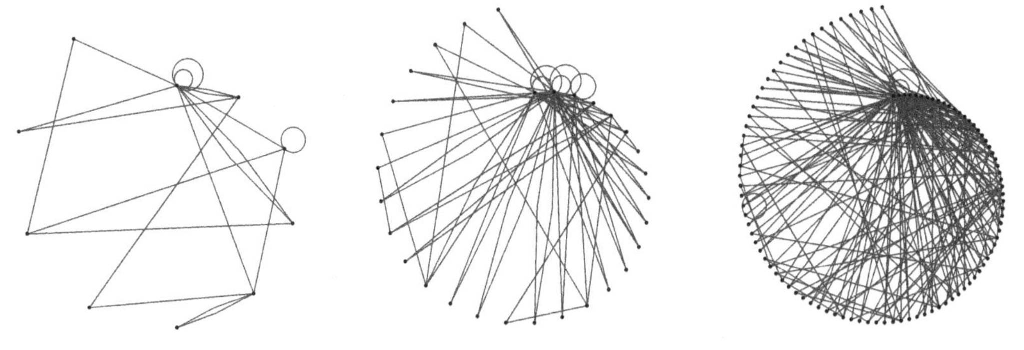

Figure 8.3 Preferential attachment random graph with $m = 2$ and $\delta = -1$ of sizes 10, 30 and 100.

first, and for $m = 1$ and $\delta \geq -1$, self-loops do not occur. We denote this variation by $\big(\mathrm{PA}_t^{(m,\delta)}(b)\big)_{t\geq 2}$ and sometimes refer to this model by *model (b)*. To define $\mathrm{PA}_t^{(1,\delta)}(b)$, we let $\mathrm{PA}_2^{(1,\delta)}(b)$ consist of two vertices $v_1^{(1)}$ and $v_2^{(1)}$ with two edges between them, and we replace the growth rule in (8.2.1) by the rule that, for all $i \in [t]$ and $t \geq 2$,

$$\mathbb{P}\big(v_{t+1}^{(1)} \to v_i^{(1)} \mid \mathrm{PA}_t^{(1,\delta)}(b)\big) = \frac{D_i(t)+\delta}{t(2+\delta)}. \tag{8.2.2}$$

The advantage of this model is that it leads to a *connected graph*. We again define the model with $m \geq 2$ and $\delta \geq -m$ in terms of $\big(\mathrm{PA}_{mt}^{(1,\delta/m)}(b)\big)_{t\geq 2}$ as below (8.2.1). We also note that the differences between $\big(\mathrm{PA}_t^{(m,\delta)}\big)_{t\geq 1}$ and $\big(\mathrm{PA}_t^{(m,\delta)}(b)\big)_{t\geq 2}$ are minor, since the probability of a self-loop in $\mathrm{PA}_t^{(m,\delta)}$ is quite small when t is large. Thus, most of the results we shall prove in this chapter for $\big(\mathrm{PA}_t^{(m,\delta)}\big)_{t\geq 1}$ shall also apply to $\big(\mathrm{PA}_t^{(m,\delta)}(b)\big)_{t\geq 2}$, but we do not state these extensions explicitly.

Interestingly, the above model with $\delta \geq 0$ can be viewed as an interpolation between the models with $\delta = 0$ and $\delta = \infty$. We show this for $m = 1$, the statement for $m \geq 2$ can again be seen by collapsing blocks of m vertices. We again let the graph at time 2 consist of two vertices with two edges between them. We fix $\alpha \in [0, 1]$. Then, we first draw a Bernoulli random variable I_{t+1} with success probability $1-\alpha$. The random variables $(I_t)_{t\geq 1}$ are independent. When $I_{t+1} = 0$, then we attach the $(t+1)$st edge to a *uniform* vertex in $[t]$. When $I_{t+1} = 1$, then we attach the $(t+1)$st edge to vertex $i \in [t]$ with probability $D_i(t)/(2t)$. We denote this model by $\big(\mathrm{PA}_t^{(1,\alpha)}(c)\big)_{t\geq 1}$. When $\alpha \geq 0$ is chosen appropriately, then this is precisely the above preferential attachment model (see also Exercise 8.7).

8.3 Degrees of Fixed Vertices

We will be interested in the degree structure in the preferential attachment model. We start by investigating the degrees of fixed vertices as $t \to \infty$, i.e., we study $D_i(t)$ for fixed i as $t \to \infty$. To formulate our results, we define the *Gamma-function* $t \mapsto \Gamma(t)$ for $t > 0$ by

$$\Gamma(t) = \int_0^\infty x^{t-1} \mathrm{e}^{-x} dx. \tag{8.3.1}$$

We also make use of the recursion formula (see e.g. Exercise 8.10)

$$\Gamma(t+1) = t\Gamma(t). \tag{8.3.2}$$

The main result in this section is the following result about the growth of the degrees of vertices:

Theorem 8.2 (Degrees of fixed vertices) *Fix $m = 1$ and $\delta > -1$. Then, $D_i(t)/t^{1/(2+\delta)}$ converges almost surely to a random variable ξ_i as $t \to \infty$, and*

$$\mathbb{E}[D_i(t)+\delta] = (1+\delta)\frac{\Gamma(t+1)\Gamma(i-1/(2+\delta))}{\Gamma(t+\frac{1+\delta}{2+\delta})\Gamma(i)}. \tag{8.3.3}$$

In Section 8.7, we considerably extend the result in Theorem 8.2. For example, we also prove the almost-sure convergence of the *maximal* degree.

Proof Fix $m = 1$ and let $t \geq i$. We compute that

$$\begin{aligned}
\mathbb{E}[D_i(t+1)+\delta \mid D_i(t)] &= D_i(t)+\delta+\mathbb{E}[D_i(t+1)-D_i(t) \mid D_i(t)] \\
&= D_i(t)+\delta+\frac{D_i(t)+\delta}{(2+\delta)t+1+\delta} \\
&= (D_i(t)+\delta)\frac{(2+\delta)t+2+\delta}{(2+\delta)t+1+\delta} \\
&= (D_i(t)+\delta)\frac{(2+\delta)(t+1)}{(2+\delta)t+1+\delta}. \qquad (8.3.4)
\end{aligned}$$

(In fact, here we rely on Exercise 8.9.) Using also that

$$\begin{aligned}
\mathbb{E}[D_i(i)+\delta] &= 1+\delta+\frac{1+\delta}{(2+\delta)(i-1)+1+\delta} \\
&= (1+\delta)\frac{(2+\delta)(i-1)+2+\delta}{(2+\delta)(i-1)+1+\delta} \\
&= (1+\delta)\frac{(2+\delta)i}{(2+\delta)(i-1)+1+\delta}, \qquad (8.3.5)
\end{aligned}$$

we obtain (8.3.3). In turn, again using (8.3.4), the sequence $(M_i(t))_{t\geq i}$ given by

$$M_i(t) = \frac{D_i(t)+\delta}{1+\delta}\prod_{s=i-1}^{t-1}\frac{(2+\delta)s+1+\delta}{(2+\delta)(s+1)} \qquad (8.3.6)$$

is a non-negative martingale with mean 1. As a consequence of the martingale convergence theorem (Theorem 2.24), $M_i(t)$ converges almost surely to a limiting random variable M_i as $t \to \infty$.

We next extend this result to almost-sure convergence of $D_i(t)/t^{1/(2+\delta)}$. For this, we compute that

$$\prod_{s=i-1}^{t-1}\frac{(2+\delta)s+1+\delta}{(2+\delta)s+2+\delta} = \prod_{s=i-1}^{t-1}\frac{s+\frac{1+\delta}{2+\delta}}{s+1} = \frac{\Gamma(t+\frac{1+\delta}{2+\delta})\Gamma(i)}{\Gamma(t+1)\Gamma(i-1/(2+\delta))}, \qquad (8.3.7)$$

so that

$$M_i(t) = \frac{D_i(t)+\delta}{1+\delta}\frac{\Gamma(t+\frac{1+\delta}{2+\delta})\Gamma(i)}{\Gamma(t+1)\Gamma(i-1/(2+\delta))}. \qquad (8.3.8)$$

Using Stirling's formula, it is not hard to see that (see Exercise 8.11) as $t \to \infty$ and for a fixed,

$$\frac{\Gamma(t+a)}{\Gamma(t)} = t^a(1+O(1/t)). \qquad (8.3.9)$$

Therefore,

$$\begin{aligned}
\frac{D_i(t)+\delta}{t^{1/(2+\delta)}} &= M_i(t)\frac{(2+\delta)\Gamma(i-1/(2+\delta))}{\Gamma(i)}(1+O(1/t)) \qquad (8.3.10) \\
&\xrightarrow{a.s.} M_i\frac{(2+\delta)\Gamma(i-1/(2+\delta))}{\Gamma(i)} \equiv \xi_i.
\end{aligned}$$

Since $1/(2+\delta) > 0$, the same conclusion follows for $t^{-1/(2+\delta)} D_i(t)$. In particular, the degrees of the first i vertices at time t is at most of order $t^{1/(2+\delta)}$. Note, however, that we do not yet know whether $\mathbb{P}(\xi_i = 0) = 0$ or not! □

We can extend the above result to the case where $m \geq 1$ by using the relation between $\mathrm{PA}_t^{(m,\delta)}$ and $\mathrm{PA}_t^{(1,\delta/m)}$. This relation in particular implies that

$$\mathbb{E}_m^{\delta}[D_i(t)] = \sum_{s=1}^{m} \mathbb{E}_1^{\delta/m}[D_{m(i-1)+s}(mt)], \tag{8.3.11}$$

where we have added a subscript m and a superscript δ to the expectation to denote the values of m and δ involved. Exercises 8.12 and 8.13 investigate $m \geq 2$ in more detail, while Exercise 8.14 studies extensions to model (b).

We close this section by giving a heuristic explanation for the occurrence of a power-law degree sequence in preferential attachment models. Theorem 8.2 (in conjunction with Exercise 8.13) implies that there exists an $a_{m,\delta}$ such that, for i, t large, and any $m \geq 1$,

$$\mathbb{E}_m^{\delta}[D_i(t)] \sim a_{m,\delta}\big(t/i\big)^{1/(2+\delta/m)}. \tag{8.3.12}$$

When the graph indeed has a power-law degree sequence, then the number of vertices with degree at least k will be close to $ctk^{-(\tau-1)}$ for some $\tau > 1$ and some $c > 0$. The number of vertices with degree at least k at time t is equal to $N_{\geq k}(t) = \sum_{i=1}^{t} \mathbb{1}_{\{D_i(t)\geq k\}}$. Now, assume that in the above formula, we are allowed to replace $\mathbb{1}_{\{D_i(t)\geq k\}}$ by $\mathbb{1}_{\{\mathbb{E}_m^{\delta}[D_i(t)]\geq k\}}$ (that is a big leap of faith, and the actual proof will be quite different as well as considerably more involved). Then we would obtain that

$$\begin{aligned} N_{\geq k}(t) &\sim \sum_{i=1}^{t} \mathbb{1}_{\{\mathbb{E}_m^{\delta}[D_i(t)]\geq k\}} \sim \sum_{i=1}^{t} \mathbb{1}_{\{a_{m,\delta}(t/i)^{1/(2+\delta/m)}\geq k\}} \\ &= \sum_{i=1}^{t} \mathbb{1}_{\{i \leq t a_{m,\delta}^{2+\delta/m} k^{-(2+\delta/m)}\}} = t a_{m,\delta}^{2+\delta/m} k^{-(2+\delta/m)}, \end{aligned} \tag{8.3.13}$$

so that we obtain a power-law with exponent $\tau - 1 = 2 + \delta/m$. This suggests that the preferential attachment model has a power-law degree sequence with exponent τ satisfying $\tau = 3 + \delta/m$. The above heuristic is made precise in the following section, but the proof is quite a bit more subtle than the above heuristic!

8.4 Degree Sequences of Preferential Attachment Models

The main result in this section establishes the scale-free nature of preferential attachment graphs. In order to state it, we need some notation. We write

$$P_k(t) = \frac{1}{t}\sum_{i=1}^{t} \mathbb{1}_{\{D_i(t)=k\}} \tag{8.4.1}$$

for the (random) proportion of vertices with degree k at time t. For $m \geq 1$ and $\delta > -m$, we define $(p_k)_{k\geq 0}$ by $p_k = 0$ for $k = 0, \ldots, m-1$ and, for $k \geq m$,

$$p_k = (2 + \delta/m)\frac{\Gamma(k+\delta)\Gamma(m+2+\delta+\delta/m)}{\Gamma(m+\delta)\Gamma(k+3+\delta+\delta/m)}. \tag{8.4.2}$$

Below, we shall show that $(p_k)_{k\geq 0}$ is a probability mass function. For $m = 1$, (8.4.2) reduces to

$$p_k = (2+\delta)\frac{\Gamma(k+\delta)\Gamma(3+2\delta)}{\Gamma(k+3+2\delta)\Gamma(1+\delta)}. \tag{8.4.3}$$

Also, when $\delta = 0$ and $k \geq m$, (8.4.2) simplifies to

$$p_k = \frac{2\Gamma(k)\Gamma(m+2)}{\Gamma(k+3)\Gamma(m)} = \frac{2m(m+1)}{k(k+1)(k+2)}. \tag{8.4.4}$$

The probability mass function $(p_k)_{k\geq 0}$ arises as the limiting degree distribution for $\mathrm{PA}_t^{(m,\delta)}$, as shown in the following theorem:

Theorem 8.3 (Degree sequence in preferential attachment model) *Fix $m \geq 1$ and $\delta > -m$. There exists a constant $C = C(m, \delta) > 0$ such that, as $t \to \infty$,*

$$\mathbb{P}\Big(\max_k |P_k(t) - p_k| \geq C\sqrt{\frac{\log t}{t}}\Big) = o(1). \tag{8.4.5}$$

Theorem 8.3 proves that $P_k(t) \xrightarrow{\mathbb{P}} p_k$ for any k fixed. Therefore, we refer to $(p_k)_{k\geq 0}$ as the *asymptotic degree distribution* of $\mathrm{PA}_t^{(m,\delta)}$.

We next investigate properties of the sequence $(p_k)_{k\geq 0}$. We start by proving that $(p_k)_{k\geq 0}$ is a *probability mass function* by noting that, by (8.3.2),

$$\frac{\Gamma(k+a)}{\Gamma(k+b)} = \frac{1}{b-a-1}\Big(\frac{\Gamma(k+a)}{\Gamma(k-1+b)} - \frac{\Gamma(k+1+a)}{\Gamma(k+b)}\Big). \tag{8.4.6}$$

Applying (8.4.6) to $a = \delta$, $b = 3 + \delta + \delta/m$, we obtain that, for $k \geq m$,

$$p_k = \frac{\Gamma(m+2+\delta+\delta/m)}{\Gamma(m+\delta)}\Big(\frac{\Gamma(k+\delta)}{\Gamma(k+2+\delta+\delta/m)} - \frac{\Gamma(k+1+\delta)}{\Gamma(k+3+\delta+\delta/m)}\Big). \tag{8.4.7}$$

Using that $p_k = 0$ for $k < m$, and by a telescoping sum identity,

$$\sum_{k\geq 0} p_k = \sum_{k\geq m} p_k = \frac{\Gamma(m+2+\delta+\delta/m)}{\Gamma(m+\delta)}\frac{\Gamma(m+\delta)}{\Gamma(m+2+\delta+\delta/m)} = 1. \tag{8.4.8}$$

Thus, since also $p_k \geq 0$, we obtain that $(p_k)_{k\geq 0}$ indeed is a probability mass function. We can give a simple description in terms of classical random variables for the probability mass function $(p_k)_{k\geq 0}$. For this, we recall that X has a *negative binomial distribution* with parameters r and p if

$$\mathbb{P}(X = k) = \frac{\Gamma(r+k)}{k!\Gamma(r)}p^r(1-p)^{k-r}. \tag{8.4.9}$$

When r is integer valued, X describes the time of the rth success in a sequence of independent experiments with success probability p. Then,

$$p_k = \mathbb{E}[\mathbb{P}(X = k)], \tag{8.4.10}$$

where X has a negative binomial distribution with parameters $r = m+\delta$ and random success probability $U^{1/(2+\delta/m)}$, where U has a uniform distribution on $[0, 1]$ (see Exercise 8.16).

We next investigate the scale-free properties of $(p_k)_{k\geq 0}$ by investigating the asymptotics of p_k for k large. By (8.4.2) and (8.3.9), as $k \to \infty$,

$$p_k = c_{m,\delta} k^{-\tau}(1 + O(1/k)), \tag{8.4.11}$$

where

$$\tau = 3 + \delta/m > 2, \qquad \text{and} \qquad c_{m,\delta} = (2 + \delta/m)\frac{\Gamma(m + 2 + \delta + \delta/m)}{\Gamma(m + \delta)}. \tag{8.4.12}$$

Therefore, by Theorem 8.3 and (8.4.11), the asymptotic degree sequence of $\mathrm{PA}_t^{(m,\delta)}$ is close to a power law with exponent $\tau = 3 + \delta/m$. We note that any exponent $\tau > 2$ is possible by choosing $\delta > -m$ and $m \geq 1$ appropriately. The power-law degree sequence can clearly be observed in a simulation; see Figure 8.4, where a realization of the degree sequence of $\mathrm{PA}_t^{(m,\delta)}$ is shown for $m = 2$, $\delta = 0$ and $t = 300{,}000$ and $t = 1{,}000{,}000$.

The important feature of the preferential attachment model is that, unlike the configuration model and the generalized random graph, the power law in $\mathrm{PA}_t^{(m,\delta)}$ is *explained* by giving a model for the growth of the graph that produces power-law degrees. Therefore, preferential attachment offers a possible explanation as to *why* power-law degree sequences occur. As Barabási (2002) puts it,

> ...the scale-free topology is evidence of organizing principles acting at each stage of the network formation. ...No matter how large and complex a network becomes, as long as preferential attachment and growth are present it will maintain its hub-dominated scale-free topology.

This may be overstating the fact, as we will see later that power laws are intimately related to *affine* preferential attachment functions as in (8.2.1). Many more possible explanations have been given for *why* power laws occur in real networks, and many adaptations of the

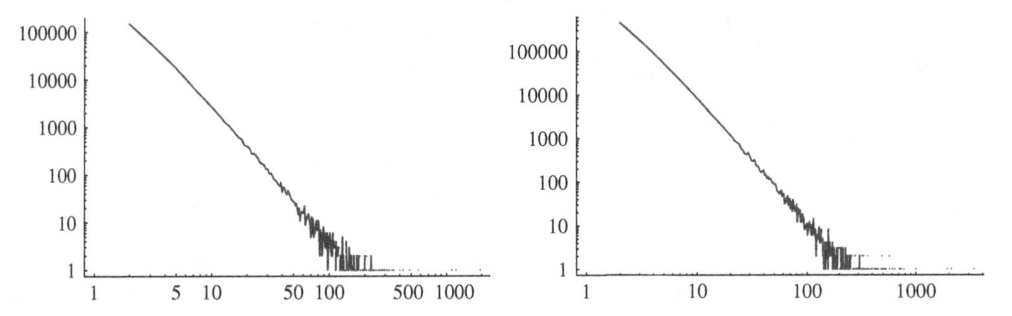

Figure 8.4 The degree sequences of a preferential attachment random graph with $m = 2$, $\delta = 0$ of sizes 300,000 and 1,000,000 in log-log scale.

above simple preferential attachment model have been studied in the literature, many giving rise to power-law degrees. See Section 8.9 for an overview.

The next two sections are devoted to the proof of Theorem 8.3, which is divided into two main parts. In Section 8.5, we prove that the degree sequence is concentrated around its mean, and in Section 8.6 we identify the mean of the degree sequence. In the course of the proof, we also prove extensions of Theorem 8.3.

8.5 Concentration of the Degree Sequence

In this section, we prove that the (random) degree sequence is strongly concentrated around its expectation. We use a martingale argument by Bollobás, Riordan, Spencer and Tusnády (2001), where it first appeared in this context, and that has been used in basically all subsequent works proving power-law degree sequences for preferential attachment models. The argument is very pretty and general, and we spend some time explaining it in detail.

We start by stating the main result in this section. In its statement, we use the notation

$$N_k(t) = \sum_{i=1}^{t} \mathbb{1}_{\{D_i(t)=k\}} = t P_k(t) \tag{8.5.1}$$

for the total number of vertices with degree k at time t.

Proposition 8.4 (Concentration of the degrees) *Fix $\delta \geq -m$ and $m \geq 1$. Then, for any $C > m\sqrt{8}$, as $t \to \infty$,*

$$\mathbb{P}\Big(\max_k |N_k(t) - \mathbb{E}[N_k(t)]| \geq C\sqrt{t \log t}\Big) = o(1). \tag{8.5.2}$$

We note that Theorem 8.3 predicts that $N_k(t) \approx tp_k$. Thus, at least for k for which p_k is not too small, i.e., $tp_k \gg \sqrt{t \log t}$, Proposition 8.4 suggests that the number of vertices with degree equal to k is very close to its expected value. Needless to say, in order to prove Theorem 8.3, we still need to investigate $\mathbb{E}[N_k(t)]$, and prove that it is quite close to tp_k. This is the second main ingredient in the proof of Theorem 8.3 and is formulated in Proposition 8.7 in the next section. In the remainder of this section, we prove Proposition 8.4. We start with two preparatory lemmas investigating an appropriate Doob martingale in the preferential attachment model.

We rely on a martingale argument and Azuma–Hoeffding's inequality (Theorem 2.27). For $n = 0, \ldots, t$, we denote the conditional expected number of vertices with degree k at time t, conditionally on the graph $\mathrm{PA}_n^{(m,\delta)}$ at time $n \in \{0, \ldots, t\}$, by

$$M_n = \mathbb{E}\big[N_k(t) \mid \mathrm{PA}_n^{(m,\delta)}\big]. \tag{8.5.3}$$

The graph $\mathrm{PA}_n^{(m,\delta)}$ can be viewed as a discrete random variable $(X_{ij}(n))_{i,j\in[n]}$ where $X_{ij}(n)$ denotes the number of edges between vertices $i, j \in [t]$ by time t, and as such, the conditional expectation given $\mathrm{PA}_n^{(m,\delta)}$ is well defined and satisfies all the usual rules.[1] We start by showing that $(M_n)_{n=0}^t$ is a martingale:

[1] Formally, $M_n = \mathbb{E}\big[N_k(t) \mid \mathscr{F}_n\big]$, where $\mathscr{F}_n$ denotes the σ-algebra generated by $(\mathrm{PA}_s^{(m,\delta)})_{s\in[n]}$.

Lemma 8.5 (Doob martingale in PAM) *The process* $(M_n)_{n=0}^t$ *given by* $M_n = \mathbb{E}\big[N_k(t) \mid \mathrm{PA}_n^{(m,\delta)}\big]$ *is a martingale w.r.t.* $\big(\mathrm{PA}_s^{(m,\delta)}\big)_{s=1}^n$*. Further,* $M_0 = \mathbb{E}[N_k(t)]$ *and* $M_t = N_k(t)$*.*

Proof Throughout this proof, we fix m and δ and abbreviate $\mathrm{PA}_n = \mathrm{PA}_n^{(m,\delta)}$. Firstly, since $N_k(t)$ is bounded by the total number of vertices at time t, we have $N_k(t) \leq t$, so that

$$\mathbb{E}[|M_n|] = \mathbb{E}[M_n] = \mathbb{E}[N_k(t)] \leq t < \infty. \tag{8.5.4}$$

Secondly, by the tower property of conditional expectations, and the fact that PA_n can be deduced from PA_{n+1}, for all $n \leq t-1$,

$$\begin{aligned}\mathbb{E}[M_{n+1}|\mathrm{PA}_n] &= \mathbb{E}\Big[\mathbb{E}\big[N_k(t) \mid \mathrm{PA}_{n+1}\big]\big|\mathrm{PA}_n\Big]\\ &= \mathbb{E}\big[N_k(t) \mid \mathrm{PA}_n\big] = M_n,\end{aligned} \tag{8.5.5}$$

so that $(M_n)_{n=0}^t$ satisfies the conditional expectation requirement for a martingale. In fact, $(M_n)_{n=0}^t$ is a so-called *Doob martingale* (see also Exercise 2.25). We conclude that $(M_n)_{n=0}^t$ is a martingale process with respect to $(\mathrm{PA}_n)_{n=0}^t$.

To identify M_0, we note that

$$M_0 = \mathbb{E}\big[N_k(t) \mid \mathrm{PA}_0\big] = \mathbb{E}[N_k(t)], \tag{8.5.6}$$

since PA_0 is the empty graph. Furthermore, M_t is trivially identified as

$$M_t = \mathbb{E}\big[N_k(t) \mid \mathrm{PA}_t\big] = N_k(t), \tag{8.5.7}$$

since one can determine $N_k(t)$ from PA_t. This completes the proof of Lemma 8.5. □

To prove concentration of $M_t = N_k(t)$, we aim to use the Azuma–Hoeffding inequality (Theorem 2.27). For this, we need to prove bounds on the martingale differences $|M_n - M_{n-1}|$, as formulated in the next lemma:

Lemma 8.6 (Bounded martingale differences in PAM) *The process* $(M_n)_{n=0}^t$ *given by* $M_n = \mathbb{E}\big[N_k(t) \mid \mathrm{PA}_n^{(m,\delta)}\big]$ *satisfies that* $|M_n - M_{n-1}| \leq 2m$ *almost surely and for every* $n \in [t]$*.*

Proof Throughout this proof, we fix m and δ, and again abbreviate $\mathrm{PA}_n = \mathrm{PA}_n^{(m,\delta)}$. We note that

$$M_n = \mathbb{E}[N_k(t) \mid \mathrm{PA}_n] = \sum_{i=1}^t \mathbb{P}(D_i(t) = k \mid \mathrm{PA}_n), \tag{8.5.8}$$

and, similarly,

$$M_{n-1} = \sum_{i=1}^t \mathbb{P}(D_i(t) = k \mid \mathrm{PA}_{n-1}). \tag{8.5.9}$$

For $s \geq 1$, we define PA'_s by $\mathrm{PA}'_s = \mathrm{PA}_s$ for $s \leq n-1$, while $s \mapsto \mathrm{PA}'_s$ evolves *independently* of $(\mathrm{PA}_s)_{s \geq n-1}$ for $s \geq n-1$ and according to the preferential attachment evolution rules in (8.2.1). Thus, both processes $(\mathrm{PA}_s)_{s\geq 1}$ and $(\mathrm{PA}'_s)_{s\geq 1}$ have the *same* marginal distribution and agree up to time $n-1$, while they evolve independently after time $n-1$. Then,

$$M_{n-1} = \sum_{i=1}^{t} \mathbb{P}(D'_i(t) = k \mid \mathrm{PA}_n), \tag{8.5.10}$$

where $D'_i(t)$ denotes the degree of vertex i in PA'_t. Indeed, since the evolution of $s \mapsto \mathrm{PA}'_s$ is independent of that of $(\mathrm{PA}_s)_{s\geq n-1}$ for $s \geq n-1$, it makes no difference whether we condition on PA_{n-1} or on PA_n in (8.5.10). Therefore, we arrive at

$$M_n - M_{n-1} = \sum_{i=1}^{t} \big[\mathbb{P}(D_i(t) = k \mid \mathrm{PA}_n) - \mathbb{P}(D'_i(t) = k \mid \mathrm{PA}_n)\big]. \tag{8.5.11}$$

Next, since the evolution of the degree $t \mapsto D_i(t)$ for $t \geq n$ only depends on $D_i(n)$ (recall Exercise 8.9),

$$\mathbb{P}(D_i(t) = k \mid \mathrm{PA}_n) = \mathbb{P}(D_i(t) = k \mid D_i(n)). \tag{8.5.12}$$

Similarly,

$$\mathbb{P}(D'_i(t) = k \mid \mathrm{PA}_n) = \mathbb{E}\big[\mathbb{P}(D'_i(t) = k \mid D'_i(n)) \mid \mathrm{PA}_n\big], \tag{8.5.13}$$

Therefore, also using that $D_i(n)$ is measurable w.r.t. PA_n, we arrive at

$$M_n - M_{n-1} = \sum_{i=1}^{t} \mathbb{E}\big[\mathbb{P}(D_i(t) = k \mid D_i(n)) - \mathbb{P}(D'_i(t) = k \mid D'_i(n)) \mid \mathrm{PA}_n\big]. \tag{8.5.14}$$

The crucial observation is that $\mathbb{P}(D_i(t) = k \mid D_i(n)) = \mathbb{P}(D'_i(t) = k \mid D'_i(n))$ whenever $D_i(n) = D'_i(n)$, since the two graphs evolve according to the same rules. Therefore,

$$\big|\mathbb{P}(D_i(t) = k \mid D_i(n)) - \mathbb{P}(D'_i(t) = k \mid D'_i(n))\big| \leq \mathbb{1}_{\{D_i(n)\neq D'_i(n)\}}. \tag{8.5.15}$$

We conclude that

$$\begin{aligned} &|M_n - M_{n-1}| \\ &\leq \sum_{i=1}^{t} \mathbb{E}\Big[\big|\mathbb{P}(D_i(t) = k \mid D_i(n)) - \mathbb{P}(D'_i(t) = k \mid D'_i(n))\big| \;\Big|\; \mathrm{PA}_n\Big] \\ &\leq \sum_{i=1}^{t} \mathbb{E}\big[\mathbb{1}_{\{D_i(n)\neq D'_i(n)\}} \mid \mathrm{PA}_n\big] = \mathbb{E}\Big[\sum_{i=1}^{t} \mathbb{1}_{\{D_i(n)\neq D'_i(n)\}} \mid \mathrm{PA}_n\Big]. \end{aligned} \tag{8.5.16}$$

Since, $D_i(n-1) = D'_i(n-1)$ for every $i \in [n-1]$ by construction, and since we only add m edges at time n both in PA_n and in PA'_n, a.s.,

$$\sum_{i=1}^{t} \mathbb{1}_{\{D_i(n)\neq D'_i(n)\}} \leq 2m. \tag{8.5.17}$$

This completes the proof of Lemma 8.6. □

We are now ready to prove Proposition 8.4:

Proof of Proposition 8.4 We start by reducing the proof. First of all, $N_k(t) = 0$ when $k > m(t+1)$. Therefore,

$$\begin{aligned}
&\mathbb{P}\Big(\max_k |N_k(t) - \mathbb{E}[N_k(t)]| \geq C\sqrt{t\log t}\Big) \\
&\quad = \mathbb{P}\Big(\max_{k \leq m(t+1)} |N_k(t) - \mathbb{E}[N_k(t)]| \geq C\sqrt{t\log t}\Big) \\
&\quad \leq \sum_{k=1}^{m(t+1)} \mathbb{P}\Big(|N_k(t) - \mathbb{E}[N_k(t)]| \geq C\sqrt{t\log t}\Big).
\end{aligned} \tag{8.5.18}$$

Lemmas 8.5 and 8.6, combined with the Azuma–Hoeffding Inequality (Theorem 2.27), yield that, for any $a > 0$,

$$\mathbb{P}\Big(|N_k(t) - \mathbb{E}[N_k(t)]| \geq a\Big) \leq 2\mathrm{e}^{-a^2/(8m^2t)}. \tag{8.5.19}$$

Taking $a = C\sqrt{t\log t}$ for C with $C^2 > 8m^2$ then proves that

$$\mathbb{P}\Big(|N_k(t) - \mathbb{E}[N_k(t)]| \geq C\sqrt{t\log t}\Big) \leq 2\mathrm{e}^{-(\log t)C^2/(8m^2)} = o(1/t). \tag{8.5.20}$$

This completes the proof of Proposition 8.4. □

The above proof is rather general, and can also be used to prove concentration around the mean of other graph properties that are related to the degrees (see Exercise 8.18). An example is the following. Denote by

$$N_{\geq k}(t) = \sum_{l=k}^{\infty} N_l(t) \tag{8.5.21}$$

the total number of vertices with degrees at least k. Then we can also prove that $N_{\geq k}(t)$ concentrates. Indeed, for $C > \sqrt{8m^2}$,

$$\mathbb{P}\big(|N_{\geq k}(t) - \mathbb{E}[N_{\geq k}(t)]| \geq C\sqrt{t\log t}\big) = o(1/t). \tag{8.5.22}$$

The proof uses the same ingredients as given above for $N_{\geq k}(t)$, where now we use the martingale

$$M'_n = \mathbb{E}[N_{\geq k}(t) \mid \mathrm{PA}_n^{(m,\delta)}]. \tag{8.5.23}$$

8.6 Expected Degree Sequence

In this section, we investigate the expected number of vertices with degree equal to k. We denote the expected number of vertices of degree k in $\mathrm{PA}_t^{(m,\delta)}$ by

$$\bar{N}_k(t) = \mathbb{E}\big[N_k(t)\big] = \mathbb{E}\big[tP_k(t)\big]. \tag{8.6.1}$$

The main aim is to prove that $\bar{N}_k(t)$ is close to $p_k t$, where p_k is defined in (8.4.2). This is the content of the following proposition:

Proposition 8.7 (Expected degree sequence) *Fix $m \geq 1$ and $\delta > -m$. Then, there exists a constant $C = C(\delta, m)$ such that, for all $t \geq 1$ and all $k \in \mathbb{N}$,*

$$|\bar{N}_k(t) - p_k t| \leq C. \tag{8.6.2}$$

Propositions 8.4 and 8.7 allow us to complete the proof of Theorem 8.3:

Proof of Theorem 8.3 By Proposition 8.7,

$$\max_k |\mathbb{E}[N_k(t)] - p_k t| \le C. \tag{8.6.3}$$

Therefore, by Proposition 8.4, with C being the maximum of the constants in Propositions 8.7 and 8.4,

$$\mathbb{P}\Big(\max_k |N_k(t) - p_k t| \ge C(1 + \sqrt{t \log t})\Big) = o(1), \tag{8.6.4}$$

which, since $P_k(t) = N_k(t)/t$, implies that

$$\mathbb{P}\Big(\max_k |P_k(t) - p_k| \ge \frac{C}{t}(1 + \sqrt{t \log t})\Big) = o(1). \tag{8.6.5}$$

Equation (8.6.5) in turn implies Theorem 8.3. □

The proof of Proposition 8.7 is split into two separate cases. We first prove the claim for $m = 1$ in Section 8.6.1, and extend the proof to $m > 1$ in Section 8.6.2. The latter proof is more involved.

8.6.1 Expected Degree Sequence for Preferential Attachment Trees

In this section, we study the expected degree sequence when $m = 1$. We adapt the argument by Bollobás, Borgs, Chayes and Riordan (2003), by deriving a recursion relation for $\bar{N}_k(t)$, and showing that the solution to this recursion relation is close to $p_k t$.

We start by writing

$$\mathbb{E}\big[N_k(t+1) \mid \mathrm{PA}_t^{(1,\delta)}] = N_k(t) + \mathbb{E}[N_k(t+1) - N_k(t) \mid \mathrm{PA}_t^{(1,\delta)}]. \tag{8.6.6}$$

Conditionally on $\mathrm{PA}_t^{(1,\delta)}$, there are four ways how $N_k(t+1) - N_k(t)$ can be unequal to zero:

(a) The end vertex of the (unique) edge incident to vertex $v_{t+1}^{(1)}$ had degree $k-1$, so that its degree is increased to k, which occurs with probability $\frac{k-1+\delta}{t(2+\delta)+(1+\delta)}$, and there are $N_{k-1}(t)$ end vertices with degree $k-1$ at time t that can be chosen.

(b) The end vertex of the (unique) edge incident to vertex $v_{t+1}^{(1)}$ had degree k, so that its degree is increased to $k+1$, which occurs with probability $\frac{k+\delta}{t(2+\delta)+(1+\delta)}$, and there are $N_k(t)$ end vertices with degree k at time t that can be chosen.

(c) Another contribution to $k = 1$ arises from vertex $v_{t+1}^{(1)}$. The degree of vertex $v_{t+1}^{(1)}$ is one, so that $N_1(t)$ is increased by one, precisely when the end vertex of the (unique) edge incident to vertex $v_{t+1}^{(1)}$ is not $v_{t+1}^{(1)}$, which occurs with probability $1 - \frac{1+\delta}{t(2+\delta)+(1+\delta)}$.

(d) The final contribution is that from vertex $v_{t+1}^{(1)}$ to $k = 2$. Indeed, the degree of vertex $v_{t+1}^{(1)}$ is equal to two, so that $N_2(t)$ is increased by one, precisely when the end vertex of the (unique) edge incident to vertex $v_{t+1}^{(1)}$ is equal to $v_{t+1}^{(1)}$, which occurs with probability $\frac{1+\delta}{t(2+\delta)+(1+\delta)}$.

The changes in the degree sequence in cases (a) and (b) arise due to the attachment of the edge (thus, the degree of one of the vertices $v_1^{(1)}, \ldots, v_t^{(1)}$ is changed), whereas in cases (c) and (d) we determine the degree of the added vertex $v_{t+1}^{(1)}$.

Taking all these cases into account, we arrive at the key identity

$$\begin{aligned}\mathbb{E}\big[N_k(t+1) - N_k(t) \mid \mathrm{PA}_t^{(1,\delta)}\big] &= \frac{k-1+\delta}{t(2+\delta)+(1+\delta)} N_{k-1}(t) \\ &\quad - \frac{k+\delta}{t(2+\delta)+(1+\delta)} N_k(t) \\ &\quad + \mathbb{1}_{\{k=1\}}\Big(1 - \frac{1+\delta}{t(2+\delta)+(1+\delta)}\Big) \\ &\quad + \mathbb{1}_{\{k=2\}} \frac{1+\delta}{t(2+\delta)+(1+\delta)}. \end{aligned} \tag{8.6.7}$$

Here $k \geq 1$, and for $k = 0$, by convention, we define

$$N_0(t) = 0. \tag{8.6.8}$$

By taking the expectation on both sides of (8.6.7), we obtain

$$\begin{aligned}\mathbb{E}[N_k(t+1)] &= \mathbb{E}[N_k(t)] + \mathbb{E}[N_k(t+1) - N_k(t)] \\ &= \mathbb{E}[N_k(t)] + \mathbb{E}\big[\mathbb{E}[N_k(t+1) - N_k(t) \mid \mathrm{PA}_t^{(1,\delta)}]\big]. \end{aligned} \tag{8.6.9}$$

Now using (8.6.7) gives us the explicit recurrence relation that, for $k \geq 1$,

$$\begin{aligned}\bar{N}_k(t+1) = \bar{N}_k(t) &+ \frac{k-1+\delta}{t(2+\delta)+(1+\delta)} \bar{N}_{k-1}(t) \\ &- \frac{k+\delta}{t(2+\delta)+(1+\delta)} \bar{N}_k(t) \\ &+ \mathbb{1}_{\{k=1\}}\big(1 - \frac{1+\delta}{t(2+\delta)+(1+\delta)}\big) \\ &+ \mathbb{1}_{\{k=2\}} \frac{1+\delta}{t(2+\delta)+(1+\delta)}. \end{aligned} \tag{8.6.10}$$

Equation (8.6.10) is the key to the proof of Proposition 8.7 for $m = 1$. We start by explaining its relation to (8.4.3). Indeed, when $\bar{N}_k(t) \approx tp_k$, then one might expect that $\bar{N}_k(t+1) - \bar{N}_k(t) \approx p_k$. Substituting these approximations into (8.6.10), and approximating $t/[t(2+\delta)+(1+\delta)] \approx 1/(2+\delta)$ and $(1+\delta)/[t(2+\delta)+(1+\delta)] \approx 0$, we arrive at the fact that p_k must satisfy the recurrence relation, for $k \geq 1$,

$$p_k = \frac{k-1+\delta}{2+\delta} p_{k-1} - \frac{k+\delta}{2+\delta} p_k + \mathbb{1}_{\{k=1\}}, \tag{8.6.11}$$

where we define $p_0 = 0$. In the following lemma, we show that the recursion in (8.6.11) has (8.4.3) as a unique solution:

Lemma 8.8 (Solution recurrence relation degree sequence) *The unique solution to* (8.6.11) *is, for* $k \geq 1$,

$$p_k = (2+\delta) \frac{\Gamma(k+\delta)\Gamma(3+2\delta)}{\Gamma(k+3+2\delta)\Gamma(1+\delta)}. \tag{8.6.12}$$

Proof We rewrite

$$p_k = \frac{k-1+\delta}{k+2+2\delta} p_{k-1} + \frac{2+\delta}{k+2+2\delta} \mathbb{1}_{\{k=1\}}. \tag{8.6.13}$$

When $k = 1$, and using that $p_0 = 0$, we obtain

$$p_1 = \frac{2+\delta}{3+2\delta}. \tag{8.6.14}$$

On the other hand, when $k > 1$, we arrive at

$$p_k = \frac{k-1+\delta}{k+2+2\delta} p_{k-1}. \tag{8.6.15}$$

Therefore, using (8.3.2) repeatedly,

$$\begin{aligned} p_k &= \frac{\Gamma(k+\delta)\Gamma(4+2\delta)}{\Gamma(k+3+2\delta)\Gamma(1+\delta)} p_1 \\ &= \frac{(2+\delta)\Gamma(k+\delta)\Gamma(4+2\delta)}{(3+2\delta)\Gamma(k+3+2\delta)\Gamma(1+\delta)} \\ &= (2+\delta)\frac{\Gamma(k+\delta)\Gamma(3+2\delta)}{\Gamma(k+3+2\delta)\Gamma(1+\delta)}, \end{aligned} \tag{8.6.16}$$

which agrees with (8.4.3) and thus proves the claim. □

Lemma 8.8 explains, at least intuitively, what the relation is between (8.6.12) and the recursion relation in (8.6.10). The next step is to use (8.6.10) and (8.6.11) to prove Proposition 8.7 for $m = 1$, by showing that $\bar{N}_k(t)$ is close to tp_k. To this end, we define

$$\varepsilon_k(t) = \bar{N}_k(t) - tp_k. \tag{8.6.17}$$

Then, in order to prove Proposition 8.7 for $m = 1$, we are left to prove that there exists a constant $C = C(\delta)$ such that

$$\sup_k |\varepsilon_k(t)| \le C. \tag{8.6.18}$$

The value of C will be determined in the course of the proof.

We use induction in t. We note that we can rewrite (8.6.11) as

$$\begin{aligned} (t+1)p_k &= tp_k + p_k \\ &= tp_k + \frac{k-1+\delta}{2+\delta} p_{k-1} - \frac{k+\delta}{2+\delta} p_k + \mathbb{1}_{\{k=1\}} \\ &= tp_k + \frac{k-1+\delta}{t(2+\delta)+(1+\delta)} tp_{k-1} - \frac{k+\delta}{t(2+\delta)+(1+\delta)} tp_k + \mathbb{1}_{\{k=1\}} \\ &\quad + \Big(\frac{1}{2+\delta} - \frac{t}{t(2+\delta)+(1+\delta)}\Big)(k-1+\delta)p_{k-1} \\ &\quad - \Big(\frac{1}{2+\delta} - \frac{t}{t(2+\delta)+(1+\delta)}\Big)(k+\delta)p_k. \end{aligned} \tag{8.6.19}$$

We abbreviate

$$\kappa_k(t) = -\Big(1/(2+\delta) - \frac{t}{t(2+\delta)+(1+\delta)}\Big) \times \Big((k+\delta)p_k - (k-1+\delta)p_{k-1}\Big), \tag{8.6.20}$$

$$\gamma_k(t) = \frac{1+\delta}{t(2+\delta)+(1+\delta)}\big(\mathbb{1}_{\{k=2\}} - \mathbb{1}_{\{k=1\}}\big). \tag{8.6.21}$$

Then, subtracting (8.6.19) from (8.6.10) leads to

$$\varepsilon_k(t+1) = \Big(1 - \frac{k+\delta}{t(2+\delta)+(1+\delta)}\Big)\varepsilon_k(t) + \frac{k-1+\delta}{t(2+\delta)+(1+\delta)}\varepsilon_{k-1}(t) + \kappa_k(t) + \gamma_k(t). \tag{8.6.22}$$

We prove the bounds on $\varepsilon_k(t)$ in (8.6.18) by induction on $t \geq 1$. We start by initializing the induction hypothesis. When $t = 1$, we have that $\mathrm{PA}_{1,\delta}(1)$ consists of a vertex with a single self-loop. Thus,

$$\bar{N}_k(1) = \mathbb{1}_{\{k=2\}}. \tag{8.6.23}$$

Therefore, since also $p_k \leq 1$, we arrive at the estimate that, uniformly in $k \geq 1$,

$$|\varepsilon_k(1)| = |\bar{N}_k(1) - p_k| \leq \max\{\bar{N}_k(1), p_k\} \leq 1. \tag{8.6.24}$$

We have initialized the induction hypothesis for $t = 1$ in (8.6.18) for any $C \geq 1$.

We next advance the induction hypothesis. We start with $k = 1$:

Lemma 8.9 (Expected number of vertices with degree 1) *Fix $m = 1$ and $\delta > -1$. Then, there exists a constant $C_1 > 0$ such that*

$$\sup_{t\geq 1} |\bar{N}_1(t) - tp_1| \leq C_1. \tag{8.6.25}$$

Proof We note that $\varepsilon_0(t) = N_0(t) - p_0 = 0$ by convention, so that (8.6.22) reduces to

$$\varepsilon_1(t+1) = \Big(1 - \frac{1+\delta}{t(2+\delta)+(1+\delta)}\Big)\varepsilon_1(t) + \kappa_1(t) + \gamma_1(t). \tag{8.6.26}$$

We note that

$$1 - \frac{1+\delta}{t(2+\delta)+(1+\delta)} \geq 0, \tag{8.6.27}$$

so that

$$|\varepsilon_1(t+1)| \leq \Big(1 - \frac{1+\delta}{t(2+\delta)+(1+\delta)}\Big)|\varepsilon_1(t)| + |\kappa_1(t)| + |\gamma_1(t)|. \tag{8.6.28}$$

Using the explicit forms in (8.6.20) and (8.6.21), it is not hard to see that there are universal constants $C_\kappa = C_\kappa(\delta)$ and $C_\gamma = C_\gamma(\delta)$ such that, uniformly in $k \geq 1$,

$$|\kappa_k(t)| \leq C_\kappa(t+1)^{-1}, \qquad |\gamma_k(t)| \leq C_\gamma(t+1)^{-1}. \tag{8.6.29}$$

Exercise 8.19 shows that $C_\gamma = 1$ and $C_\kappa = \frac{(1+\delta)(2+\delta)}{3+2\delta}$ do the job.

Using the induction hypothesis (8.6.18), as well as (8.6.29), we arrive at

$$|\varepsilon_1(t+1)| \le C_1\Big(1 - \frac{1+\delta}{t(2+\delta)+(1+\delta)}\Big) + \frac{C_\kappa + C_\gamma}{t+1}. \tag{8.6.30}$$

Next, we use that $t(2+\delta)+(1+\delta) \le (t+1)(2+\delta)$, so that

$$|\varepsilon_1(t+1)| \le C_1 - (t+1)^{-1}\Big(C_1\frac{1+\delta}{2+\delta} - (C_\kappa + C_\gamma)\Big). \tag{8.6.31}$$

Therefore, $|\varepsilon_1(t+1)| \le C_1$ whenever

$$C_1 \ge \frac{2+\delta}{1+\delta}(C_\kappa + C_\gamma). \tag{8.6.32}$$

This advances the induction hypothesis in $t \ge 1$ for $k = 1$, and thus completes the proof of Lemma 8.9. □

We now extend the argument to $k \ge 2$:

Lemma 8.10 (Expected number of vertices with degree k) *Fix $m = 1$ and $\delta > -1$. Then, there exists a constant $C > 0$ such that*

$$\sup_{k\ge 1}\sup_{t\ge 1} |\bar{N}_k(t) - tp_k| \le C. \tag{8.6.33}$$

Proof We use induction on k, where the induction hypothesis is the statement that $\sup_{t\ge 1} |\bar{N}_k(t) - tp_k| \le C$. We determine the value of C in the course of the proof.

Lemma 8.9 initiates the induction for $k = 1$. We next advance the induction on $k \ge 1$. For this, we assume that (8.6.33) is true for $k-1$, and prove it for k. For this, in turn, we use induction on t, with the result for $t = 1$ following from (8.6.24). To advance the induction on t, we again use (8.6.22). We note that

$$1 - \frac{k+\delta}{t(2+\delta)+(1+\delta)} \ge 0 \quad \text{as long as} \quad k \le t(2+\delta)+1. \tag{8.6.34}$$

We assume (8.6.34) for the time being, and deal with $k > t(2+\delta)+1$ later.

By (8.6.22) and (8.6.34), for $k \ge 2$ and $\delta > -1$ so that $k-1+\delta \ge 0$, it follows that

$$|\varepsilon_k(t+1)| \le \Big(1 - \frac{k+\delta}{t(2+\delta)+(1+\delta)}\Big)|\varepsilon_k(t)| + \frac{k-1+\delta}{t(2+\delta)+(1+\delta)}|\varepsilon_{k-1}(t)| + |\kappa_k(t)| + |\gamma_k(t)|. \tag{8.6.35}$$

Again using the induction hypothesis (8.6.18), as well as (8.6.29), we arrive at

$$\begin{aligned} |\varepsilon_k(t+1)| &\le C\Big(1 - \frac{k+\delta}{t(2+\delta)+(1+\delta)}\Big) + C\frac{k-1+\delta}{t(2+\delta)+(1+\delta)} + \frac{C_\kappa + C_\gamma}{t+1} \\ &= C\Big(1 - \frac{1}{t(2+\delta)+(1+\delta)}\Big) + \frac{C_\kappa + C_\gamma}{t+1}. \end{aligned} \tag{8.6.36}$$

As before,

$$t(2+\delta)+(1+\delta) \le (t+1)(2+\delta), \tag{8.6.37}$$

so that

$$|\varepsilon_k(t+1)| \leq C - (t+1)^{-1}\Big(\frac{C}{2+\delta} - (C_\kappa + C_\gamma)\Big) \leq C, \tag{8.6.38}$$

the last inequality being valid whenever

$$C \geq (2+\delta)(C_\kappa + C_\gamma). \tag{8.6.39}$$

This advances the induction hypothesis in the case where $k \leq t(2+\delta)+1$.

Finally, we deal with the case that $k > t(2+\delta)+1$. Note that then $k \geq t(2+\delta)+2 > t+2$, since $\delta > -1$. Since the maximal degree of $\mathrm{PA}_t^{(1,\delta)}$ is $t+2$ (which happens precisely when all edges are connected to the initial vertex), we have that $\bar{N}_k(t+1) = 0$ for $k \geq t(2+\delta)+2$. Therefore, for $k \geq t(2+\delta)+2$,

$$|\varepsilon_k(t+1)| = (t+1)p_k. \tag{8.6.40}$$

By (8.4.11) and (8.4.12), uniformly for $k \geq t(2+\delta)+2 \geq t+2$ for $\delta \geq -1$, there exists a $C_p = C_p(\delta)$ such that

$$p_k \leq C_p(t+1)^{-(3+\delta)}. \tag{8.6.41}$$

Therefore, we conclude that for $\delta > -1$, and again uniformly for $k \geq t+2$,

$$(t+1)p_k \leq C_p(t+1)^{-(2+\delta)} \leq C_p. \tag{8.6.42}$$

Therefore, if $C \geq C_p$, then also the claim follows for $k \geq t(2+\delta)+2$. Comparing to (8.6.32) and (8.6.39), we choose

$$C = \max\Big\{(2+\delta)(C_\kappa + C_\gamma), \frac{(2+\delta)(C_\kappa + C_\gamma)}{1+\delta}, C_p\Big\}. \tag{8.6.43}$$

This advances the induction hypothesis in t for $k \geq 2$, and completes the proof of Lemma 8.10. □

Lemma 8.10 implies Proposition 8.7 when $m = 1$ and $\delta > -1$.

8.6.2 Expected Degree Sequence for $m \geq 2$*

In this section, we prove Proposition 8.7 for $m > 1$. We adapt the argument in Section 8.6.1 above. There, we have been rather explicit in the derivation of the recursion relation in (8.6.10), which in turn gives the explicit recursion relation on the errors $\varepsilon_k(t)$ in (8.6.22). In this section, we make the derivation more abstract, since the explicit derivations become too involved when $m > 1$. The argument presented in this section is rather flexible, and can, e.g., be extended to different preferential attachment models.

We make use of the fact that we add precisely m edges in a preferential way to go from $\mathrm{PA}_t^{(m,\delta)}$ to $\mathrm{PA}_{t+1}^{(m,\delta)}$. This process can be described in terms of certain operators. For a sequence of numbers $Q = (Q_k)_{k\geq 1}$, we define the operator $T_{t+1}\colon \mathbb{R}^{\mathbb{N}} \mapsto \mathbb{R}^{\mathbb{N}}$ by

$$(T_{t+1}Q)_k = \Big(1 - \frac{k+\delta'}{t(2+\delta')+(1+\delta')}\Big)Q_k + \frac{k-1+\delta'}{t(2+\delta')+(1+\delta')}Q_{k-1}, \tag{8.6.44}$$

where we recall that $\delta' = \delta/m$. Writing $\bar{N}(t) = (\bar{N}_k(t))_{k\geq 1}$ with $\bar{N}_k(t) = \mathbb{E}[N_k(t)]$ denoting the expected number of vertices with degree k, we can rewrite (8.6.10) when $m = 1$ (for which $\delta' = \delta$) as

$$\bar{N}_k(t+1) = (T_{t+1}\bar{N}(t))_k + \mathbb{1}_{\{k=1\}}\Big(1 - \frac{1+\delta}{t(2+\delta)+(1+\delta)}\Big) \qquad (8.6.45)$$
$$+ \mathbb{1}_{\{k=2\}}\frac{1+\delta}{t(2+\delta)+(1+\delta)}.$$

Thus, as remarked above (8.6.7), the operator T_{t+1} describes the effect to the sequence $\bar{N}(t) = (\bar{N}_k(t))_{k\geq 1}$ of a single addition of the $(t+1)$st edge, apart from the degree of the newly added vertex. The latter degree is equal to 1 with probability $1 - \frac{1+\delta}{t(2+\delta)+(1+\delta)}$, and equal to 2 with probability $\frac{1+\delta}{t(2+\delta)+(1+\delta)}$. This explains the origin of each of the terms appearing in (8.6.10).

In the case when $m > 1$, every vertex has m edges that are sequentially connected in a preferential way. Therefore, we need to investigate the effect of attaching m edges in sequel. Due to the fact that we update the degrees after attaching an edge, the effect of attaching the $(j+1)$st edge is described by applying the operator T_j to $\bar{N}(j)$. When we add the edges incident to the tth vertex, this corresponds to attaching the edges $m(t-1)+1, \ldots, mt$ in sequel with intermediate updating.

The effect on the degrees of vertices $v_1^{(m)}, \ldots, v_t^{(m)}$ is described precisely by applying first T_{mt+1} to describe the effect of the addition of the first edge, followed by T_{mt+2} to describe the effect of the addition of the second edge, etc. After attaching the m edges, we further inspect the degree of the newly added vertex. Therefore, the recurrence relation of the expected number of vertices with degree k in (8.6.45) is changed to

$$\bar{N}_k(t+1) = (T_{t+1}^{(m)}\bar{N}(t))_k + \alpha_k(t), \qquad (8.6.46)$$

where

$$T_{t+1}^{(m)} = T_{m(t+1)} \circ \cdots \circ T_{mt+1}, \qquad (8.6.47)$$

denotes the m-fold application of successive T_j's, and where, for $k = m, \ldots, 2m$, $\alpha_k(t)$ is the probability that the degree of the $(t+1)$st added vertex is precisely equal to k.

When t grows large, the probability distribution $k \mapsto \alpha_k(t)$ is such that $\alpha_m(t)$ is very close to 1, while $\alpha_k(t)$ is close to zero when $k > m$. Indeed, for $k > m$, in contributions to $\alpha_m(t)$, at least one of the m edges should be connected to vertex $t+1$, so that

$$\sum_{k=m+1}^{2m} \alpha_k(t) \leq \frac{m^2(1+\delta')}{mt(2+\delta')+(1+\delta')}. \qquad (8.6.48)$$

We define

$$\gamma_k(t) = \alpha_k(t) - \mathbb{1}_{\{k=m\}}, \qquad (8.6.49)$$

then we obtain from (8.6.48) that there exists a constant $C_\gamma = C_\gamma(\delta, m)$ such that, for all $k > m$,

$$|\gamma_k(t)| \leq \frac{C_\gamma}{t+1}. \qquad (8.6.50)$$

For $k = m$, the same bound follows since $\sum_{k\geq m} \alpha_m(t) = 1$, so that $|\alpha_m(t) - 1| = \sum_{k>m} \alpha_k(t)$. The bound in (8.6.50) replaces the bound on $|\gamma_k(t)|$ for $m = 1$ in (8.6.29).

Denote the operator $S^{(m)}$ on sequences of numbers $Q = (Q_k)_{k\geq 1}$ by

$$(S^{(m)}Q)_k = m\frac{k-1+\delta}{2m+\delta}Q_{k-1} - m\frac{k+\delta}{2m+\delta}Q_k. \tag{8.6.51}$$

Then, for $m = 1$, we have that (8.6.11) is equivalent to

$$p_k = (S^{(1)}p)_k + \mathbb{1}_{\{k=1\}}. \tag{8.6.52}$$

For $m > 1$, we replace the above recursion on p by $p_k = 0$ for $k < m$ and, for $k \geq m$,

$$p_k = (S^{(m)}p)_k + \mathbb{1}_{\{k=m\}}. \tag{8.6.53}$$

Again, we can explicitly solve for $p = (p_k)_{k\geq 0}$. The solution is given in the following lemma:

Lemma 8.11 (Solution recursion for $m > 1$) *Fix $\delta > -1$ and $m \geq 1$. Then, the solution to* (8.6.53) *is given by* (8.4.2).

Proof We start by noting that $p_k = 0$ for $k < m$, and identify p_m using the recursion for $k = m$

$$p_m = -m\frac{m+\delta}{2m+\delta}p_m + 1, \tag{8.6.54}$$

so that

$$p_m = \frac{2m+\delta}{m(m+\delta)+2m+\delta} = \frac{2+\delta/m}{(m+\delta)+2+\delta/m}. \tag{8.6.55}$$

For $k > m$, the recursion relation in (8.6.53) becomes

$$p_k = \frac{m(k-1+\delta)}{m(k+\delta)+2m+\delta}p_{k-1} = \frac{k-1+\delta}{k+\delta+2+\delta/m}p_{k-1}. \tag{8.6.56}$$

As a result, again repeatedly using (8.3.2),

$$\begin{aligned} p_k &= \frac{\Gamma(k+\delta)\Gamma(m+3+\delta+\delta/m)}{\Gamma(m+\delta)\Gamma(k+3+\delta+\delta/m)}p_m \\ &= \frac{\Gamma(k+\delta)\Gamma(m+3+\delta+\delta/m)}{\Gamma(m+\delta)\Gamma(k+3+\delta+\delta/m)} \frac{(2+\delta/m)}{(m+\delta+2+\delta/m)} \\ &= (2+\delta/m)\frac{\Gamma(k+\delta)\Gamma(m+2+\delta+\delta/m)}{\Gamma(m+\delta)\Gamma(k+3+\delta+\delta/m)}. \end{aligned} \tag{8.6.57}$$

This is the same as (8.4.2), as required. □

Similarly to (8.6.19), we can rewrite (8.6.53) as

$$\begin{aligned} (t+1)p_k &= tp_k + p_k = tp_k + (S^{(m)}p)_k + \mathbb{1}_{\{k=m\}} \\ &= (T^{(m)}_{t+1}tp)_k + \mathbb{1}_{\{k=m\}} - \kappa_k(t), \end{aligned} \tag{8.6.58}$$

where, writing I for the identity operator,

$$\kappa_k(t) = -\big(\big[S^{(m)} + t(I - T^{(m)}_{t+1})\big]p\big)_k. \tag{8.6.59}$$

While (8.6.59) is not very explicit, it can be effectively used to bound $\kappa_k(t)$ in a similar way as in (8.6.29):

Lemma 8.12 (A bound on $\kappa_k(t)$) *Fix $\delta \geq -1$ and $m \geq 1$. Then there exists a constant $C_\kappa = C_\kappa(\delta, m)$ such that*

$$|\kappa_k(t)| \leq \frac{C_\kappa}{t+1}. \tag{8.6.60}$$

Proof We start with

$$\begin{aligned} T_{t+1}^{(m)} &= T_{m(t+1)} \circ \cdots \circ T_{mt+1} \\ &= \big(I + (T_{m(t+1)} - I)\big) \circ \cdots \circ \big(I + (T_{mt+1} - I)\big). \end{aligned} \tag{8.6.61}$$

By (8.6.44),

$$\big((T_{t+1} - I)Q\big)_k = -\frac{k+\delta}{t(2+\delta') + (1+\delta')} Q_k + \frac{k-1+\delta}{t(2+\delta') + (1+\delta')} Q_{k-1}. \tag{8.6.62}$$

When $\sup_k k|Q_k| \leq K$, there exists a constant C_T such that

$$\sup_k \Big|\big((T_{t+1} - I)Q\big)_k\Big| \leq \frac{C_T}{t+1}. \tag{8.6.63}$$

Moreover, when $\sup_k k^2|Q_k| \leq K$, there exists a constant $C = C_K$ such that, when $u, v \geq t$,

$$\sup_k \big|((T_{u+1} - I) \circ (T_{v+1} - I)Q)_k\big| \leq \frac{C_K}{(t+1)^2}. \tag{8.6.64}$$

We expand out the brackets in (8.6.61), and note that, by (8.6.64) and since the operators T_u are contractions, the terms where we have at least two factors $T_u - I$ lead to error terms. More precisely, we rewrite

$$(T_{t+1}^{(m)} Q)_k = Q_k + \sum_{a=1}^{m} \big((T_{mt+a} - I)Q\big)_k + E_k(t, Q), \tag{8.6.65}$$

where, uniformly in k and Q for which $\sup_k k^2|Q_k| \leq K$,

$$|E_k(t, Q)| \leq \frac{C_K}{(t+1)^2}. \tag{8.6.66}$$

As a result,

$$\big((I - T_{t+1}^{(m)})Q\big)_k = -\sum_{a=1}^{m} \big((T_{mt+a} - I)Q\big)_k - E_k(t, Q). \tag{8.6.67}$$

Furthermore, for every $a = 1, \ldots, m$,

$$\big((T_{mt+a} - I)Q\big)_k = \frac{1}{mt}(S^{(m)}Q)_k + F_{k,a}(t, Q), \tag{8.6.68}$$

where, uniformly in k, Q for which $\sup_k k|Q_k| \leq K$ and $a = 1, \ldots, m$,

$$|F_{k,a}(t, Q)| \leq \frac{C_K'}{(t+1)^2}. \tag{8.6.69}$$

Therefore also

$$\sum_{a=1}^{m} \big((T_{mt+a} - I)Q\big)_k = \frac{1}{t}(S^{(m)}Q)_k + F_k(t, Q), \tag{8.6.70}$$

where

$$F_k(t, Q) = \sum_{a=1}^{m} F_{k,a}(t, Q). \tag{8.6.71}$$

We summarize from (8.6.67) and (8.6.70) that

$$\big([S^{(m)} + t(I - T_{t+1}^{(m)})]Q\big)_k = -tF_k(t, Q) - tE_k(t, Q), \tag{8.6.72}$$

so that

$$\kappa_k(t) = -\big([S^{(m)} + t(I - T_{t+1}^{(m)})]p\big)_k = tF_k(t, p) + tE_k(t, p). \tag{8.6.73}$$

By (8.4.11) and (8.4.12), p satisfies that there exists a $K > 0$ such that

$$\sup_k k^2 p_k \leq K, \tag{8.6.74}$$

so that

$$\begin{aligned} \|\kappa(t)\|_\infty &= \sup_k \Big|\big([S^{(m)} + t(I - T_{t+1}^{(m)})p]\big)_k\Big| \leq \sup_k t\big(|E_k(t, p)| + |F_k(t, p)|\big) \\ &\leq \frac{t(C_K + C_K')}{(t+1)^2} \leq \frac{C_K + C_K'}{t+1}. \end{aligned} \tag{8.6.75}$$

□

We continue with the proof of Proposition 8.7 for $m > 1$ by investigating $\bar{N}_k(t) - tp_k$. We define, for $k \geq m$,

$$\varepsilon_k(t) = \bar{N}_k(t) - tp_k. \tag{8.6.76}$$

Subtracting (8.6.58) from (8.6.46) and writing $\varepsilon(t) = (\varepsilon_k(t))_{k\geq 1}$ leads to

$$\varepsilon_k(t+1) = (T_{t+1}^{(m)}\varepsilon(t))_k + \kappa_k(t) + \gamma_k(t). \tag{8.6.77}$$

To study the recurrence relation (8.6.77) in more detail, we investigate the properties of the operator $T_t^{(m)}$. We let $Q = (Q_k)_{k\geq 1}$ be a sequence of real numbers, and we let $\mathcal{Q} = \mathbb{R}^\infty$ denote the set of all such sequences. For $Q \in \mathcal{Q}$, we define the supremum-norm to be

$$\|Q\|_\infty = \sup_{k\geq 1} |Q_k|. \tag{8.6.78}$$

Thus, in functional analytic terms, we consider the ℓ^∞ norm on $\mathcal{Q} = \mathbb{R}^\infty$.

Furthermore, we let $\mathcal{Q}_m(t) \subseteq \mathcal{Q}$ be the subset of sequences for which $Q_k = 0$ for $k > m(t+1)$, i.e.,

$$\mathcal{Q}_m(t) = \{Q \in \mathcal{Q}\colon Q_k = 0 \;\; \forall k > m(t+1)\}. \tag{8.6.79}$$

Clearly, $\bar{N}(t) \in \mathcal{Q}_m(t)$. We regard $T_{t+1}^{(m)}$ in (8.6.47) as an operator on $\mathcal{Q}$. We now show that $T_{t+1}^{(m)}$ acts as a *contraction* on elements of $\mathcal{Q}_m(t)$.

Lemma 8.13 (A contraction property) *Fix $\delta \geq -1$ and $m \geq 1$. Then $T^{(m)}_{t+1}$ maps $\mathcal{Q}_m(t)$ into $\mathcal{Q}_m(t+1)$ and, for every $Q \in \mathcal{Q}_m(t)$,*

$$\|T^{(m)}_{t+1}Q\|_\infty \leq \Big(1 - \frac{1}{t(2m+\delta)+(m+\delta)}\Big)\|Q\|_\infty. \tag{8.6.80}$$

Proof We recall that

$$T^{(m)}_{t+1} = T_{m(t+1)} \circ \cdots \circ T_{mt+1}, \tag{8.6.81}$$

Thus, the fact that $T^{(m)}_{t+1}$ maps $\mathcal{Q}_m(t)$ into $\mathcal{Q}_m(t+1)$ follows from the fact that T_{t+1} maps $\mathcal{Q}_1(t)$ into $\mathcal{Q}_1(t+1)$. This proves the first claim in Lemma 8.13.

To prove that the contraction property of $T^{(m)}_{t+1}$ in (8.6.80) holds, we first prove that, for all $Q \in \mathcal{Q}_1(mt+a-1)$, $a = 1, \ldots, m$, $\delta > -m$ and $\delta' = \delta/m > -1$,

$$\|T_{mt+a}Q\|_\infty \leq \Big(1 - \frac{1}{t(2+\delta')+(1+\delta')}\Big)\|Q\|_\infty. \tag{8.6.82}$$

For this, we recall from (8.6.44) that

$$\begin{aligned}(T_{mt+a}Q)_k &= \Big(1 - \frac{k+\delta'}{(mt+a-1)(2+\delta')+(1+\delta')}\Big)Q_k \\ &\quad + \frac{k-1+\delta'}{(mt+a-1)(2+\delta')+(1+\delta')}Q_{k-1}.\end{aligned} \tag{8.6.83}$$

When $Q \in \mathcal{Q}_1(mt+a-1)$, then, for all k for which $Q_k \neq 0$,

$$1 - \frac{k+\delta'}{(mt+a-1)(2+\delta')+(1+\delta')} \in [0,1], \tag{8.6.84}$$

and, for $k \geq 2$, also

$$\frac{k-1+\delta'}{(mt+a-1)(2+\delta')+(1+\delta')} \in [0,1]. \tag{8.6.85}$$

As a consequence,

$$\begin{aligned}\|T_{mt+a}Q\|_\infty &\leq \sup_k \Big[\Big(1 - \frac{k+\delta'}{(mt+a-1)(2+\delta')+(1+\delta')}\Big)\|Q\|_\infty \\ &\quad + \frac{k-1+\delta'}{(mt+a-1)(2+\delta')+(1+\delta')}\|Q\|_\infty\Big] \\ &= \Big(1 - \frac{1}{(mt+a-1)(2+\delta')+(1+\delta')}\Big)\|Q\|_\infty.\end{aligned} \tag{8.6.86}$$

Now, by (8.6.86), the application of T_{mt+a} to an element Q of $\mathcal{Q}_1(mt+a-1)$ reduces its norm. By (8.6.81), we therefore conclude that, for every $Q \in \mathcal{Q}_m(t)$,

$$\begin{aligned}\|T^{(m)}_{t+1}Q\|_\infty \leq \|T_{mt+1}Q\|_\infty &\leq \Big(1 - \frac{1}{mt(2+\delta')+(1+\delta')}\Big)\|Q\|_\infty \\ &= \Big(1 - \frac{1}{t(2m+\delta)+(m+\delta)}\Big)\|Q\|_\infty,\end{aligned} \tag{8.6.87}$$

since $\delta' = \delta/m$. This completes the proof of Lemma 8.13. □

Lemmas 8.12 and 8.13, as well as (8.6.50), allow us to complete the proof of Proposition 8.7:

Proof of Proposition 8.7 We start with the recurrence relation in (8.6.77). We define the truncated sequence $\varepsilon'(t) = (\varepsilon'_k(t))_{k\geq 1}$ by

$$\varepsilon'_k(t) = \varepsilon_k(t)\mathbb{1}_{\{k\leq m(t+1)\}}. \tag{8.6.88}$$

Then, by construction, $\varepsilon'(t) \in \mathcal{Q}_m(t)$. Therefore, by Lemma 8.13,

$$\begin{aligned}\|\varepsilon(t+1)\|_\infty &\leq \|T^{(m)}_{t+1}\varepsilon'(t)\|_\infty + \|\varepsilon'(t+1) - \varepsilon(t+1)\|_\infty + \|\kappa(t)\|_\infty + \|\gamma(t)\|_\infty \\ &\leq \Big(1 - \frac{1}{(2m+\delta)+(m+\delta)}\Big)\|\varepsilon'(t)\|_\infty \\ &\quad + \|\varepsilon'(t+1) - \varepsilon(t+1)\|_\infty + \|\kappa(t)\|_\infty + \|\gamma(t)\|_\infty.\end{aligned} \tag{8.6.89}$$

Equation (8.6.50) and Lemma 8.12, respectively, imply that

$$\|\gamma(t)\|_\infty \leq \frac{C_\gamma}{t+1} \quad \text{and} \quad \|\kappa(t)\|_\infty \leq \frac{C_\kappa}{t+1}. \tag{8.6.90}$$

It is not hard to see that

$$\|\varepsilon'(t+1) - \varepsilon(t+1)\|_\infty \leq C_{\varepsilon'}(t+1)^{-(\tau-1)}, \tag{8.6.91}$$

where $\tau > 2$ is defined in (8.4.12). See (8.6.41)–(8.6.42) for the analogous proof for $m = 1$.
Therefore,

$$\|\varepsilon(t+1)\|_\infty \leq \Big(1 - \frac{1}{t(2m+\delta)+(m+\delta)}\Big)\|\varepsilon(t)\|_\infty + \frac{C_\gamma + C_\kappa + C_{\varepsilon'}}{t+2}. \tag{8.6.92}$$

Using further that, for $m \geq 1$ and $\delta > -m$,

$$t(2m+\delta) + (m+\delta) \leq (2m+\delta)(t+1) \tag{8.6.93}$$

we arrive at

$$\|\varepsilon(t+1)\|_\infty \leq \Big(1 - \frac{1}{(t+1)(2m+\delta)}\Big)\|\varepsilon(t)\|_\infty + \frac{C_\gamma + C_\kappa + C_{\varepsilon'}}{t+1}. \tag{8.6.94}$$

Now we can advance the induction hypothesis

$$\|\varepsilon(t)\|_\infty \leq C. \tag{8.6.95}$$

For some $C > 0$ sufficiently large, this statement trivially holds for $t = 1$. To advance it, we use (8.6.94), to see that

$$\|\varepsilon(t+1)\|_\infty \leq \Big(1 - \frac{1}{(2m+\delta)(t+1)}\Big)C + \frac{C_\gamma + C_\kappa + C_{\varepsilon'}}{t+1} \leq C, \tag{8.6.96}$$

whenever

$$C \geq (2m+\delta)(C_\gamma + C_\kappa + C_{\varepsilon'}). \tag{8.6.97}$$

This advances the induction hypothesis, and completes the proof that the inequality $\|\varepsilon(t)\|_\infty = \sup_{k\geq 0}|\bar{N}_k(t) - p_k t| \leq C$ holds for all $m \geq 2$. □

8.7 Maximal Degree in Preferential Attachment Models

In this section we investigate the maximal degree in the graph $\mathrm{PA}_t^{(m,\delta)}$. In order to state the results on the maximal degree, we denote

$$M_t = \max_{i\in[t]} D_i(t). \tag{8.7.1}$$

The main result on the maximal degree is the following theorem:

Theorem 8.14 (Maximal degree of $\mathrm{PA}_t^{(m,\delta)}$) *Fix $m \geq 1$ and $\delta > -m$. Then, there exists a random variable μ with with $\mathbb{P}(\mu = 0) = 0$ such that*

$$M_t t^{-1/(2+\delta/m)} \xrightarrow{a.s.} \mu. \tag{8.7.2}$$

By Theorem 8.3 and (8.4.11), $\mathrm{PA}_t^{(m,\delta)}$ has a power-law asymptotic degree sequence with power-law exponent $\tau = 3 + \delta/m$. The maximum of t i.i.d. random variables with such a power-law distribution is of the order $t^{1/(\tau-1)} = t^{1/(2+\delta/m)}$. Thus, Theorem 8.14 shows that the maximal degree is of the same order of magnitude.

The proof of Theorem 8.14 further reveals that $\mu = \sup_{i\geq 1} \xi_i$, where ξ_i is the a.s. limit of $D_i(t)t^{-1/(2+\delta/m)}$ as identified in Theorem 8.2. The proof reveals much more, and, for example, allows us to compute moments of the a.s. limits of $D_i(t)t^{-1/(2+\delta/m)}$. In turn, this for example allows us to show that $\mathbb{P}(\xi_i = 0) = 0$ for every $i \geq 1$.

We now start with the proof of Theorem 8.14. We fix $m = 1$ for the time being, and extend the results to $m \geq 2$ at the end of this section. Let $X_j(t) = D_j(t) + \delta$ be the weight of vertex j at time t, let $\Delta_j(t) = X_j(t+1) - X_j(t)$. If $j \leq t$, then the conditional probability that $\Delta_j(t) = 1$, given $\mathrm{PA}_t^{(1,\delta)}$, is equal to

$$\mathbb{P}\left(\Delta_j(t) = 1 \mid \mathrm{PA}_t^{(1,\delta)}\right) = X_j(t)/n(t), \tag{8.7.3}$$

where $n(t) = (2+\delta)t + 1 + \delta$ is the total weight of all the vertices at time t. From this, we get

$$\mathbb{E}\left(X_j(t+1) \mid \mathrm{PA}_t^{(1,\delta)}\right) = X_j(t)\left(1 + \frac{1}{n(t)}\right), \tag{8.7.4}$$

so that $(c_t X_j(t))_{t\geq 1}$ is a martingale with respect to $(\mathrm{PA}_t^{(1,\delta)})_{t\geq 1}$ if and only if $c_{t+1}/c_t = n(t)/(n(t)+1)$.

Anticipating the definition of a larger collection of martingales, we let

$$c_k(t) = \frac{\Gamma(t + \frac{1+\delta}{2+\delta})}{\Gamma(t + \frac{k+1+\delta}{2+\delta})}, \qquad t, k \geq 1. \tag{8.7.5}$$

For fixed $k \geq 0$, by (8.3.9),

$$c_k(t) = t^{-k/(2+\delta)}(1 + o(1)) \quad \text{as } t \to \infty. \tag{8.7.6}$$

By the recursion $\Gamma(r) = (r-1)\Gamma(r-1)$,

$$\frac{c_k(t+1)}{c_k(t)} = \frac{t + \frac{1+\delta}{2+\delta}}{t + \frac{k+1+\delta}{2+\delta}} = \frac{n(t)}{n(t)+k}. \tag{8.7.7}$$

In particular, it follows that $c_1(t)X_j(t)$ is a martingale for $t \geq j$. Being a positive martingale it converges a.s. to a random variable ξ_j, as discussed in full detail in Theorem 8.2 and its proof. We now extend this line of arguments considerably, by studying various joint distributions of degrees.

To study the joint distribution of the $X_j(t)$ we make use of a whole class of martingales. We first introduce some notation. For $a, b > -1$ with $a - b > -1$, where a, b are not necessarily integers, we write

$$\binom{a}{b} = \frac{\Gamma(a+1)}{\Gamma(b+1)\Gamma(a-b+1)}. \tag{8.7.8}$$

The restriction on a, b is such that the arguments of the Gamma-function are all strictly positive, which makes the Gamma-function well defined. Then the following proposition identifies a whole class of useful martingales related to the degrees of the vertices:

Proposition 8.15 (A rich class of degree martingales) *Let $r \geq 0$ be an integer, $k_1, k_2, \ldots, k_r > -\max\{1, 1+\delta\}$, and $1 \leq j_1 < \cdots < j_r$ be integers. Then, with $k = \sum_{i=1}^r k_i$,*

$$Z_{\vec{j},\vec{k}}(t) = c_k(t) \prod_{i=1}^r \binom{X_{j_i}(t) + k_i - 1}{k_i} \tag{8.7.9}$$

is a martingale for $t \geq \max\{j_r, 1\}$.

The restriction $k_i > -\max\{1, 1+\delta\}$ is to satisfy the restrictions $a, b, a - b > -1$ in (8.7.8), since $X_j(t) \geq 1 + \delta$. Since $\delta > -1$, this means that Proposition 8.15 also holds for certain $k_i < 0$, a fact that we will make convenient use of later on.

Proof By considering the two cases $\Delta_j(t) = 0$ or $\Delta_j(t) = 1$, and using (8.7.8) and $\Gamma(r) = (r-1)\Gamma(r-1)$, it is easy to check that, for all k,

$$\begin{aligned}\binom{X_j(t+1) + k - 1}{k} &= \binom{X_j(t) + k - 1}{k} \frac{\Gamma(X_j(t+1)+k)}{\Gamma(X_j(t)+k)} \frac{\Gamma(X_j(t))}{\Gamma(X_j(t+1))} \\ &= \binom{X_j(t) + k - 1}{k} \left(1 + \frac{k\Delta_j(t)}{X_j(t)}\right).\end{aligned} \tag{8.7.10}$$

Since $m = 1$, at most one $X_j(t)$ can change at time $t + 1$, so that

$$\prod_{i=1}^r \left(1 + \frac{k_i \Delta_{j_i}(t)}{X_{j_i}(t)}\right) = 1 + \sum_{i=1}^r \frac{k_i \Delta_{j_i}(t)}{X_{j_i}(t)}. \tag{8.7.11}$$

Together, (8.7.10) and (8.7.11) imply that

$$\prod_{i=1}^r \binom{X_{j_i}(t+1) + k_i - 1}{k_i} = \left(1 + \sum_{i=1}^r \frac{k_i \Delta_{j_i}(t)}{X_{j_i}(t)}\right) \prod_{i=1}^r \binom{X_{j_i}(t) + k_i - 1}{k_i}. \tag{8.7.12}$$

Since

$$\mathbb{P}\big(\Delta_j(t) = 1 \mid \mathrm{PA}_t^{(1,\delta)}\big) = X_j(t)/n(t), \tag{8.7.13}$$

using the definition of $Z_{\vec{j},\vec{k}}(t)$ and taking expected value,

$$\mathbb{E}\left(Z_{\vec{j},\vec{k}}(t+1) \mid \mathrm{PA}_t^{(1,\delta)}\right) = Z_{\vec{j},\vec{k}}(t) \cdot \frac{c_k(t+1)}{c_k(t)}\left(1 + \frac{\sum_{i=1}^r k_i}{n(t)}\right) = Z_{\vec{j},\vec{k}}(t), \tag{8.7.14}$$

where $k = \sum_{i=1}^r k_i$ and the last equality follows from (8.7.7). □

Being a non-negative martingale, $Z_{\vec{j},\vec{k}}(t)$ converges by the Martingale Convergence Theorem (Theorem 2.24). From the form of the martingale, the convergence result for the factors and the asymptotics for the normalizing constants in (8.7.6), the limit must be $\prod_{i=1}^r \xi_i^{k_i}/\Gamma(k_i+1)$, where we recall that ξ_i is the almost sure limit of $D_i(t)t^{-1/(2+\delta)}$ from Theorem 8.2. Here we make use of the fact that $D_i(t) \xrightarrow{a.s.} \infty$ (see Exercise 8.8), together with (8.3.9), which implies that

$$\binom{X_j(t)+k-1}{k} = \frac{X_j(t)^k}{\Gamma(k+1)}(1 + O(1/X_j(t))). \tag{8.7.15}$$

Our next step is to check that the martingale converges in L^2. We use the L^2-Martingale Convergence Theorem (Theorem 2.25), for which we rely on the following L^2-boundedness result:

Lemma 8.16 (L^2-boundedness) *For $m = 1$ and $\delta > -1$. For all $r \in \mathbb{N}$, for all $j \geq 1$ integer and $k \in \mathbb{R}$ with $k \geq -\max\{1, 1+\delta\}/2$, there exists a constant $K = K_{r,k,\vec{j}}$ such that*

$$\mathbb{E}\big[Z_{\vec{j},\vec{k}}(t)^2\big] \leq K. \tag{8.7.16}$$

Proof We begin by observing that (8.7.6) implies $c_m(t)^2/c_{2m}(t) \to 1$ and

$$\binom{x+k-1}{k}^2 = \left(\frac{\Gamma(x+k)}{\Gamma(x)\Gamma(k+1)}\right)^2 = \frac{\Gamma(x+k)}{\Gamma(x)}\frac{\Gamma(x+k)}{\Gamma(x)\Gamma(k+1)^2}. \tag{8.7.17}$$

Now we use that $x \mapsto \Gamma(x+k)/\Gamma(x)$ is increasing for $k \geq 0$, to obtain $\Gamma(x+k)/\Gamma(x) \leq \Gamma(x+2k)/\Gamma(x+k)$. Substitution yields that

$$\binom{x+k-1}{k}^2 \leq \frac{\Gamma(x+2k)}{\Gamma(x+k)}\frac{\Gamma(x+k)}{\Gamma(x)\Gamma(k+1)^2} = \binom{x+2k-1}{2k}\cdot\binom{2k}{k}. \tag{8.7.18}$$

From this it follows that

$$Z_{\vec{j},\vec{k}}(t)^2 \leq C_{\vec{k}} Z_{\vec{j},2\vec{k}}(t), \tag{8.7.19}$$

where

$$C_{\vec{k}} = \prod_{i=1}^r \binom{2k_i}{k_i}. \tag{8.7.20}$$

Note that $k \geq -\max\{1, 1+\delta\}/2$, so that $2k \geq -\max\{1, 1+\delta\}$. By Proposition 8.15, $Z_{\vec{j},2\vec{k}}(t)$ is a martingale with a finite expectation that is independent of t. This completes the proof. □

By Lemma 8.16, $Z_{\vec{j},\vec{k}}(t)$ is an L^2-bounded martingale, and hence converges in L^2, and thus also in L^1. Taking $r = 1$ we have, for all $j \geq 1$ integer and $k \in \mathbb{R}$ with $k \geq -\max\{1, 1+\delta\}/2$,

$$\mathbb{E}[\xi_j^k/\Gamma(k+1)] = \lim_{t\to\infty} \mathbb{E}[Z_{j,k}(t)] = \mathbb{E}[Z_{j,k}(j)] = c_k(j-1)\binom{k+\delta}{k}. \tag{8.7.21}$$

Recalling that $c_k(j-1) = \frac{\Gamma(j-\frac{1}{2+\delta})}{\Gamma(j+\frac{k-1}{2+\delta})}$, we thus arrive at the fact that, for all j non-negative integers, and all k non-negative,

$$\mathbb{E}[\xi_j^k] = \frac{\Gamma(j-\frac{1}{2+\delta})}{\Gamma(j+\frac{k-1}{2+\delta})} \frac{\Gamma(k+1+\delta)}{\Gamma(1+\delta)}. \tag{8.7.22}$$

The above moments identify the distribution, see Exercises 8.23 and 8.24.

We next show that $\mathbb{P}(\xi_j = 0) = 0$ for all $j \geq 1$:

Lemma 8.17 (No atom at zero for ξ_j) *Fix $m = 1$ and $\delta > -1$. Then, $\mathbb{P}(\xi_j = 0) = 0$ for all $j \geq 1$.*

Proof We use (8.3.9) and (8.7.15), which imply that for $k > -\max\{1, 1+\delta\}$ and some constant $A_k < \infty$,

$$\limsup_{t\to\infty} \mathbb{E}\Big[\Big(\frac{X_j(t)}{t^{1/(2+\delta)}}\Big)^k\Big] \leq A_k \limsup_{t\to\infty} \mathbb{E}[Z_{j,k}(t)] < \infty. \tag{8.7.23}$$

Since $\delta > -1$, we have $-1-\delta < 0$, so that a negative moment of $X_j(t)/t^{1/(2+\delta)}$ remains uniformly bounded. We use that $X_j(t)/t^{1/(2+\delta)} \xrightarrow{a.s.} \xi_j$, which implies that $X_j(t)/t^{1/(2+\delta)} \xrightarrow{d} \xi_j$, so that, using the Markov inequality (Theorem 2.17), for every $\varepsilon > 0$ and $k \in (-\max\{1, 1+\delta\}, 0)$,

$$\begin{aligned}\mathbb{P}(\xi_j \leq \varepsilon) &= \limsup_{t\to\infty} \mathbb{P}\Big(X_j(t)/t^{1/(2+\delta)} \leq \varepsilon\Big) \\ &\leq \limsup_{t\to\infty} \varepsilon^{|k|}\mathbb{E}\Big[\Big(\frac{X_j(t)}{t^{1/(2+\delta)}}\Big)^k\Big] = O(\varepsilon^{|k|}).\end{aligned} \tag{8.7.24}$$

Letting $\varepsilon \downarrow 0$, we obtain that $\mathbb{P}(\xi_j = 0) = 0$ for every $j \geq 1$. □

With the above results in hand, we are now ready to complete the proof of Theorem 8.14:

Proof of Theorem 8.14 We start by proving Theorem 8.14 for $m = 1$. Let M_t denote the maximal degree in our random graph after t steps, and, for $t \geq j$, let

$$M_j(t) = \max_{1\leq i\leq j} Z_{i,1}(t). \tag{8.7.25}$$

Note that $M_t(t) = c_1(t)(M_t + \delta)$. We will prove that $M_t(t) \xrightarrow{a.s.} \sup_{j\geq 1} \xi_j$. Being a maximum of martingales, $(M_t(t))_{t\geq 1}$ is a non-negative submartingale. Therefore, $M_t(t) \xrightarrow{a.s.} \mu$ for some limiting random variable μ, and we are left to prove that $\mu = \sup_{j\geq 1} \xi_j$.

Since $Z_{j,1}(t)^k$ is a submartingale for every $k \geq 1$, $t \mapsto \mathbb{E}[Z_{j,1}(t)^k]$ is non-decreasing. Further, by the L^2-Martingale Convergence Theorem (Theorem 2.25) and the bound in (8.7.23), $Z_{j,1}(t)^k$ converges in L^2 to ξ_j^k, so that

$$\mathbb{E}[Z_{j,1}(t)^k] \leq \mathbb{E}[\xi_j^k]. \tag{8.7.26}$$

Then, using the trivial inequality

$$M_t(t)^k = \max_{1 \leq j \leq t} Z_{j,1}(t)^k \leq \sum_{j=1}^{t} Z_{j,1}(t)^k, \tag{8.7.27}$$

and (8.7.26), we obtain

$$\mathbb{E}[M_t(t)^k] \leq \sum_{j=1}^{t} \mathbb{E}[Z_{j,1}(t)^k] \leq \sum_{j=1}^{\infty} \mathbb{E}[\xi_j^k] = \Gamma(k+1)\binom{k+\delta}{k} \sum_{j=1}^{\infty} c_k(j), \tag{8.7.28}$$

which is finite by (8.7.6) if $k > 2 + \delta$. Thus $M_t(t)$ is bounded in L^k for every integer $k > 2 + \delta$, and hence bounded and convergent in L^p for any $p \geq 1$.

We conclude that in order to prove that $\mu = \sup_{j \geq 0} \xi_j$, we are left to show that $M_t(t)$ converges to $\sup_{j \geq 0} \xi_j$ in L^k for some k.

Let $k > 2 + \delta$ be fixed. Then, by a similar inequality as in (8.7.27),

$$\mathbb{E}\big[(M_t(t) - M_j(t))^k\big] \leq \sum_{i=j+1}^{t} \mathbb{E}[Z_{i,1}(t)^k] \tag{8.7.29}$$

Since $M_j(t)$ is a *finite* maximum of martingales, it is a non-negative submartingale. Since each $M_j(t) \xrightarrow{a.s.} \xi_j$, and the convergence also holds in L^k for any $k > 2 + \delta$, also $\max_{1 \leq i \leq j} Z_{j,1}(t) \xrightarrow{a.s.} \max_{1 \leq i \leq j} \xi_i = \mu_j$, Therefore, the limit of the left-hand side of (8.7.29) is

$$\mathbb{E}\Big[\Big(\lim_{t \to \infty} t^{-1/(2+\delta)} M_t - \mu_j\Big)^k\Big], \tag{8.7.30}$$

while the right-hand side of (8.7.29) increases to (compare to (8.7.26))

$$\sum_{i=j+1}^{\infty} \mathbb{E}[\xi_i^k] = k!\binom{k+\delta}{k} \sum_{i=j+1}^{\infty} c_k(i), \tag{8.7.31}$$

which is small if j is large by (8.7.6). Since $t^{-1/(2+\delta)} M_t \xrightarrow{a.s.} \mu$, we obtain

$$\lim_{j \to \infty} \mathbb{E}\Big[\big(\mu - \mu_j\big)^k\Big] = 0. \tag{8.7.32}$$

Hence $t^{-1/(2+\delta)} M_t \xrightarrow{a.s.} \sup_{j \geq 1} \xi_j$ as claimed.

We next extend the argument to $m \geq 2$. By Exercise 8.13,

$$D_i(t)(mt)^{-1/(2+\delta/m)} \xrightarrow{a.s.} \xi_i', \tag{8.7.33}$$

where

$$\xi_i' = \sum_{j=(i-1)m+1}^{mi} \xi_j, \tag{8.7.34}$$

and ξ_j is the almost-sure limit of $D_j(t)$ in $(\mathrm{PA}_t^{(1,\delta/m)})_{t\geq 1}$. This implies that $M_t \xrightarrow{a.s.} \mu = \sup_{j\geq 1} \xi_j'$. We omit the details.

We next show that $\mathbb{P}(\mu = 0) = 0$. For $m = 1$ and by Lemma 8.17, $\mathbb{P}(\xi_1 = 0) = 0$, and we conclude that $\mathbb{P}(\mu = 0) = \mathbb{P}(\sup_{j=1}^{\infty} \xi_j = 0) \leq \mathbb{P}(\xi_1 = 0) = 0$. Since $\xi_1' \geq \xi_1$, this also holds for $m \geq 2$. □

8.8 Related Results for Preferential Attachment Models

In this section, we collect some related results on preferential attachment models.

The Limit Law of the Maximal Degree

There are more precise results concerning the maximal degrees in preferential attachment models, particularly about the precise limiting distribution in Theorem 8.14. Fix $m \geq 1$ and $\delta = 0$. Let $(X_i)_{i\geq 1}$ be the points of a Poisson process on $[0, \infty)$ with rate m, so, setting $X_0 = 0$, the variables $X_i - X_{i-1}$ are i.i.d. exponentials with mean $1/m$. Let $Y_i = \sqrt{X_{mi}}$ and let $\mu = \mu_m = \max_{i\geq 1}(Y_i - Y_{i-1})$. (It can be seen that this maximum exists with probability one.) Then Bollobás and Riordan (2003a) prove that $\max_i D_i(t)/(2m\sqrt{t}) \xrightarrow{d} \mu$.

Second Degrees

Fix $m = 1$ and $\delta = 0$. We define the *second degree* of vertex i at time t by

$$D_i^{(2)}(t) = \#\{kl : k, l \neq i, ik, kl \in \mathrm{PA}_t^{(1,0)}\}, \tag{8.8.1}$$

i.e., $D_i^{(2)}(t)$ is the number of vertices at graph distance 2 away from i. We let $N_d^{(2)}(t)$ denote the number of vertices i with $D_i^{(2)}(t) = d$. Then Ostroumova and Grechnikov (2012) prove that

$$\mathbb{E}[N_d^{(2)}(t)] = \frac{4t}{d^2}\big[1 + O((\log d)^2/d) + O(d^2/t)\big], \tag{8.8.2}$$

and there exists a sequence $b_t = o(t)$ such that

$$\mathbb{P}\Big(\big|N_d^{(2)}(t) - \mathbb{E}[N_d^{(2)}(t)]\big| \geq b_t/d^2\Big) = o(1). \tag{8.8.3}$$

It would be of interest to obtain an exact expression for $\lim_{t\to\infty} \mathbb{E}[N_d^{(2)}(t)]/t$ in the general case where $m \geq 1$ and $\delta > -m$.

Assortativity Coefficient of the PAM

Computations such as the ones leading to the above also allow one to identify the *assortativity coefficient* of the PAM. We start with some notation. Let $G = (V, E)$ be an undirected random graph. For a *directed* edge $e = (u, v)$, we write $\underline{e} = u, \overline{e} = v$ and we denote the set of directed edges in E by E' (so that $|E'| = 2|E|$), and D_v is the degree of vertex $v \in V$.

Recall (1.5.8). The *assortativity coefficient* of G is defined by Newman (see Newman, 2002, (4)) as

$$\rho(G) = \frac{\frac{1}{|E'|}\sum_{(u,v)\in E'} D_u D_v - \Big(\frac{1}{|E'|}\sum_{(u,v)\in E'} \frac{1}{2}(D_u + D_v)\Big)^2}{\frac{1}{|E'|}\sum_{(u,v)\in E'} \frac{1}{2}(D_u^2 + D_v^2) - \Big(\frac{1}{|E'|}\sum_{(u,v)\in E'} \frac{1}{2}(D_u + D_v)\Big)^2}, \tag{8.8.4}$$

so that the assortativity coefficient in (8.8.4) is equal to the sample correlation coefficient of the sequence of variables $((D_u, D_v))_{(u,v)\in E'}$. Then, one can explicitly compute that

$$\rho(\mathrm{PA}_t^{(m,\delta)}) \xrightarrow{\mathbb{P}} \begin{cases} 0 & \text{if } \delta \le m, \\ \rho & \text{if } \delta > m, \end{cases} \tag{8.8.5}$$

where, abbreviating $a = \delta/m$,

$$\rho = \frac{(m-1)(a-1)[2(1+m) + a(1+3m)]}{(1+m)[2(1+m) + a(5+7m) + a^2(1+7m)]}. \tag{8.8.6}$$

Peculiarly, the degrees across edges are always *uncorrelated* when $m = 1$, i.e., for preferential attachment trees. We have no convincing explanation for this fact. The fact that $\rho(\mathrm{PA}_t^{(m,\delta)}) \xrightarrow{\mathbb{P}} 0$ for $\delta \in (-m, m)$ is an example of a general feature, explored in more detail in Litvak and van der Hofstad (2013) and van der Hofstad and Litvak (2014), that the assortativity coefficient converges to zero when the graph sequence has third moment degrees that converge to infinity. Infinite third moment degrees correspond to $\delta \le m$, which explains that the assortativity coefficient vanishes in (8.8.5) in this case.

The proof of (8.8.5)–(8.8.6) relies on a more general result about the convergence of the local topology of $\mathrm{PA}_t^{(m,\delta)}$. Indeed, let $N_{k,l}(t)$ denote the number of oriented edges (u, v) for which $D_u(t) = k$, $D_v(t) = l$. Then, there exists a limiting (joint) probability mass function $q = (q_{k,l})$ such that, for every $k, l \ge m$,

$$N_{k,l}(t)/(2mt) \xrightarrow{\mathbb{P}} q_{k,l}. \tag{8.8.7}$$

Note that $\sum_l q_{k,l} = kp_k/(2m)$, and $\sum_l N_{k,l}(t) = kN_k(t)$. The limiting law can be identified as

$$q_{k,l} = \tfrac{1}{2}(p_{k,l} + p_{l,k}), \tag{8.8.8}$$

where $(p_{k,l})_{k,l\ge 1}$ is the limiting distribution of $(D_{U_1}(t), D_{U_2}(t))$, where $U_1 \in [t]$ is a uniform vertex in $[t]$, while U_2 is one of the m vertices to which U_1 has attached one of its m edges, chosen uniformly at random among them. For $k \ge m$ and $l \ge m+1$, $p_{k,l}$ can be identified as (with $b = 2m + \delta$)

$$p_{k,l} = (2+\delta/m)(3+\delta/m)\frac{\Gamma(j+b+2)}{\Gamma(j+1)\Gamma(b+1)}\frac{\Gamma(k+b)}{\Gamma(k+1)\Gamma(b)} \times \int_0^1 (1-v)^{j-m} v^{b+2+\delta/m} \int_v^1 (1-u)^{k-m-1} u^b du dv. \tag{8.8.9}$$

The above result follows from a detailed description of local neighborhoods in $\mathrm{PA}_t^{(m,\delta)}$ that relies on de Finetti's theorem and Pólya urn schemes that we describe in more detail in [II, Chapter 6].

8.9 Related Preferential Attachment Models

There are numerous related preferential attachment models in the literature. Here we discuss a few of them:

A Directed Preferential Attachment Model

Bollobás, Borgs, Chayes and Riordan (2003) investigate a directed preferential attachment model and prove that the degrees obey a power law similar to the one in Theorem 8.3. We first describe the model. Let G_0 be any fixed initial directed graph with t_0 edges, where t_0 is some arbitrary positive integer.

We next define $G(t)$. Fix some non-negative parameters α, β, γ, δ_{in} and δ_{out}, where $\alpha + \beta + \gamma = 1$. We say that we choose a vertex *according to* $f_i(t)$ when we choose vertex i with probability

$$\frac{f_i(t)}{\sum_j f_j(t)}. \tag{8.9.1}$$

Thus, the probability that we choose a vertex i is proportional to the value of the function $f_i(t)$. Also, we denote the in-degree of vertex i in $G(t)$ by $D_i^{(\text{in})}(t)$, and the out-degree of vertex i in $G(t)$ by $D_i^{(\text{out})}(t)$.

We let $G(t_0) = G_0$, where t_0 is chosen appropriately, as we will indicate below. For $t \geq t_0$, we form $G(t+1)$ from $G(t)$ according to the following growth rules:

(A) With probability α, we add a new vertex v together with an edge from v to an existing vertex w which is chosen according to $D_w^{(\text{in})}(t) + \delta_{\text{in}}$.
(B) With probability β, we add an edge between the existing vertices v and w, where v and w are chosen independently, v according to $D_v^{(\text{in})}(t) + \delta_{\text{in}}$ and w according to $D_w^{(\text{out})}(t) + \delta_{\text{out}}$.
(C) With probability γ, we add a vertex w and an edge from an existing vertex v to w according to $D_v^{(\text{out})}(t) + \delta_{\text{out}}$.

The above growth rule produces a graph process $(G(t))_{t\geq t_0}$ where $G(t)$ has precisely t edges. The number of vertices in $G(t)$ is denoted by $T(t)$, where $T(t) \sim \mathsf{Bin}(t, \alpha+\gamma)$.

It is not hard to see that if $\alpha\delta_{\text{in}} + \gamma = 0$, then all vertices outside of G_0 have in-degree zero, while if $\gamma = 1$ all vertices outside of G_0 have in-degree one. Similar trivial graph processes arise when $\gamma\delta_{\text{out}} + \alpha = 0$ or $\alpha = 1$.

We exclude the above cases. Then, Bollobás et al. (2003) show that both the in-degree and the out-degree of the graph converge, in the sense that we will explain now. Denote by $(X_k(t))_{k\geq 0}$ the in-degree sequence of $G(t)$, so that

$$X_k(t) = \sum_{v\in G(t)} \mathbb{1}_{\{D_v^{(\text{in})}=k\}}, \tag{8.9.2}$$

and, similarly, let $(Y_k(t))_{k\geq 0}$ be the out-degree sequence of $G(t)$, so that

$$Y_k(t) = \sum_{v\in G(t)} \mathbb{1}_{\{D_v^{(\text{out})}=k\}}. \tag{8.9.3}$$

Denote

$$\tau_{\rm in} = 1 + \frac{1+\delta_{\rm in}(\alpha+\beta)}{\alpha+\beta}, \qquad \tau_{\rm out} = 1 + \frac{1+\delta_{\rm out}(\gamma+\beta)}{\gamma+\beta}. \tag{8.9.4}$$

Then (Bollobás et al., 2003, Theorem 3.1) shows that there exist probability distributions $p = (p_k)_{k\geq 0}$ and $q = (q_k)_{k\geq 0}$ such that

$$X_k(t) - p_k t = o_{\mathbb{P}}(t), \qquad Y_k(t) - q_k t = o_{\mathbb{P}}(t), \tag{8.9.5}$$

where, for $k \to \infty$,

$$p_k = C_{\rm in} k^{-\tau_{\rm in}}(1+o(1)), \qquad q_k = C_{\rm out} k^{-\tau_{\rm out}}(1+o(1)). \tag{8.9.6}$$

In fact, the probability distributions p and q are determined explicitly, as in (8.4.2) above, and p and q have a similar shape as p in (8.4.2). Also, since $\delta_{\rm in}, \delta_{\rm out} \geq 0$, and $\alpha+\beta, \gamma+\beta \leq 1$, we again have that $\tau_{\rm in}, \tau_{\rm out} \in (2, \infty)$.

A General Preferential Attachment Model

A quite general version of preferential attachment models is presented by Cooper and Frieze (2003). In this paper, an undirected graph process is defined. At time 0, there is a single initial vertex v_0. Then, to go from $G(t)$ to $G(t+1)$, either a new vertex or a number of edges between existing vertices can be added. The first case is called NEW, the second OLD. With probability α, we choose to apply the procedure OLD, and with probability $1-\alpha$ we apply the procedure NEW.

In the procedure NEW, we add a single vertex, and let $f = (f_i)_{i\geq 1}$ be such that f_i is the probability that the new vertex generates i edges. With probability β, the end vertices of these edges are chosen *uniformly* among the vertices, and, with probability $1-\beta$, the end vertices of the added edges are chosen proportionally to the degree.

In the procedure OLD, we choose a single old vertex. With probability δ, this vertex is chosen uniformly, and with probability $1-\delta$, it is chosen with probability proportionally to the degree. We let $g = (g_i)_{i\geq 1}$ be such that g_i is the probability that the old vertex generates i edges. With probability γ, the end vertices of these edges are chosen *uniformly* among the vertices, and, with probability $1-\gamma$, the end vertices of the added edges are chosen proportionally to the degree.

The main result by Cooper and Frieze (2003) states that the empirical degree distribution converges to a probability distribution which obeys a power law with a certain exponent τ that depends on the parameters of the model. More precisely, a result such as in Theorem 8.3 is proved, at least for $k \leq t^{1/21}$. Also, a version of Proposition 8.7 is proved, where the error term $\mathbb{E}[N_k(t)] - tp_k$ is proved to be at most $Mt^{1/2}\log t$. For this result, some technical conditions need to be made on the first moment of f, as well as on the distribution g. The result is nice, because it is quite general. The precise bounds are a bit weaker than the ones presented here.

Interestingly, also the maximal degree is investigated, and it is shown that the maximal degree is of order $\Theta_{\mathbb{P}}(t^{1/(\tau-1)})$ as one would expect. This result is proved as long as $\tau < 3$.[2]

[2] In Cooper and Frieze (2003, Page 318), it is mentioned that when the power law holds with power-law exponent τ, this suggests that the maximal degree should grow like $t^{1/\tau}$. However, when the degrees are independent and identically distributed with a power-law exponent equal to τ, then the maximal degree should grow like $\Theta(t^{1/(\tau-1)})$, which is precisely what is proved in Cooper and Frieze (2003, Theorems 2 and 5).

Finally, results close to those that we present here are given by Aiello et al. (2002). In fact, the error bound in Proposition 8.7 is proved there for $m = 1$ for several models. The result for $m > 1$ is, however, not contained there.

Non-Linear Preferential Attachment

There is also work on preferential attachment models where the probability of connecting to a vertex with degree k depends in a non-linear way on k. In Krapivsky et al. (2000), the attachment probabilities have been chosen proportional to k^γ for some γ. The linear case was non-rigorously investigated by Krapivsky and Redner (2001), and the cases where $\gamma \neq 1$ by Krapivsky et al. (2000). As one can expect, the results depend dramatically in the choice of γ. When $\gamma < 1$, the degree sequence is predicted to have a power law with a certain stretched exponential cut-off. Indeed, the number of vertices with degree k at time t is predicted to be roughly equal to $t\alpha_k$, where

$$\alpha_k = \frac{\mu}{k^\gamma} \prod_{j=1}^{k} \frac{1}{1 + \frac{\mu}{j^\gamma}}, \tag{8.9.7}$$

and where μ satisfies the implicit equation that $\sum_k \alpha_k = 1$. When $\gamma > 1$, Krapivsky and Redner (2001) predict that there is a single vertex that is connected to nearly all the other vertices. In more detail, when $\gamma \in (1+\frac{1}{m+1}, 1+\frac{1}{m})$, it is predicted that there are only finitely many vertices that receive at least $m+1$ links, while there are, asymptotically, infinitely many vertices that receive at least m links. This was proved rigorously by Oliveira and Spencer (2005).

Rudas, Tóth and Valko (2007) study random trees with possibly non-linear preferential attachment by relating them to continuous-time branching processes and using properties of such branching processes. Their analysis can be seen as a way to make the heuristic in Section 1.7.2 precise. To explain their results, let $f(k)$ be the weight of a vertex of degree k. The function $k \mapsto f(k)$ will act as the preferential attachment function. The random tree evolves, conditionally on the tree at time t, by attaching the $(t+1)$st vertex to vertex i with probability proportional to $f(D_i(t) - 1)$. Let λ^* be the solution, if it exists, of the equation

$$1 = \sum_{n=1}^{\infty} \prod_{i=0}^{n-1} \frac{f(i)}{f(i) + \lambda}. \tag{8.9.8}$$

Then, it is proved in Rudas et al. (2007) that the degree distribution converges to $(p_k)_{k \geq 0}$, where[3]

$$p_k = \frac{\lambda^*}{f(k) + \lambda^*} \prod_{i=0}^{k-1} \frac{f(i)}{f(i) + \lambda^*}. \tag{8.9.9}$$

For linear preferential attachment models where $f(i) = i + 1 + \delta$, $\lambda^* = \delta$, so that (8.9.9) reduces to (8.4.3) (see Exercise 8.27).

[3] The notion of degree used in Rudas et al. (2007) is slightly different since Rudas et al. (2007) makes use of the in-degree only. For trees, the degree is the in-degree plus 1, which explains the apparent difference in our formula.

Interestingly, Rudas, Tóth and Valko (2007) study not only the degree of a uniformly chosen vertex, but also its neighborhood. We refrain from describing these results here. These analyses are extended beyond the tree case by Bhamidi (In preparation 2007).

Preferential Attachment with Fitness

The models studied by Bianconi and Barabási (2001a,b) and by Ergün and Rodgers (2002) include preferential attachment models with *random fitness*. In general, in such models, the vertex v_i which is added at time i is given a random fitness (ζ_i, η_i). The later vertex v_t at time $t > i$ connects to vertex v_i with a conditional probability which is proportional to $\zeta_i D_i(t) + \eta_i$. The variable ζ_i is called the *multiplicative fitness*, and η_i is the *additive fitness*. The case of additive fitness only was introduced by Ergün and Rodgers (2002), the case of multiplicative fitness was introduced by Bianconi and Barabási (2001a,b) and studied further by Borgs, Chayes, Daskalis and Roch (2007). Bhamidi (In preparation 2007) finds the exact degree distribution both for the additive and multiplicative models.

Preferential Attachment and Power-Law Exponents in $(1, 2)$

In all models, and similarly to Theorem 8.3, the power-law exponents τ are limited to the range $(2, \infty)$. It would be of interest to find simple examples where the power law exponent can lie in the interval $(1, 2)$. A possible solution to this is presented by Deijfen et al. (2009), where a preferential attachment model is presented in which a *random* number of edges can be added which is, unlike in the work by Cooper and Frieze (2003), not bounded. In this case, when the number of edges obeys a power law, there is a cross-over between a preferential attachment power law and the power law from the edges, the one with the smallest exponent winning. Unfortunately, the case where the weights have degrees with power-law exponent in $(1, 2)$ is not entirely analyzed. The conjecture by Deijfen et al. (2009) in this case is partially proved in Bhamidi (In preparation 2007, Theorem 40).

Universal Techniques to Study Preferential Attachment Models

Bhamidi (In preparation 2007) investigates various preferential attachment models using universal techniques from continuous-time branching processes (see Aldous (1991) and the works by Jagers and Nerman (1984, 1996); Nerman and Jagers (1984)) to prove powerful results for preferential attachment graphs. Models that can be treated within this general methodology include fitness models (Bianconi and Barabási (2001a,b); Ergün and Rodgers (2002)), competition-induced preferential attachment models (Berger et al. (2004, 2005a)), linear preferential attachment models as studied in this chapter, but also sublinear preferential attachment models and preferential attachment models with a cut-off. Bhamidi is able to prove results for (1) the degree distribution of the graph; (2) the maximal degree; (3) the degree of the initial root; (4) the local neighborhoods of vertices; (5) the height of various preferential attachment trees; and (6) properties of percolation on the graph, where we erase the edges independently and with equal probability.

Preferential Attachment Models with Conditionally Independent Edges

A preferential attachment model with conditionally independent edges is investigated by Dereich and Mörters (2009, 2011, 2013). Fix a preferential attachment function $f\colon \mathbb{N}_0 \mapsto (0, \infty)$. Then, the graph evolves as follows. Start with $G(1)$ being a graph containing one vertex v_1 and no edges. At each time $t \geq 2$, we add a vertex v_t. Conditionally on $G(t-1)$, and independently for every $i \in [t-1]$, we connect this vertex to i by a directed edge with probability

$$\frac{f(D_i(t-1))}{t-1}, \tag{8.9.10}$$

where $D_i(t-1)$ is the in-degree of vertex i at time $t-1$. This creates the random graph $G(t)$. Note that the number of edges in the random graph process $(G(t))_{t\geq 1}$ is *not* fixed, and equals a random variable. In particular, it makes a difference whether we use the in-degree in (8.9.10).

We consider functions $f\colon \mathbb{N} \mapsto (0,\infty)$ that satisfy that $f(k+1) - f(k) < 1$ for every $k \geq 0$. Under this assumption and when $f(0) \leq 1$, Mörters and Dereich show that the empirical degree sequence converges as $t \to \infty$, i.e.,

$$P_k(t) \equiv \frac{1}{t}\sum_{i\in[t]} \mathbb{1}_{\{D_i(t)=k\}} \xrightarrow{\mathbb{P}} p_k, \qquad \text{where} \qquad p_k = \frac{1}{1+f(k)}\prod_{l=0}^{k-1}\frac{f(l)}{1+f(l)}. \tag{8.9.11}$$

Note the similarity with (8.9.9). In particular, $\log(1/p_k)/\log(k) \to 1+1/\gamma$ when $f(k)/k \to \gamma \in (0,1)$, while $\log(1/p_k) \sim k^{1-\alpha}/(\gamma(1-\alpha))$ when $f(k) \sim \gamma k^{\alpha}$ for some $\alpha \in (0,1)$. Interestingly, Mörters and Dereich also show that when $\sum_{k\geq 1} 1/f(k)^2 < \infty$, then there exists a persistent hub, i.e., a vertex that has maximal degree for all but finitely many times. When $\sum_{k\geq 1} 1/f(k)^2 = \infty$, this does not happen.

8.10 Notes and Discussion

Notes on Section 8.2

There are various ways of modeling the Rich-get-Richer or preferential attachment phenomenon.

Notes on Section 8.3

The degrees of fixed vertices plays a crucial role in the analysis of preferential attachment models, see, e.g., the first mathematical work on the topic by Bollobás, Riordan, Spencer and Tusnády (2001). Szymański (2005) computes several moments of the degrees for the Barabási–Albert model, including the result in Theorem 8.2 and several extensions.

For $m = 1$ and $\delta = 0$, Peköz, Röllin and Ross (2013) investigate the explicit law of the different vertices, and identify the density to be related to confluent hypergeometric functions of the second kind (also known as the Kummer U function). Interestingly, the precise limit is different (though related) whether we allow for self-loops or not. They also provide bounds of order $1/\sqrt{t}$ for the total variation distance between the law of $D_i(t)/\sqrt{\mathbb{E}[D_i(t)^2]}$ and the limiting law, using Stein's method.

Notes on Section 8.4

Most papers on specific preferential attachment models prove that the degree sequences obey a power law. We shall refer in more detail to the various papers on the topic when we discuss the various different ways of proving Proposition 8.7. General results in this direction can be found for example in the work by Bhamidi (In preparation 2007). The fact that the limiting degree distribution is a mixed-negative binomial distribution as stated in (8.4.10) is due to Ross (2013).

Notes on Section 8.5

The proof of Theorem 8.3 relies on two key propositions, namely, Propositions 8.4 and 8.7. Proposition 8.4 is a key ingredient in the investigation of the degrees in preferential attachment models, and is used in many related results for other models. The first version, as far as we know, of this proof is by Bollobás et al. (2001).

Notes on Section 8.6

The proof of the expected empirical degree sequence in Proposition 8.7 is new, and proves a stronger result than the one for $\delta = 0$ appearing in Bollobás et al. (2001). The proof of Proposition 8.7 is also quite flexible. For example, instead of the growth rule in (8.2.1), we could attach the m edges of the newly added vertex $v_{t+1}^{(m)}$ each independently and with equal probability to a vertex $i \in [t]$ with probability proportional to $D_i(t) + \delta$. More precisely, this means that, for $t \geq 3$,

$$\mathbb{P}\big(v_{t+1}^{(m)} \to v_i^{(m)} \big| \mathrm{PA}_t^{(m,\delta)}\big) = \frac{D_i(t) + \delta}{t(2m + \delta)} \qquad \text{for } i \in [t], \tag{8.10.1}$$

and, conditionally on $\mathrm{PA}_t^{(m,\delta)}$, the attachment of the edges is *independent*. We can define $\mathrm{PA}_2^{(m,\delta)}$ to consist of 2 vertices connected by m edges.

It is not hard to see that the proof of Proposition 8.4 applies verbatim (see Exercise 8.28).

It is also not hard to see that the proof of Proposition 8.7 applies by making the obvious changes. In fact, the limiting degree sequence remains unaltered. A second slightly different model, in which edges are added independently without intermediate updating, is studied by Jordan (2006).

The original proof by Bollobás et al. (2001) of the asymptotics of the expected empirical degree sequence for $\delta = 0$ makes use of an interesting relation between this model and so-called *n-pairings*. An n-pairing is a partition of the set $\{1, \ldots, 2n\}$ into pairs. We can think about the pairs as being points on the x-axis, and the pairs as chords joining them. This allows us to speak of the left and right endpoints of the pairs.

The link between an n-pairing and the preferential attachment model with $\delta = 0$ and $m = 1$ is obtained as follows. We start from the left, and merge all left endpoints up to and including the first right endpoint into the single vertex v_1. Then, we merge all further left endpoints up to the next right endpoint into vertex v_2, etc. For the edges, we replace each pair by a directed edge from the vertex corresponding to its right endpoint to the vertex corresponding to its left endpoint. Then, as noted by Bollobás and Riordan (2004), the resulting graph has the same distribution as $\mathrm{PA}_t^{(1,0)}$. The proof by Bollobás et al. (2001) then uses explicit computations to prove that for $k \leq t^{1/15}$,

$$\mathbb{E}[N_k(t)] = tp_k(1 + o(1)). \tag{8.10.2}$$

The advantage of the current proof is that the restriction on k in $k \leq t^{1/15}$ is absent, that the error term in (8.10.2) is bounded uniformly by a constant, and that the proof applies to $\delta = 0$ and $\delta \neq 0$.

The approach by Hagberg and Wiuf (2006) is closest to ours. In it, the authors assume that the model is a preferential attachment model, where the expected number of vertices of degree k in the graph at time $t+1$, conditionally on the graph at time t solves

$$\mathbb{E}[N_k(t+1) \mid N(t)] = (1 - \frac{a_k}{t})N_k(t) - \frac{a_{k-1}}{t}N_{k-1}(t) + c_k, \tag{8.10.3}$$

where $N_k(t)$ is the number of vertices of degree k at time t, $N(t) = (N_k(t))_{k\geq 0}$ and it is assumed that $a_{-1} = 0$, and where $c_k \geq 0$ and $a_k \geq a_{k-1}$. Also, it is assumed that $|N_k(t) - N_k(t-1)|$ is uniformly bounded. This is almost true for the model considered in this chapter. Finally, $(N(t))_{t\geq 0}$ is assumed to be a Markov process, starting at some time t_0 in a configuration $N(t_0)$. Then, with

$$\alpha_k = \sum_{j=0}^{k} \frac{c_j}{1+a_j} \prod_{i=j+1}^{\infty} \frac{a_{i-1}}{1+a_i}, \tag{8.10.4}$$

it is shown that $N_k(t)/t$ converges to α_k. Exercise 8.29 investigates the monotonicity properties in t of $\sup_{j\leq k} |\mathbb{E}[N_j(t) - t\alpha_j|$.

Notes on Section 8.7

The beautiful martingale description in Proposition 8.15 is due to Móri (2002) (see also Móri (2005)). We largely follow the presentation in (Durrett, 2007, Section 4.3), adapting it to the setting of the preferential attachment models introduced in Section 8.2. The fact that Proposition 8.15 also holds for *non-integer* k_i is, as far as we know, new. This is relevant, since it identifies *all* moments of the limiting random variables ξ_j.

We have reproduced Mori's argument, applied to a slightly different model. See also (Durrett, 2007, Section 4.3). Móri (2002) studies a setting in which the graph at time 1 consists of two vertices, 0 and 1, connected by a single edge. In the attachment scheme, no self-loops are created, so that the resulting graph is a tree. The proof generalizes easily to other initial configurations and attachment rules, and we have adapted the argument here to the usual preferential attachment model in which self-loops *do* occur and $\mathrm{PA}_1^{(1,\delta)}$ consists of one vertex with a single self-loop. Lancelot James corrected an error in Exercise 8.24.

Notes on Section 8.8

The results on second degrees were obtained by Ostroumova and Grechnikov (2012). The explicit limiting distribution of $\max_i D_i(t)/(2m\sqrt{t})$ was proved by Bollobás and Riordan (2003a). For $m = 1$, the precise form of the limiting distribution was also obtained by Peköz, Röllin and Ross (Preprint (2014)), who generalized it to the setting where the graph at time 2 consists of two vertices with one edge in between, and has no self-loops. The key observation in this proof is that the sequence $(D_i(t)/2\sqrt{t})_{i=1}^k$, for each $k \geq 1$, converges jointly in distribution to the random vector $(Y_1, Y_2-Y_1, \ldots, Y_k-Y_{k-1})$, where $Y_i = \sqrt{X_i}$. This is performed by bounding the total variation distance between the distributions $(D_i(t)/2\sqrt{t})_{i=1}^k$ and $(Y_1, Y_2 - Y_1, \ldots, Y_k - Y_{k-1})$ by $C(k)/\sqrt{t}$ using Stein's method.

The convergence of the assortativity coefficient is performed rigorously by van der Hofstad and Litvak (2014), and it was already predicted by Dorogovtsev, Ferreira, Goltsev and Mendes (2010). The formula for $p_{k,l}$ in (8.8.9) is proved by Berger, Borgs, Chayes and Saberi (see Berger et al., 2014, Lemma 5.3) in the case where $\delta \geq 0$. The latter paper is an extension of Berger et al. (2005b), where the spread of an infection on preferential attachment graphs is investigated, and also contains local weak convergence results.

Notes on Section 8.9

The results on directed preferential attachment models are due to Bollobás, Borgs, Chayes and Riordan (2003). The proof in Bollobás et al. (2003) is similar to the one chosen here. Again the proof is split into a concentration result as in Proposition 8.4, and a determination of the expected empirical degree sequence in Proposition 8.7. In fact, the proof Proposition 8.7 is adapted after the proof in Bollobás et al. (2003), which is also based on the recurrence relation in (8.6.22), but analyses it in a different way, by performing induction on k, rather than on t as we do in Sections 8.6.1 and 8.6.2. As a result, the result proved in Proposition 8.7 is slightly stronger. A related result on a directed preferential attachment model can be found in Buckley and Osthus (2004). In this model, the preferential attachment probabilities only depend on the in-degrees, rather than on the total degree, and power-law in-degrees are proved. In Bollobás et al. (2003), there is also a result on the joint distribution of the in- and out-degrees of $G(t)$, which we shall not state here.

Preferential attachment models with conditionally independent edges are studied by Dereich and Mörters (2009, 2011), where the degree evolution is discussed. The giant component is investigated by Dereich and Mörters (2013), while vulnerability is studied by Eckhoff and Mörters (2014). There, also the vulnerability of various other models (such as the configuration model and generalized random graphs) is investigated.

8.11 Exercises for Chapter 8

Exercise 8.1 (A dynamic formulation of $\mathrm{ER}_n(\lambda/n)$) Give a dynamical model for the Erdős–Rényi random graph, where at each time n we add a single individual, and where at time n the graph is equal to $\mathrm{ER}_n(\lambda/n)$. See also the dynamic description of the Norros–Reittu model on page 208.

Exercise 8.2 (Non-negativity of $D_i(t)+\delta$) Fix $m=1$. Verify that $D_i(t)\geq 1$ for all i and t with $t\geq i$, so that $D_i(t)+\delta\geq 0$ for all $\delta\geq -1$.

Exercise 8.3 (Attachment probabilities sum up to one) Verify that the probabilities in (8.2.1) sum up to one.

Exercise 8.4 (Total degree) Prove that the total degree of $\mathrm{PA}_t^{(m,\delta)}$ equals $2mt$.

Exercise 8.5 (Collapsing vs. growth of the PA model) Prove that the alternative definition of $(\mathrm{PA}_t^{(m,\delta)})_{t\geq 1}$ is indeed equal to the one obtained by collapsing m consecutive vertices in $(\mathrm{PA}_t^{(1,\delta/m)})_{t\geq 1}$.

Exercise 8.6 (Graph topology for $\delta=-1$) Show that when $\delta=-1$, the graph $\mathrm{PA}_t^{(1,\delta)}$ consists of a self-loop at vertex $v_1^{(1)}$, and each other vertex is connected to $v_1^{(1)}$ with precisely one edge. What is the implication of this result for $m>1$?

Exercise 8.7 (Alternative formulation of $\mathrm{PA}_t^{(1,\delta)}$) For $\alpha = \frac{\delta}{2+\delta}$, show that the law of $\big(\mathrm{PA}_t^{(1,\alpha)}(c)\big)_{t\geq 2}$ is equal to the one of $\big(\mathrm{PA}_t^{(1,\delta)}(b)\big)_{t\geq 2}$. For the original PA model $\big(\mathrm{PA}_t^{(1,\delta)}\big)_{t\geq 2}$ a similar identity holds, with the only difference that the coin probability $\alpha = \alpha_t = \delta(t+1)/[(2t+1)+\delta(t+1)]$ depends slightly on t. Note that, for large t, α_t is asymptotic to $\delta/(2+\delta)$, as for $\big(\mathrm{PA}_t^{(1,\delta)}(b)\big)_{t\geq 2}$.

Exercise 8.8 (Degrees grow to infinity a.s.) Fix $m = 1$ and $i \geq 1$. Prove that $D_i(t) \xrightarrow{a.s.} \infty$, by using that $\sum_{s=i}^{t} I_s \preceq D_i(t)$, where $(I_t)_{t\geq i}$ is a sequence of independent Bernoulli random variables with $\mathbb{P}(I_t = 1) = (1+\delta)/(t(2+\delta)+1+\delta)$. What does this imply for $m > 1$?

Exercise 8.9 (Degree Markov chain) Prove that the degree $(D_i(t))_{t\geq i}$ forms a (time-inhomogeneous) Markov chain. Compute its transition probabilities for $m = 1$.

Exercise 8.10 (Recursion formula for the Gamma function) Prove (8.3.2) using partial integration, and also prove that $\Gamma(n) = (n-1)!$ for $n = 1, 2, \ldots$.

Exercise 8.11 (Asymptotics for ratio $\Gamma(t+a)/\Gamma(t)$) Prove (8.3.9), using Stirling's formula (see Gradshteyn and Ryzhik, 1965, 8.327) in the form

$$\mathrm{e}^{-t} t^{t+\frac{1}{2}} \sqrt{2\pi} \leq \Gamma(t+1) \leq \mathrm{e}^{-t} t^{t+\frac{1}{2}} \sqrt{2\pi}\, \mathrm{e}^{\frac{1}{12t}}. \tag{8.11.1}$$

Exercise 8.12 (Mean degree for $m \geq 2$) Prove (8.3.11) and use it to compute $\mathbb{E}_m^{\delta}[D_i(t)]$.

Exercise 8.13 (A.s. limit of degrees for $m \geq 2$) Prove that, for $m \geq 2$ and any $i \geq 1$, $D_i(t)(mt)^{-1/(2+\delta/m)} \xrightarrow{a.s.} \xi_i'$, where

$$\xi_i' = \sum_{j=(i-1)m+1}^{mi} \xi_j, \tag{8.11.2}$$

and ξ_j is the almost-sure limit of $D_j(t)t^{-1/(2+\delta/m)}$ in $\big(\mathrm{PA}_t^{(1,\delta/m)}\big)_{t\geq 1}$.

Exercise 8.14 (Mean degree for model (b)) Prove that for $\mathrm{PA}_t^{(1,\delta)}(b)$, (8.3.3) becomes

$$\mathbb{E}[D_i(t)+\delta] = (1+\delta)\frac{\Gamma(t+1/(2+\delta))\Gamma(i)}{\Gamma(t)\Gamma(i+1/(2+\delta))}. \tag{8.11.3}$$

Exercise 8.15 (The degree of a uniform vertex) Prove that Theorem 8.3 implies that the degree at time t of a *uniform* vertex in $[t]$ converges in distribution to a random variable with probability mass function $(p_k)_{k\geq 0}$.

Exercise 8.16 (A negative binomial representation of the degree distribution) Let X have a negative binomial distribution with parameters $r = m+\delta$ and random success probability $U^{1/(2+\delta/m)}$, where U has a uniform distribution on $[0, 1]$. Prove (8.4.10) by showing that

$p_k = \mathbb{E}[\mathbb{P}(X = k)]$, where the expectation is over U. [Hint: Use the integral representation of the beta-function.]

Exercise 8.17 (Degree sequence uniform recursive tree Janson (2005)) In a uniform recursive tree we attach each vertex to a uniformly chosen old vertex. This can be seen as the case where $m = 1$ and $\delta = \infty$ of $\big(\mathrm{PA}_t^{(m,\delta)}(b)\big)_{t\geq 2}$. Show that Theorem 8.3 remains true, but now with $p_k = 2^{-(k+1)}$.

Exercise 8.18 (Concentration of the number of vertices of degree at least k) Prove (8.5.22) by adapting the proof of Proposition 8.4.

Exercise 8.19 (Formulas for C_γ and C_κ) Consider (8.6.29). Show that $C_\gamma = 1$ does the job, and $C_\kappa = \sup_{k\geq 1}(k+\delta)p_k = (1+\delta)p_1 = (1+\delta)(2+\delta)/(3+2\delta)$.

Exercise 8.20 (The total degree of high degree vertices) Use Propositions 8.7 and 8.4 to prove that for $l = l(t) \to \infty$ as $t \to \infty$ such that $tl^{2-\tau} \geq K\sqrt{t\log t}$ for some $K > 0$ sufficiently large, there exists a constant $B > 0$ such that with probability exceeding $1 - o(t^{-1})$, for all such l,

$$\sum_{i:\, D_i(t)\geq l} D_i(t) \geq Btl^{2-\tau}. \tag{8.11.4}$$

Exercise 8.21 (Dommers et al. (2010)) Fix $m = 1$ and $\delta > -1$. Then, prove that for all $t \geq i$

$$\mathbb{P}(D_i(t) = j) \leq C_j \frac{\Gamma(t)\Gamma(i + \frac{1+\delta}{2+\delta})}{\Gamma(t + \frac{1+\delta}{2+\delta})\Gamma(i)}, \tag{8.11.5}$$

where $C_1 = 1$ and

$$C_j = \frac{j-1+\delta}{j-1} C_{j-1}. \tag{8.11.6}$$

Exercise 8.22 (Martingale mean) Use Proposition 8.15 to show that, for all $t \geq \max\{j_r, 1\}$,

$$\mathbb{E}[Z_{\vec{j},\vec{k}}(t)] = \prod_{i=1}^{r} \frac{c_{K_i}(j_i - 1)}{c_{K_{i-1}}(j_i - 1)} \binom{k_i + \delta}{k_i}, \tag{8.11.7}$$

where $K_i = \sum_{a=1}^{i} k_a$.

Exercise 8.23 (Uniqueness of limit) Prove that the moments in (8.7.22) identify the distribution of ξ_j uniquely. Prove also that $\mathbb{P}(\xi_j > x) > 0$ for every $x > 0$, so that ξ_j has unbounded support.

Exercise 8.24 (A.s. limit of $D_j(t)$ in terms of limit $D_1(t)$) Use the method of moments to show that ξ_j has the same distribution as

$$\xi_1 \prod_{k=1}^{j-1} B_k, \tag{8.11.8}$$

where B_k has a Beta$((2+\delta)k-1, 1)$-distribution.

Exercise 8.25 (Martingales for alternative construction PA model Móri (2002)) Prove that when the graph at time 0 is given by two vertices with a single edge between them, and we do not allow for self-loops, then (8.7.21) remains valid when we instead define

$$c_k(t) = \frac{\Gamma(t + \frac{\delta}{2+\delta})}{\Gamma(t + \frac{k+\delta}{2+\delta})} \qquad t \geq 1, k \geq 0. \tag{8.11.9}$$

Exercise 8.26 (Special cases directed PA model) Prove that if $\alpha\delta_{in} + \gamma = 0$ in the directed preferential attachment model, then all vertices outside of G_0 have in-degree zero, while if $\gamma = 1$ all vertices outside of G_0 will have in-degree one.

Exercise 8.27 (The affine preferential attachment case) Prove that, when $\lambda^* = \delta$ and $f(i) = i + 1 + \delta$, (8.9.9) reduces to (8.4.3).

Exercise 8.28 (Adaptation concentration degree sequence) Adapt the proof of Proposition 8.4 showing the concentration of the degrees to the preferential attachment model defined in (8.10.1).

Exercise 8.29 (Monotonicity error Hagberg and Wiuf (2006)) Show that, under the assumption (8.10.3) by Hagberg and Wiuf,

$$\max_{j=1}^{k} |\mathbb{E}[N_j(t)] - \alpha_j t| \tag{8.11.10}$$

is non-increasing in t.

Appendix

Some Facts about Measure and Integration

In this section, we give some classical results from the theory of measure and integration, which will be used in the course of the proofs. For details and proofs of these results, we refer to the books by Billingsley (1995); Feller (1971); Dudley (2002); Halmos (1950). We treat three results, namely, Lebesgue's dominated convergence theorem, the monotone convergence theorem and Fatou's lemma. We also treat some slight adaptations that will prove to be useful in the text.

A.1 Lebesgue's Dominated Convergence Theorem

The statement of Lebesgue's dominated convergence theorem in probabilistic terms is as follows:

Theorem A.1 (Lebesgue's dominated convergence theorem) *Let $(X_n)_{n\geq 1}$ and Y satisfy $\mathbb{E}[Y] < \infty$, $X_n \xrightarrow{a.s.} X$, and $|X_n| \leq Y$ almost surely. Then*

$$\mathbb{E}[X_n] \to \mathbb{E}[X] \tag{A.1.1}$$

and $\mathbb{E}[|X|] < \infty$.

A proof of Theorem A.1 can be found in the classic of Feller (1971, Page 111).

We also make use of a slight extension, where almost sure convergence is replaced with convergence in distribution:

Theorem A.2 (Lebesgue's dominated convergence theorem (cont.)) *Let $(X_n)_{n\geq 1}$ and Y satisfy that $|X_n| \leq Y$ a.s. for all $n \geq 1$, where $\mathbb{E}[Y] < \infty$. Further, $X_n \xrightarrow{d} X$. Then*

$$\mathbb{E}[X_n] \to \mathbb{E}[X], \tag{A.1.2}$$

and $\mathbb{E}[|X|] < \infty$.

We give the simple proof of Theorem A.2, as it relies on the useful notion of *uniform integrability*, which we next define:

Definition A.3 (Uniform integrability) The sequence of random variables $(X_n)_{n\geq 1}$ is called *uniformly integrable* when

$$\limsup_{K\to\infty} \limsup_{n\to} \mathbb{E}[|X_n|\mathbb{1}_{\{|X_n|>K\}}] = 0. \tag{A.1.3}$$

Proof of Theorem A.2 First of all, the sequence $(X_n)_{n\geq 1}$ is uniformly integrable, since $|X_n| \leq Y$ implies that $|X_n|\mathbb{1}_{\{|X_n|>K\}} \leq Y\mathbb{1}_{\{Y>K\}}$, so that

$$\mathbb{E}[|X_n|\mathbb{1}_{\{|X_n|>K\}}] \leq \mathbb{E}[Y\mathbb{1}_{\{Y>K\}}] \to 0, \tag{A.1.4}$$

as $K \to \infty$, since $\mathbb{E}[Y] < \infty$. Thus, it suffices to prove Theorem A.2 when we replace the a.s. bound $|X_n| \leq Y$ by the requirement that $(X_n)_{n\geq 1}$ is uniformly integrable. Then, we can split

$$\mathbb{E}[X_n] = \mathbb{E}[X_n\mathbb{1}_{\{|X_n|\leq K\}}] + \mathbb{E}[X_n\mathbb{1}_{\{|X_n|>K\}}]. \tag{A.1.5}$$

The first expectation converges to $\mathbb{E}[X\mathbb{1}_{\{|X|\leq K\}}]$ by Theorem A.1, which converges to $\mathbb{E}[X]$ when $K \to \infty$. The second term vanishes when first $n \to \infty$ followed by $K \to \infty$. □

We close the discussion of Lebesgue's dominated convergence theorem by discussing a version involving *stochastic domination*:

Theorem A.4 (Lebesgue's dominated convergence theorem (cont.)) *Let $(X_n)_{n\geq 1}$ and Y satisfy that $|X_n| \preceq Y$ for all $n \geq 1$, where $\mathbb{E}[Y] < \infty$. Further, $X_n \xrightarrow{d} X$. Then*

$$\mathbb{E}[X_n] \to \mathbb{E}[X] \tag{A.1.6}$$

and $\mathbb{E}[|X|] < \infty$.

Proof Let F_{X_n}, F_X, F_Y denote the distribution functions of X_n, X, Y, respectively. Let U be uniform on $[0, 1]$, and let $\hat{X}_n = F_{X_n}^{-1}(U)$, $\hat{X} = F_X^{-1}(U)$ and $\hat{Y} = F_Y^{-1}(U)$.

Note that $\mathbb{E}[X_n] = \mathbb{E}[\hat{X}_n]$, $\mathbb{E}[X] = \mathbb{E}[\hat{X}]$ and $\mathbb{E}[Y] = \mathbb{E}[\hat{Y}]$, so it suffices to prove the statement for $\hat{X}_n$, $\hat{X}$ and $\hat{Y}$. However, since $X_n \preceq Y$, we have that $\hat{X}_n \leq \hat{Y}$ a.s. Further, $\hat{X}_n \xrightarrow{a.s.} \hat{X}$ since $X_n \xrightarrow{d} X$. Therefore, the claim follows from Theorem A.1. □

A.2 Monotone Convergence Theorem

We continue with the *monotone convergence theorem:*

Theorem A.5 (Monotone convergence theorem) *Let $(X_n)_{n\geq 1}$ be a monotonically increasing sequence, i.e., $X_n \leq X_{n+1}$ a.s. Assume that $\mathbb{E}[|X_n|] < \infty$ for every $n \geq 1$. Then $X_n(\omega) \nearrow X(\omega)$ for all ω and some limiting random variable X (that is possibly degenerate), and*

$$\mathbb{E}[X_n] \nearrow \mathbb{E}[X]. \tag{A.2.1}$$

In particular, when $\mathbb{E}[X] = \infty$, then $\mathbb{E}[X_n] \nearrow \infty$.

A proof of Theorem A.5 can again be found in (Feller, 1971, Page 110).

We again formulate a stochastic domination version:

Theorem A.6 (Monotone convergence theorem (cont.)) *Let X_n be a stochastically monotonically increasing sequence, i.e., $X_n \preceq X_{n+1}$ for all $n \geq 1$. Assume that $\mathbb{E}[|X_n|] < \infty$. Then $X_n \xrightarrow{d} X$ for some limiting random variable X (that is possibly degenerate), and*

$$\mathbb{E}[X_n] \nearrow \mathbb{E}[X]. \tag{A.2.2}$$

The proof of Theorem A.6 again follows from Theorem A.5 by instead considering $\hat{X}_n = F_{X_n}^{-1}(U)$, where U is a uniform random variable.

A.3 Fatou's Lemma

The following result, known under the name *Fatou's lemma*, shows that mass may run away to infinity, but it cannot appear out of nowhere:

Theorem A.7 (Fatou's lemma) *If $X_n \geq 0$ and $\mathbb{E}[X_n] < \infty$, then*

$$\mathbb{E}[\liminf_{n\to\infty} X_n] \leq \liminf_{n\to\infty} \mathbb{E}[X_n]. \tag{A.3.1}$$

In particular, if $X_n \xrightarrow{a.s.} X$, then

$$\mathbb{E}[X] \leq \liminf_{n\to\infty} \mathbb{E}[X_n]. \tag{A.3.2}$$

A proof of Theorem A.7 can also be found in Feller (1971, Page 110).

Glossary

$G = (V, E)$	Graph with vertex set V and edge set E.	2
T	The total progeny of a branching process.	91
$X^\star$	The size-biased version of a (non-negative) random variable X.	67
$Z_{\geq k}$	The number of vertices in connected component of size at least k.	125
$[n]$	Vertex set $[n] = \{1, \ldots, n\}$.	2
$\mathsf{Be}(p)$	Bernoulli random variable with success probability p.	49
$\mathsf{Bin}(n, p)$	Binomial random variable with parameters n and success probability p.	49
$\mathrm{CL}_n(\boldsymbol{w})$	Chung-Lu model with n vertices and weight sequence $\boldsymbol{w} = (w_i)_{i\in[n]}$.	207
$\mathrm{CM}_n(\boldsymbol{d})$	Configuration model with n vertices and degree sequence $\boldsymbol{d} = (d_i)_{i\in[n]}$.	219
$\mathscr{C}_{\max}$	The connected component of maximal size.	118
$\mathrm{ER}_n(p)$	Erdős-Rényi random graph with n vertices and edge probability p.	118
$\mathrm{GRG}_n(\boldsymbol{w})$	Generalized random graph with n vertices and weight sequence $\boldsymbol{w} = (w_i)_{i\in[n]}$.	184
$\mathrm{NR}_n(\boldsymbol{w})$	Norros-Reittu model with n vertices and weight sequence $\boldsymbol{w} = (w_i)_{i\in[n]}$.	208
$\mathrm{PA}_t^{(m,\delta)}$	Preferential attachment model with t vertices. At each time step, one vertex and m edges are added and the attachment function equals the degree plus δ.	260
$\mathsf{Poi}(\lambda)$	Poisson random variable with parameter λ.	49
$\preceq$	X is stochastically smaller than Y is written as $X \preceq Y$.	66
$\mathscr{C}(v)$	The connected component of $v \in [n]$.	118
$\xrightarrow{a.s.}$	Convergence almost surely.	58
$\xrightarrow{d}$	Convergence in distribution.	58
$\xrightarrow{\mathbb{P}}$	Convergence in probability.	58
ℓ_n	Total vertex weight $\ell_n = \sum_{i\in[n]} w_i$ for GRG, total degree $\ell_n = \sum_{i\in[n]} d_i$ for CM.	184, 218
η	The extinction probability of a branching process.	88

ζ	The survival probability of a branching process.	90
d_{TV}	Total variation distance.	63
$\mathrm{UG}_n(\boldsymbol{d})$	Uniform random graph with n vertices and degree sequence $\boldsymbol{d} = (d_i)_{i\in[n]}$.	237

References

Achlioptas, D., Clauset, A., Kempe, D., and Moore, C. 2005. On the bias of traceroute sampling or, power-law degree distributions in regular graphs. In: *STOC'05: Proceedings of the 37th Annual ACM Symposium on Theory of Computing*.

Adamic, L. A. 1999. The small world web. Pages 443–454 of: *Lecture Notes in Computer Science*, vol. **1696**. Springer.

Addario-Berry, L., Broutin, N., and Goldschmidt, C. 2010. Critical random graphs: limiting constructions and distributional properties. *Electron. J. Probab.*, **15**(25), 741–775.

Aiello, W., Chung, F., and Lu, L. 2002. Random evolution in massive graphs. Pages 97–122 of: *Handbook of massive data sets*. Massive Comput., vol. **4**. Dordrecht: Kluwer Acad. Publ.

Aigner, M., and Ziegler, G. 2014. *Proofs from The Book*. Fifth edn. Springer-Verlag, Berlin. Including illustrations by Karl H. Hofmann.

Albert, R., and Barabási, A.-L. 2002. Statistical mechanics of complex networks. *Rev. Modern Phys.*, **74**(1), 47–97.

Albert, R., Jeong, H., and Barabási, A.-L. 1999. Internet: diameter of the World-Wide Web. *Nature*, **401**, 130–131.

Albert, R., Jeong, H., and Barabási, A.-L. 2001. Error and attack tolerance of complex networks. *Nature*, **406**, 378–382.

Aldous, D. *Random graphs and complex networks*. Transparencies available from `http://www.stat.berkeley.edu/~aldous/Talks/net.ps`.

Aldous, D. 1991. Asymptotic fringe distributions for general families of random trees. *Ann. Appl. Probab.*, **1**(2), 228–266.

Aldous, D. 1993. Tree-based models for random distribution of mass. *J. Stat. Phys.*, **73**, 625–641.

Aldous, D. 1997. Brownian excursions, critical random graphs and the multiplicative coalescent. *Ann. Probab.*, **25**(2), 812–854.

Alili, L., Chaumont, L., and Doney, R. A. 2005. On a fluctuation identity for random walks and Lévy processes. *Bull. London Math. Soc.*, **37**(1), 141–148.

Alon, N., and Spencer, J. 2000. *The probabilistic method*. Second edn. Wiley-Interscience Series in Discrete Mathematics and Optimization. New York: John Wiley & Sons.

Amaral, L. A. N., Scala, A., Barthélémy, M., and Stanley, H. E. 2000. Classes of small-world networks. *Proc. Natl. Acad. Sci. USA*, **97**, 11149–11152.

Angel, O., Hofstad, R. van der, and Holmgren, C. Preprint (2016). *Limit laws for self-loops and multiple edges in the configuration model*. Available from `http://arxiv.org/abs/1603.07172`.

Anthonisse, J. 1971. *The rush in a graph*. Technical Report University of Amsterdam Mathematical Center.

Arratia, R., and Liggett, T. 2005. How likely is an i.i.d. degree sequence to be graphical? *Ann. Appl. Probab.*, **15**(1B), 652–670.

Athreya, K., and Ney, P. 1972. *Branching processes*. New York: Springer-Verlag. Die Grundlehren der mathematischen Wissenschaften, Band 196.

Austin, T. L., Fagen, R. E., Penney, W. F., and Riordan, J. 1959. The number of components in random linear graphs. *Ann. Math. Statist*, **30**, 747–754.

Azuma, K. 1967. Weighted sums of certain dependent random variables. *Tohoku Math. J.*, **3**, 357–367.

Backstrom, L., Boldi, P., Rosa, M., Ugander, J., and Vigna, S. 2012. Four degrees of separation. Pages 33–42 of: *Proceedings of the 3rd Annual ACM Web Science Conference*. ACM.

Bak, P. 1996. *How Nature Works: The Science of Self-Organized Criticality*. New York: Copernicus.

Ball, F., Sirl, D., and Trapman, P. 2009. Threshold behaviour and final outcome of an epidemic on a random network with household structure. *Adv. in Appl. Probab.*, **41**(3), 765–796.

Ball, F., Sirl, D., and Trapman, P. 2010. Analysis of a stochastic SIR epidemic on a random network incorporating household structure. *Math. Biosci.*, **224**(2), 53–73.

Barabási, A.-L. 2002. *Linked: the new science of networks*. Cambridge, Massachusetts: Perseus Publishing.

Barabási, A.-L., and Albert, R. 1999. Emergence of scaling in random networks. *Science*, **286**(5439), 509–512.

Barabási, A.-L., Albert, R., and Jeong, H. 2000. Scale-free characteristics of random networks: the topology of the world-wide web. *Phys. A*, **311**, 69–77.

Barabási, A.-L., Jeong, H., Néda, Z., Ravasz, E., Schubert, A., and Vicsek, T. 2002. Evolution of the social network of scientific collaborations. *Phys. A*, **311**(3-4), 590–614.

Barraez, D., Boucheron, S., and Fernandez de la Vega, W. 2000. On the fluctuations of the giant component. *Combin. Probab. Comput.*, **9**(4), 287–304.

Bavelas, A. 1950. Communication patterns in task-oriented groups. *Journ. Acoust. Soc. Amer.*, **22**(64), 725–730.

Behrisch, M., Coja-Oghlan, A., and Kang, M. 2014. Local limit theorems for the giant component of random hypergraphs. *Combin. Probab. Comput.*, **23**(3), 331–366.

Bender, E.A., and Canfield, E.R. 1978. The asymptotic number of labelled graphs with given degree sequences. *Journal of Combinatorial Theory (A)*, **24**, 296–307.

Bender, E.A., Canfield, E.R., and McKay, B.D. 1990. The asymptotic number of labeled connected graphs with a given number of vertices and edges. *Random Structures Algorithms*, **1**(2), 127–169.

Bennet, G. 1962. Probability inequaltities for the sum of independent random variables. *J. Amer. Statist. Assoc.*, **57**, 33–45.

Berger, N., Borgs, C., Chayes, J. T., D'Souza, R. M., and Kleinberg, R. D. 2004. Competition-induced preferential attachment. Pages 208–221 of: *Automata, languages and programming*. Lecture Notes in Comput. Sci., vol. **3142**. Berlin: Springer.

Berger, N., Borgs, C., Chayes, J. T., D'Souza, R. M., and Kleinberg, R. D. 2005a. Degree distribution of competition-induced preferential attachment graphs. *Combin. Probab. Comput.*, **14**(5-6), 697–721.

Berger, N., Borgs, C., Chayes, J.T., and Saberi, A. 2005b. On the spread of viruses on the Internet. Pages 301–310 of: *SODA '05: Proceedings of the sixteenth annual ACM-SIAM symposium on Discrete algorithms*. Philadelphia, PA, USA: Society for Industrial and Applied Mathematics.

Berger, N., Borgs, C., Chayes, J., and Saberi, A. 2014. Asymptotic behavior and distributional limits of preferential attachment graphs. *Ann. Probab.*, **42**(1), 1–40.

Bertoin, J. 2006. *Random fragmentation and coagulation processes*. Cambridge Studies in Advanced Mathematics, vol. **102**. Cambridge: Cambridge University Press.

Bhamidi, S. In preparation (2007). *Universal techniques to analyze preferential attachment trees: global and local analysis*. Available from `http://www.unc.edu/~bhamidi/preferent.pdf`.

Bhamidi, S., Hofstad, R. van der, and Hooghiemstra, G. 2010. First passage percolation on random graphs with finite mean degrees. *Ann. Appl. Probab.*, **20**(5), 1907–1965.

Bhamidi, S., Hofstad, R. van der, and Hooghiemstra, G. *Universality for first passage percolation on sparse random graphs*. Preprint (2012). To appear in *Ann. Probab.*

Bianconi, G., and Barabási, A.-L. 2001a. Bose–Einstein condensation in complex networks. *Phys. Rev. Lett.*, **86**(24), 5632–5635.

Bianconi, G., and Barabási, A.-L. 2001b. Competition and multiscaling in evolving networks. *Europhys. Lett.*, **54**, 436–442.

Billingsley, P. 1968. *Convergence of probability measures*. New York: John Wiley and Sons.

Billingsley, P. 1995. *Probability and measure*. Third edn. Wiley Series in Probability and Mathematical Statistics. New York: John Wiley & Sons Inc. A Wiley-Interscience Publication.

Boldi, P., and Vigna, S. 2014. Axioms for centrality. *Internet Math.*, **10**(3-4), 222–262.

Bollobás, B. 1980. A probabilistic proof of an asymptotic formula for the number of labelled regular graphs. *European J. Combin.*, **1**(4), 311–316.
Bollobás, B. 1981. Degree sequences of random graphs. *Discrete Math.*, **33**(1), 1–19.
Bollobás, B. 1984a. The evolution of random graphs. *Trans. Amer. Math. Soc.*, **286**(1), 257–274.
Bollobás, B. 1984b. The evolution of sparse graphs. Pages 35–57 of: *Graph theory and combinatorics (Cambridge, 1983)*. London: Academic Press.
Bollobás, B. 1998. *Modern graph theory*. Graduate Texts in Mathematics, vol. **184**. Springer-Verlag, New York.
Bollobás, B. 2001. *Random graphs*. Second edn. Cambridge Studies in Advanced Mathematics, vol. **73**. Cambridge: Cambridge University Press.
Bollobás, B., and Riordan, O. 2003a. Mathematical results on scale-free random graphs. Pages 1–34 of: *Handbook of graphs and networks*. Wiley-VCH, Weinheim.
Bollobás, B., and Riordan, O. 2003b. Robustness and vulnerability of scale-free random graphs. *Internet Math.*, **1**(1), 1–35.
Bollobás, B., and Riordan, O. 2004. The diameter of a scale-free random graph. *Combinatorica*, **24**(1), 5–34.
Bollobás, B., Riordan, O., Spencer, J., and Tusnády, G. 2001. The degree sequence of a scale-free random graph process. *Random Structures Algorithms*, **18**(3), 279–290.
Bollobás, B., Borgs, C., Chayes, J., and Riordan, O. 2003. Directed scale-free graphs. Pages 132–139 of: *Proceedings of the Fourteenth Annual ACM-SIAM Symposium on Discrete Algorithms (Baltimore, MD, 2003)*. New York: ACM.
Bollobás, B., Janson, S., and Riordan, O. 2007. The phase transition in inhomogeneous random graphs. *Random Structures Algorithms*, **31**(1), 3–122.
Bonato, A. 2008. *A course on the web graph*. Graduate Studies in Mathematics, vol. **89**. Providence, RI: American Mathematical Society.
Borgs, C., Chayes, J. T., Kesten, H., and Spencer, J. 1999. Uniform boundedness of critical crossing probabilities implies hyperscaling. *Random Structures Algorithms*, **15**(3-4), 368–413.
Borgs, C., Chayes, J. T., Kesten, H., and Spencer, J. 2001. The birth of the infinite cluster: finite-size scaling in percolation. *Comm. Math. Phys.*, **224**(1), 153–204. Dedicated to Joel L. Lebowitz.
Borgs, C., Chayes, J., Hofstad, R. van der, Slade, G., and Spencer, J. 2005a. Random subgraphs of finite graphs. I. The scaling window under the triangle condition. *Random Structures Algorithms*, **27**(2), 137–184.
Borgs, C., Chayes, J., Hofstad, R. van der, Slade, G., and Spencer, J. 2005b. Random subgraphs of finite graphs. II. The lace expansion and the triangle condition. *Ann. Probab.*, **33**(5), 1886–1944.
Borgs, C., Chayes, J., Hofstad, R. van der, Slade, G., and Spencer, J. 2006. Random subgraphs of finite graphs. III. The phase transition for the n-cube. *Combinatorica*, **26**(4), 395–410.
Borgs, C., Chayes, J. T., Daskalis, C., and Roch, S. 2007. First to market is not everything: an analysis of preferential attachment with fitness. Pages 135–144 of: *STOC '07: Proceedings of the thirty-ninth annual ACM symposium on Theory of computing*. New York, NY, USA: ACM Press.
Breiman, L. 1992. *Probability*. Classics in Applied Mathematics, vol. **7**. Society for Industrial and Applied Mathematics (SIAM), Philadelphia, PA. Corrected reprint of the 1968 original.
Bressloff, P. 2014. *Waves in neural media*. Lecture Notes on Mathematical Modelling in the Life Sciences. Springer, New York. From single neurons to neural fields.
Brin, S., and Page, L. 1998. The anatomy of a large-scale hypertextual Web search engine. Pages 107–117 of: *Computer Networks and ISDN Systems*, vol. **33**.
Britton, T., Deijfen, M., and Martin-Löf, A. 2006. Generating simple random graphs with prescribed degree distribution. *J. Stat. Phys.*, **124**(6), 1377–1397.
Broder, A., Kumar, R., Maghoul, F., Raghavan, P., Rajagopalan, S., Stata, R., Tomkins, A., and Wiener, J. 2000. Graph structure in the Web. *Computer Networks*, **33**, 309–320.
Buckley, P. G., and Osthus, D. 2004. Popularity based random graph models leading to a scale-free degree sequence. *Discrete Math.*, **282**(1-3), 53–68.
Cayley, A. 1889. A theorem on trees. *Q. J. Pure Appl. Math.*, **23**, 376–378.
Champernowne, D.G. 1953. A model of income distribution. *Econ. J.*, **63**, 318.

Chen, N., and Olvera-Cravioto, M. 2013. Directed random graphs with given degree distributions. *Stoch. Syst.*, **3**(1), 147–186.

Chernoff, H. 1952. A measure of asymptotic efficiency for tests of a hypothesis based on the sum of observations. *Ann. Math. Statistics*, **23**, 493–507.

Choudum, S. A. 1986. A simple proof of the Erdős-Gallai theorem on graph sequences. *Bull. Austral. Math. Soc.*, **33**(1), 67–70.

Chung, F., and Lu, L. 2002a. The average distances in random graphs with given expected degrees. *Proc. Natl. Acad. Sci. USA*, **99**(25), 15879–15882 (electronic).

Chung, F., and Lu, L. 2002b. Connected components in random graphs with given expected degree sequences. *Ann. Comb.*, **6**(2), 125–145.

Chung, F., and Lu, L. 2003. The average distance in a random graph with given expected degrees. *Internet Math.*, **1**(1), 91–113.

Chung, F., and Lu, L. 2006a. *Complex graphs and networks*. CBMS Regional Conference Series in Mathematics, vol. **107**. Published for the Conference Board of the Mathematical Sciences, Washington, DC.

Chung, F., and Lu, L. 2006b. Concentration inequalities and martingale inequalities: a survey. *Internet Math.*, **3**(1), 79–127.

Chung, F., and Lu, L. 2006c. The volume of the giant component of a random graph with given expected degrees. *SIAM J. Discrete Math.*, **20**, 395–411.

Clauset, A., and Moore, C. Preprint (2003). *Traceroute sampling makes random graphs appear to have power law degree distributions*. Available from `https://arxiv.org/abs/cond-mat/0312674`.

Clauset, A., and Moore, C. 2005. Accuracy and scaling phenomena in Internet mapping. *Phys. Rev. Lett.*, **94**, 018701: 1–4.

Clauset, A., Shalizi, C., and Newman, M. E. J. 2009. Power-law distributions in empirical data. *SIAM review*, **51**(4), 661–703.

Cohen, R., Erez, K., ben Avraham, D., and Havlin, S. 2000. Resilience of the Internet to random breakdowns. *Phys. Rev. Letters*, **85**, 4626.

Cohen, R., Erez, K., ben Avraham, D., and Havlin, S. 2001. Breakdown of the Internet under intentional attack. *Phys. Rev. Letters*, **86**, 3682.

Cooper, C., and Frieze, A. 2003. A general model of web graphs. *Random Structures Algorithms*, **22**(3), 311–335.

Cooper, C., and Frieze, A. 2004. The size of the largest strongly connected component of a random digraph with a given degree sequence. *Combin. Probab. Comput.*, **13**(3), 319–337.

Corless, R.M., Gonnet, G.H., Hare, D.E.G., Jeffrey, D.J., and Knuth, D.E. 1996. On the Lambert *W* function. *Adv. Comput. Math.*, **5**, 329–359.

Coupechoux, E., and Lelarge, M. 2014. How clustering affects epidemics in random networks. *Adv. in Appl. Probab.*, **46**(4), 985–1008.

De Castro, R., and Grossman, J.W. 1999a. Famous trails to Paul Erdős. *Rev. Acad. Colombiana Cienc. Exact. Fí s. Natur.*, **23**(89), 563–582. Translated and revised from the English.

De Castro, R., and Grossman, J.W. 1999b. Famous trails to Paul Erdős. *Math. Intelligencer*, **21**(3), 51–63. With a sidebar by Paul M. B. Vitanyi.

Deijfen, M., Esker, H. van den, Hofstad, R. van der, and Hooghiemstra, G. 2009. A preferential attachment model with random initial degrees. *Ark. Mat.*, **47**(1), 41–72.

Dembo, A., and Zeitouni, O. 1998. *Large deviations techniques and applications*. 2nd Applications of Mathematics (New York), vol. **38**. New York: Springer-Verlag.

Dereich, S., and Mörters, P. 2009. Random networks with sublinear preferential attachment: Degree evolutions. *Electronic Journal of Probability*, **14**, 122–1267.

Dereich, S., and Mörters, P. 2011. Random networks with concave preferential attachment rule. *Jahresber. Dtsch. Math.-Ver.*, **113**(1), 21–40.

Dereich, S., and Mörters, P. 2013. Random networks with sublinear preferential attachment: the giant component. *Ann. Probab.*, **41**(1), 329–384.

Dodds, P., Muhamad, R., and Watts, D. 2003. An experimental study of search in global social networks. *Science*, **301**(5634), 827–829.

Dommers, S., Hofstad, R. van der, and Hooghiemstra, G. 2010. Diameters in preferential attachment graphs. *Journ. Stat. Phys.*, **139**, 72–107.

Dorogovtsev, S.N. 2010. *Lectures on complex networks*. Oxford Master Series in Physics, vol. **20**. Oxford University Press, Oxford. Oxford Master Series in Statistical Computational, and Theoretical Physics.

Dorogovtsev, S.N., Ferreira, A.L., Goltsev, A.V., and Mendes, J.F.F. 2010. Zero Pearson coefficient for strongly correlated growing trees. *Phys. Rev. E*, **81**(3), 031135.

Dudley, R. M. 2002. *Real analysis and probability*. Cambridge Studies in Advanced Mathematics, vol. **74**. Cambridge: Cambridge University Press. Revised reprint of the 1989 original.

Durrett, R. 2007. *Random graph dynamics*. Cambridge Series in Statistical and Probabilistic Mathematics. Cambridge: Cambridge University Press.

Durrett, R. 2010. *Probability: theory and examples*. Fourth edn. Cambridge Series in Statistical and Probabilistic Mathematics. Cambridge University Press, Cambridge.

Dwass, M. 1962. A fluctuation theorem for cyclic random variables. *Ann. Math. Statist.*, **33**, 1450–1454.

Dwass, M. 1968. A theorem about infinitely divisible distributions. *Z. Wahrscheinleikheitsth.*, **9**, 206–224.

Dwass, M. 1969. The total progeny in a branching process and a related random walk. *J. Appl. Prob.*, **6**, 682–686.

Easley, D., and Kleinberg, J. 2010. *Networks, crowds, and markets: Reasoning about a highly connected world*. Cambridge University Press.

Ebel, H., Mielsch, L.-I., and Bornholdt, S. 2002. Scale-free topology of e-mail networks. *Phys. Rev. E*, **66**, 035103.

Eckhoff, M., and Mörters, P. 2014. Vulnerability of robust preferential attachment networks. *Electron. J. Probab.*, **19**, no. 57, 47.

Embrechts, P., Klüppelberg, C., and Mikosch, T. 1997. *Modelling extremal events*. Applications of mathematics (New York), vol. **33**. Berlin: Springer-Verlag. For insurance and finance.

Erdős, P. 1947. Some remarks on the theory of graphs. *Bull. Amer. Math. Soc.*, **53**, 292–294.

Erdős, P., and Gallai, T. 1960. Graphs with points of prescribed degrees. (Hungarian). *Mat. Lapok*, **11**, 264–274.

Erdős, P., and Rényi, A. 1959. On random graphs. I. *Publ. Math. Debrecen*, **6**, 290–297.

Erdős, P., and Rényi, A. 1960. On the evolution of random graphs. *Magyar Tud. Akad. Mat. Kutató Int. Közl.*, **5**, 17–61.

Erdős, P., and Rényi, A. 1961a. On the evolution of random graphs. *Bull. Inst. Internat. Statist.*, **38**, 343–347.

Erdős, P., and Rényi, A. 1961b. On the strength of connectedness of a random graph. *Acta Math. Acad. Sci. Hungar.*, **12**, 261–267.

Erdős, P., and Wilson, R. J. 1977. On the chromatic index of almost all graphs. *J. Combinatorial Theory Ser. B*, **23**(2–3), 255–257.

Ergün, G., and Rodgers, G. J. 2002. Growing random networks with fitness. *Phys. A*, **303**, 261–272.

Esker, H. van den, Hofstad, R. van der, and Hooghiemstra, G. 2008. Universality for the distance in finite variance random graphs. *J. Stat. Phys.*, **133**(1), 169–202.

Faloutsos, C., Faloutsos, P., and Faloutsos, M. 1999. On power-law relationships of the internet topology. *Computer Communications Rev.*, **29**, 251–262.

Feld, S.L. 1991. Why your friends have more friends than you do. *American Journal of Sociology*, **96**(6), 1464–1477.

Feller, W. 1968. *An introduction to probability theory and its applications.* Volume I. 3rd edn. New York: Wiley.

Feller, W. 1971. *An introduction to probability theory and its applications.* Volume II. 2nd edn. New York: Wiley.

Fortunato, S. 2010. Community detection in graphs. *Physics Reports*, **486**(3), 75–174.

Freeman, L. 1977. A set of measures in centrality based on betweenness. *Sociometry*, **40**(1), 35–41.

Gao, P., and Wormald, N. 2016. Enumeration of graphs with a heavy-tailed degree sequence. *Adv. Math.*, **287**, 412–450.

Gilbert, E. N. 1959. Random graphs. *Ann. Math. Statist.*, **30**, 1141–1144.

Gladwell, M. 2006. *The tipping point: How little things can make a big difference*. Hachette Digital, Inc.

Gnedin, A., Hansen, B., and Pitman, J. 2007. Notes on the occupancy problem with infinitely many boxes: general asymptotics and power laws. *Probab. Surv.*, **4**, 146–171.

Gradshteyn, I. S., and Ryzhik, I. M. 1965. *Table of integrals, series, and products.* Fourth edition prepared by Ju. V. Geronimus and M. Ju. Ceĭtlin. Translated from the Russian by Scripta Technica, Inc. Translation edited by Alan Jeffrey. New York: Academic Press.

Granovetter, M.S. 1973. The strength of weak ties. *American Journal of Sociology*, 1360–1380.

Granovetter, M.S. 1995. *Getting a job: A study of contacts and careers.* University of Chicago Press.

Grimmett, G. 1999. *Percolation.* 2nd edn. Berlin: Springer.

Grimmett, G.R., and Stirzaker, D.R. 2001. *Probability and random processes.* Third edn. New York: Oxford University Press.

Grossman, J.W. 2002. The evolution of the mathematical research collaboration graph. Pages 201–212 of: *Proceedings of the Thirty-third Southeastern International Conference on Combinatorics, Graph Theory and Computing (Boca Raton, FL, 2002)*, vol. **158**.

Hagberg, O., and Wiuf, C. 2006. Convergence properties of the degree distribution of some growing network models. *Bull. Math. Biol.*, **68**, 1275–1291.

Halmos, P. 1950. *Measure theory.* D. Van Nostrand Company, Inc., New York, N. Y.

Harary, F. 1969. *Graph theory.* Addison-Wesley Publishing Co., Reading, Mass.-Menlo Park, Calif.-London.

Harris, T. 1963. *The theory of branching processes.* Die Grundlehren der Mathematischen Wissenschaften, Bd. 119. Berlin: Springer-Verlag.

Heuvel, M. van den, and Sporns, O. 2011. Rich-club organization of the human connectome. *The Journal of Neuroscience*, **31**(44), 15775–15786.

Hirate, Y., Kato, S., and Yamana, H. 2008. Web structure in 2005. Pages 36–46 of: *Algorithms and models for the web-graph.* Springer.

Hoeffding, W. 1963. Probability inequalities for sums of bounded random variables. *J. Amer. Statist. Assoc.*, **58**, 13–30.

Hofstad, R. van der. 2010. Percolation and random graphs. Pages 173–247 of: *New perspectives in stochastic geometry.* Oxford Univ. Press, Oxford.

Hofstad, R. van der. 2015. *Random graphs and complex networks. Vol. II.* In preparation, see `http://www.win.tue.nl/~rhofstad/NotesRGCNII.pdf`.

Hofstad, R. van der, Kager, W., and Müller, T. 2009. A local limit theorem for the critical random graph. *Electron. Commun. Probab.*, **14**, 122–131.

Hofstad, R. van der, and Keane, M. 2008. An elementary proof of the hitting time theorem. *Amer. Math. Monthly*, **115**(8), 753–756.

Hofstad, R. van der, Leeuwaarden, J.H.S. van, and Stegehuis, C. Preprint (2015). *Hierarchical configuration model.*

Hofstad, R. van der, and Litvak, N. 2014. Degree-degree dependencies in random graphs with heavy-tailed degrees. *Internet Math.*, **10**(3-4), 287–334.

Hofstad, R. van der, and Spencer, J. 2006. Counting connected graphs asymptotically. *European J. Combin.*, **27**(8), 1294–1320.

Hollander, F. den. 2000. *Large deviations.* Fields Institute Monographs, vol. **14**. Providence, RI: American Mathematical Society.

Hoorn, P. van der, and Litvak, N. 2015. *Upper bounds for number of removed edges in the Erased Configuration Model.* Proceedings of the 12th International Workshop on Algorithms and Models for the Web-Graph, WAW 2015, 10–11 Dec 2015, Eindhoven, pp. 54–65. Lecture Notes in Computer Science 9479. Springer International Publishing. ISSN 0302-9743 ISBN 978-3-319-26783-8.

Huffaker, B., Fomenkov, M., and Claffy, K. 2012 (May). *Internet topology data comparison.* Tech. rept. Cooperative Association for Internet Data Analysis (CAIDA).

Jagers, P. 1975. *Branching processes with biological applications.* London: Wiley-Interscience [John Wiley & Sons]. Wiley Series in Probability and Mathematical Statistics—Applied Probability and Statistics.

Jagers, P., and Nerman, O. 1984. The growth and composition of branching populations. *Adv. in Appl. Probab.*, **16**(2), 221–259.

Jagers, P., and Nerman, O. 1996. The asymptotic composition of supercritical multi-type branching populations. Pages 40–54 of: *Séminaire de Probabilités, XXX*. Lecture Notes in Math., vol. **1626**. Berlin: Springer.

Janson, S. 2005. Asymptotic degree distribution in random recursive trees. *Random Structures Algorithms*, **26**(1-2), 69–83.

Janson, S. 2007. Monotonicity, asymptotic normality and vertex degrees in random graphs. *Bernoulli*, **13**(4), 952–965.

Janson, S. 2009. The probability that a random multigraph is simple. *Combin. Probab. Comput.*, **18**(1-2), 205–225.

Janson, S. 2010. Asymptotic equivalence and contiguity of some random graphs. *Random Structures Algorithms*, **36**(1), 26–45.

Janson, S. 2011. *Probability asymptotics: notes on notation*. Available at `http://arxiv.org/pdf/1108.3924.pdf`.

Janson, S. 2014. The probability that a random multigraph is simple. II. *J. Appl. Probab.*, **51A** (Celebrating 50 Years of The Applied Probability Trust), 123–137.

Janson, S., Knuth, D.E., Łuczak, T., and Pittel, B. 1993. The birth of the giant component. *Random Structures Algorithms*, **4**(3), 231–358. With an introduction by the editors.

Janson, S., Łuczak, T., and Rucinski, A. 2000. *Random graphs*. Wiley-Interscience Series in Discrete Mathematics and Optimization. Wiley-Interscience, New York.

Janson, S., and Spencer, J. 2007. A point process describing the component sizes in the critical window of the random graph evolution. *Combin. Probab. Comput.*, **16**(4), 631–658.

Járai, A., and Kesten, H. 2004. A bound for the distribution of the hitting time of arbitrary sets by random walk. *Electron. Comm. Probab.*, **9**, 152–161 (electronic).

Jin, S., and Bestavros, A. 2006. Small-world characteristics of Internet topologies and implications on multicast scaling. *Computer Networks*, **50**, 648–666.

Jordan, J. 2006. The degree sequences and spectra of scale-free random graphs. *Random Structures Algorithms*, **29**(2), 226–242.

Karinthy, F. 1929. Chains. In: *Everything is different*. Publisher unknown.

Karlin, S. 1967. Central limit theorems for certain infinite urn schemes. *J. Math. Mech.*, **17**, 373–401.

Karp, R.M. 1990. The transitive closure of a random digraph. *Random Structures Algorithms*, **1**(1), 73–93.

Kemperman, J. H. B. 1961. *The passage problem for a stationary Markov chain*. Statistical Research Monographs, Vol. I. The University of Chicago Press, Chicago, Ill.

Kesten, H., and Stigum, B. P. 1966a. Additional limit theorems for indecomposable multidimensional Galton–Watson processes. *Ann. Math. Statist.*, **37**, 1463–1481.

Kesten, H., and Stigum, B. P. 1966b. A limit theorem for multidimensional Galton–Watson processes. *Ann. Math. Statist.*, **37**, 1211–1223.

Kesten, H., and Stigum, B. P. 1967. Limit theorems for decomposable multi-dimensional Galton–Watson processes. *J. Math. Anal. Appl.*, **17**, 309–338.

Kleinberg, J. M. 1999. Authoritative sources in a hyperlinked environment. *J. ACM*, **46**(5), 604–632.

Kleinberg, J. M. 2000a. Navigation in a small world. *Nature*, **406**, 845.

Kleinberg, J. M. 2000b (May). The small-world phenomenon: an algorithm perspective. Pages 163–170 of: *Proc. of the twenty-third annual ACM symposium on Principles of distributed computing*.

Kleinberg, J.M., Kumar, R., Raghavan, P., Rajagopalan, S, and Tomkins, A. 1999. The Web as a graph: measurements, models, and methods. Pages 1–17 of: *Computing and Combinatorics: 5th Annual International Conference, COCOON'99, Tokyo, Japan, July 1999. Proceedings*. Lecture Notes in Computer Science.

Konstantopoulos, T. 1995. Ballot theorems revisited. *Statist. Probab. Lett.*, **24**(4), 331–338.

Krapivsky, P. L., and Redner, S. 2001. Organization of growing random networks. *Phys. Rev. E*, **63**, 066123.

Krapivsky, P. L., and Redner, S. 2005. Network growth by copying. *Phys. Rev. E*, **71**(3), 036118.

Krapivsky, P. L., Redner, S., and Leyvraz, F. 2000. Connectivity of growing random networks. *Phys. Rev. Lett.*, **85**, 4629.

Krioukov, D., Kitsak, M., Sinkovits, R., Rideout, D., Meyer, D., and Boguñá, M. 2012. Network cosmology. *Scientific reports*, **2**.

Kumar, R., Raghavan, P., Rajagopalan, S, and Tomkins, A. 1999. Trawling the Web for emerging cyber communities. *Computer Networks*, **31**, 1481–1493.

Kumar, R., Raghavan, P., Rajagopalan, S., Sivakumar, D., Tomkins, A., and Upfal, E. 2000. Stochastic models for the web graph. Pages 57–65 of: *42st Annual IEEE Symposium on Foundations of Computer Science*.

Lakhina, A., Byers, J.W., Crovella, M., and Xie, P. 2003. Sampling biases in IP topology measurements. Pages 332–341 of: *Proceedings of IEEE INFOCOM 1*.

Lengler, J., Jug, F., and Steger, A. 2013. Reliable neuronal systems: the importance of heterogeneity. *PLoS ONE*, **8**(12), 1–10.

Leskovec, J., Kleinberg, J., and Faloutsos, C. 2005. Graphs over time: densification laws, shrinking diameters and possible explanations. Pages 177–187 of: *Proceedings of the eleventh ACM SIGKDD international conference on Knowledge discovery in data mining*. ACM.

Leskovec, J., Kleinberg, J., and Faloutsos, C. 2007. Graph evolution: Densification and shrinking diameters. *ACM Transactions on Knowledge Discovery from Data (TKDD)*, **1**(1), 2.

Leskovec, J., Lang, K., Dasgupta, A., and Mahoney, M. 2009. Community structure in large networks: Natural cluster sizes and the absence of large well-defined clusters. *Internet Math.*, **6**(1), 29–123.

Leskovec, J., Lang, K., and Mahoney, M. 2010. Empirical comparison of algorithms for network community detection. Pages 631–640 of: *Proceedings of the 19th International Conference on World Wide Web*. WWW '10. New York, NY, USA: ACM.

Liljeros, F., Edling, C. R., Amaral, L. A. N., and Stanley, H. E. 2001. The web of human sexual contacts. *Nature*, **411**, 907.

Lindvall, T. 2002. *Lectures on the coupling method*. Dover Publications, Inc., Mineola, NY. Corrected reprint of the 1992 original.

Lint, J. H. van, and Wilson, R. M. 2001. *A course in combinatorics*. 2nd Cambridge: Cambridge University Press.

Litvak, N., and Hofstad, R. van der. 2013. Uncovering disassortativity in large scale-free networks. *Phys. Rev. E*, **87**(2), 022801.

Lotka, A.J. 1926. The frequency distribution of scientific productivity. *Journal of the Washington Academy of Sciences*, **16**(12), 317–323.

Lovász, L. 2012. *Large networks and graph limits*. American Mathematical Society Colloquium Publications, vol. **60**. American Mathematical Society, Providence, RI.

Lu, L. 2002. *Probabilistic methods in massive graphs and Internet computing*. Ph.D. thesis, University of California, San Diego. Available at `http://math.ucsd.edu/~llu/thesis.pdf`.

Łuczak, T. 1990a. Component behavior near the critical point of the random graph process. *Random Structures Algorithms*, **1**(3), 287–310.

Łuczak, T. 1990b. On the number of sparse connected graphs. *Random Structures Algorithms*, **1**(2), 171–173.

Łuczak, T., Pittel, B., and Wierman, J. 1994. The structure of a random graph at the point of the phase transition. *Trans. Amer. Math. Soc.*, **341**(2), 721–748.

Lyons, R., Pemantle, R., and Peres, Y. 1995. Conceptual proofs of $L \log L$ criteria for mean behavior of branching processes. *Ann. Probab.*, **23**(3), 1125–1138.

Martin-Löf, A. 1986. Symmetric sampling procedures, general epidemic processes and their threshold limit theorems. *J. Appl. Probab.*, **23**(2), 265–282.

Martin-Löf, A. 1998. The final size of a nearly critical epidemic, and the first passage time of a Wiener process to a parabolic barrier. *J. Appl. Probab.*, **35**(3), 671–682.

Milgram, S. 1967. The small world problem. *Psychology Today*, **May**, 60–67.

Miller, J.C. 2009. Percolation and epidemics in random clustered networks. *Phys. Rev. E*, **80**(Aug), 020901.

Mitzenmacher, M. 2004. A brief history of generative models for power law and lognormal distributions. *Internet Math.*, **1**(2), 226–251.

Molloy, M., and Reed, B. 1995. A critical point for random graphs with a given degree sequence. *Random Structures Algorithms*, **6**(2-3), 161–179.

Molloy, M., and Reed, B. 1998. The size of the giant component of a random graph with a given degree sequence. *Combin. Probab. Comput.*, **7**(3), 295–305.

Móri, T. F. 2002. On random trees. *Studia Sci. Math. Hungar.*, **39**(1-2), 143–155.

Móri, T. F. 2005. The maximum degree of the Barabási–Albert random tree. *Combin. Probab. Comput.*, **14**(3), 339–348.

Nachmias, A., and Peres, Y. 2010. The critical random graph, with martingales. *Israel J. Math.*, **176**, 29–41.

Nerman, O., and Jagers, P. 1984. The stable double infinite pedigree process of supercritical branching populations. *Z. Wahrsch. Verw. Gebiete*, **65**(3), 445–460.

Newman, M. E. J. 2000. Models of the small world. *J. Stat. Phys.*, **101**, 819–841.

Newman, M. E. J. 2001. The structure of scientific collaboration networks. *Proc. Natl. Acad. Sci. USA*, **98**, 404.

Newman, M. E. J. 2002. Assortative mixing in networks. *Phys. Rev. Lett.*, **89(20)**, 208701.

Newman, M. E. J. 2003. The structure and function of complex networks. *SIAM Rev.*, **45**(2), 167–256 (electronic).

Newman, M. E. J. 2005. Power laws, Pareto distributions and Zipf's law. *Contemporary Physics*, **46**(5), 323–351.

Newman, M. E. J. 2009. Random graphs with clustering. *Phys. Rev. Lett.*, **103**(Jul), 058701.

Newman, M. E. J. 2010. *Networks: an introduction*. Oxford University Press.

Newman, M. E. J., Strogatz, S., and Watts, D. 2002. Random graph models of social networks. *Proc. Nat. Acad. Sci.*, **99**, 2566–2572.

Newman, M. E. J., Watts, D. J., and Barabási, A.-L. 2006. *The structure and dynamics of networks*. Princeton Studies in Complexity. Princeton University Press.

Norros, I., and Reittu, H. 2006. On a conditionally Poissonian graph process. *Adv. in Appl. Probab.*, **38**(1), 59–75.

O'Connell, N. 1998. Some large deviation results for sparse random graphs. *Probab. Theory Related Fields*, **110**(3), 277–285.

Okamoto, M. 1958. Some inequalities relating to the partial sum of binomial probabilities. *Ann. Inst. Statist. Math.*, **10**, 29–35.

Oliveira, R., and Spencer, J. 2005. Connectivity transitions in networks with super-linear preferential attachment. *Internet Math.*, **2**(2), 121–163.

Olivieri, E., and Vares, M.E. 2005. *Large deviations and metastability*. Encyclopedia of Mathematics and its Applications. Cambridge: Cambridge University Press.

Ostroumova, L., and Grechnikov, E. 2012. The distribution of second degrees in the Bollobás–Riordan random graph model. *Mosc. J. Comb. Number Theory*, **2**(2), 85–110.

Otter, R. 1949. The multiplicative process. *Ann. Math. Statist.*, **20**, 206–224.

Pansiot, J.-J., and Grad, D. 1998. On routes and multicast trees in the Internet. *ACM SIGCOMM Computer Communication Review*, **28**(1), 41–50.

Pareto, V. 1896. *Cours d'economie politique*. Geneva, Switserland: Droz.

Pastor-Satorras, R., and Vespignani, A. 2007. *Evolution and structure of the Internet: A statistical physics approach*. Cambridge University Press.

Peköz, E., Röllin, A., and Ross, N. 2013. Degree asymptotics with rates for preferential attachment random graphs. *Ann. Appl. Probab.*, **23**(3), 1188–1218.

Peköz, E., Röllin, A., and Ross, N. Preprint (2014). *Joint degree distributions of preferential attachment random graphs*. Available from `http://arxiv.org/pdf/1402.4686.pdf`.

Pitman, J., and Yor, M. 1997. The two-parameter Poisson–Dirichlet distribution derived from a stable subordinator. *Ann. Probab.*, **25**(2), 855–900.

Pittel, B. 1990. On tree census and the giant component in sparse random graphs. *Random Structures Algorithms*, **1**(3), 311–342.

Pittel, B. 2001. On the largest component of the random graph at a nearcritical stage. *J. Combin. Theory Ser. B*, **82**(2), 237–269.

Pool, I. de S., and Kochen, M. 1978. Contacts and influence. *Social Networks*, **1**, 5–51.

Rényi, A. 1959. On connected graphs. I. *Magyar Tud. Akad. Mat. Kutató Int. Közl.*, **4**, 385–388.

Resnick, S. 2007. *Heavy-tail phenomena*. Springer Series in Operations Research and Financial Engineering. Springer, New York. Probabilistic and statistical modeling.

Ross, N. 2013. Power laws in preferential attachment graphs and Stein's method for the negative binomial distribution. *Adv. in Appl. Probab.*, **45**(3), 876–893.

Rudas, A., Tóth, B., and Valkó, B. 2007. Random trees and general branching processes. *Random Structures Algorithms*, **31**(2), 186–202.

Seneta, E. 1969. Functional equations and the Galton-Watson process. *Advances in Appl. Probability*, **1**, 1–42.

Sierksma, G., and Hoogeveen, H. 1991. Seven criteria for integer sequences being graphic. *J. Graph Theory*, **15**(2), 223–231.

Siganos, G., Faloutsos, M., Faloutsos, P., and Faloutsos, C. 2003. Power laws and the AS-level Internet topology. *IEEE/ACM Trans. Netw.*, **11**(4), 514–524.

Simon, H. A. 1955. On a class of skew distribution functions. *Biometrika*, **42**, 425–440.

Solomonoff, R., and Rapoport, A. 1951. Connectivity of random nets. *Bull. Math. Biophys.*, **13**, 107–117.

Spencer, J. 1997. Enumerating graphs and Brownian motion. *Comm. Pure Appl. Math.*, **50**(3), 291–294.

Spitzer, F. 1956. A combinatorial lemma and its application to probability theory. *Trans. Amer. Math. Soc.*, **82**, 323–339.

Spitzer, F. 1976. *Principles of random walk.* 2nd edn. New York: Springer.

Sporns, O. 2011. *Networks of the Brain.* MIT Press.

Strassen, V. 1965. The existence of probability measures with given marginals. *Ann. Math. Statist.*, **36**, 423–439.

Strogatz, S. 2001. Exploring complex networks. *Nature*, **410**(8), 268–276.

Szemerédi, E. 1978. Regular partitions of graphs. Pages 399–401 of: *Problèmes combinatoires et théorie des graphes (Colloq. Internat. CNRS, Univ. Orsay, Orsay, 1976).* Colloq. Internat. CNRS, vol. **260**. CNRS, Paris.

Szymański, J. 2005. Concentration of vertex degrees in a scale-free random graph process. *Random Structures Algorithms*, **26**(1-2), 224–236.

Thorisson, H. 2000. *Coupling, stationarity, and regeneration.* Probability and its applications (New York). New York: Springer-Verlag.

Trapman, P. 2007. On analytical approaches to epidemics on networks. *Theoretical Population Biology*, **71**(2), 160–173.

Travers, J., and Milgram, S. 1969. An experimental study of the small world problem. *Sociometry*, **32**, 425–443.

Ugander, J., Karrer, B., Backstrom, L., and Marlow, C. Preprint (2011). The anatomy of the Facebook social graph. Available from `http://arxiv.org/pdf/1111.4503.pdf`.

Vega-Redondo, F. 2007. *Complex social networks.* Econometric Society Monographs, vol. **44**. Cambridge University Press, Cambridge.

Viswanath, B., Mislove, A., Cha, M., and Gummadi, K. 2009. On the evolution of user inter-action in Facebook. Pages 37–42 of: *Proceedings of the 2Nd ACM Workshop on Online Social Networks.* WOSN '09. New York, NY, USA: ACM.

Watts, D. J. 1999. *Small worlds. The dynamics of networks between order and randomness.* Princeton Studies in Complexity. Princeton, NJ: Princeton University Press.

Watts, D. J. 2003. *Six degrees. The science of a connected age.* New York: W. W. Norton & Co. Inc.

Watts, D. J., and Strogatz, S. H. 1998. Collective dynamics of 'small-world' networks. *Nature*, **393**, 440–442.

Wendel, J. G. 1975. Left-continuous random walk and the Lagrange expansion. *Amer. Math. Monthly*, **82**, 494–499.

Williams, D. 1991. *Probability with martingales.* Cambridge Mathematical Textbooks. Cambridge: Cambridge University Press.

Willinger, W., Alderson, D., and Doyle, J.C. 2009. Mathematics and the Internet: A source of enormous confusion and great potential. *Notices of the American Mathematical Society*, **56**(5), 586–599.

Willinger, W., Govindan, R., Jamin, S., Paxson, V., and Shenker, S. 2002. Scaling phenomena in the Internet: Critically examining criticality. *Proc. Natl. Acad. Sci.*, **99**, 2573–2580.

Wilson, R., Gosling, S., and Graham, L. 2012. A review of Facebook research in the social sciences. *Perspectives on Psychological Science*, **7**(3), 203–220.

Wright, E. M. 1977. The number of connected sparsely edged graphs. *J. Graph Theory*, **1**(4), 317–330.

Wright, E. M. 1980. The number of connected sparsely edged graphs. III. Asymptotic results. *J. Graph Theory*, **4**(4), 393–407.

Yook, S.-H., Jeong, H., and Barabási, A.-L. 2002. Modeling the Internet's large-scale topology. *Proc. Natl. Acad. Sci.*, **99**(22), 13382–13386.

Yule, G. U. 1925. A mathematical theory of evolution, based on the conclusions of Dr. J. C. Willis, F.R.S. *Phil. Trans. Roy. Soc. London, B*, **213**, 21–87.

Zipf, G.K. 1929. Relative frequency as a determinant of phonetic change. *Harvard Studies in Classical Philology*, **15**, 1–95.

Index

Printed in the United States
by Baker & Taylor Publisher Services